Neuropathogenic Viruses
and Immunity

INFECTIOUS AGENTS AND PATHOGENESIS

Series Editors: Mauro Bendinelli, *University of Pisa*
Herman Friedman, *University of South Florida*

COXSACKIEVIRUSES
A General Update
Edited by Mauro Bendinelli and Herman Friedman

MYCOBACTERIUM TUBERCULOSIS
Interactions with the Immune System
Edited by Mauro Bendinelli and Herman Friedman

NEUROPATHOGENIC VIRUSES AND IMMUNITY
Edited by Steven Specter, Mauro Bendinelli, and Herman Friedman

VIRUS-INDUCED IMMUNOSUPPRESSION
Edited by Steven Specter, Mauro Bendinelli, and Herman Friedman

Neuropathogenic Viruses and Immunity

Edited by

Steven Specter
University of South Florida
Tampa, Florida

Mauro Bendinelli
University of Pisa
Pisa, Italy

and

Herman Friedman
University of South Florida
Tampa, Florida

Plenum Press • New York and London

Library of Congress Cataloging-in-Publication Data

Neuropathogenic viruses and immunity / edited by Steven Specter, Mauro
Bendinelli, and Herman Friedman.
    p.   cm. -- (Infectious agents and pathogenesis)
  Includes bibliographical references and index.
  ISBN-13: 978-1-4684-5888-6   e-ISBN-13: 978-1-4684-5886-2
  DOI: 10.1007/978-1-4684-5886-2
    1. Nervous system--Infections--Immunological aspects. 2. Virus
diseases--Immunological aspects.   I. Specter, Steven.
II. Bendinelli, Mauro. III. Friedman, Herman, 1931-   .
IV. Series.
    [DNLM: 1. Nervous System Diseases--etiology. 2. Nervous System
Diseases--immunology. 3. Viruses--immunology. 4. Viruses-
-pathogenicity.   WL 100 N495173]
RC346.5.N49 1992
616.8'04194--dc20
DNLM/DLC
for Library of Congress   92-3281
                                                              CIP

ISBN-13: 978-1-4684-5888-6

©1992 Plenum Press, New York
Softcover reprint of the hardcover 1st edition 1992

A Division of Plenum Publishing Corporation
233 Spring Street, New York, N.Y. 10013

Contributors

ÓLAFUR S. ANDRÉSSON • Institute for Experimental Pathology, University of Iceland, Keldur, IS-128 Reykjavík, Iceland

MELVYN J. BALL • Division of Neuropathology, Oregon Health Sciences University, Portland, Oregon 97201

MAURO BENDINELLI • Department of Biomedicine, University of Pisa, I-56127 Pisa, Italy

RICHARD I. CARP • New York State Institute for Basic Research in Developmental Disabilities, Staten Island, New York 10314

MAURO C. DAL CANTO • Departments of Neurology and Pathology (Neuropathology), Northwestern University Medical School, Chicago, Illinois 60611

ANNE M. DEATLY • Department of Microbiology, University of Minnesota, Minneapolis, Minnesota 55455

HERMAN FRIEDMAN • Department of Microbiology and Immunology, University of South Florida, College of Medicine, Tampa, Florida 33612-4799

MURRAY GARDNER • Department of Pathology, School of Medicine, University of California, Davis, California 95616

HOWARD E. GENDELMAN • HIV-Immunopathogenesis Program, Department of Cellular Immunology, Walter Reed Army Institute of Research, Rockville, Maryland 20850, and Henry M. Jackson Foundation for the Advancement of Military Medicine, Uniformed Services University Health Science Center, Bethesda, Maryland 20814

SEYMOUR GENDELMAN • Department of Neurology, Mt. Sinai Medical Center, New York, New York 10028

GUDMUNDUR GEORGSSON • Institute for Experimental Pathology, University of Iceland, Keldur, IS–128 Reykjavík, Iceland

MICHAEL C. GRAVES • Department of Neurology, Reed Neurological Research Center, UCLA Medical Center, Los Angeles, California 90024-1769

DIANE E. GRIFFIN • Departments of Medicine and Neurology, The Johns Hopkins University School of Medicine, Baltimore, Maryland 21205

ASHLEY T. HAASE • Department of Microbiology, University of Minnesota, Minneapolis, Minnesota 55455

LESLEY M. HALLICK • Department of Microbiology and Immunology, Oregon Health Sciences University, Portland, Oregon 97201

SIDNEY A. HOUFF • Laboratory of Viral and Molecular Pathogenesis, National Institute of Neurological Disorders and Stroke, National Institutes of Health, Bethesda, Maryland 20892. *Present address*: Department of Neurology, Washington Veterans Administration Hospital, Washington, DC 20422

RICHARD T. JOHNSON • Department of Neurology, Johns Hopkins University School of Medicine, Baltimore, Maryland 21205

ANDREW LACKNER • California Primate Research Center, University of California, Davis, California 95616. *Present address:* New Mexico Regional Primate Research Laboratory, New Mexico State University, Holloman Air Force Base, New Mexico 88330-1027

HOWARD L. LIPTON • Department of Neurology, Northwestern University Medical School, Chicago, Illinois 60611. *Present address:* Department of Neurology, University of Colorado School of Medicine, Denver, Colorado 80262

J. MATTHIAS LÖHR • Department of Internal Medicine, University of Erlangen, D–8520 Erlangen, Germany

LINDA LOWENSTINE • California Primate Research Center, University of California, Davis, California 95616

EUGENE O. MAJOR • Laboratory of Viral and Molecular Pathogenesis, National Institute of Neurological Disorders and Stroke, National Institutes of Health, Bethesda, Maryland 20892

JOSEPH L. MELNICK • Division of Molecular Virology, Baylor College of Medicine, Houston, Texas 77030

ANTHONY A. NASH • Department of Pathology, University of Cambridge, Cambridge CB2 1QP, England

CLAES ÖRVELL • Department of Virology, National Bacteriological Laboratory and Karolinska Institute, School of Medicine, S-105 21 Stockholm, Sweden

GUDMUNDUR PÉTURSSON • Institute for Experimental Pathology, University of Iceland, Keldur, IS-128 Reykjavík, Iceland

STEVEN SPECTER • Department of Microbiology and Immunology, University of South Florida, College of Medicine, Tampa, Florida 33612-4799

JANICE R. STEVENS • Neuropsychiatry Branch, National Institute of Mental Health, Neuroscience Center at St. Elizabeths Hospital, Washington, DC 20032, and Departments of Psychiatry and Neurology, Oregon Health Sciences University, Portland, Oregon 97201

WILLIAM R. TYOR • Department of Neurology, Johns Hopkins University School of Medicine, Baltimore, Maryland 21205

DOMINICK A. VACANTE • Laboratory of Viral and Molecular Pathogenesis, National Institute of Neurological Disorders and Stroke, National Institutes of Health, Bethesda, Maryland 20892. *Present address*: Microbiological Associates, Rockville, Maryland 20852

HARRY V. VINTERS • Department of Pathology (Neuropathology) and Brain Research Institute, UCLA Medical Center, Los Angeles, California 90024-1732

H. E. WEBB • Department of Neurovirology, UMDS, The Rayne Institute, St. Thomas' Hospital, London SE1 7EH, England

Preface to the Series

The mechanisms of disease production by infectious agents are presently the focus of an unprecedented flowering of studies. The field has undoubtedly received impetus from the considerable advances recently made in the understanding of the structure, biochemistry, and biology of viruses, bacteria, fungi, and other parasites. Another contributing factor is our improved knowledge of immune responses and other adaptive or constitutive mechanisms by which hosts react to infection. Furthermore, recombinant DNA technology, monoclonal antibodies, and other newer methodologies have provided the technical tools for examining questions previously considered too complex to be successfully tackled. The most important incentive of all is probably the regenerated idea that infection might be the initiating event in many clinical entities presently classified as idiopathic or of uncertain origin.

Infectious pathogenesis research holds great promise. As more information is uncovered, it is becoming increasingly apparent that our present knowledge of the pathogenic potential of infectious agents is often limited to the most noticeable effects, which sometimes represent only the tip of the iceberg. For example, it is now well appreciated that pathological processes caused by infectious agents may emerge clinically after an incubation of decades and may result from genetic, immunologic, and other indirect routes more than from the infecting agent itself. Thus, there is a general expectation that continued investigation will lead to the isolation of new agents of infection, the identification of hitherto unsuspected etiologic correlations, and eventually, more effective approaches to prevention and therapy.

Studies on the mechanisms of disease caused by infectious agents demand a breadth of understanding across many specialized areas, as well as much cooperation between clinicians and experimentalists. The series *Infectious Agents and Pathogenesis* is intended not only to document the state of the art in this fascinating and challenging field but also to help lay bridges among diverse areas and people.

M. Bendinelli
H. Friedman

Preface

There has been a tremendous increase in interest in the neuropathogenicity of viruses during the past decade as we have come to recognize that the human immunodeficiency virus, which causes the acquired immunodeficiency syndrome (AIDS), can infect glial cells and cause neurological disease. Yet this increase has not been limited to AIDS but has extended to viruses that infect either or both the central and peripheral nervous systems. The changes examined here include both neurological and psychological diseases or syndromes. Moreover, the chapters in this volume review the interaction of the host immune system with the viruses examined and how such interactions may increase or decrease the neuropathogenicity of the viruses.

Questions regarding viral neuropathogenesis include: (1) What is the mode of transmission of virus to the nervous system? (2) What types of cells are infected, and do they contain receptors for the virus? (3) What is the extent of damage that results from viral infection? (4) What are the immunologic mechanisms by which damage is mediated or limited? Many of these questions remain unanswered, but this volume delves into efforts to provide some answers.

There is an overall increase in awareness that many neurological and psychological conditions have a physiological basis. In many cases it has not been possible to detect any reason for such disease, but in recent years research has drawn us nearer to solving some of these mysteries. One approach has been to look for virological causes of such disease, and in several cases the path has led to the discovery of etiologies that are likely virus induced (e.g., human T-lymphotropic virus I is linked to tropical spastic paraparesis). Yet, other diseases elude the ability of the scientific community to detect a cause. In some cases there is evidence that an antecedent viral infection may have occurred in the host or even in neural tissue, but there is no evidence of virus being present at the time that active disease is seen. In certain circumstances it has been shown that the earlier viral infection may have led to activation of the immune system, resulting in damage to neural tissues long after the viral infection has cleared. Alternatively, neural tissue may be persistently infected by latent viruses that may express viral antigens on their surface resulting in destruction of neural tissue because of an immunologic attack against the foreign antigens displayed on the cell surface. Thus, developing an understanding of the role of both viruses and immune responses in eliciting neuropathogenesis has been a major goal of this volume.

Our understanding of such viral and immune interactions with neural tissue has been acquired through the study of animal models of viral neuropathogenesis, and it is for this reason that a portion of this volume is devoted to animal models. Because of the tremendous interest in and support of AIDS research, retroviruses, especially lentiviruses, have been examined more extensively than any other animal models of neuropathogenic viruses.

In recent years viruses have come under increasing scrutiny as etiologic agents in psychosis. Again, much of this can be attributed to the AIDS epidemic and our increasing knowledge of virus-induced dementia. For the past decade or so there has been the suggestion that even a more severe disease, schizophrenia, may have a viral component to its etiology, and this is examined.

As we realize more convincingly that virally induced neurological disease is more widespread than we previously thought, it becomes clearer that our approaches to dealing with such infections are dependent on a more thorough understanding of the nature and mechanisms by which these agents attack the nervous tissues and cause disease. The advent of biotechnology and newer, more sophisticated methods to examine disease and affected tissues will surely assist our progress toward understanding neuropathogenic viruses and the diseases they cause. This will take a concerted effort of a multidisciplinary nature, involving molecular and cellular biologists, virologists, immunologists, neurologists, psychiatrists, and others. This volume is designed to provide an integration of both the current understanding of the virological and immunologic components of these neuropathogenic diseases.

Steven Specter
Mauro Bendinelli
Herman Friedman

Contents

1. Viruses and Neuropsychiatric Disorders: Facts and Suppositions

STEVEN SPECTER, MAURO BENDINELLI,
and HERMAN FRIEDMAN

1. The Facts 1
   1.1. Acute Diseases Resulting Directly from Viral Invasion 2
   1.2. Acute Parainfectious Syndromes 6
   1.3. Congenital Defects 8
   1.4. Subacute and Chronic Diseases 8
2. The Suppositions 10
   References 11

PART I   GENERAL INFORMATION

2. Immune Responses and the Central Nervous System

WILLIAM R. TYOR and RICHARD T. JOHNSON

1. Introduction 15
2. General Overview of the Immune System 16
   2.1. Anatomy of the Immune System 16
   2.2. Cellular Development 19
   2.3. Activation of the Immune System 22
3. Immune Responses to Viral Infection 27
   3.1. Humoral Interaction 27
   3.2. Cell-Mediated Interaction 28
4. Immune Responses in the Central Nervous System 28
   4.1. Immune Cells Normally Present in the Central Nervous
        System 28
   4.2. Cerebrospinal Fluid 31

4.3. Response to Viral Infection 31
4.4. Viral Clearance 33
5. Sindbis Virus as a Model of Encephalitis 33
6. Summary 37
   References 37

3. Virus Infection of Peripheral Nerve

MICHAEL C. GRAVES and HARRY V. VINTERS

1. Introduction 41
2. Virus Infection of Neurons: Herpesvirus Latency and Reactivation
   in Sensory Ganglia 42
   2.1. Herpes Simplex Virus Type 1 42
   2.2. Herpes Simplex Virus Type 2 43
   2.3. Varicella–Zoster Virus 44
3. Other Herpesviruses Associated with Neuropathy: Marek's Disease
   Virus, Cytomegalovirus, and Epstein–Barr Virus 45
4. Hepatitis B Virus 47
5. Human Immunodeficiency Virus and Peripheral Neuromuscular
   Syndromes 48
   5.1. Inflammatory Neuropathies that Usually Occur Early in the
        Course of Infection 48
   5.2. Axonal Neuropathies that Usually Occur Late in the Course
        of HIV Infection 50
   5.3. Morphological Studies of Nerves 54
   5.4. Autonomic Neuropathy 54
   5.5. Other Neuromuscular Syndromes 57
   References 57

PART II ANIMAL MODELS

4. Visna, a Lentiviral Disease of Sheep

GUDMUNDUR PÉTURSSON, ÓLAFUR S. ANDRÉSSON,
and GUDMUNDUR GEORGSSON

1. Visna Virus: Biology and Structure 63
   1.1. Introduction 63
   1.2. Neurotropism 64
   1.3. Virus Structure and Genome Organization 65
   1.4. Virus Replication in Vitro 65

2. Clinical Features and Pathology 66
   2.1. Transmission and Incubation Period 66
   2.2. Clinical Symptoms 66
   2.3. Pathology 67
3. Immune Response 71
   3.1. Humoral Immune Response 71
   3.2. Cell-Mediated Immune Response 72
   3.3. Interferon 72
   3.4. Vaccination Trials 72
4. Pathogenesis 73
   References 74

5. Theiler's Virus-Induced Demyelinating Disease in Mice:
   Picornavirus Animal Model

   HOWARD L. LIPTON and MAURO C. DAL CANTO

1. Introduction 79
2. Two Neurovirulence Groups 81
3. Clinical Signs 82
4. Pathological Features 82
5. Virus-Specific Immunity 83
6. Sites of Virus Persistence 84
7. Immune-Mediated Mechanism of Demyelination 85
8. Multigenic Control of Demyelinating Disease 87
   References 88

6. Neurotropic Retroviruses of Mice, Cats, Macaques, and Humans

   MURRAY GARDNER, ANDREW LACKNER,
   and LINDA LOWENSTINE

1. Introduction 93
2. Neurotropic Murine Leukemia Virus 94
3. Neurotropic Type D Retrovirus of Macaques 97
4. Simian Immunodeficiency Virus 98
5. Comparison of Animal Models of CNS Retroviral Infection
   with Central Nervous System Infections Caused by Human
   Retroviruses 100
   5.1. Human T-Lymphotropic Virus I 100
   5.2. Human Immunodeficiency Virus 101
6. Spumaviruses 101
7. Feline Immunodeficiency Virus 102
8. Summary 104
   References 104

7. Scrapie: Unconventional Infectious Agent

RICHARD I. CARP

1. Scrapie Biology 111
2. Diseases Associated with Nervous System Infection 113
3. Pathogenesis 116
    3.1. Pathogenesis and Route of Injection 116
    3.2. The Cell Type Involved in Early Steps of Pathogenesis 116
    3.3. The Spread of Infectivity in the Infected Host 118
    3.4. The Concept of Clinical Target Areas 119
    3.5. Genetic Aspects of Pathogenesis 120
4. The Absence of an Immunopathological Effect in Scrapie 121
5. Agent Persistence 123
6. Treatment 123
7. Miscellaneous 125
    References 129

PART III   HUMAN INFECTIONS OF THE CNS

8. Enteroviruses

JOSEPH L. MELNICK

1. Introduction 139
2. Description and Classification 139
3. Epidemiology 140
    3.1. General Epidemiology of Enteroviruses 140
    3.2. Epidemiology of Poliomyelitis 141
4. Clinical Diseases Caused by Enteroviruses 142
    4.1. Poliomyelitis 142
    4.2. Meningitis and Mild Paresis Caused by Nonpolio
        Enteroviruses 143
    4.3. Other Diseases Caused by Enteroviruses 143
5. Pathogenesis and Immunity 144
6. Persistence 147
7. Control of Enteroviral Diseases 147
    7.1. Control of Paralytic Poliomyelitis 147
    7.2. Control of Other Enteroviral Diseases 150
    References 151

9. Pathogenesis and Immunology of Herpesvirus Infections of the
Nervous System

ANTHONY A. NASH and J. MATTHIAS LÖHR

1. Introduction 155
2. Properties of Herpes Simplex Virus 156
3. The Relationship between Herpes Simplex Virus and the
   Nervous System 156
   3.1. Primary Infection 156
   3.2. Latency 159
   3.3. Reactivation/Recurrence 160
4. Neurological Diseases Arising from Herpes Simplex Virus
   Infections 161
5. The Host Response to Herpes Simplex Virus Infections 162
   5.1. Induction and Activity of T Cells during a Herpes Simplex
        Virus Infection 162
   5.2. T Cells in Recurrent Infection 164
   5.3. Significance of Antibody in Herpes Simplex Virus
        Infections 164
6. Immunopathology Associated with Herpes Infections of the
   Nervous System 166
7. Toward the Prevention of Primary and Recurrent Herpes
   Infections 166
   References 168

10. Paramyxoviruses

CLAES ÖRVELL

1. Introduction 177
2. Classification and Structural Properties of Paramyxoviruses 178
3. Pathogenesis of CNS Infections Caused by Paramyxoviruses 180
   3.1. Modes of Spread of Infection to the CNS 180
   3.2. Models for Immunobiological Characterization of
        Paramyxoviruses Involved in Brain Infection 182
   3.3. Persistent Infections with Paramyxoviruses in Vitro 182
   3.4. Persistent Paramyxovirus Infection in the Brain 183
   3.5. Local Antibody Production Caused by Paramyxovirus
        Infections in the CNS 186
   3.6. Cellular Immune Response in CNS Caused by
        Paramyxovirus Infections 188
4. Diseases of the CNS Caused by Paramyxoviruses 189
   4.1. Diseases of the CNS Caused by Viruses Belonging to the
        Paramyxovirus Genus 189

4.2. Diseases of the CNS Caused by Viruses Belonging to the
     Morbillivirus Genus 192
4.3. Other Less-Common Viruses Causing Diseases of the CNS
     of Animals 196
5. Summary and Concluding Remarks 197
   References 198

11. Human Papovaviruses: JC Virus, Progressive Multifocal
    Leukoencephalopathy, and Model Systems for Tumors of the
    Central Nervous System

EUGENE O. MAJOR, DOMINICK A. VACANTE,
and SIDNEY A. HOUFF

1. Introduction 207
2. Biology of JC Virus 208
   2.1. Host Range Properties for Growth 208
   2.2. Genetics of the Viral Genome 209
   2.3. Protein Products of the Viral Genome 211
3. Pathology of JC Virus in the Human Nervous System: Progressive
   Multifocal Leukoencephalopathy 213
   3.1. Clinical Features 213
   3.2. Description of the Pathology 214
   3.3. Mechanisms of Pathogenesis 214
4. Oncology of JC Virus 216
   4.1. Animal Models for Tumor Induction with JC Virus 216
   4.2. Animal Models for Tumor Induction with BK Virus 218
   4.3. Involvement of JC Virus in Human Brain Tumors 219
   4.4. Involvement of BK Virus in Human Tumors 220
5. Summary 221
   References 222

12. Neurological Aspects of Human Immunodeficiency Virus Infection

HOWARD E. GENDELMAN and SEYMOUR GENDELMAN

1. Introduction 229
2. Lentivirus–Host Interactions: An Overview 230
3. HIV Infection during Disease: Modes of Immunosuppression 231
4. Role of Monocytes/Macrophages in the Persistence and
   Dissemination of Lentivirus Infections 232
5. Monocyte/Macrophages and HIV Infection 234
6. Neurological Manifestations of HIV Infection: Clinical Aspects 234
   6.1. Clinical and Neuropathological Observations: Introduction ... 234
   6.2. Neurological Manifestations of Primary HIV Infection:
        Aseptic Meningitis 235

     6.3. AIDS–Dementia Complex (Subacute Encephalopathy) 237
     6.4. Vacuolar Myelopathy 238
     6.5. Neuromuscular Syndromes Associated with HIV Infection ... 239
  7. Biology and Pathogenesis of CNS Disease 240
  8. Conclusion 245
    References 246

13. Alphaviruses, Flaviviruses, and Bunyaviruses

DIANE E. GRIFFIN

  1. Introduction 255
  2. Alphaviruses 256
     2.1. Virus Biology 256
     2.2. Clinical Disease 258
     2.3. Pathogenesis and Pathology 258
     2.4. Persistence 259
     2.5. Prevention and Treatment 259
  3. Flaviviruses 259
     3.1. Virus Biology 259
     3.2. Clinical Disease 263
     3.3. Pathogenesis and Pathology 264
     3.4. Persistence 265
     3.5. Prevention and Treatment 266
  4. Bunyaviruses 266
     4.1. Virus Biology 266
     4.2. Clinical Disease 268
     4.3. Pathogenesis and Pathology 268
     4.4. Persistence 268
     4.5. Prevention and Treatment 268
    References 269

PART IV PERSPECTIVES

14. Antiglycolipid Immunity: Possible Viral Etiology of Multiple Sclerosis

H. E. WEBB

  1. Introduction 277
  2. Epidemiologic Evidence 277
  3. Evidence from Antiviral Antibody Studies and Virus Isolation 278
  4. General Mechanisms by Which Virus Damage Could Occur 281

5. Molecular Mimicry and Autoimmunity Related to Viruses
   and Their Possible Role in the Immunopathogenesis
   of Demyelination 281
6. Experimental Allergic Encephalitis, Experimental Allergic Neuritis,
   Viruses, and Demyelination 283
7. Nervous System Glycolipids and Multiple Sclerosis 285
8. Evidence that Antiglycolipid Activity Occurs in Multiple Sclerosis .. 286
9. Evidence that Host-Derived Envelope Membrane in Viruses Is
   Antigenic 286
10. Is There Evidence that Antiglycolipid Activity Produces
    Demyelination of the Central or Peripheral Nervous System? 287
11. Can Virus Infections of the Nervous System Present Glycolipids in
    Such a Way as to Be Immunogenic? 288
12. Conclusion 290
13. Summary of the Evidence that Virus-Induced Antiglycolipid
    Activity May Be Important in Multiple Sclerosis 292
    References 295

15. Viruses and Schizophrenia

JANICE R. STEVENS and LESLEY M. HALLICK

1. Introduction 303
2. Evidence for a Viral Origin 304
   2.1. Course and Pathology of the Illness 304
   2.2. Epidemiology 305
   2.3. Season of Onset and Season of Birth 307
   2.4. Immunologic Factors 307
   2.5. Search for the Virus 308
3. Conclusions 311
   References 312

16. The Role of Viruses in Dementia

ANNE M. DEATLY, ASHLEY T. HAASE, and MELVYN J. BALL

1. Introduction 317
2. Viruses or Virus-Like Agents, Dementia, and Chronic Neurological
   Disease 318
   2.1. Viruses Known to Cause Dementia 318
   2.2. Viruses and Chronic Neurological Disease 320
3. The Role of Viruses in the Etiology of Alzheimer's Disease 324
4. Neurological Disorders with a Possible Viral Etiology 327
   4.1. Tropical Spastic Paraparesis 327
   4.2. Postencephalitic Parkinsonism 328

5. Discussion 329
   References 330

17. Continuing the Investigation: Viruses and Neurological Disorders

STEVEN SPECTER, MAURO BENDINELLI,
and HERMAN FRIEDMAN

1. Unanswered Questions 339
2. The Need for Models 342
3. Unconventional Thinking 342
   References 343

INDEX 345

Viruses and Neuropsychiatric Disorders
Facts and Suppositions

STEVEN SPECTER, MAURO BENDINELLI, and HERMAN FRIEDMAN

1. THE FACTS

The list of human viruses that exhibit a well-documented though varied degree of neurotropism and neuropathogenicity is impressive. It includes RNA and DNA viruses of many different families and genera (Tables 1-1 and 1-2). In fact, there are few human viruses that appear entirely incapable of producing diseases at the level of the central (CNS) or peripheral (PNS) nervous system. To the list of well-characterized viruses we must add a still undefined number of poorly understood but clearly neurotropic transmissible agents commonly, though not solely, referred to as "unconventional viruses." These agents appear to share with viruses little more than the ability to replicate and be transmitted (Table 1-3).

The list of virus-induced diseases of the CNS and PNS is also a long one. These diseases can be loosely grouped as in the following sections.

STEVEN SPECTER and HERMAN FRIEDMAN • Department of Microbiology and Immunology, University of South Florida, College of Medicine, Tampa, Florida 33612-4799.   MAURO BENDINELLI • Department of Biomedicine, University of Pisa, I-56127 Pisa, Italy.

Neuropathogenic Viruses and Immunity, edited by Steven Specter *et al.* Plenum Press, New York, 1992.

TABLE 1-1
Major DNA Viruses Known to Affect the Nervous System of Man

Family	Genus	Virus[a]	Neurological involvement Frequency[b]	Localization[c]
Adenoviridae	*Mastadenovirus*	Adenovirus	±	M, E
Herpesviridae	*Simplexvirus*	HSV-1	+++	N, E, M
		HSV-2	+++	N, E, M
		B	+++	E
	Varicellavirus	VZV	++	N, M?, E?
	Cytomegalovirus	CMV	±	E
	Lymphocryptovirus	EBV	±	M?, E?
Polyomaviridae	*Polyomavirus*	JC	±	E
		BK	?	

[a]CMV, cytomegalovirus; EBV, Epstein–Barr virus; HSV, herpes simplex virus; VZV, varicella–zoster virus.
[b]Relative to total number of infections with the indicated virus.
[c]E, brain and/or spinal cord; M, leptomeninges; N, nerves.

1.1. Acute Diseases Resulting Directly from Viral Invasion

Viral invasion of the CNS has long been known to lead to acute diseases, which are primarily the result of direct cell damage or destruction by the infecting virus, although inflammatory and immunopathologically mediated injury of infected tissue usually also contributes significantly to the disease process. Because neurons and supporting cells have limited or no capacity to regenerate, the consequences may be devastating.

Viruses are responsible for most cases of acute meningitis and essentially all cases of acute encephalitis and encephalomyelitis in man. Although the etiologic agent of most such cases remains undetermined, the viruses that are more frequently found associated with encephalitis are herpes simplex virus (HSV), togaviruses, bunyaviruses, and enteroviruses (Fig. 1-1). Whereas togaviral, bunyaviral, and enteroviral forms are epidemic in nature and present clear geographic and temporal variations in their incidence (Fig. 1-2), HSV-induced cases occur sporadically.

The incidence of at least some epidemic forms of viral encephalitis has decreased recently in developed countries as a result of control measures to limit biological vectors (togaviruses), improved environmental sanitation (enteroviruses), and widespread use of vaccinations against polio and other viruses. Nevertheless, even in Western countries, the global public health impact of these diseases remains significant. The reported annual average 1500 cases of primary encephalitis and 10,000 cases of aseptic meningitis (although aseptic meningitis is not always viral meningitis, the great majority of cases are virus-induced) in the United States (Table 1-4) are gross underestimates because of marked under-reporting, and the true occurrence of these infections is probably 6 to 12 times higher.[3] In third-world countries preventive measures have been much less

TABLE 1-2
Major RNA Viruses Known to Affect the Nervous System of Man

Family	Genus	Virus[a]	Neuropathogenicity Frequency[b]	Neuropathogenicity Localization[c]
Arenaviridae	*Arenavirus*	LCMV	±	M, E
		Junin	±	M?
		Lassa	±	M?
Bunyaviridae	*Bunyavirus*	California encephalitis	+ +	E, M
		LaCrosse encephalitis	+ +	E
		Tahyna	±	E
	Phlebovirus	Rift Valley fever	±	E
		Toscana	±	M
Filoviridae		Marburg	±	E
Paramyxoviridae	*Paramyxovirus*	Mumps	±	M, E
	Morbillivirus	Measles	±	E
Picornaviridae	*Enterovirus*	Polio	+ +	M, E
		Coxsackie A	+	M, E
		Coxsackie B	+	M, E
		Echo	+	M
		Entero 70	+	M, E
Reoviridae	*Orbivirus*	Colorado tick fever	±	M, E
Retroviridae	*Oncovirus*	HTLV-1	?	
	Lentivirus	HIV-1	+ +	E, N
Rhabdoviridae	*Lyssavirus*	Rabies	+ + +	E
Togaviridae	*Alphavirus*	EEE	+ + +	E
		VEE	+ + +	E
		WEE	+ + +	E
	Rubivirus	Rubella	±	E
Flaviviridae	*Flavivirus*	Japanese encephalitis	+ +	E
		St. Louis encephalitis	+ +	E
		Murray Valley encephalitis	+ +	E
		Rocio	+	E
		West Nile encephalitis	+ +	E
		Central European TBE	+ +	E
		RSSE	+ +	E
		Louping Ill	+	E
		Powassan	+	E

[a]EEE, Eastern equine encephalitis; HIV, human immunodeficiency virus; HTLV, human T-lymphotropic virus; LCMV, lymphocytic choriomeningitis virus; RSSE, Russian spring–summer encephalitis; TBE, tick-borne encephalitis; VEE, Venezuelan equine encephalitis; WEE, Western equine encephalitis.
[b]Relative to total number of infections with the indicated virus.
[c]E, brain and/or spinal cord; M, leptomeninges.

S. SPECTER *et al.*

TABLE 1-3
Diseases Produced by "Unconventional
Viruses" in Man and Animals

Human
  Kuru
  Creutzfeldt–Jakob disease
  Gerstmann–Straussler syndrome
Animal
  Scrapie
  Transmissible mink encephalopathy
  Chronic wasting disease of mule deer and elk
  Bovine spongiform encephalopathy

effective to date, and even poliomyelitis and rabies, which have become very rare in developed countries (Figs. 1-3 and 1-4), still represent serious scourges. On the other hand, attempts to develop vaccines against important causes of human encephalitis such as Japanese encephalitis virus are still under way.

Although viral meningitis is usually a mild illness, viral encephalitides are very serious diseases, with death rates ranging between 2% and 50% depending on the etiologic agent and residual mental and motor disabilities in a significant proportion of cases. The recent introduction of effective chemotherapy has markedly reduced the gravity of HSV encephalitis, which previously was a devastating disease with up to 70% lethality and some sequelae in 25–50% of those who survived. Many such cases occur in children and adolescents.

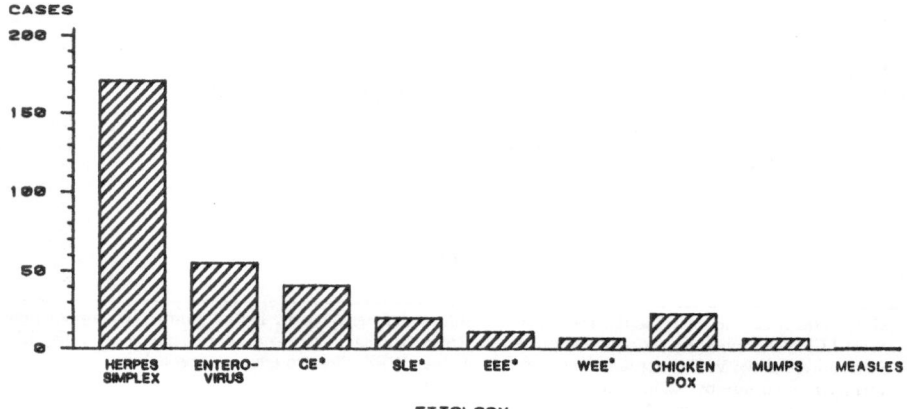

FIGURE 1-1. Reported cases of encephalitis by etiology in the United States in 1983. Cases of indeterminate etiology in the same year were 1,401. CE, California encephalitis; SLE, St. Louis encephalitis; EEE, Eastern equine encephalitis; WEE, Western equine encephalitis. (From Centers for Disease Control.[1])

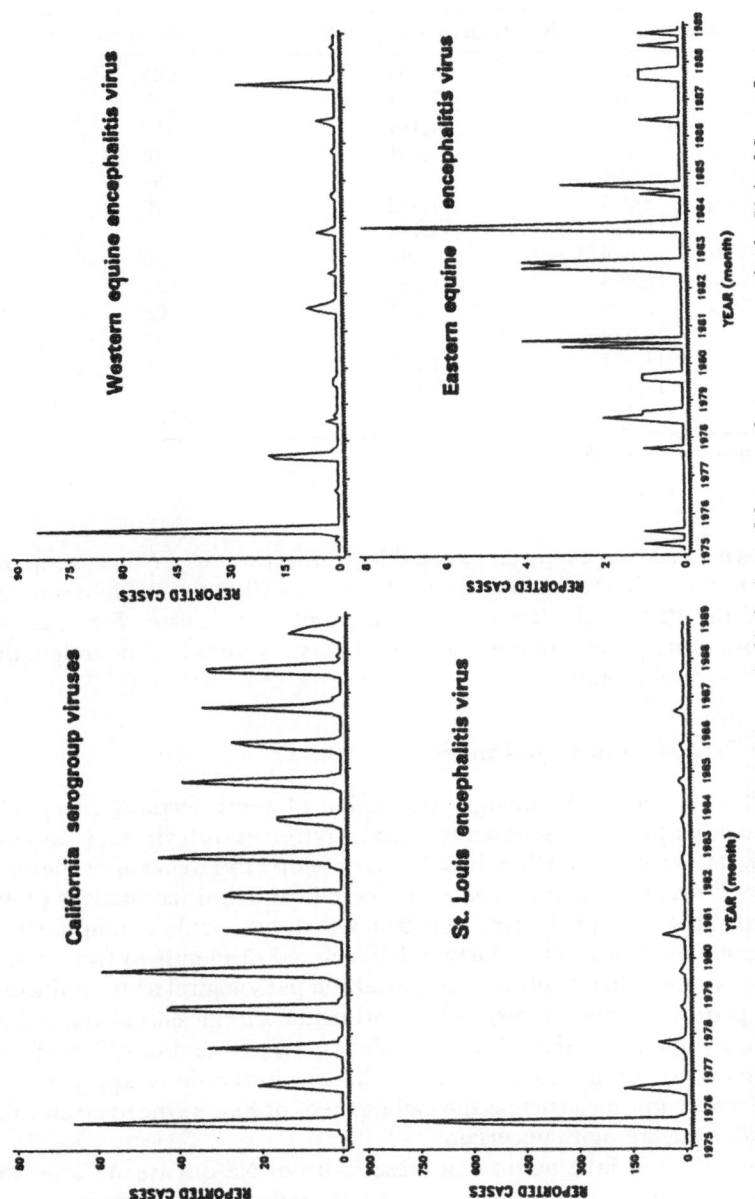

FIGURE 1-2. Reported cases of bunyaviral and togaviral infections of the central nervous system in the United States from 1975 to 1988. (From Centers for Disease Control.[2])

TABLE 1-4
Reported Cases of Aseptic Meningitis, Primary Encephalitides,
and Postinfectious Encephalitis, United States, 1975–1988[a]

Year	Aseptic meningitis	Primary encephalitis	Postinfectious encephalitis
1975	4,475	4,064	237
1976	3,510	1,651	175
1977	4,789	1,414	119
1978	6,573	1,351	78
1979	8,754	1,504	84
1980	8,028	1,362	40
1981	9,547	1,492	43
1982	9,680	1,464	36
1983	12,696	1,761	34
1984	8,326	1,257	108
1985	10,619	1,376	161
1986	11,374	1,302	124
1987	11,487	1,418	121
1988	7,234	882	121

[a]Data from Centers for Disease Control.[2]

Fewer viruses are recognized as capable of infecting the PNS as compared with those that invade the brain and spinal cord (see Chapter 3). However, viral invasion of the PNS is also associated with significant disease. For example, varicella zoster virus replication in peripheral nerves results in a neuralgia that may last for several months.

1.2. Acute Parainfectious Syndromes

The clinical and epidemiologic association of postinfectious encephalomyelitis and other parainfectious neurological syndromes with viruses is also well established. It was soon noted that these forms develop 1 to 4 weeks after the onset of a variety of viral infections and also after certain antiviral vaccinations (Table 1-5). Postinfectious encephalomyelitis is believed to be mainly autoimmune in nature because no virus can be consistently recovered or identified from neural tissues and because it shares substantial clinical and pathological features (including similar patterns of myelin loss) with experimental allergic encephalomyelitis. In addition, a high proportion of patients show antibody- and/or cell-mediated immune reactivity to neuroantigens. Similar considerations apply to the Guillain–Barré syndrome, whereas the pathogenesis of Reye's syndrome and Von Economo's disease are more uncertain.

Although there is little doubt that these forms of NS disease are somehow related to a preceding exposure to viruses or viral antigens, how so many widely diverse agents can trigger the same immunopathological response remains an enigma. An answer might lie in the complex immunologic dysregulations produced by many viral infections,[5] but it has also been suggested that the activation

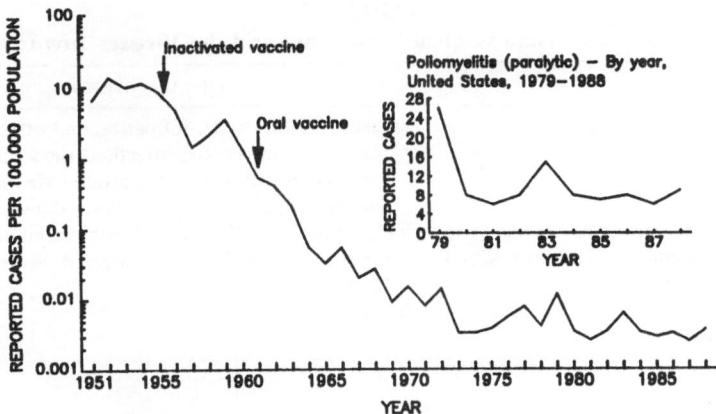

FIGURE 1-3. Reported cases of paralytic poliomyelitis in the United States from 1951 to 1987. (From Centers for Disease Control.[2])

of a second hitherto unidentified virus might be involved. Today, the Guillain–Barré syndrome is frequently associated with human immunodeficiency virus (HIV) infection, where it generally occurs when there still is little if any evidence of immunosuppression. Current hypotheses to explain this complication of HIV infection include a direct action of the virus or neurotropic HIV variants on nerve cells, autoimmune mechanisms, and circulating neurotoxins. It also seems likely that genetic and other cofactors contribute to the genesis of parainfectious neurological diseases. Epidemiologic studies have confirmed prior reports of an

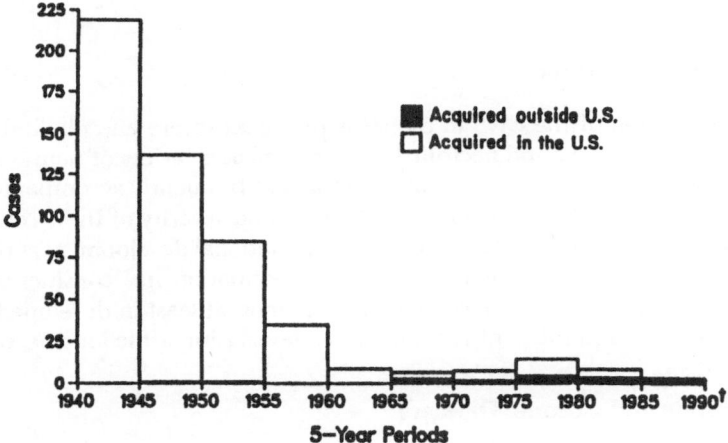

FIGURE 1-4. Reported cases of human rabies in the United States from 1940 to 1988, by 5-year periods. (From Centers for Disease Control.[4]) †Four-year period only.

TABLE 1-5
Parainfectious Neurological Syndromes and the Viruses Involved

Syndromes	Triggering viruses
Postinfectious encephalomyelitis	Measles, VZV, rubella, influenza, and other respiratory infections; infectious mononucleosis; vaccination with live or inactivated viruses (smallpox, yellow fever, measles, rabies)
Guillain–Barré syndrome (acute immune-mediated polyneuropathy)	VZV, mumps, CMV, HIV, EBV, rubella, enteroviruses; vaccination with live or inactivated viruses
Reye's syndrome	VZV, influenza
Von Economo's disease (encephalitis lethargica)	Influenza

association between ingestion of aspirin during antecedent viral illness and subsequent development of Reye's syndrome.[6] Whatever the underlying mechanisms, it is clear that a wide spectrum of viruses can initiate an immune-mediated attack on the CNS, the PNS, or both.

The discontinuation of smallpox vaccination and the continuing practice of vaccination against exanthematous diseases of childhood have considerably reduced the incidence of postinfectious encephalitis in developed countries (Table 1-4), where it is now mainly seen in conjunction with upper respiratory tract infections and varicella.[7] In countries where measles vaccination has not yet become widespread, measles remains the major cause of postinfectious encephalitis. Its frequency is approximately one case per 1000 cases of measles,[8] and so worldwide incidence might be as high as 100,000 cases per year. Case-fatality rates of postinfectious encephalitis may be high, up to 50%, depending on the triggering event. However, most patients who survive the acute phase of the disease recover completely.

1.3. Congenital Defects

Viral invasion of the CNS in the fetus produces severe encephalitides with extensive inflammatory and necrotic lesions as well as a variety of neural malformations and functional deficits (Table 1-6) that are frequently accompanied by a number of other teratogenetic effects. The type and severity of these conditions depend on the etiologic agent as well as on the gestational development at the time of infection.[9,10] The introduction of rubella vaccination has considerably decreased the frequency of these congenital infections, at least in developed countries, but a limited number of cases of these devastating afflictions persist.

1.4. Subacute and Chronic Diseases

More recently, a number of well-defined subacute and chronic neurological disorders have been recognized to be caused by persistent infection with conven-

TABLE 1-6
Major Neurological Malformations and
Deficits Associated with Viral Infections of the Fetus

Diseases	Viruses involved
Microcephaly, hydrocephalus, limb hypoplasia, microgyria, cerebral calcifications, visual, auditory, motor, and mental deficits, retardation	Rubella, CMV, HIV, HSV, enteroviruses, others

tional and unconventional viruses (Table 1-7). These disorders include both "degenerative" noninflammatory diseases and demyelinating inflammatory diseases. These conditions are generally rare, but some are opportunistic diseases and develop preferentially in immunocompromised subjects; their number appears likely to increase in association with the increasing numbers of immunocompromised individuals. An augmented incidence of papovavirus-induced progressive multifocal leukoencephalopathy has already been noted as a consequence of the HIV pandemic. It also has been suggested that in HIV-immunosuppressed patients papovavirus activation may not only cause damage by directly infecting oligodendroglia but may also cause additional damage by attracting HIV-infected macrophages.[11] As discussed below, the importance of the recognition of these NS diseases as being of viral etiology lies also in the expectations that the etiology of other apparently idiopathic chronic diseases of the CNS may be similarly demonstrated to be related to detectable infectious agents.

In conclusion, despite a number of significant victories that have reduced morbidity and mortality from NS infections in the last two to three decades, the burden in suffering, premature death, and long-term sequelae of the neuropathogenic potential of viruses, as presently understood, is a very heavy one.

TABLE 1-7
Subacute and Chronic Neuropathies Caused by Viruses

Disease	Virus
Subacute sclerosing panencephalitis	Measles
Chronic progressive panencephalitis	Rubella
Progressive congenital encephalomyelitis	Rubella
Congenital cytomegalic inclusion disease	CMV
AIDS–dementia complex and peripheral neuropathies	HIV
Spongiform encephalopathies	Unconventional viruses
Recurrent meningitis[a]	Echo
Subacute encephalitis[a]	Measles
Progressive multifocal leukoencephalopathy[a]	JC polyomavirus

[a]Occurring in immunocompromised individuals only.

2. THE SUPPOSITIONS

As mentioned above, certain rare chronic neuropathies of man long considered idiopathic in nature are now known to be caused by viruses, and others by unprecedented infectious agents—discovered much as a result of the work of Carleton D. Gajdusek—that are much more elusive than conventional viruses. Interestingly the latter agents do not appear to evoke an inflammatory or immune response (see Chapter 7). In addition, animal models have clearly shown that the nervous system is a target for persistence by a wide spectrum of viruses and that neurotropic viruses can cause significant behavioral changes without producing obvious histopathological alterations.[12] These findings have aroused suspicions (and hopes) that other and more widespread unexplained neurological and psychiatric disorders of man may be of similar etiologies. There is now consistent, albeit circumstantial, evidence that the human T lymphotropic (retro)virus type 1 (HTLV-1) is implicated in the genesis of tropical spastic paraparesis, a disease that markedly resembles multiple sclerosis and other neuropathies[13] (see Chapter 12). Because HTLV-1 isolated from tropical spastic paraparesis does not differ significantly from the leukemogenic prototypes, it has been speculated that the virus might induce either lymphoproliferative disease or chronic neuromyelopathy depending on as yet unknown cofactors.[14] Since experimental infection with certain viruses (adenoviruses, polyomaviruses, and retroviruses) can induce tumors of varied histotype in the CNS of rodents and other animals, it is also suspected that some brain tumors of man might have a viral origin (Table 1-8).

Because the suspicion of a viral etiology for "idiopathic" nervous system diseases is of the utmost importance, this area of research is currently one of great excitement. Several chapters of the present volume describe the intensive research pursued with regard to the role of viruses, virus-like elements, and virus-related genetic elements in a number of important neuropsychiatric diseases of unknown origin. The difficulties encountered in these studies are many, and progress is inevitably slow. Frustration in this area is exemplified by the fact that over the years some 20 different viruses have been suggested as possible etiologic agents of multiple sclerosis (see Chapter 14). Thus, today it is impossible to predict whether and to what extent current suspects will be corroborated.

One point seems, however, easy to forecast. Several years ago Professor Richard T. Johnson pointed out that, "in the etiopathogenesis of neuropathies it is often unclear where the field of virology ends and that of immunology begins."[15] This has become even more evident with the recognition that complex bidirectional functional and regulatory interactions are operative between the nervous and immune systems. In light of such interactions, it seems highly likely that viruses and other agents that primarily affect one of the systems may also have a profound impact on the other. It is, therefore, to be expected that future advances will mainly stem from collaborative efforts in the areas of molecular virology, viral immunobiology, neurobiology, and neuroimmunology. The newly recognized HIV-associated neuropsychiatric disorders have already taught a great deal in this direction. Mimicking the contrast between the severe gener-

TABLE 1-8
Some Neurological and Psychiatric Disorders
for Which a Viral Origin Has Been Suggested

Demyelinating
  Multiple sclerosis and its variants (neuromyelitis optica, concentric sclerosis, etc.)
  Tropical spastic paraparesis (HTLV-1?)
  Behcet's disease
  Other chronic myelopathies
Degenerative
  Amyotrophic lateral sclerosis
  Parkinson's disease
  Alzheimer's disease
  Other presenile dementias
Vascular
  Arteriosclerosis
  Angiitis
Psychiatric
  Schizophrenia
Proliferative
  Meningioma
  Glioma
  Von Recklinghausen's disease
  Other tumors
Others
  Myalgic encephalomyelitis (or postviral fatigue syndrome)

alized immunodeficiency and the low proportion of HIV-expressing immunocytes found in AIDS (1/10,000 to 1/100,000), the number of HIV-infected cells in
neural tissues appears inadequate to explain the neuropsychological changes,
including dementia and cortical atrophy, present in many patients (see Chapter
12). This substantiates a long history of speculations about possible mechanisms,
both immunologically and nonimmunologically mediated, of indirect viral damage to the nervous system. Why should we presume that such disparity is limited
to HIV?

REFERENCES

1. Centers for Disease Control, 1984, Annual summary 1983: Reported morbidity and mortality in the United States, *MMWR* **32**:1–125.
2. Centers for Disease Control, 1989, Summary of notifiable diseases, United States 1988,
   MWWR **37**:1–57.
3. Beghi, E., Nicolosi, A., Kurland, L. T., Mulder, D. W., Hauser, A., and Shuster, L., 1984,
   Encephalitis and aseptic meningitis, Olmsted County, Minnesota, 1950–1981: I. Epidemiology, *Ann. Neurol.* **16**:283–294.

4. Centers for Disease Control, 1988, Rabies surveillance, United States, 1988, *MMWR* **38** (Suppl. 1):1–21.
5. Specter, S., Bendinelli, M., and Friedman, H. (Eds.), 1989, *Virus-Induced Immunosuppression*, Plenum Press, New York.
6. Pinsky, P. F., Hurwitz, E. S., Schonberger, L. B., and Gunn, W. J., 1988, Reye's syndrome and aspirin: Evidence for a dose–response effect, *JAMA* **260**:657–661.
7. Johnson, R. T., Griffin, D. E., and Gendelman, H. E., 1985, Postinfectious encephalomyelitis, *Semin. Neurol.* **5**:180–190.
8. Miller, H. G., Stanton, J. B., and Gibbons, J. L., 1956, Parainfectious encephalomyelitis and related syndromes, *J. Med.* **25**:427–505.
9. Hanshaw, J. B., Dudgeon, J. A., and Marshall, W. C., 1985, *Viral Diseases of the Fetus and Newborn*, 2nd ed. W. B. Saunders, Philadelphia.
10. Johnson, R. T., and Lyon, G. (Eds.), 1988, *Virus Infections and the Developing Nervous System*, Kluwer, Dordrecht.
11. Wiley, C. A., Grafe, M., Kennedy, C., and Nelson, J. A., 1988, Human immunodeficiency virus and JC virus in acquired immunodeficiency syndrome patients with progressive multifocal leukoencephalopathy, *Acta Neuropathol. (Berl.)* **76**:338–346.
12. Kristensson, K., and Norrby, E., 1986, Persistence of RNA viruses in the central nervous system, *Annu. Rev. Microbiol.* **40**:159–184.
13. Roman, G. C., Vernant, J.-C., and Osame, M. (Eds.), 1989, *HTLV-1 and the Nervous System*, Alan R. Liss, New York.
14. Gessain, A., Saal, F., Morozov, V., Lasmeret, J., Vilette, D., Gout, O., Emanoil-Ravier, R., Sigaux, F., deThe, G., and Peries, J., 1989. Characterization of HTLV-1 isolates and T lymphoid cell lines derived from French West Indian patients with tropical spastic paraparesis, *Int. J. Cancer* **43**:327–333.
15. Vinken, P. J., and Bruyn, G. B. (Eds.), 1978, *Handbook of Clinical Neurology. Volume 34. Infections of the Nervous System*, Part II, North-Holland, Amsterdam.

I

General
Information

2

Immune Responses and the Central Nervous System

WILLIAM R. TYOR and RICHARD T. JOHNSON

1. INTRODUCTION

The central nervous system (CNS) is relatively isolated from systemic immune responses in the absence of disease. Within the normal CNS, there is no mechanism for antibody production, no lymphatic system, and few if any phagocytic cells. The CNS has been described as an "immunologically privileged site" because of the paucity of normal immune surveillance. Consequently, when a virus penetrates the blood–brain barriers that exclude most infectious agents, the same barriers may deter viral clearance.

Low levels of immunoglobulins are found in normal cerebrospinal fluid (CSF), but these immunoglobulins are derived solely from the blood.[1] Small numbers of lymphocytes are also present in the CSF, and they normally mirror the T helper/T suppressor–cytotoxic cell ratios in the blood. The circulation or function of these cells is unknown; they presumably enter from the blood through the arachnoid or choroid plexus vessels and exit at the cervical lymphatics.[2]

Immune responses in the CNS during infection are recruited from the systemic circulation in a relatively selective and specific fashion. Cells and antibodies found in the nervous system during infections differ from those that follow nonspecific rupture in the blood–brain barrier such as occurs after a traumatic injury (e.g., a stab wound). In traumatic lesions the transudate of serum contains antibodies, and cells of all types enter, but with a predominance of monocytes that differentiate into macrophages. During viral infections, although an early in-

WILLIAM R. TYOR and RICHARD T. JOHNSON • Department of Neurology, Johns Hopkins University School of Medicine, Baltimore, Maryland 21205.

Neuropathogenic Viruses and Immunity, edited by Steven Specter *et al.* Plenum Press, New York, 1992.

crease in permeability of vessels allows transudation of serum proteins, cell entry is immunologically specific, and the cells that enter have specific kinetics and do not simply mirror the proportions of cell phenotypes in the blood. These cells in turn are caused to replicate or differentiate within the CNS, and B lymphocytes mature into antibody-forming cells that may persist for long periods of time within the CNS.[3]

Therefore, to review immune responses in the CNS it is necessary first to discuss the anatomy and physiology of extraneural immune responses and then to discuss how the responses are recruited into and evolve within the CNS.

2. GENERAL OVERVIEW OF THE IMMUNE SYSTEM

2.1. Anatomy of the Immune System

The general anatomy of the immune system can be divided into the lymph nodules, lymph nodes, the spleen, the thymus, and the reticuloendothelial system (RE system). The cellular constituents of the immune system include T cells, B cells, monocyte–macrophages, dendritic cells, and endothelial cells.

Lymph nodules are collections of lymphoid cells that occur in submucosal layers in the gastrointestinal (i.e., Peyer's patches), respiratory, and genitourinary tracts. The B cells that lie within these nodules secrete relatively high amounts of IgA and IgE, which contribute to the defense against pathogens in the external environment (i.e., on the mucosal surfaces of these tracts). If pathogens breach this barrier, they may then enter the lymphatic system in which they are carried to the lymph nodes, which are concentrated in various locations throughout the body. The lymph nodes that receive most of the drainage from the CNS are the anterior and the posterior cervical lymph nodes.[4]

Lymph nodes are surrounded by a connective tissue capsule through which afferent and efferent lymphatic vessels pass, as well as arteries and veins. From the capsule (Fig. 2-1) trabeculae extend between germinal centers, which contain actively dividing lymphocytes (primarily B cells) and blood vessels. Toward the center of the lymph node is the thymic-dependent area (primarily T cells), which is highly vascularized, allowing passage of lymphocytes to and from the blood. Lymph nodes serve as filters in the lymphatics as well as providing a source for circulating lymphocytes in blood and the lymphatics.[4]

If pathogens gain access to the blood, they will encounter the spleen and/or the RE system. The spleen serves a number of functions. It removes undesirable elements from the blood including abnormal or senescent cells as well as pathogenic microbes. Overwhelming bacterial infections can occur when the spleen is removed or when its function is impaired, as in sickle cell anemia. In addition, the spleen converts hemoglobin to bilirubin, recirculates iron, and serves as a major source for the production of lymphocytes. The spleen has a connective tissue capsule (Fig. 2-2) with trabeculae extending into the tissue. The white pulp contains lymph nodules with germinal follicles that produce B cells. These follicles are adjacent to T-cell areas through which arterioles course, providing a

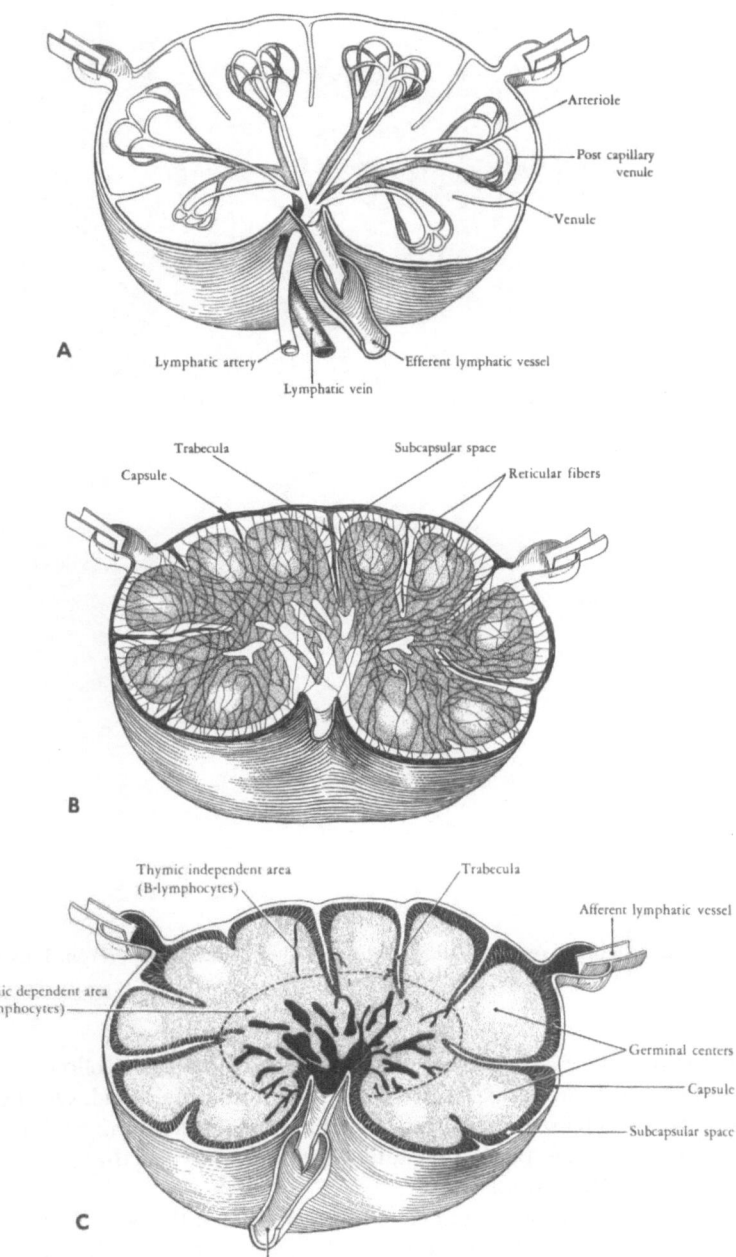

FIGURE 2-1. Structure of lymph node, schematic: A, circulation; B, supporting structures (reticular fibers); C, general areas of thymic-dependent (T-cell) and -independent (B-cell) areas. (From Bellanti,[4] p. 37.)

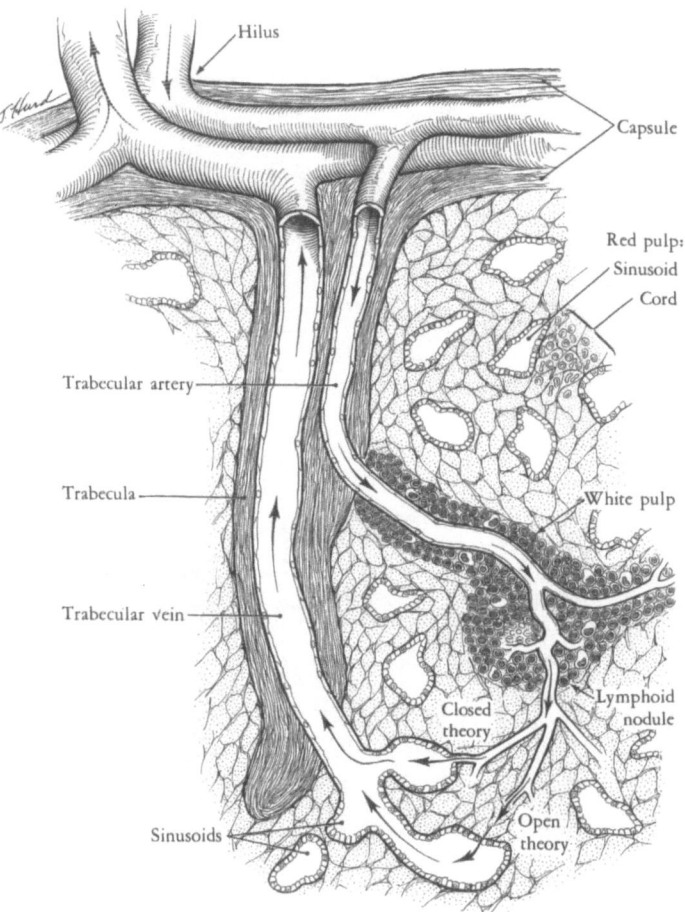

FIGURE 2-2. Schematic representation of the structure of the spleen. (From Bellanti,[4] p. 39.)

means for passage of lymphocytes to and from the blood. Following antigenic stimulation, B cells are released from the spleen into the blood. On the margins of the T- and B-cell areas are significant numbers of macrophages. The red pulp surrounds this white pulp area and contains many erythrocytes as well as elements of the RE system.[5]

The RE system is present in most of the organs of the body. It is a division of the mononuclear-phagocyte system, which includes the immature precursors of the monocytes that are found in the bone marrow, monocytes in the peripheral blood, infiltrating macrophages, and tissue or resident macrophages.[6] The RE system is comprised of the tissue or resident macrophages, the particular organs in which they reside, and possibly dendritic and endothelial cells. The RE system serves a role both as a scavenger and as a signaler to the immune system, alerting it to the presence of pathogens. The cell types that have been identified with the highest degree of certainty as being tissue macrophages are alveolar macrophages, peritoneal macrophages, and Kupffer cells (Table 2-1).[7] When tissue macrophages are activated by invading pathogens, they produce a number of soluble factors (monokines) and process and present antigens as described later.

Dendritic cells are often considered to be a part of the RE system. These cells are found in lymphoid tissue but are also found in the skin in the form of Langerhans' cells and in most other organs, except brain, in the form of interstitial dendritic cells. These cells are irregularly shaped with a variety of cell processes, including dendrites and pseudopods. Dendritic cells do not phagocytose particles *in vitro*; however, some evidence indicates that they perform this activity *in vivo*. They constitutively express class I and class II molecules of the major histocompatibility complex (MHC) as well as complement receptors. Class I and II molecules are involved in signaling T cells about the presence of a foreign antigen. Dendritic cells function as potent accessory cells in T-cell-dependent immune responses; therefore, they are important in primary immune responses as antigen-presenting cells[8] (discussed in Section 2.3). Vascular endothelial cells may also function in a role similar to that of dendritic cells. They are able to present antigen to T cells in an MHC-restricted fashion *in vitro*; however, it is unclear if they present antigen *in vivo*.[9]

2.2. Cellular Development

T-lymphocyte precursors develop from stem cells in the bone marrow (BM). They then migrate to the thymus, where they are termed thymocytes, and further maturation occurs. The thymus is composed of a connective tissue capsule with septa extending into the gland, dividing its two lobes into multiple lobules (Fig. 2-3). In addition to T-cell lymphopoiesis, the thymus may serve an endocrine

TABLE 2-1
Components of the Reticuloendothelial System

Cell type	Tissue location
Kupffer cells	Liver
Alveolar macrophages	Lung
Langerhans' cells	Skin
Histiocytes	Connective tissue, skin
Osteoclasts	Bone
Peritoneal and pleural macrophages	Serous cavities
Tissue or resident macrophages	Lymph nodes, spleen, bone marrow, and others
Dendritic cells (?)	Lymphoid tissues such as spleen, lymph nodes, and lymph nodules
Vascular endothelial cells (?)	All tissues
Microglia (?)	Brain

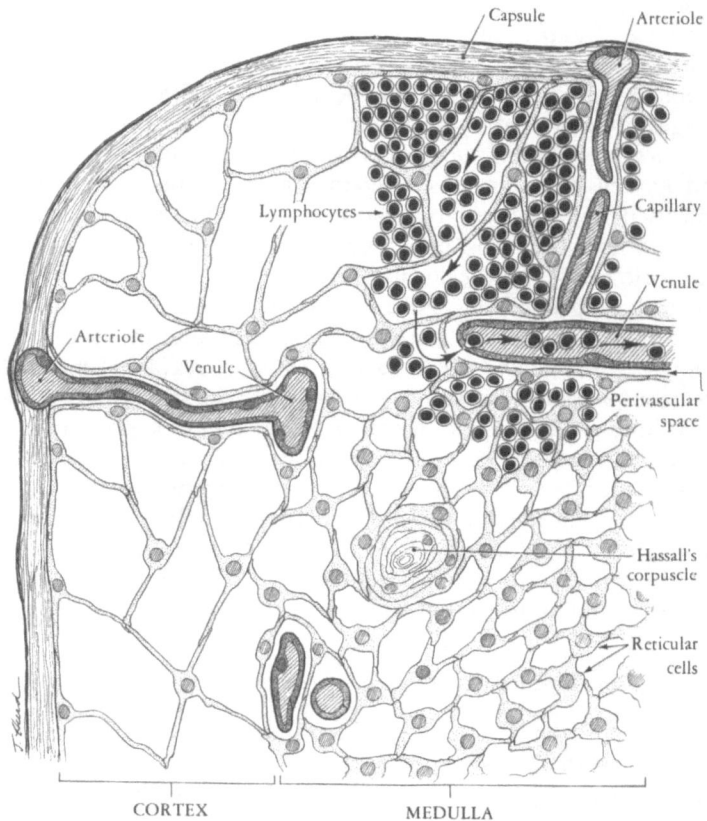

FIGURE 2-3. Thymus gland. Schematic representation of the perivascular epithelium surrounding blood vessels in the cortex. Note the barrier provided by this sheath and the pathways of lymphocutes formed in the cortex into the blood vessels. (From Bellanti,[4] p. 44.)

function with the production of hormones including thymosin, which may serve to modulate the immune system.[4,5]

The further development of T-lymphocyte precursors from the BM begins in the peripheral portion or cortex of the thymus. Here a subpopulation of immature T cells is selected and expanded by mechanisms that are not fully understood. Many immature thymocytes that are not selected for expansion will die. Further maturation of these expanded populations of immature thymocytes is characterized by changes in their surface antigens (Table 2-2) and functional properties as they migrate toward the central portion or medulla of the thymus. It is thought that these immature thymocytes first encounter antigen-presenting cells at the corticomedullary junction, and tolerance to self-antigens is induced. In humans, three stages of development have been defined with the use of various

TABLE 2-2
Cell Surface Antigens of Lymphocytes and Macrophages

Cell type	MHC		CD		Complement receptor	FC receptor
	Class I	Class II	4	8		
T-helper cells	+	±	++	−	−	−
T-suppressor cells/cytotoxic	+	±	−	++	−	−
B cells	+	++	−	−	+	+
Monocyte/macrophages	+	±	+	−	+	+

monoclonal antibodies. The first two stages are characterized by the addition of certain cell surface molecules that can be identified using specific monoclonal antibodies and immunohistochemical techniques. During the third stage two distinct populations of mature thymocytes can be identified. These are CD4+ thymocytes and CD8+ thymocytes. These cells are eventually released into the blood; the CD4+ cells comprise approximately 60% of the total blood T lymphocytes, and the CD8+ cells comprise 20–30% of the total T-lymphocyte population in the blood. The CD4+ cells roughly correspond to the helper/inducer population of T cells, and the CD8+ cells include the cytotoxic and the suppressor populations of T cells. Circulating T cells also have T 11 (the sheep erythrocyte-binding protein) and a receptor for antigen termed the Ti–T3 complex. The Ti–T3 receptor complex is composed of the Ti αβ or γδ heterodimer and three molecules constituting T3. This receptor complex is discussed later.[4,5,10]

B lymphocytopoiesis occurs in the BM when multipotent stem cells enter the B-cell pathway and begin to produce cytoplasmic μ heavy chains. These pre-B cells then express surface IgM and will subsequently coexpress IgD. The fascinating sequence of immunoglobulin gene rearrangement and isotype switching is beyond the scope of this text. Suffice it to say that light and heavy chain rearrangement is complete by the time these immature B cells express IgM and IgD on their surfaces. At this time, these cells are antigen committed although not yet antigen stimulated. When immature B cells leave the bone marrow, they are able to respond to antigen stimulation. At this time they can migrate to peripheral lymph organs such as the spleen and lymph nodes. In addition to surface immunoglobulin, B cells have been found to vary in their expression of complement receptors, interferon (IFN) receptors, IgG, IgD, IgM, and IgE receptors, B-cell-stimulating factor-1 receptors, B-cell-stimulating factor-2 receptors, transferrin receptors, interleukin-2 (IL-2) receptors, and class II molecules.[4,11]

Mononuclear phagocyte (monocyte/macrophage) development begins in the BM. The monoblast is the most immature cell of this line and is derived from the colony-forming unit granulocyte–monocyte cell. With division of the monoblast the promonocyte is formed, which is the direct precursor of the monocyte. Monocytes are able to enter the circulation (their half-life is approximately 3 days), where they may then migrate to various tissues and become resident macrophages.[12] The macrophage population within a tissue is primarily dependent on monocyte influx rather than on local production of new macrophages from the existing resident macrophages. The life span of tissue or resident macrophages is on the order of months. Whether these resident macrophages are able to migrate to other tissues is unclear, but the recirculation of macrophages in peripheral blood is minimal. Nonetheless, the migration of macrophages to nearby lymph nodes does occur, and there is evidence that they may then perish there. Macrophages have Fc receptors, complement receptors, a low concentration of CD4 molecules, and may express class I and class II molecules as well.[13]

2.3. Activation of the Immune System

Once pathogens enter the body, they may be taken up and processed by a number of different cell types that are a part of the RE system. After the pathogen has been processed within the cell, antigen is expressed on the cell membrane in association with immune-associated (Ia) molecules, also called class II molecules.[14] Class II molecules are coded for by genes in the MHC. Specifically in humans, this is designated the human leukocyte antigen D (HLA-D) region. This region is polymorphic, providing for a great deal of genetic variability in the expression of class II molecules.[15] Class II molecules are constitutively expressed on dendritic cells of the RE system and mature B cells. They are also expressed on a minority of macrophages and T cells normally; however, during immune activation all of these cell types can up-regulate class II molecule expression.[4]

Pathogens invading a particular tissue may be phagocytized by resident macrophages. Alternatively, the pathogen may infect or be taken up by cells such as dendritic or possibly vascular endothelial cells. The processed pathogen or antigen, when expressed on the surface of these cell types in association with class II molecules, can be recognized by specific T-helper cell clones. These specific T-helper cell clones are then signaled to proliferate and induce other cellular components of the immune system through the elaboration of cytokines (see below). Generally T-helper cells are only able to recognize antigen if it is associated with class II molecules on an antigen-presenting cell (APC) such as macrophages and dendritic cells.[4,8–10,13]

Class I molecules are also coded for by genes in the MHC. Specifically in humans these are the HLA-A, HLA-B, and HLA-C regions, each of which has multiple alleles (like the HLA-D region, they are polymorphic). Class I molecules are found on most nucleated cells. They were originally described as major transplantation antigens involved in graft rejection. Cytotoxic (CD8+) T cells are only able to recognize antigen if it is associated with class I molecules on the surface of cells that have been infected or that have taken up the antigen. Classically this has been demonstrated in the context of virally infected cells that are lysed by cytotoxic T cells that recognize the viral antigen on the surface of the infected cell in association with class I molecules.[4]

Many immune cells, when activated, can produce cytokines. Cytokines function as amplifiers of the immune system. Many of these cytokines and their biological properties are listed in Table 2-3 along with the cell types that produce them.[16,17]

TABLE 2-3
Relevant Cytokines

Cytokine	Cell production	Biological properties
Interleukin-1 (IL-1)	Most cells, including T and B lymphocytes, macrophages, natural killer cells, epithelial cells, mesangial cells, vascular endothelial cells, astrocytes, microglia, and skin keratinocytes	Activates T cells, endothelial cells, and macrophages; induces sleep, fever, release of ACTH, cortisol, and insulin, synthesis of lymphokines and collagen; cofactor for hematopoietic growth factors; mediates inflammation and acute-phase responses
Interleukin-2 (IL-2)	T-helper cells	Stimulates the clonal expansion of T cells and B cells; induces the synthesis of other lymphokines; enhances cytolytic activity of natural killer cells
Interleukin-3 (IL-3)	Activated T cells	Stimulates multilineage bone marrow stem cell and mast cell growth
Interleukin-4 (IL-4; B-cell-stimulating factor 1)	Activated helper T cells	Stimulates growth of B cells, T cells, thymocytes, and macrophages; stimulates Ia, IgG_1, and IgE; activates macrophages
Interleukin-5 (IL-5; B-cell growth factor II, eosinophil–CSF)	Activated helper T cells	Stimulates B-cell and eosinophil growth; enhances IgA, IgM, and IL-4-induced IgE production; enhances IL-2-mediated killer cell induction
Interleukin-6 (IL-6; B-cell-stimulating, β_2-stimulating factor)	T-helper cells, macrophages, vascular endothelial cells, fibroblasts, bone marrow stromal cells	Induces proliferating B cells to differentiate into plasma cells; increases synthesis of hepatic factor 2, interferon, and hepatocyte acute-phase proteins
Tumor necrosis factor (TNF, cachectin), α and β	Macrophages, actived T cells	Induces sleep, fever, and acute-phase response; cytotoxic for some tumor cells; stimulates synthesis of cytokines and collagen; activates macrophages and endothelial cells; mediates inflammation and septic shock
Colony-stimulating factors granulocyte–macrophage	T cells, vascular endothelial cells, fibroblasts, bone marrow stromal cells	Stimulates growth of neutrophils, eosinophils, and macrophage colonies in bone marrow; enhances granulocyte functions
Granulocyte	Macrophages, vascular endothelial cells, fibroblasts, bone marrow stromal cells	Stimulates growth of neutrophil colonies
Macrophage	Macrophages, vascular endothelial cells, bone marrow stromal cells, fibroblasts	Stimulates growth of macrophage colonies

Continued

TABLE 2-3 (*Continued*)

Cytokine	Cell production	Biological properties
Interferon γ (IFN γ; immune interferon)	Activated T cells	Induction of class I and class II molecules; activates macrophages and endothelial cells; synergistic or antagonist interaction with other cytokines; antiviral activity; enhances natural killer cell activity
Interferon α (IFN α; leukocyte interferon)	Lymphocytes, macrophages	Antiviral and antiproliferative activity; increases class I molecules on lymphocytes; enhances natural killer cell activity
Interferon β (IFN β; fibroblast interferon)	Fibroblasts, epithelial cells, and macrophages	Antiviral and antiproliferative activity

Interleukin-1 (IL-1) may be produced by any nucleated cell type in response to foreign antigen, toxin, injury, or inflammation. It is a general stimulator of the immune system. It accomplishes this primarily by stimulating the production of other cytokines such as IL-2, IL-3, IL-4, and interferon (IFN) γ. The systemic effects of IL-1 are similar to those of tumor necrosis factor (TNF), and these factors may act synergistically to effect such responses as tumor necrosis, hypotension, fever, and inflammation.[16,18] Interleukin-2 is produced by T cells and serves primarily to stimulate IL-2 receptor formation and the proliferation of activated T-helper cells, cytotoxic/suppressor T cells, and B cells. Interleukin-3, IL-4, and IL-5 are also produced by activated T cells. Interleukin-3 stimulates hematopoiesis; IL-4 and IL-5 primarily serve as stimulators of B-cell growth. Interleukin-6 is produced by a number of cell types including macrophages and T cells; like IL-3 it serves to stimulate hematopoiesis.[16,17]

Interferon α and β are primarily induced by viral infection. Interferon α is produced by lymphocytes, and IFN β is produced by fibroblasts, epithelial cells, and macrophages. They have potent antiproliferative and antiviral properties. Interferon γ is produced by activated T cells, induces class I and class II molecule expression, activates macrophages, exerts antiviral activity, and variously enhances or inhibits other cytokine activities.[16]

Colony-stimulating factors are produced during immune activation by lymphocytes. Granulocyte–macrophage colony-stimulating factor is the best characterized, and this cytokine supports growth and differentiation of granulocytes and monocytes.[16,17]

The complement (C) system is composed of a large number of plasma and cell membrane proteins that interact on a sequential basis during complement activation. There are two pathways of complement activation: (1) The classical pathway, which usually involves antibody-bound cell-associated antigen or antigen–antibody complexes, and (2) the alternative pathway, which may be activated in the absence of antibody, provided a suitable target is present.[4,19]

The classical pathway typically is activated by antibody (IgG_1, IgG_2, IgG_3, or IgM)–antigen complexes. The antigen may be free in the plasma or cell-associated. The classical pathway can also be activated by bacterial lipopolysaccharide, retroviruses, and C-reactive protein, which may be bound to *Streptococcus pneumoniae*. The first step involves binding of C-1 to two or more Fc regions of the antibody (or antibodies) bound to antigen. This activates C-1 and begins a cascade of events that is next highlighted by activation of C-3, the first event common to both pathways. The alternative pathway is activated by polysaccharides, yeast cell walls, fungi, bacterial cell wall components, and certain viruses. This pathway begins with C-3 activation. Once C-3 is activated, the series of events that follow is common to both pathways. A number of proteins are involved in sequential enzymatic reactions that result in the formation of by-products, the most important of which are the membrane attack complex, C-3a and C-5a. C-3a and C-5a have important immunoregulatory effects. They act as anaphylatoxins causing mast cells to degranulate. C-5a also acts as a chemotactic agent for neutrophils. The membrane attack complex that is generated by the full complement cascade on the cell surface is a cylindrical structure through which small ions can pass. Once this structure is formed, the cell cannot maintain its osmotic equilibrium and is disrupted.[4,19]

Complement activation may cause severe tissue injury, especially in the context of autoantibody activation and immune-complex formation. Autoantibodies may arise during infection or with tissue damage. The binding of antibody to host tissue can cause activation of the complement system and subsequent tissue damage. Immune-complex deposition in the walls of small vessels may also activate complement and result in tissue damage.[19]

During inflammatory processes, increasing numbers of monocytes migrate into affected tissues and undergo differentiation into macrophages. Macrophages phagocytize or take up antigen and catabolize it to varying degrees in the lysosomes of the cytoplasm.[20] These processed antigens are degraded or altered and may be distributed on the cell membrane in association with class I or II molecules. If the processed antigen is associated with a class I or class II molecule on the macrophage membrane, T-suppressor/cytotoxic or T-helper cells, respectively, can be signaled to the presence of this antigen and undergo clonal proliferation.[21] Resting tissue macrophages and new infiltrating macrophages may then be activated through release of cytokines such as IFN γ and granulocyte–macrophage colony-stimulating factor from T-helper cells. Once activated, macrophages increase their ability to phagocytize particles and to present processed antigen in association with MHC molecules. They increase certain IgG Fc receptors and thus increase their ability to pick up antibody-coated antigens. They have increased ability to ingest and kill *Mycobacterium*, *Listeria*, *Toxoplasma*, fungi, organic particles, and tissue debris and to enhance tumor rejection. The functions of macrophages in tumor rejection include secretion of products such as lysozymes, cytolytic proteases, IL-1, prostaglandins, IFNβ, and TNF in addition to their enhanced phagocytic capacities. Activated macrophages migrate more vigorously. In certain situations, the final stage of macrophage development is the multinucleated giant cell. The precise function of this cell is unclear, although it appears to carry on most, if not all, of the activities of the activated macrophage.[13,22]

T lymphocytes recognize antigen in association with class I or class II molecules on the surface of antigen-presenting cells such as macrophages. Cytotoxic T cells (CD8+ cells) recognize antigen in the context of class I molecules and are able to lyse specific target cells such as tumor cells or cells infected with virus. Helper T cells (CD4+ cells) recognize antigen in the context of class II molecules and act as inducers of other T cells, B cells, macrophages, and other cell types.[4,10,21]

Both CD4+ and CD8+ cells express the T-cell receptor (Ti–T3 complex). The CD4 and CD8 molecules are believed to play a role in binding to class II and class I molecules, respectively. The Ti portion of the T-cell receptor appears to form a binding site for antigen as well as the MHC molecule (i.e., class II or class I molecules). The T3 component is involved in signal transduction. The T cell is triggered by the interaction of the MHC molecule and antigen on the surface of the antigen-presenting cell with the T-cell receptor complex. This results in the induction of surface IL-2 receptors as well as IL-2 secretion. DNA synthesis and cell mitosis begin to occur in the particular T-cell clones that have been stimulated.[10]

T-helper cells, in addition to producing IL-2 and IL-2 receptors, produce a number of other cytokines whose actions are listed in Table 2-2. There are at least two sets of T-helper cells: (1) helper effectors for B cells and other immune components and (2) helper/suppressor inducers, which provide help only for T-suppressor cells. CD8+ cytotoxic cells are capable of lysing virus-infected cells and tumor cells. CD8+ suppressor cells appear to inhibit B cell activity.[4,10]

B cells are activated by two types of processes. Nonspecific mitogens such as lipopolysaccharide (mouse) or pokeweed mitogen (humans) can stimulate B cells indiscriminately (i.e., polyclonal activation). Mitogens are able to cross-link surface immunoglobulin molecules, resulting in induction of B-cell DNA synthesis and mitosis. B cells will then typically secrete IgM.[4,11]

The second mode of B-cell activation involves binding of the antigen by surface immunoglobulin on the B cell. When activated in this manner, the B cell can respond to the specific T-cell clones that have also been activated and secrete cytokines that will promote the proliferation and differentiation of the B-cell clones (Table 2-3). Ultimately, these B cells become plasma cells that will secrete antibody of a certain immunoglobulin isotype specific for the antigen. Memory B cells are long-lived, poised to make the specific antibody quickly on a subsequent stimulation with the same antigen.[4,11]

Immunoglobulins are composed of two heavy chains (γ, α, μ, δ, or ϵ) and two light chains (κ or λ). The Fc fragment is involved in binding of complement or may be bound by Fc receptors on such cells as macrophages. The Fab fragment is the portion involved in antigen binding. The immunoglobulin isotypes in man are IgG (IgG$_1$, IgG$_2$, IgG$_3$, and IgG$_4$), IgA, IgM, IgD, and IgE. Of the immunoglobulin isotypes, IgG is found in the highest concentrations in serum and extracellular fluid. IgG is transported across the placenta, and its half-life is approximately 25 days. IgA is present primarily in mucous secretions but is also found in breast milk and serum. IgM is found primarily in the serum, especially early in the course of infections, and its half-life is only 2–5 days. IgE is thought to play a role in parasitic infections, asthma, and atopic reactions. Finally, IgD has primarily been recognized as a surface immunoglobulin on "immature" B lymphocytes that have not yet been antigen stimulated.[4]

Null cells are lymphocytes that do not bear the typical identifying antigens for T or B cells and comprise about 5–15% of peripheral blood lymphocytes. These cells have also been called killer cells and include a subpopulation of large granular lymphocytes including natural killer (NK) cells, which are able to lyse tumor cells spontaneously and through antibody-dependent cellular cytotoxicity. They bear Fc receptors and can lyse virus-infected cells. The NK cells are important in immune surveillance and early cell lysis of virus-infected cells prior to the induction of significant numbers of T and B cells.[4,23]

Circulating granulocytes, important in many immune responses, include neutrophils, eosinophils, and basophils. These cells, when activated, release a variety of substances including chemotactic factors, kallikreins, and vasoactive amines. Neutrophils and eosinophils are also phagocytic. Mast cells and platelets can release vasoactive amines as well.[4]

3. IMMUNE RESPONSES TO VIRAL INFECTION

3.1. Humoral Interaction

Antibody that attaches to circulating virus particles is effective in preventing infectivity of the virus if it has neutralizing capabilities. Antibody can neutralize virus by several methods. It may prevent viral attachment to the cell membrane. It may prevent penetration of the virus after it has attached to the cell and then may enhance the ability of the cell to degrade the virus within a pinocytotic vacuole. Antibody-bound virus may result in complement activation and viral lysis. Virus that is bound by antibody may also be more easily phagocytosed by macrophages.[3,24] Circulating immune complexes (i.e., virus–antibody complexes) may be thermally inactivated or deposited into tissues such as the kidney glomerulus, choroid plexus, or in arteries, resulting in indirect damage to these organs. Antibodies that arise in response to certain viruses may be complement-fixing, hemagglutination-inhibiting, and/or precipitating. These properties can be helpful diagnostically in identifying the specific virus.[3,4]

Enveloped viruses express their surface proteins on the cellular membranes of infected host cells prior to budding from the cell surface. Antibody may bind to the viral proteins that are expressed on the cell surface. This can result in the activation of complement and the lysis of the infected cell. If antibody that is attached to infected cells cross-links with other antibody that is attached to the same cell, then capping can occur. Capping can result in pinocytosis of the virus into the cell or extrusion of the viral antigen from the surface of the cell. The biological importance of this phenomenon is unclear.[3,24]

3.2. Cell-Mediated Interaction

Cell-mediated antiviral actions are primarily against virus-infected cells. After antibody attachment to viral antigens on the host cell surface, antibody-dependent cellular cytotoxicity can occur via NK cells, macrophages, or polymorphonuclear cells. Cytotoxic T cells (CD8+) can lyse a virus-infected cell without antibody attachment. Cell-mediated antiviral actions also include the elaboration of cytokines (Table 2-3) that enhance the immune response and, in the case of IFN, inhibit viral replication.[3,24]

Activation of the immune system during viral infection can be a double-edged sword. In some viral infections the immune response is responsible for more tissue damage than the virus itself. Other viruses are associated with an increased frequency of autoimmune phenomena, resulting in tissue damage secondary to an overactive immune system. Viral mimicry of self-antigens, especially in patients who may have a genetic predisposition that often is related to MHC class II typing, may be one mode for the development of autoimmune disease after viral infection.[24] In addition, viruses that directly infect immune cells are associated with immune dysregulation, which can lead to states ranging from immunodeficiency to autoimmunity.

4. IMMUNE RESPONSES IN THE CENTRAL NERVOUS SYSTEM

Normally the CNS has a relatively small population of immune cells. Lymphatic drainage within the brain has not been clearly demonstrated, and the brain has no areas of lymphocyte concentration such as lymph nodes or RE system as we normally define them. The first line of defense against viral invasion is the blood–brain barrier, which also keeps the CNS relatively isolated from the systemic immune system.[25]

Tight junctions connect CNS capillary endothelial cells, choroid plexus epithelial cells, and arachnoid cells, and under normal conditions these inhibit the passage of proteins and cells from the blood into the brain parenchyma and CSF. These tight junctions are not present between ventricular ependymal cells so that flow of substances between CSF and the extracellular spaces of the brain occurs more freely.[25]

4.1. Immune Cells Normally Present in the Central Nervous System

Animal experiments indicate that small numbers of leukocytes of hematogenous origin are normally present in brain. Primarily, these are monocytes and lymphocytes, although rare polymorphonuclear cells may be present. Lymphocytes and monocytes are able to cross the blood–brain barrier in small numbers, and these cells can move into the CSF or back into the peripheral circulation.[26] Studies suggest that some leukocytes eventually migrate to the cervical lymph nodes.[2] In this way, there appears to be a constant circulation of peripheral blood mononuclear cells through the CNS and back into the blood or peripheral lymphatics (i.e., immunologic surveillance).

Supraependymal cells resembling phagocytic macrophages have been described in animals.[3] However, it remains somewhat controversial whether the CNS contains resident macrophages and if microglia are those cells. Microglia are stellate cells in brain parenchyma (Fig. 2-4) that have variously been described as bone marrow or neuroectodermal in origin. If microglia are related to monocytes/macrophages, one would expect them to have some surface antigens in common. Some immunocytochemical studies have not demonstrated shared antigens.[27] Nevertheless, more recent studies have indicated that microglia share a number of surface markers with macrophages and are most likely bone marrow derived.[28] Microglia are ubiquitous in brain but more common in gray matter than in white, and their turnover is probably slow. They frequently have been found next to blood vessels. They express Fc and complement receptors, low levels of CD4 antigen, and class I molecules. Studies to date have not documented constitutive expression of class II molecules on microglia. A number of animal and human studies have demonstrated class II molecule expression immunocytochemically during various CNS inflammatory processes, but little if any in normal brain.[29–31] Their function during these inflammatory processes is unclear, but if they do

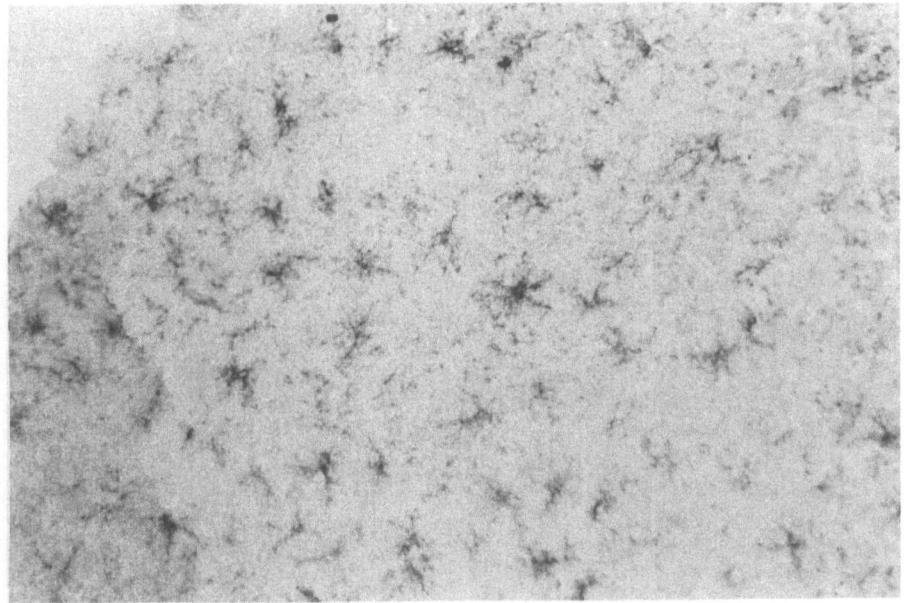

FIGURE 2-4. Microglia stained immunocytochemically with anti-Ia antibody in Sindbis-virus-infected mouse brain.

indeed function as resident macrophages in the CNS, then they could play a role in phagocytosis, antigen presentation to T cells, and possibly elaboration of cytokines and other monocyte/macrophage factors.

Astrocytes have a similar appearance to microglia (Fig. 2-5) in routine light microscopic tissue sections, although their nuclei tend to be larger. They are usually differentiated by the presence of glial fibrillary acidic protein and are present throughout the gray and white matter. Astrocytic processes often end on blood vessels, not infrequently encircling the basal membranes surrounding the endothelial lining of capillaries. There is also a concentration of astrocytes beneath the pia mater of the meninges and the ependyma of the ventricles. During inflammatory processes, astrocytes proliferate around lesions (i.e., gliosis) and can produce IL-1.[32] A few animal studies have demonstrated class II molecule expression by astrocytes *in vivo* during inflammatory processes.[31,33] Astrocytes have been demonstrated *in vitro* to express class II molecules and to present antigen in a MHC-restricted fashion to T cells, but whether antigen-presenting capacity is retained *in vivo* and occurs to any significant degree during inflammatory processes is unknown.[34] Astrocytes can also express class I molecules.

Another candidate for antigen-presenting cells in the CNS is the vascular endothelial cell.[30,31] Endothelial cells have been shown to express class II molecules and to present antigen to T cells *in vitro*.[35] It is unclear whether they present

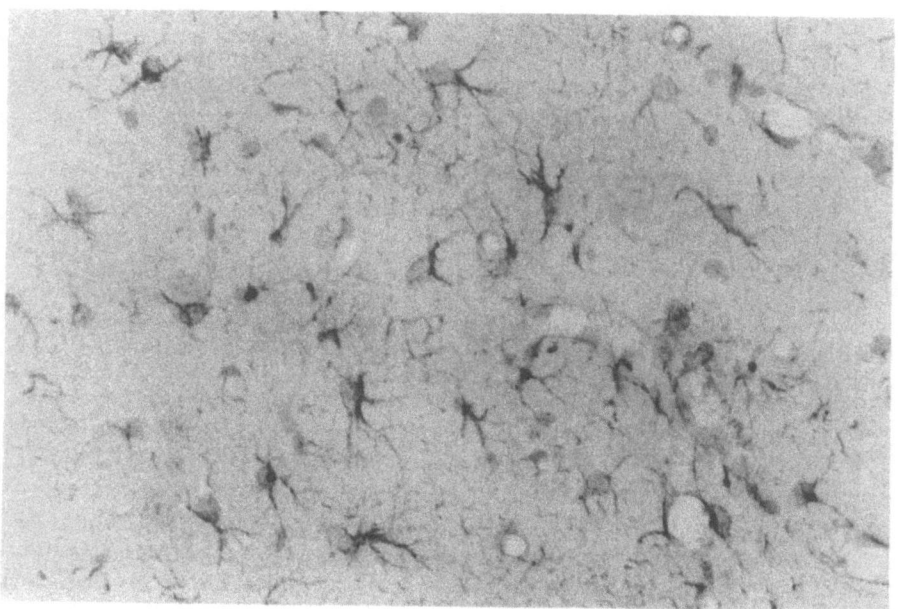

FIGURE 2-5. Astrocytes stained immunocytochemically with anti-GFAP antibody in Sindbis-virus-infected mouse brain. Nuclei are lightly counterstained with hematoxylin.

antigen to T cells *in vivo*. The results of studies have been conflicting concerning constitutive class II molecule expression on endothelial cells in the CNS, but in a few studies endothelial cells expressed class II molecules *in vivo* during CNS inflammatory processes. Class II molecule expression on endothelial cells may precede the onset of inflammation in experimental autoimmune encephalomyelitis, which suggests that endothelial cells may be important in the initiation of this autoimmune disease. However, as with astrocytes, a number of studies have failed to show class II molecule expression in endothelial cells during inflammatory processes; therefore, the significance of class II molecule expression on endothelial cells and their role in antigen presentation *in vivo* is unclear. Endothelial cells express class I molecules and when activated secrete IL-1, platelet-activating factor (activates platelets, neutrophils, and monocytes and increases vascular permeability), fibronectin (binds to cellular surfaces and promotes attachment), and granulocyte–macrophage colony-stimulating factor. Factors such as IL-1 and TNF can induce molecular adhesion of leukocytes to endothelial cells.[36]

Dendritic cells have not been demonstrated in the CNS. Dendritic cells are required for antigen presentation to resting T cells, that is, T cells not previously exposed to antigen.[8] Microglia, astrocytes, or vascular endothelial cells may possibly assume this role in the CNS, or alternatively, this function may occur only outside of the CNS. Class II molecule expression on neurons and oligodendroglia has not been demonstrated. However, these cells may express class I molecules.

4.2. Cerebrospinal Fluid

Proteins, including immunoglobulins, found in the CSF under normal conditions are derived primarily from the blood. Entry through the blood–brain barrier is related to the size of the protein and also to its charge. Passage is inversely related to size; therefore, smaller proteins enter more easily. Proteins with higher isoelectric points (more positive charge) enter more easily. Both IgG and IgA are present in about 0.2% to 0.4% of serum concentration, and IgM concentration is even lower; IgM is relatively excluded on the basis of size, and more positively charged IgG isotypes are present in relatively higher proportions.[3,37]

Cells that are found normally in CSF are of hematogenous origin, although they are not present in the same proportions as in blood. As previously mentioned, the few cells that are able to enter through the blood–brain barrier under normal conditions are primarily monocytes and lymphocytes. Plasma cells are not normally found in the CNS, and polymorphonuclear leukocytes found in the CSF are considered abnormal.[37]

4.3. Response to Viral Infection

Once a virus has infected cells in the brain and significant numbers of viral particles are replicated, then macrophages, T cells, and NK cells may be activated. Expression of viral proteins on the infected cell surfaces along with class I molecules may enable recognition of the viral infection by cytotoxic T cells. Phagocytosis of viral particles by macrophages and possibly microglia can lead to further MHC-restricted signaling of T cells via antigen presentation. This process may be contributed to or possibly even initiated by astrocytes or endothelial cells. The activated macrophages contribute to the release of IL-1 and can also release a number of other factors and cytokines that can enhance the inflammatory response. The inflammatory response is further enhanced by the cytokines released by activated T-helper cells.

The blood–brain barrier becomes "leaky," and increased protein transudation occurs. The CSF protein in this circumstance is essentially in proportion to serum protein. Coincident with this protein transudation is the increase in mononuclear cells entering the perivascular areas of brain parenchyma, meninges, subependymal areas, and CSF. Natural killer cells, neutrophils, macrophages, and T cells constitute the initial mononuclear infiltrate, with B cells arriving relatively late.[3,26] In addition to chemotactic factors that are released by activated inflammatory cells, the entry of cells into the CNS is also facilitated by activation of endothelial cells and expression of molecules on their cell surface that make them "sticky" and allow circulating mononuclear cells to adhere to the vessel wall before entering. In addition, T cells, when activated, also appear to express adhesion molecules on their cell surfaces.[38] B cells are known to home specifically to certain tissues such as lymph nodes or Peyer's patches in the walls of the intestines.[39] Whether or not cells specifically home to the CNS is unknown.

Inflammatory responses to viruses within the CNS are immunologically specific and dependent on sensitized T cells. The proportion of cytotoxic or helper T cells as well as other types of inflammatory cells entering the CNS varies with the type of viral infection. In addition, inflammatory cell types that predominate in the brain parenchyma may not necessarily predominate in the meninges or CSF. Later, usually after the first week, a significant number of plasma cells are present within the CNS. Immunoglobulins are synthesized in the CNS and are found elevated in the CSF as manifest by distorted CSF/serum ratios of immunoglobulin and the development of oligoclonal bands (i.e., evidence of local clonal expansion and secretion of immunoglobulin by plasma cells). As a result of local synthesis of antibody, the concentration of virus-specific antibody increases. IgG is most commonly elevated, but IgM and IgA can be found as well. Usually the CSF antibody levels begin to return to normal within a few weeks, but in certain viral infections antibody may persist for years. Plasma cells have been found to persist for months in the brain.[3,26]

Nevertheless, most inflammatory cells leave relatively rapidly after virus clearance. The blood–brain barrier function usually normalizes within a week or two, as reflected by the normalization of CSF albumin concentration, CSF pleocytosis, and the perivascular inflammatory response. Inflammatory cells may pass back into the circulation or enter the cervical lymph nodes. The pathway to the cervical lymph nodes has not been elucidated. In chronic viral infections of the CNS, inflammatory cells may persist in the brain parenchyma and CSF, and levels of immunoglobulin may remain elevated. Levels of specific IgG in the CSF may become higher than those in serum and can persist for the duration of the encephalitis. CD8+ T cells can remain chronically elevated in the CSF during human immunodeficiency virus (HIV) infection, cytomegalovirus encephalitis, and subacute sclerosing panencephalitis.[3,26]

4.4. Viral Clearance

Virus is normally cleared within the first 1 to 2 weeks by the mechanisms described above. Alternatively, infection may resolve because of the virus' limited capability of replication in the CNS or other nonspecific factors. The failure to clear virus, on the other hand, can result from a number of factors.[3]

Immunodeficiency states, whether they are genetic, iatrogenic such as cancer chemotherapy, related to cancer, or secondary to infections such as HIV, can lead to unusual and persistent viral infections. The normal immune mechanisms necessary for viral clearance are not available under these conditions. Viruses that normally would not invade the CNS may produce opportunistic infections in these individuals and are often persistent.[40]

There are viruses such as herpesviruses and HIV that are capable of escaping immune detection for a period of time in ganglion cells and immune cells, respectively. During these periods no viral antigen is presented to the immune system. Visna virus and probably HIV persist in part through a propensity to generate mutants that are not neutralized by previously formed antibodies to the original infecting strain.[41] Some viruses go undetected via spread from cell to cell by bridging cytoplasmic membranes. In this way, they are never exposed to immune surveillance. Finally, the immune system can be tolerized to antigens and will then fail to recognize viral polypeptides as foreign.[3]

5. SINDBIS VIRUS AS A MODEL OF ENCEPHALITIS

The inflammatory reaction in the CNS to Sindbis virus (SV) has been studied extensively as a model for acute viral encephalomyelitis. Sindbis virus is an alphavirus related to Western equine encephalitis virus. When injected intracerebrally into weanling mice, SV causes an acute nonfatal encephalitis.[26,43]

Viral antigen of a neuroadapted SV is found in ependymal cells as early as one day after inoculation. Subependymal areas are affected by day 2. Subsequently, further spread into the brain parenchyma occurs with involvement of neurons and, to a lesser extent, glial cells. Virus is present primarily in gray matter areas by days 3 through 5, including the cerebral cortex, subcortical nuclei, brainstem nuclei, cerebellar cortex, deep cerebellar nuclei, and spinal cord gray matter, especially the ventral horns. Virus content in brain is maximal by day 2, and infectious virus can no longer be recovered after 7 to 8 days (Fig. 2-6). Small amounts of virus are found transiently in blood after intracerebral inoculation.[43]

Coincident with viral replication, blood–brain barrier changes are evident 2 to 3 days after inoculation.[26] These are reflected by increased protein in the CSF by day 2 and the development of CSF pleocytosis and perivascular inflammatory infiltrates (Fig. 2-7) by day 3. Mononuclear cells also appear in the meninges and

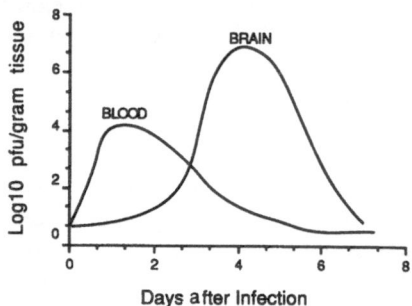

FIGURE 2-6. Sindbis virus titers in brain and serum as measured over the course of the encephalitis in mice.

brain parenchyma by day 3.[44] Class II molecule expression on perivascular mononuclear cells can be detected as early as day 2, which is coincident with the development of a specific T-cell response to SV.[44] Class II molecule expression steadily increases, and by day 7 40% of the perivascular inflammatory cells are class II positive. Class II molecule expression in the spleen and in peripheral blood mononuclear cells also increases by day 8. Microglia (Fig. 2-4) that are class II positive are found near blood vessels and in brain parenchyma as early as day 3 and are numerous by day 7. Astrocytes do not express Ia during SV encephalitis; however, rare perivascular fusiform cells, probably endothelial cells or pericytes, are found to be class II positive after day 3. Macrophages and microglia are the predominant class-II-positive cell types found during SV encephalitis and may therefore be the most important population of cells responsible for antigen presentation to T-helper cells.

Further immunocytochemical analysis of mononuclear cell types in the CSF and perivascular inflammatory cuffs is depicted in Fig. 2-7. In the CSF T-helper and -suppressor cells can be found by day 2. A very small percentage of B cells and macrophages are found, and a large percentage of CSF cells are unidentified by immunocytochemical analysis with markers for T-helper cells, T-suppressor cells, macrophages, or B cells. Natural killer cell activity has been shown to be high during this time, and it is presumed that some of the unidentified CSF cells are NK cells.[45] The percentage of T-helper and -suppressor cells increases in the CSF during the 2 weeks after inoculation of SV. Early on, a majority of the perivascular inflammatory cells are unstained by markers for the cell types mentioned above. Again, some of these are presumed to be NK cells. However, macrophages and T-helper cells each represent approximately 20% of the cells initially detected in the inflammatory response of the encephalitis, and they increase over the 14-day period of encephalitis that has been examined. T-suppressor/cytotoxic cells remain a relatively small percentage (5–10%) of perivascular inflammatory cells, and B cells are initially found in small numbers but constitute about 20% of the

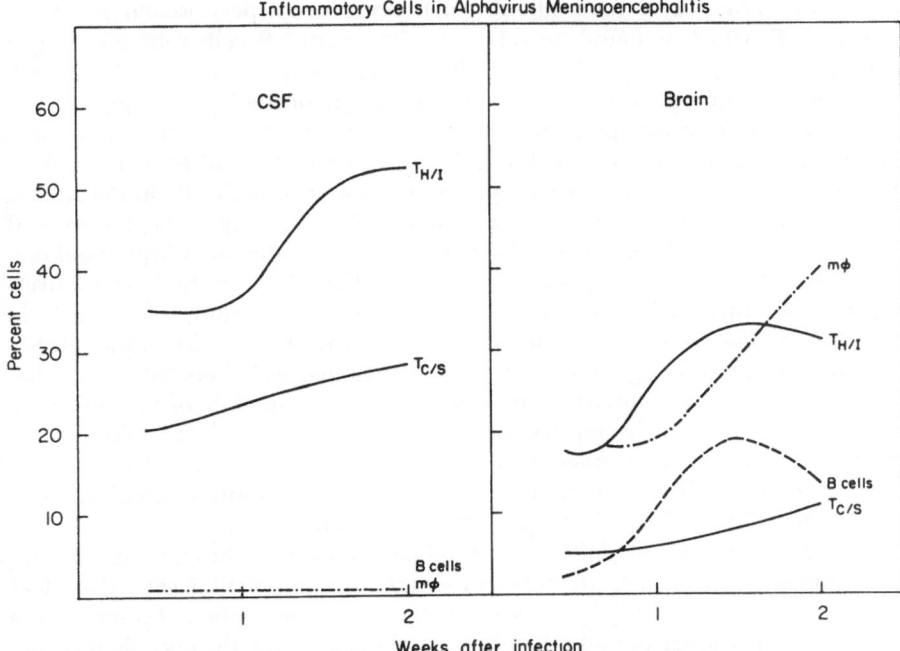

FIGURE 2-7. The percentages of various types of mononuclear cells in CSF and perivascular inflammatory cuffs during Sindbis virus encephalitis in mice. $T_{H/I}$ (T-helper/inducer), $T_{C/S}$ (T-cytoxic/suppressor), MO (macrophages/monocytes), and B cells were stained immunocytochemically.

perivascular infiltrate by day 10. The discrepancy between the number of T-suppressor cells, macrophages and B cells in the CSF versus brain parenchyma is unclear.[44]

The neuroadapted form of SV causes a fatal encephalitis in weanling mice when inoculated intracerebrally. Treatment of these mice with immune serum (i.e., antibody) protects a high percentage from fatal infection and reduces neuroadapted SV titers in the brain.[46] In addition, monoclonal antibody to neuroadapted SV, when given to mice soon after virus inoculation, will protect them from fatal encephalitis.[47] Neutralizing antibody of IgM and IgG isotypes to SV (the nonneuroadapted strain) can be detected in serum as early as day 4. However, in neuroadapted SV infection, protection does not seem to correlate with virus-neutralizing ability of the antibody. To further characterize antibody production during SV encephalitis IgM, IgG_1, IgG_{2a}, IgG_{2b}, IgG_3, and IgA isotypes were measured in CSF and serum. All isotypes begin to be elevated by day 4 and are maximum by days 8 through 15. SV-specific IgG can be detected in serum on day 8 and in CSF on day 15.[48] The above antibody isotypes were examined immunocytochemically for their presence on perivascular B cells as well as IgD, which is found on relatively "immature" B cells (not yet antigen stimulated).[49] Early (days 3 through 5) B cells express IgM or both IgM and IgD; later (days 10 through 14), most B cells express one of the IgG isotypes of IgA.

The pattern of isotype expression in brain is reflected in spleen and blood mononuclear cells. Isotype switching of B cells from IgM and IgM/IgD cells to IgG and IgA cells probably occurs primarily outside of the CNS, in the spleen, but may also occur in the brain. Of the immunoglobulin isotypes expressed on B cells, IgG_{2a}-positive B cells increase the most relative to the other immunoglobulin isotypes found on B cells during the encephalitis. However, IgG_{2a} is relatively underrepresented in CSF. This may be because of local consumption of IgG_{2a} in the perivascular areas, thereby limiting its concentration in CSF. Although SV-specific IgG_{2a} immunoglobulin has not been studied in SV encephalitis, it has been shown to be an important murine isotype in a number of systemic viral infections of mice, including SV infection. Another finding of interest is the relatively increased percentage of IgA-positive B cells in perivascular cuffs and IgA concentration in CSF versus the blood. The reason for this preponderance of IgA in the CNS during SV encephalitis is unknown.[49]

Sindbis virus encephalitis in T-cell-deficient nude mice has also been examined. Although these mice are deficient in immunocompetent T cells, they clear virus and recover normally from SV encephalitis.[50] Transudation of protein from serum to CSF is lower in nude than in normal mice during the encephalitis, and perivascular inflammatory cells are about one-tenth as numerous. Although class II molecules are expressed on the same types of cells in nude mice, the percentage of cells expressing class II in perivascular inflammatory cuffs is approximately one-third that in normal mice. B-cell isotype switching is known to be impaired in nude mice. They respond to infection with normal IgM production but low levels of IgG and IgA. This is reflected in the brain during SV encephalitis. The switch from predominantly IgM- and IgM/IgD-positive perivascular B cells to IgG- and IgA-positive B cells occurs several days later in nude mice as compared with normal mice. The studies outlined above have suggested a less important role for T cells in SV encephalitis. This is further supported by data in which neuroadapted SV-infected mice are not protected by sensitized lymph node cells, in contrast to the protection afforded by immune serum.[46]

In summary, SV, when injected intracerebrally into mice, replicates primarily in ependymal cells and neurons, with viral titers peaking by day 2 and virus being cleared from brain by days 7–8. Class II molecule expression on perivascular mononuclear cells is coincident with protein transudation into CSF by day 2. By day 3 SV-specific T cells appear in perivascular cuffs along with macrophages and a few B cells. Over the course of 2 weeks the cell types increase, with T-helper cells and macrophages predominating. However, in CSF, macrophages and B cells are for the most part excluded, although CSF antibody is readily found. Several studies suggest that T cells are less important than the production of antibody by B cells with respect to the immunologic reaction to SV encephalitis. In addition, early on in infection, NK cells and lymphokines such as IFN γ probably play a role in viral clearance.

6. SUMMARY

Different viruses evoke different types of immune response. As in SV encephalitis in mice, B cells may comprise a relatively important population of immune cells. On the other hand, in other viral infections T cells may be primarily important in viral clearance, and within the general T-cell population, T-suppressor/cytotoxic cells may predominate over T-helper cells. Also CNS pathology and the clinical condition of the organism may be caused by the immunologic reaction to the virus more than by the effects of the virus itself. There is a continuum of immune reaction to viral disease ranging from mild immune reaction to severe reaction, where relative amounts of CNS pathology are caused either by virus destruction or by immune-associated destruction. Within this continuum there are varying contributions to this process by different components of the immune system (i.e., cellular versus humoral versus other factors such as cytokines). At the other end of the continuum are the CNS parainfectious and autoimmune diseases in which the immune system, once triggered, is completely responsible for the pathogenesis of disease. In all of these situations there is a complex interaction of the immune system with the CNS and the virus. The interaction is affected by intrinsic properties of the virus, the intactness of the immune system, and the genetic makeup particular to each individual.

REFERENCES

1. Cutler, R. W. P., Watters, G. V., and Hammerstad, J. P., 1969, The origin and turnover rates of cerebrospinal fluid albumin and gamma-globulin in man, *J. Neurol. Sci.* **10**:259–268.
2. Oehmichen, M., Domasch, D., and Wutholter, H., 1982, Origin, proliferation and fate of cerebrospinal fluid cells, *J. Neurol.* **227**:145–150.
3. Johnson, R. T., 1982, *Viral Infections of the Nervous System*, Raven Press, New York.
4. Bellanti, J. A. (ed.), 1985, *Immunology III*, W. B. Saunders, Philadelphia.
5. Oppenheim, J. J., Rosenstreich, D. L., and Potter, M. (eds.), 1981, *Cellular Functions in Immunity and Inflammation*, Elsevier/North-Holland, New York.
6. van Furth, R., 1982, Current view on the mononuclear phagocyte system, *Immunobiology* **161**: 178–185.
7. Robbins, S. L., Cotran, R. S., and Kuman, V. (eds.), 1984, *Textbook of Pathology*, W. B. Saunders, Philadelphia, pp. 59–60.
8. Austyn, J. M., 1987, Lymphoid dendritic cells, *Immunology* **62**:161–170.
9. Hirschberg, H., Breathen, L. R., and Thorsby, E., 1982, Antigen presentation by vascular endothelial cells and epidermal Langerhans cells: The role of the HLA-DR, *Immunol. Rev.* **66**:57–78.
10. Roger, H. D., and Reinherz, E. L., 1987, Current concepts: T lymphocytes: Ontogeny, function, and relevance to clinical disorders, *N. Engl. J. Med.* **317**:1136–1142.
11. Cooper, M. D., 1987, Current concepts: B lymphocytes, normal development and function, *N. Engl. J. Med.* **317**:1452–1456.
12. Dannenberg, A. M., 1980, Macrophages and monocytes, in: *Fundamentals of Clinical Hematology* (J. L. Spivak ed.), Harper & Row, Hagerstown, MD, pp. 137–153.
13. Johnston, R. B., 1988, Current concepts: Immunology, monocytes and macrophages, *N. Engl. J. Med.* **318**:747–752.

14. Bures, S., Sette, A., and Greg, H. M., 1987, The interaction between protein-derived immunogenic peptides and Ia, *Immunol. Rev.* **98**:115–141.
15. Benacerraf, B., 1985, Significance and biological function of class II MHC molecules, *Am. J. Pathol.* **120**:334–343.
16. Dinarello, C. A., and Mier, J. W., 1987, Current concepts: Lymphokines, *N. Engl. J. Med.* **317**: 940–945.
17. Miyajima, A., Miyatake, S., Schreurs, J., DeVries, J., Arai, N., Yokota, T., and Arai, K., 1988, Coordinate regulation of immune and inflammatory responses by T cell-derived lymphokines, *FASEB J.* **2**:2462–2473.
18. Movat, H. Z., 1987, Tumor necrosis factor and interleukin-1: Role in acute inflammation and microvascular injury, *J. Lab. Clin. Med.* **110**:668–681.
19. Frank, M. M., 1987, Current concepts: Complement in the pathophysiology of human disease, *N. Engl. J. Med.* **316**:1525–1530.
20. Unanue, E. R., and Allen, P. M., 1986, Biochemistry and biology of antigen presentation by macrophages, *Cell. Immunol.* **99**:3–6.
21. Braciale, T. J., Morrison, L. A., Sweetser, M. T., Sambrook, J., Gething, M. J., and Braciale, V. L., 1987, Antigen presentation pathways to class I and class II MHC-restricted T lymphocytes, *Immunol. Rev.* **98**:95–114.
22. Unanue, E. R., and Allen, P. M., 1987, The immunoregulatory role of the macrophage, *Hosp. Pract.* **22**:87–104.
23. Kaplan, J., 1986, NK lineage and target specificity: A unifying concept, *Immunol. Today* **7**:10–12.
24. Notkins, A. L., and Oldstone, M. B. A., 1984, *Concepts in Viral Pathogenesis*, Springer-Verlag, New York.
25. Goldstein, G. W., and Bety, A. L., 1986, Blood vessels and blood–brain barrier, in: *Diseases of the Nervous System* (A. K. Asbury, G. M. McKhann, and W. I. McDonald, eds.), W. B. Saunders, Philadelphia, pp. 172–184.
26. Griffin, D. E., Hess, J. L., and Moench, T. R., 1987, Immune responses in the central nervous system, *Toxicol. Pathol.* **15**:294–302.
27. Perry, V. H., and Gordon, S., 1988, Macrophages and microglia in the nervous system, *Trends Neurosci.* **11**:273–277.
28. Hickey, W. F., and Kimura, H., 1988, Perivascular microglial cells of the CNS are bone marrow-derived and present antigen *in vivo*, *Science* **239**:290–292.
29. Lassmann, H., Vass, K., Brunner, C., and Seitelberger, F., 1986, Characterization of inflammatory infiltrates in experimental allergic encephalomyelitis, in: *Progress in Neuropathology*, Vol. 6 (H. M. Zimmerman, ed.), Raven Press, New York, pp. 33–62.
30. Traugott, U., Scheinberg, L. C., and Raine, C. S., 1985, On the presence of Ia-positive endothelial cells and astrocytes in multiple sclerosis lesions and its relevance to antigen presentation, *J. Neuroimmunol.* **8**:1–14.
31. Rodriguez, M., Pierce, M. L., and Howie, E. A., 1987, Immune response gene products (Ia antigens) on glial and endothelial cells in virus-induced demyelination, *J. Immunol.* **138**: 3438–3442.
32. Adams, J. H., Corsellis, J. A. N., and Duchen, L. W. (eds.), 1984, *Greenfield's Neuropathology*, 4th ed., John Wiley & Sons, New York.
33. Hickey, W. F., Osborn, J. P., and Kirby, W. M., 1985, Expression of Ia molecules by astrocytes during acute experimental allergic encephalomyelitis in the Lewis rat, *Cell. Immunol.* **91**: 528–535.
34. Fierz, W., Endler, B., Reske, K., Wekerle, H., and Fontana, A., 1985, Astrocytes as antigen-presenting cells. I. Induction of Ia antigen expression on astrocytes by T cells via immune interferon and its effect on antigen presentation, *J. Immunol.* **134**:3785–3793.
35. McCarron, R. M., Spatz, M., Kempski, O., Hogan, R. N., Muehl, L., and McFarlin, D. E., 1986, Interaction between myelin basic protein-sensitized T lymphocytes and murine cerebral vascular endothelial cells, *J. Immunol.* **137**:3428–3435.

36. Ryan, U. S., 1986, The endothelial surface and responses to injury, *Fed. Proc.* **45:**101–107.
37. Griffin, D. E., Hess, J. L., and Mokhtarian, F., 1984, Entry of proteins and cells into the normal and virus-infected central nervous system, in: *Neuroimmunology* (P. O. Behan and F. Spreafico, eds.), Raven Press, New York, pp. 193–206.
38. Hemler, M. E., 1988, Adhesive protein receptors on hematopoietic cells, *Immunol. Today* **9:** 109–113.
39. Jalkanen, S., Reichart, R. A., Gallatin, W. M., Bargatze, R. F., Weissman, I. L., and Butcher, E. C., 1986, Homing receptors and the control of lymphocyte migration, *Immunol. Rev.* **91:** 39–60.
40. Johnson, R. T., 1984, Failure of viral clearance in central nervous system infections, in: *Neuroimmunology* (P. O. Behan and F. Spreafico, eds.), Raven Press, New York, pp. 229–236.
41. Johnson, R. T., McArthur, J. C., and Narayan, O., 1988, The neurobiology of human immunodeficiency virus infections, *FASEB J.* **2:**2970–2981.
42. Griffin, D. E., 1986, Alphavirus pathogenesis and immunity, in: *The Togaviridae and Flaviviridae* (S. Schlesinger and M. J. Schlesinger, eds.), Plenum Press, New York, pp. 209–249.
43. Jackson, A. C., Moench, T. R., Griffin, D. E., and Johnson, R. T., 1987, The pathogenesis of spinal cord involvement in the encephalomyelitis of mice caused by neuroadapted Sindbis virus infection, *Lab. Invest.* **56:**418–423.
44. Moench, T. R., and Griffin, D. E., 1984, Immunocytochemical identification and quantitation of the mononuclear cells in the cerebrospinal fluid, meninges, and brain during acute viral meningoencephalitis, *J. Exp. Med.* **159:**77–88.
45. Griffin, D. E., and Hess, J. L., 1986, Cells with natural killer activity in the cerebrospinal fluid of normal mice and athymic nude mice with acute Sindbis virus encephalitis, *J. Immunol.* **136:** 1841–1845.
46. Griffin, D. E., and Johnson, R. T., 1977, Role of the immune response in recovery from Sindbis virus encephalitis in mice, *J. Immunol.* **118:**1070–1075.
47. Stanley, J., Cooper, S. J., and Griffin, D. E., 1986, Monoclonal antibody cure and prophylaxis of lethal Sindbis virus encephalitis in mice, *J. Virol.* **58:**107–115.
48. Griffin, D. E., 1981, Immunoglobulins in the cerebral spinal fluid: Changes during acute viral encephalitis in mice. *J. Immunol.* **126:**27–31.
49. Tyor, W. R., Moench, T. R., and Griffin, D. E., 1989, Characterization of the local and systemic B cell response of normal and athymic nude mice with Sindbis virus encephalitis, *J. Neuroimmunol.* **24:**27–215.
50. Hirsch, R. L., and Griffin, D. E., 1979, The pathogenesis of Sindbis virus infection in athymic nude mice, *J. Immunol.* **123:**1215–1218.

3

Virus Infection of Peripheral Nerve

MICHAEL C. GRAVES and HARRY V. VINTERS

1. INTRODUCTION

This chapter reviews the peripheral nerve diseases associated with herpes simplex virus (HSV), varicella–zoster virus (VZV), hepatitis B virus, human cytomegalovirus (HCMV), Epstein–Barr virus (EBV), and human immunodeficiency virus (HIV). In contrast to the many viral infections of brain and spinal cord,[1] peripheral nerve has not been generally recognized as a target tissue for viruses. Acute inflammatory demyelinating polyneuropathy (AIDP), like the analogous central nervous system (CNS) disease, postinfectious encephalomyelitis, is an autoimmune disease.[1,2] Both may follow a systemic viral infection, but the mechanism for the autosensitization is unknown. The mechanism is better understood in hepatitis B virus neuropathy, where circulating immune complexes have been detected.[3] Probably a wide variety of mechanisms are responsible for the peripheral nerve disorders that are associated with viral infection.

The latent infections of the sensory ganglia by HSV and VZV have long been recognized in humans, although largely as a cause of dermatologic disease. When these viruses reactivate from their latent state in neurons, they migrate down the axon or dendrite, produce skin lesions, and occasionally cause a motor or sensory neuropathy.[1,4] Modern awareness that other viruses may infect the peripheral nerve began with the study of animal model systems.

MICHAEL C. GRAVES • Department of Neurology, Reed Neurological Research Center, UCLA Medical Center, Los Angeles, California 90024-1769.   HARRY V. VINTERS • Department of Pathology (Neuropathology) and Brain Research Institute, UCLA Medical Center, Los Angeles, California 90024-1732.

Neuropathogenic Viruses and Immunity, edited by Steven Specter *et al.* Plenum Press, New York, 1992.

Marek's disease virus (MDV) is a herpesvirus that causes an inflammatory neuropathy in chickens very similar to that of AIDP in humans. These chickens have an autoimmune response to peripheral nerves, but in addition the Schwann cells harbor latent virus.[5] The latter observation prompted the hypothesis that a human herpesvirus such as EBV or CMV might establish latency in the Schwann cell and that reactivation provides the sensitization to nerves.[6] In another experimental system, it was established that CMV can infect the Schwann cell and can cause a latent infection in peripheral nerve of the mouse.[7]

The most promising clinical setting for the study of virus and peripheral nerve disease today is in patients with HIV infection, including those with acquired immunodeficiency syndrome (AIDS). The many types of peripheral nerve involvement in these patients are currently in the stage of clinical description and classification, and there is little information about pathogenesis. Possible causes include nerve infection by HIV or other opportunistic viruses, a disturbance in the immune system with resultant autoimmunity, or a depletion of tropic growth factors needed for maintenance of nerves.[8–14]

2. VIRUS INFECTION OF NEURONS: HERPESVIRUS LATENCY AND REACTIVATION IN SENSORY GANGLIA

Herpes simplex virus types 1 and 2 and VZV all form latent infections in sensory ganglia of man, and reactivation is characterized by lesions in skin or mucous membranes in the distribution of the sensory nerve root involved.[1,4] With all three of these viruses, patients may feel numbness and pain in a nerve distribution prior to or during the eruption, but in most cases no lasting neurological impairment ensues. The encephalitis frequently caused by HSV 1 is not generally accompanied by a recurrence of oral ulcerations. Likewise, peripheral neuropathy almost never accompanies the recurrences of oral ulcerations of HSV 1.[1,4] However, neuropathy may complicate both the genital recurrences of HSV 2 and the dermatomal reactivation of VZV (shingles). Both HSV 1 and 2 may reactivate multiple times during a patient's lifetime, whereas shingles usually does not recur.[1,4]

2.1. Herpes Simplex Virus Type 1

By age 15 years, the incidence of HSV 1 seropositivity is 90%. The high incidence of HSV 1 latency in trigeminal ganglia was first demonstrated by neurosurgeons who noted herpes labialis in 90% of patients treated for trigeminal neuralgia with preganglionic sectioning of the trigeminal nerve. The virus can be recovered from human trigeminal ganglia removed at autopsy and grown in explant culture.[1,4] The detection of HSV immediate-early gene product antisense RNA in ganglia from seropositive but not seronegative cases at autopsy suggests that virtually all individuals who have been infected by this virus have a latent infection of the trigeminal ganglia.[15]

Approximately 25% of people have episodic bouts of oral ulcerations caused by HSV 1 reactivation. As mentioned above, these are almost never complicated by numbness or other neurological impairment. The syndrome of trigeminal neuralgia consists of severe bouts of stabbing pain in the face. The pain is often triggered by sensory stimulation of the face or mouth. There is little direct evidence to implicate HSV 1 as the cause of this syndrome, since both cases and controls have the same high incidence of viral latency in the trigeminal nerve.[1,4] The possibility that the painful syndrome results from an abnormal reaction to a ubiquitous virus remains. Finally, a few rare causes of peripheral nerve disease with HSV 1 have been reported, including radiculoneuropathy with isolation of HSV 1 from cerebrospinal fluid (CSF)[16] and an occasional AIDP following an HSV 1 infection.[17] Clinicians are aware that a herpetic eruption may be secondary to a structural lesion compressing the preganglionic segment of the trigeminal nerve, and these lesions should be sought before a cranial nerve palsy is attributed to HSV 1 infection.

2.2. Herpes Simplex Virus Type 2

Herpes simplex virus type 2 is the cause of recurrent bouts of genital ulcerations and is generally spread by sexual contact. The episodes of recurrence may be accompanied by pain and subjective tingling in the affected regions and by generalized malaise. A few patients suffer from definite neurological symptoms, which are accompanied by the local ulcerations. Although HSV 2 is a major cause of neonatal encephalitis, the virus causes meningitis, myelitis, and lumbosacral polyradiculoneuritis in adults.[18-23]

The HSV 2 is the established cause of a number of related syndromes, all of lumbosacral root distribution, and all tend to occur at times of either initial or recurrent genital ulcerations. This is in contrast to HSV 1 infections, where ulcerations may be misleading in diagnosis as they do not correlate with CNS infections.

Acute urinary retention may occur with an HSV 2 recurrence.[18] The patient may have some numbness in the sacral region. The ulcerations may precede or accompany the onset of retention and may be occult, occurring inside of the bladder, urethra, or vagina.[19,20]

The second neuropathic syndrome of HSV 2 is retention plus definite neurological findings: weakness, numbness, and arreflexia of the legs and impotence in males. The HSV 2 also causes aseptic meningitis and ascending myelitis.[21,22] A syndrome of sciatic pain without retention was seen in two married couples with genital herpes was termed conjugal sciatica. These patients did not have urinary retention or impotence.[23]

Most patients with genital herpes do not have these neurological complications. Retention or retention plus neuropathy was reported in 17 of 486 cases (3.5%) of anogenital herpes in a clinic for sexually transmitted diseases in England.[18] The diagnosis of neurological HSV infection is made in patients with urinary retention and lumbosacral neurological findings who also have genital herpes. The ulcerations may be occult and may require gynecological and urological examination for detection in the vagina or bladder. The condition is easily distinguished from AIDP because of the early involvement of the bladder and the presence of genital ulcerations. A myelogram, computerized tomography scan, or magnetic resonance imaging of the lumbosacral spine may be needed to rule out acute cauda equina compressive lesions in some cases, and CSF examination may be helpful. In HSV 2 meningitis there is always a pleocytosis averaging 200 cells, a protein concentration of up to 100 mg%, and usually a normal glucose. In HSV 2 lumbosacral neuropathy the CSF may be normal but usually shows a mild pleocytosis.

Although HSV 2 has been recovered by cultivation with susceptible cells of human lumbosacral ganglia from autopsy of unselected cases, pathological studies of the acute neuropathy are not available. Because of the rarity of the condition, and since the patients survive, it is unlikely that such material will be examined. However, animal models for HSV 2 genital infections and viral latency in lumbosacral ganglia have provided useful information that would be difficult to obtain from patient studies.[24,25] Martin and Suzuki reported that HSV 2 became latent in mouse lumbosacral ganglia after genital infection.[25] These ganglia and associated nerves had no detectable viral antigen, and virus could not be cultured from homogenates. It is thus remarkable that nerve inflammation was seen during latent infection in the absence of viral antigen. Possibly this inflammation lingers from a previous subclinical reactivation. Alternatively, it may be a clue that an immune response to a nonviral antigen is important in the neuropathy, analogous to that of MDV of chickens (Section 3).

2.3. Varicella–Zoster Virus

Children with chickenpox, the primary VZV infection, have virus-induced vesicles disseminated randomly over trunk, face, and limbs. With healing, the virus moves from the cutaneous lesions to the sensory nerves and, by axonal transport, to sensory ganglia, where it persists in a latent state throughout life.[1,4]

In the adult, herpes zoster or shingles is the result of reactivation of VZV in ganglia with transport down axons to the skin. There the virus causes a dermatomally localized eruption. The episodes are usually quite painful. Postherpetic neuralgia, the chronic persistence of pain, is a major clinical problem and a challenge to the development of effective therapy. From 2% to 16.5% of patients have residual pain 1 year after the healing of the vesicles, and this is more frequent in the elderly.[26]

Following recovery from zoster, areas of numbness in the distribution of the lesions are common and probably reflect damage to some sensory neurons during the inflammation. Up to 38% of patients have CSF pleocytosis, and the few histopathological studies of ganglia in this condition show inflammation, hemorrhages, and cell necrosis with extension of inflammation both into the spinal cord or brainstem and down into the distal nerve segment.[1] Although it is somewhat controversial, treatment of patients with either antiviral agents or steroids appears to lower the incidence of subsequent postherpetic neuralgia.[26] This suggests that both viral and inflammatory factors contribute to the pathology. A significant number of patients have postherpetic motor deficits in the distribution of the affected nerve.[1,4] Zoster in the distribution of the geniculate ganglion may result in vesicles and pain in the external auditory meatus, loss of taste over the anterior two-thirds of the tongue, and a facial palsy, the Ramsey–Hunt syndrome. Ophthalmic zoster may produce ocular motor palsies, ptosis, and dilation of the pupil. Otitic zoster may be complicated by hearing loss, facial palsy, and vertigo, implicating involvement of multiple cranial nerves and also probably the brainstem. Zoster of the cervical dermatomes may result in brachial motor neuropathies or unilateral diaphragmatic paralysis. Intercostal muscle denervation may follow thoracic zoster. Intestinal hypomotility and ileus have been reported in abdominal cases. Lumbar involvement may result in a persistent foot drop. Sacral zoster may leave the patient with residual urinary retention or incontinence.

3. OTHER HERPESVIRUSES ASSOCIATED WITH NEUROPATHY: MAREK'S DISEASE VIRUS, CYTOMEGALOVIRUS, AND EPSTEIN–BARR VIRUS

All three of these herpesviruses are definitely associated with peripheral neuropathy, but so far no direct causal relationship has been proven. The pathogenesis of the neuropathies associated with these three viruses is not completely understood but may involve infection of the Schwann cell and induction of immunity. The MDV induces an inflammatory neuropathy in chickens. The histopathology closely resembles human AIDP with demyelination and inflammation, and from this point of view it is the best animal model for AIDP of humans. There are two important observations in this disease. First, virus has been observed in explanted ganglia and is found earliest in nonmyelinating Schwann cells.[5] Secondly, an immune response to peripheral nerve can be demonstrated.[5] Stevens and Pepose and their co-workers proposed that human viruses similar to avian MDV might be responsible for AIDP by a similar mechanism, involving infection of the Schwann cell and subsequent induction of immunity.[5,6]

If a virus is involved in the pathogenesis of human AIDP, it is likely to be a mammalian virus analogous to MDV of chickens. Therefore, it is interesting to point out several biological similarities shared by MDV and the two human viruses, HCMV and EBV. All three are herpesviruses. In contrast to the HSV and VZV, these viruses do not infect neurons. Rather, the experimental evidence suggests Schwann cell infection under certain circumstances.[7] All three establish latency in lymphoid tissues and are associated with lymphoproliferative disorders. Since all three are latent, there usually is no infectious virus being shed for host-to-host spread of disease. In modern times CMV and EBV are transferred by blood transfusion or organ transplantation. Natural host-to-host spread is also analogous for the three: they replicate productively during the early acute stage of infection in epithelial surfaces, MDV in feather epithelium, EBV in nasopharynx, and CMV in salivary gland and urogenital excretions.[27]

Studies from my laboratory demonstrated that mouse (M) CMV grew productively in cultured Schwann cells but produced a latent infection in intact sciatic nerve *in vivo*.[7] Previously, Davis *et al.* reported MCMV replication in fibroblasts, Schwann cells, and neurons of trigeminal ganglion explants taken from mice surviving neonatal intracerebral MCMV inoculation.[28] Similarly, studies of *in vitro* infection of mouse dorsal root ganglion cultures showed neurons, neuroglia, and Schwann cells to be infected. Neurons were least susceptible to viral infection.[28–30] Thus, MCMV can infect the Schwann cell in experimentally infected mice. Plotkin and co-workers isolated HCMV from two of 20 human thoracic ganglion explants in an unsuccessful attempt to isolate VZV.[31] This suggests that latent MCMV may be present in nerve tissue of asymptomatic individuals. More direct observation has been made in the inflammatory neuropathies that occur in some AIDS patients. In some patients with these conditions, CMV inclusions have been found in Schwann cells[11–14,32] and microvascular endothelium (Fig. 3-1).[33]

All of these findings indicate that the Schwann cell can be infected by CMV. The state of viral expression, latent or replicating, depends on two factors: The state of cellular differentiation of the infected cell and the immune status of the host. Schwann cells in monolayer culture or in explanted ganglia lose their

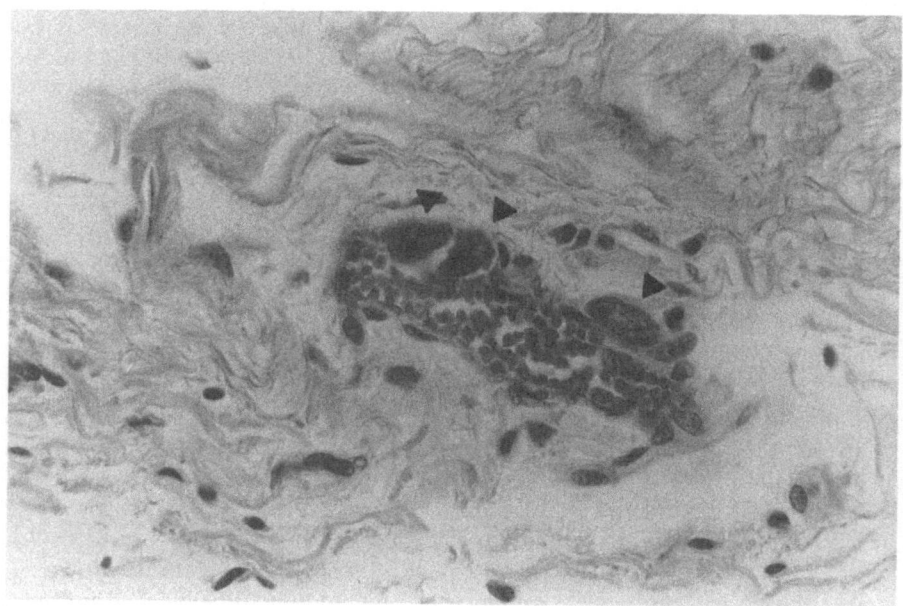

FIGURE 3-1. Section of epimysial vessel from an AIDS patient with widespread CMV infection of the CNS, peripheral nerve, and muscle. Note characteristic endothelial localization of cytomegalic cells (arrowheads), whereas endothelial cells along opposite side of lumen appear unremarkable. There is negligible inflammation in the surrounding tissues. (Hematoxylin and eosin; ×371)

association with myelin and begin to replicate. These small spindle-shaped cells are permissive for CMV *in vitro*. In contrast, the myelin-producing differentiated Schwann cell is found in intact nerve, and our data show that intact nerve can harbor latent infection. In similar studies, permissive infection in nonmyelinating and latent infection in myelinating Schwann cells were reported in MDV-infected chickens.[5] Similarly, latently infected B cells and macrophages reactivate MCMV when they are stimulated.[25,26] The reverse observation was made in MCMV infection of mouse teratoma cells. Latent infection was found in undifferentiated cells, and productive infection in differentiated cells.[34,35] These examples show that the state of differentiation of a cell determines whether CMV infection is latent or productive. The escape from the immune system is another factor favoring reactivation of virus *in vitro*. Immune suppression *in vivo* is capable of reactivating CMV in experimentally infected mice, in the clinical context of transplantation, or in patients with AIDS.

Primary virus infection or reactivation of a latent virus in the myelin-producing cell might be the mechanism of initiation of immune-mediated demyelination in some patients with AIDP. Experimental MCMV pneumonitis is particularly relevant in this regard, since immune mechanisms are required for development of the pulmonary lesion. This proves that immune-mediated pathology can occur in MCMV infection.[36] It is therefore possible that CMV initiates an immune-mediated disorder of nerves as well.

A variety of neurological complications, including peripheral neuropathies, occur in patients with primary infections with EBV and with HCMV.[37-40] Dowling and Cook reported serological evidence for HCMV infection in 33 of 220 AIDP patients and for EBV infection in eight of another series of 100 AIDP patients.[41,42] In patients with AIDS, HCMV infection of the nervous system is very common,[14,43] and these patients commonly have peripheral nerve disorders,[9-14] which may result from direct viral involvement or from immune factors. In AIDP, paralysis commonly follows exposure to any of a long list of stressful antecedent events such as fever, viral infection, trauma, or immunization.[6] How these stressful events might activate an immune response to peripheral nerve is not known, but reactivation of a latent virus is one possibility.[6]

4. HEPATITIS B VIRUS

The metabolic effects of hepatic dysfunction may cause some mild slowing of nerve conduction velocity measurements, and a small proportion of all liver disease patients have a mild clinical neuropathy.[44,45] However, a more pronounced neuropathy caused by segmental demyelination has been reported in patients with infectious hepatitis. Patients with chronic forms of hepatitis B may have a serum sickness syndrome with a rash, arthritis, glomerulonephritis, and neuropathy.[46] However, in some patients neuropathy is the most significant presenting symptom, although abnormal liver function tests and hepatitis B antigen and antibody(ies) in serum may be found in the course of evaluation of the neuropathy. These patients have a mononeuritis multiplex secondary to small vessel vasculitis. Involvement of multiple nerve twigs may coalesce to give the clinical appearance of a motor and sensory polyneuropathy with symmetrical or slightly asymmetric involvement of arms and legs. The onset may be acute or subacute, so that these patients clinically resemble either AIDP or chronic inflammatory demyelinating polyneuropathy (CIDP). Nerve biopsy has demonstrated inflammation of small vessels and a necrotizing vasculitis and periarteritis with segmental demyelination.[3,46,47] The diagnosis is made by positive hepatitis B serology in these patients.

All patients with otherwise unexplained neuropathy should have hepatitis B antigen and antibody tests. Cryoglobulins, consisting of antigen–antibody–complement complexes that precipitate in the cold, are demonstrable in the sera of many of these patients. These circulating immune complexes have been found in the small vessels of peripheral nerve and are probably responsible for the inflammation and demyelination.[3,46,47] The neuropathy may respond to treatment with corticosteroids and plasmapheresis. The ultimate outcome is dependent on the severity of the liver disease and the success of its management. Some success has been reported in treatment with interferon and adenine arabinoside 5′-monophosphate. Prolonged treatment with the latter drug may cause a painful distal sensory neuropathy.

5. HUMAN IMMUNODEFICIENCY VIRUS AND PERIPHERAL NEUROMUSCULAR SYNDROMES

Clinically apparent neurological complications are very common in patients with AIDS and related conditions (e.g., AIDS-related complex), and structural abnormalities of the nervous system discovered at autopsy are even more common.[48] It is likely that this virus invades the CNS soon after the primary human immunodeficiency virus (HIV) infection, long before the development of AIDS. Overall, neurological complications occur in approximately 70–80% of such patients, and at autopsy neuropathological lesions may be seen in almost 100% of patients.[8,48] The peripheral neuromuscular manifestations of HIV infection have been less well studied than those of the CNS, but it has been estimated that they affect 15% to 20% of infected individuals.[10–14]

A variety of neuropathies and other neuromuscular complications have been described in HIV infection[10–14] (Table 3-1), and it is likely that several different etiologies will be established by future investigations. Early in the course of infection, autoimmune mechanisms are probably important. Later, the direct or indirect effects of HIV itself or of other viruses such as HCMV, HSV 1, and HSV 2 may predominate.

5.1. Inflammatory Neuropathies that Usually Occur Early in the Course of Infection

The number and types of neuropathies seen in patients with HIV are at first bewildering, but it is reassuring that the majority of cases fit into one of two major

TABLE 3-1
Recognized Syndromes of the Peripheral Neuromuscular Manifestations of HIV Infection[a]

Inflammatory polyneuropathies, usually early
  Acute inflammatory demyelinating polyneuropathy (AIDP, Guillain–Barré syndrome)
  Chronic inflammatory demyelinating polyneuropathy (CIDP)
  Mononeuritis multiplex
  Large-fiber ataxic neuropathy (sensory ganglioneuronitis)
Axonal neuropaties, usually late
  Distal axonopathy
    Small-fiber distal sensory type
    Distal symmetrical polyneuropathy (DSPN)
  Progressive inflammatory polyradiculoneuropathy of legs with sphincter involvement (cauda
    equina syndrome)

[a]Sources in references 10–14.

diagnostic groups: one type that usually occurs early and a second that occurs later in the course of HIV infection. Early in the course of HIV infection, a mononucleosis-like syndrome may be observed, HIV may be cultured from blood and CSF, and weeks to months later HIV antibody becomes detectable in the serum. The clinician is unlikely to be aware of these early events, but the types of neuropathies listed in Table 3-1 begin to occur at the time of seroconversion. This highlights the importance of obtaining HIV serology in all patients with neuropathies.

These neuropathies also occur in the general population and are thought to have an autoimmune etiology. The onset is either acute over hours to days or subacute over weeks to months. The chronic form may have relapses and remissions. The patient experiences motor weakness of arms and legs (both proximal and distal) and some numbness and sensory loss. Electrodiagnostic testing shows slow nerve conduction velocities or conduction block, indicating that they are demyelinating neuropathies. Cornblath *et al.* reported that about 8% of a population of patients with inflammatory demyelinating polyneuropathies were HIV positive.[9] The HIV-seropositive group is distinguished by an overrepresentation of males and a CSF pleocytosis with a mean of 23 cells/mm^3. The seronegative patients classically have no cells in their CSF. Otherwise the HIV-infected patients clinically resemble the seronegative patients with neuropathies. These HIV-infected patients have a good prognosis if treated with either corticosteroids or plasmapheresis. The patients may survive up to several years before opportunistic infections, neoplasms, and other hallmarks of AIDS supervene.[9]

Mononeuritis and mononeuritis multiplex also occur early with seroconversion and may involve a segmental dermatome or any cranial or peripheral nerve. The single neuropathy may spontaneously improve or may herald the onset of widespread involvement leading to the appearance of a polyneuropathy. There is morphological evidence of inflammation around small blood vessels, and nerve infarcts may be present. Circulating immune complexes are the likely cause of the vasculitis, similar to the pathogenetic mechanism in hepatitis B. Widespread necrotizing vasculitis of the cauda equina and spinal cord has also been described in an individual with AIDS-related complex (ARC).[49]

The large-fiber sensory neuropathy or sensory ganglioneuronitis is a very rare disorder that has been reported in a patient with HIV infection.[11] A number of cases in the tropics have been associated with HTLV-I infection, and the condition may occur with no identifiable viral cause. The patients have ataxia with loss of position sense. Pathological findings included inflammation of the dorsal root ganglia and sensory roots with proliferation of ganglionic satellite cells, changes consistent with ganglioneuronitis.[50] The case reported by Dalakas and Pezeshkpour[11] did not improve with plasmapheresis and steroid medications.

The etiologies of the acute and chronic inflammatory neuropathies are thought to be autoimmune. Just as in the HIV-seronegative patients with the same diseases, the nature of the immunizing event and the relevant antigens are still unknown. Sural nerve biopsies have shown inflammatory cells around blood vessels and to some extent in the perineurium, similar to changes observed in seronegative cases. This type of inflammation is rare, however, in seronegative chronic inflammatory polyneuropathies. In a few reported cases, HIV has been isolated from nerve, but it is not known if the virus came from blood cells in the nerve, because most patients have viremia at the time of biopsy. So far, *in situ* hybridization studies of peripheral nerve using HIV genome probes have been negative: HIV-like particles have been demonstrated by electron microscopy in the nerve of one patient with AIDS-related neuropathy.[51] Studies with probes to other viruses such as HCMV and EBV are needed. In a case reported by Dalakas and Pezeshkpour,[11] HCMV inclusions were seen in the Schwann cells of peripheral nerve of an HIV-seropositive AIDP patient, and we have frequently found characteristic CMV inclusions in and around the nerves of AIDS patients with neuropathy (Fig. 3-2).

5.2. Axonal Neuropathies that Usually Occur Late in the Course of HIV Infection

After the diagnosis of AIDS is made, patients may develop numbness and burning pain in the feet, followed by distal leg weakness. Hand numbness begins only when leg numbness and weakness have progressed to the level of the knee. The electrodiagnostic features of this neuropathy are low amplitudes or loss of distal sensory nerve action potentials, findings of acute and chronic denervation on needle electromyography of distal lower extremity muscles, and relative preservation of nerve conduction velocities, especially of more proximal nerves. This neuropathy accounts for about 70% of the cases of peripheral neuropathy in AIDS and is probably underdiagnosed in the more advanced patients with other more pressing medical problems.[10–13]

The pattern of distal and symmetrical involvement is typical of toxic or metabolic neuropathies, where the most distal segments of the longest axons suffer the earliest damage, presumably related to deranged axoplasmic flow mechanisms. Two types of distal axonopathies have been reported in HIV

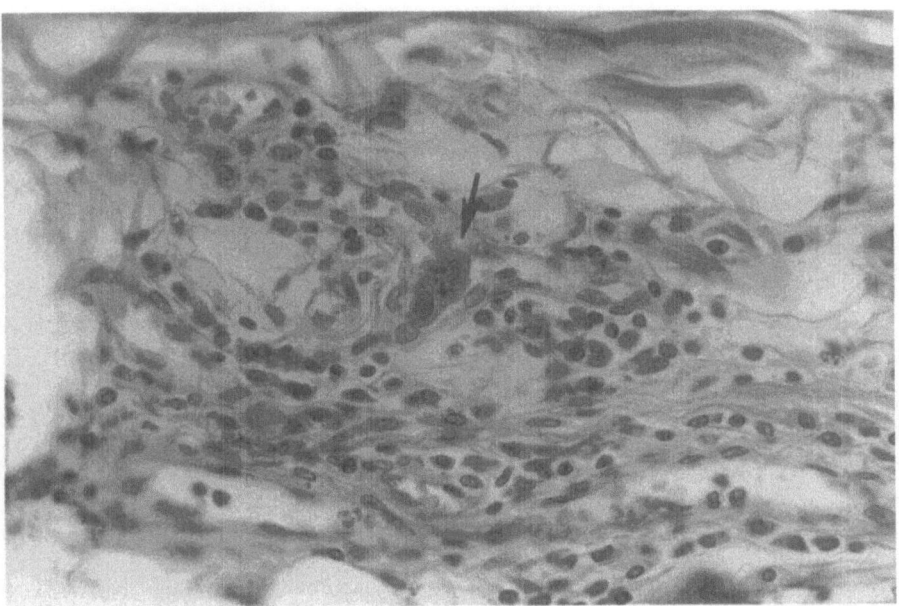

FIGURE 3-2. A section of peripheral nerve from the same patient as illustrated in Fig. 3-1 shows an isolated cytomegalic inclusion (with nuclear and cytoplasmic component, arrow) in the epineurium. Moderately severe chronic inflammation, including plasma cell infiltrate, is seen around the inclusion. (Hematoxylin and eosin; ×371)

infection: sensory and sensorimotor. In the small-fiber distal sensory neuropathy, burning dysesthesias of the feet are the major symptom. The syndrome may coexist with the AIDS–dementia complex and tends not to respond to treatment with azidothymidine.[52] Spinal cords from these patients show gracile tract degeneration with most severe involvement of upper thoracic and cervical segments.[53] Distal symmetrical polyneuropathy (DSPN) is probably a more advanced stage of the same pathological process, with distal leg weakness added to the sensory findings. Some authors lump the two together under the more inclusive term DSPN. The etiology is unknown but it probably results from a toxic product of either the HIV itself or the immune response to infection. Occasional patients have reported spontaneous improvement, and a number of patients have improved with antiviral drug treatment.[9]

Progressive inflammatory polyradiculoneuropathy, frequently affecting the legs most severely with sphincter involvement (cauda equina syndrome), is another cause of progressive weakness. A number of these cases have come to autopsy and have shown CMV infection of peripheral nerve and nerve roots.[12,13,43,54,55] Direct infection of nerve by HCMV (or perhaps HSV 2 in other cases) is the probable etiology. Of interest is the fact that, in at least one patient, CMV localized

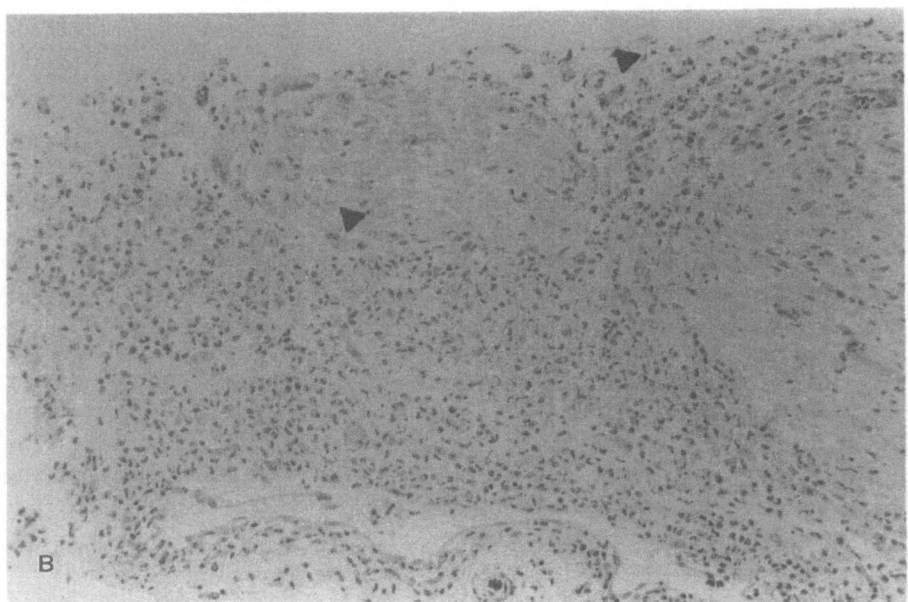

FIGURE 3-3. Severe radiculomyelopathy associated with CMV infection. Panel A shows low-power view of a dorsal nerve root infiltrated by inflammatory cells, found in clusters in some regions (arrow) or as a single cell infiltrate. Panel B shows an area with prominent radicular inflammation, including mononuclear inflammatory cells and scattered polymorphonuclear leukocytes. Even at this magnification, several cytomegalic cells can be identified (arrowheads).

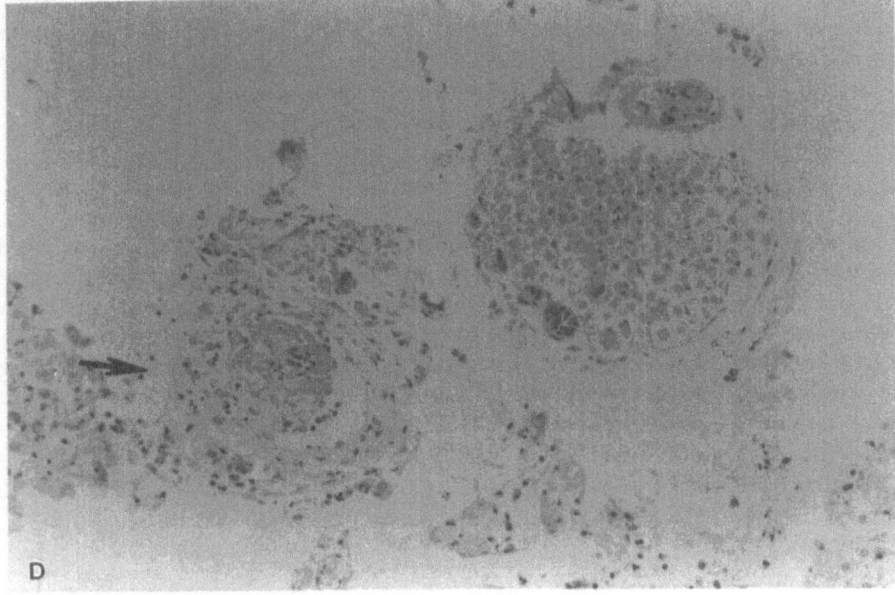

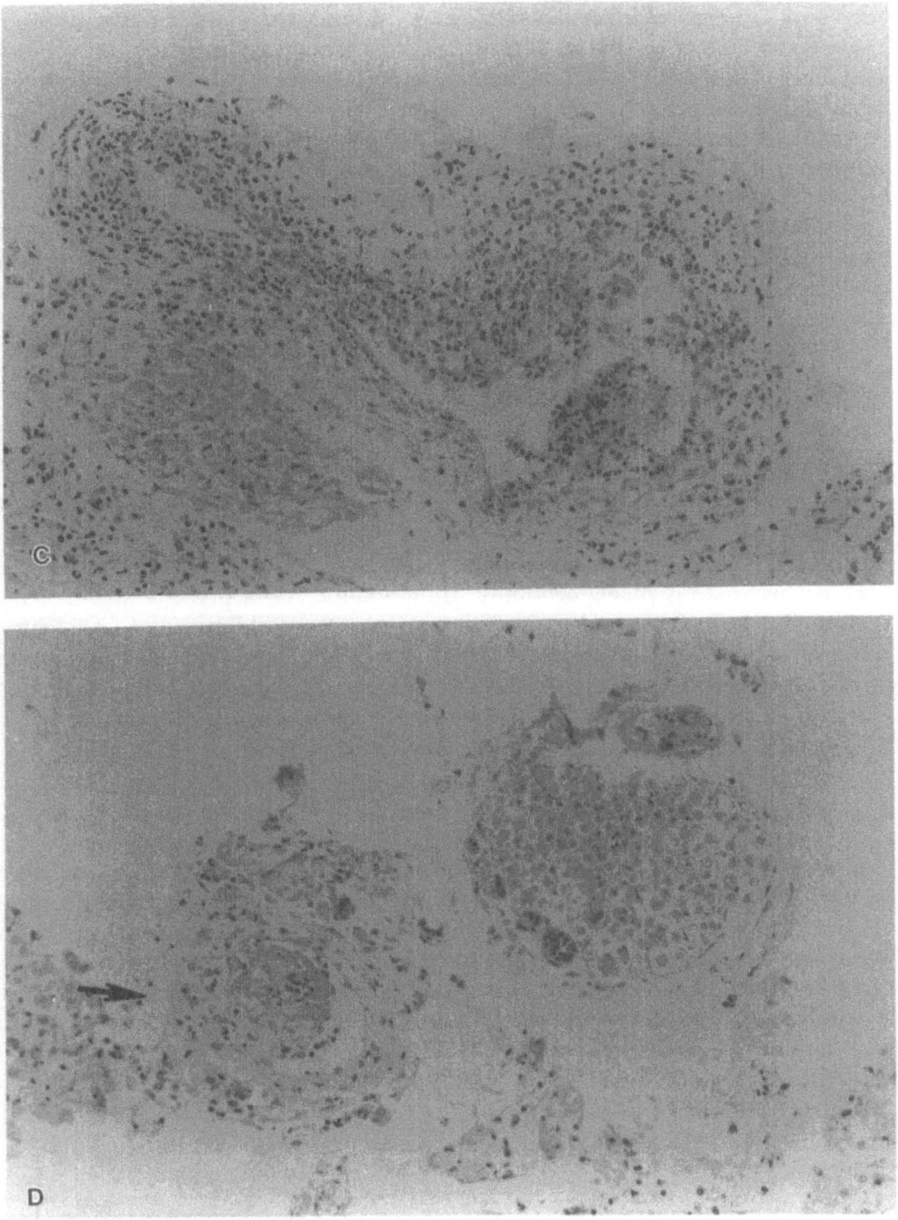

Panel C shows prominent acute and chronic inflammation of the wall of a venule in the subarachnoid space. Panel D shows a partly thrombosed microvessel (arrow) adjacent to a nerve root, seen at upper right. The wall and adventitia of the partly thrombosed microvessel show cytomegalic inclusions. Also see Vinters et al.[43] (All micrographs hematoxylin and eosin: A ×60; B ×123; C ×123; D ×123)

prominently to cells within blood vessel walls of the subarachnoid space.[43] This in turn had induced vasculitis and vascular thrombosis with resultant infarcts of the nerve roots, among which CMV inclusions were prominent (Fig. 3-3). Recognition and diagnosis of CMV infection in life is important because antivirals directed at CMV might be effective. However, the cases reported by Miller *et al.* did not respond to gancyclovir treatment, although this drug has been successful in improving HCMV chorioretinitis in some patients with both conditions.[13] One patient stabilized with plasmapheresis but did not improve. Thus, both immune and viral factors might be involved in this neuropathy. In view of the sacral neuropathies caused by HSV 2 in the non-HIV-infected population and the prevalence of HSV 2 infection in persons at risk for AIDS, it seems reasonable to suspect that some cases of this neuropathy may be caused by HSV 2.

5.3. Morphological Studies of Nerves

Biopsy and autopsy studies of large numbers of AIDS patients reveal that structural lesions of peripheral nerve are common even in clinically asymptomatic patients.[51,56,57] Findings include moderate or severe demyelination in a majority of specimens,[56] axonal degeneration, and variable degrees of mononuclear infiltration (Fig. 3-4). Inflammation is more severe in patients with chronic inflammatory demyelinating polyneuropathy. Inflammatory cells present within nerve include T lymphocytes and macrophages, with a preponderance of CD8+ cytotoxic or suppressor cells.[56] Morphometric studies of autopsy nerve specimens from patients with AIDS also reveal a decrease in the density of total myelinated fibers, with a disproportionately severe loss of large myelinated fibers (Fig. 3-5) in almost 50% of patients surveyed.[57]

5.4. Autonomic Neuropathy

The conclusion that autonomic neuropathy occurs in AIDS patients is based on physiological studies, since (as yet) no detailed morphological assessment of the autonomic nervous system in patients with AIDS has been carried out.[58–60] A syndrome resembling amyotrophic lateral sclerosis or motor neuron disease (ALS/MND) has been observed only rarely in patients with AIDS,[61,62] including one individual who had appropriate findings confirmed at autopsy. In one patient who had severe peripheral neuropathy and ALS/MND-like picture, we found structural correlates of a profound sensorimotor neuropathy, immunohistochemical evidence for HIV infection of anterior horn cells within the spinal cord, and widespread HCMV infection of nerves and muscles.[63]

---→

FIGURE 3-4. Inflammatory infiltrates within peripheral nerve. Both sections show autopsy specimens of peripheral nerve, showing variable degrees of predominantly epineurial inflammation. Elsewhere in the nerve illustrated in panel B, typical cytomegalic cells were seen, but no evidence of opportunistic infection was seen in the nerve illustrated in panel A. (Hematoxylin and eosin; both panels ×153)

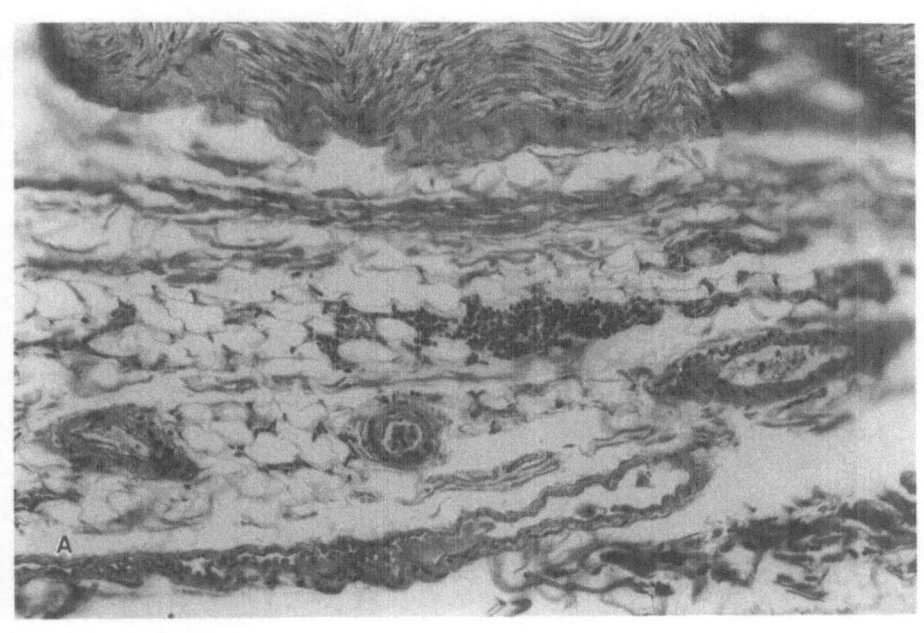

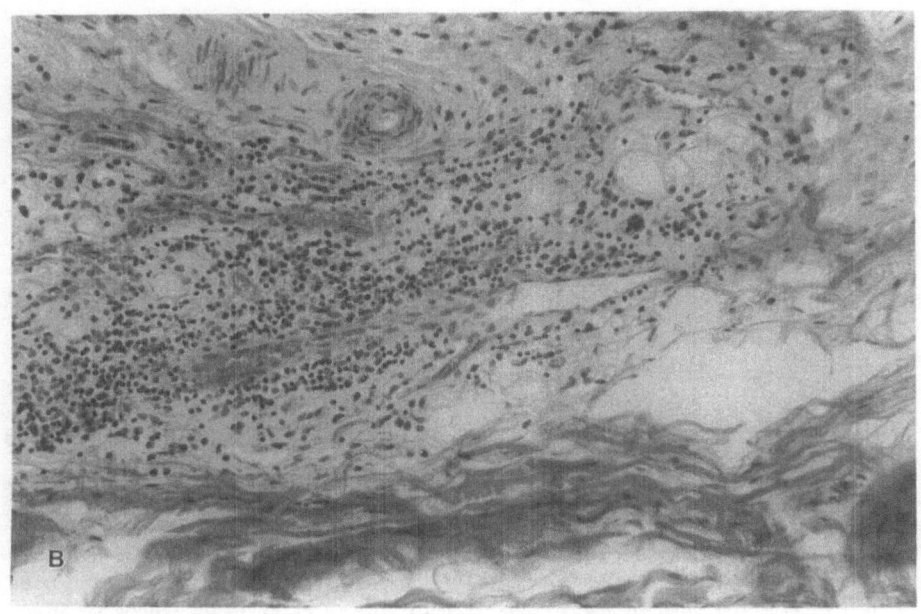

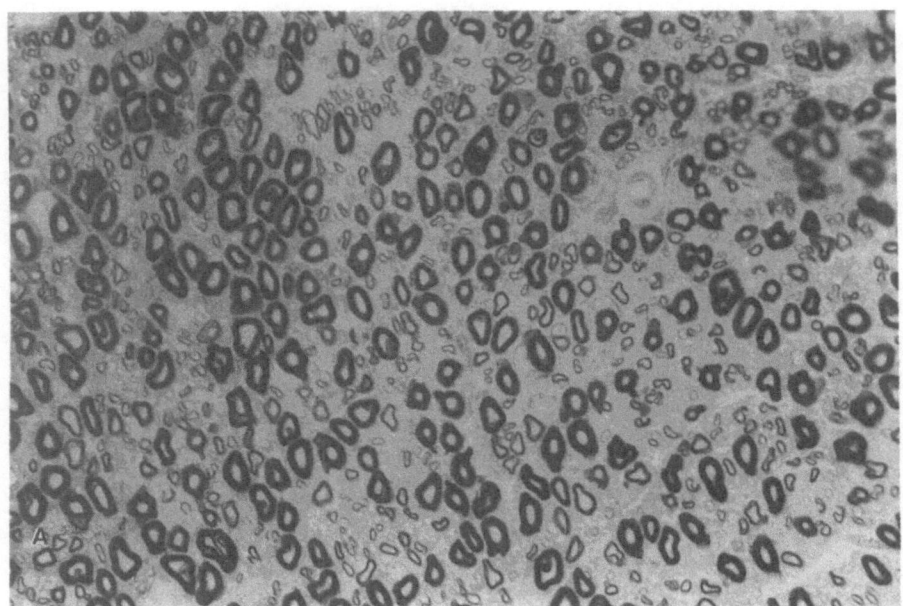

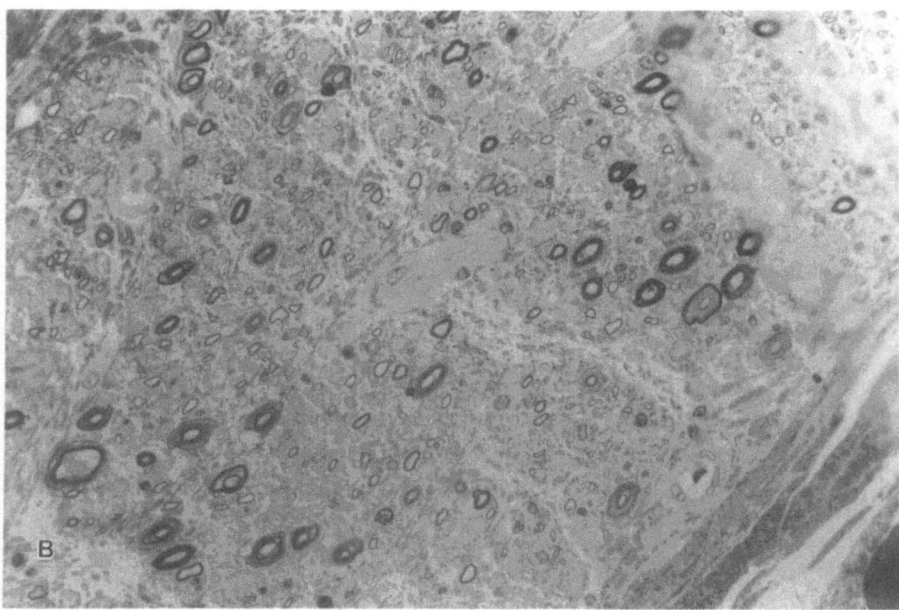

FIGURE 3-5. Plastic-embedded sections of peripheral nerve stained with toluidine blue and photographed at identical magnification. Panel A shows an approximately normal population density of large and small myelinated fibers. Panel B shows a section of the same nerve from another patient, with profound loss of large and small myelinated fibers seen throughout most of the cross-sectional area. This illustrates a severe degree of pathological findings commonly seen in appropriately examined nerves from patients with AIDS. (Magnification, both A and B, ×390)

5.5. Other Neuromuscular Syndromes

Myopathies are rare manifestations of HIV infection, occurring approximately one-tenth as often as abnormalities of peripheral nerve.[14] HIV-antigen-bearing inflammatory cells have been reported in muscle biopsies from seropositive patients with clinical polymyositis.[11] Patients from defined HIV risk groups with myopathic symptoms definitely need to be tested for HIV infection serologically, since they may be candidates for antiviral therapy.

Myopathies in patients with HIV infection are often inflammatory, resembling polymyositis.[64,65] Giant cells have been seen within the inflammatory infiltrates, but HIV antigen has not been localized to these multinucleate cells; multinucleated giant cells are a fairly reliable marker for HIV in the brain.[14] Myositis has also been observed in simian AIDS.[66,67]

Noninflammatory myopathies are also observed in patients with AIDS.[68] The most structurally intriguing of these are myopathies in which rod or cytoplasmic bodies are observed within muscle fibers.[69–72] Opportunistic pathogens can often be seen within skeletal muscle, and denervation atrophy is (as expected) a frequent finding in muscles from AIDS patients.[14,73]

REFERENCES

1. Johnson, R. T., 1982, *Virus Infections of the Nervous System*, Raven Press, New York.
2. Johnson, R. T., Griffin, D. E., Hirsch, R. L., Wolinsky, J. S., Roedenbeck, S., Lindo de Soriano, I., and Vaisberg, A., 1984, Measles encephalitis—clinical and immunologic studies, *N. Engl. J. Med.* **310:**137–141.
3. Tsukada, N., Koh, C.-S., Inoue, A., and Yanagisawa, N., 1987, Demyelinating neuropathy associated with hepatitis B virus infection. Detection of immune complexes composed of hepatitis B virus surface antigen, *J. Neurol. Sci.* **77:**203–216.
4. Tenser, R. B., 1984, Herpes simplex and herpes zoster: Nervous system involvement, *Neurol. Clin.* **2:**215–240.
5. Pepose, J. S., Stevens, J. G., Cook, M. L., and Lampert, P. W., 1981, Marek's disease as a model for the Landry–Guillain–Barré syndrome. Latent viral infection in nonneuronal cells accompanied by specific immune responses to peripheral nerve and myelin, *Am. J. Pathol.* **103:**309–320.
6. Pepose, J. S., 1982, A theory of virus-induced demyelination in the Landry–Guillain–Barré syndrome, *J. Neurol.* **227:**93–97.
7. Abols-Mantyh, I., Siegel, L., Verity, M. A., Londe, H., and Graves, M. C., 1987, Latency of mouse cytomegalovirus in sciatic nerve, *Neurology* **37:**1809–1812.
8. Elder, G. A., and Sever, J. L., 1988, AIDS and neurological disorders: An overview, *Ann. Neurol.* **23:**S4–S6.
9. Cornblath, D. R., McArthur, J. C., Kennedy, P. G. E., Witte, A. S., and Griffin, J. W., 1987, Inflammatory demyelinating peripheral neuropathies associated with human T-cell lymphotropic virus type III infection, *Ann. Neurol.* **21:**32–40.
10. Cornblath, D. R., 1988, Treatment of the neuromuscular complications of human immunodeficiency virus infection, *Ann. Neurol.* **23:**S88–S91.
11. Dalakas, M. C., and Pezeshkpour, G. H., 1988, Neuromuscular diseases associated with human immunodeficiency virus infection, *Ann. Neurol.* **23:**S38–S48.
12. Parry, G. J., 1988, Peripheral neuropathies associated with human immunodeficiency virus infections, *Ann. Neurol.* **23:**S49–S53.
13. Miller, R. G., Parry, G. J., Pfaeffl, W., Lang, W., Lippert, R., and Kiprov, D., 1988, The

spectrum of peripheral neuropathy associated with ARC and AIDS, *Muscle Nerve* **11**:857–863.

14. Vinters, H. V., and Anders, K. H., 1990, *Neuropathology of AIDS*, CRC Press, Boca Raton, Florida.
15. Stevens, J. G., Wagner, E. K., Devi-Rao, G. B., Cook, M. L., and Feldman, L. T., 1987, RNA complementary to a herpesvirus alpha gene mRNA is prominent in latently infected neurons, *Science* **235**:1056–1059.
16. Morrison, R. E., Shatsky, S. A., Holmes, G. E., Top, F. H., Jr., and Martins, A. N., 1979, Herpes simplex virus type 1 from a patient with radiculoneuropathy, *J.A.M.A.* **241**:393–394.
17. Olivarius, B. D., and Buhl, M., 1975, Herpes simplex virus and Guillain–Barré polyradiculitis, *Br. Med. J.* **1**:192–193.
18. Oates, J. K., and Greenhouse, P. R. D. H., 1978, Retention of urine in anogenital herpetic infection, *Lancet* **1**:691– 692.
19. Riehle, R. A., Jr., and Williams, J. J., 1979, Transient neuropathic bladder following herpes simplex genitalis, *J. Urol.* **122**:263–264.
20. Caplan, L. R., Kleeman, F. J., and Berg, S., 1977, Urinary retention probably secondary to herpes genitalis, *N. Engl. J. Med.* **297**:920–921.
21. Black, D., Stewart, J., and Melmed, C., 1983, Sacral nerve dysfunction plus generalized polyneuropathy in herpes simplex genitalis. *Ann. Neurol.* **14**:692.
22. Handler, C. E., Perkin, G. D., Fray, R., and Woinarski, J., 1983, Radiculomyelopathy associated with herpes simplex genitalis treated with adenosine arabinoside, *Postgrad. Med. J.* **59**:388–389.
23. Hinthorn, D. R., Baker, L. H., Romig, D. A., and Liu, C., 1976, Recurrent conjugal neuralgia caused by herpesvirus hominis type-2, *J.A.M.A.* **236**:587–588.
24. Bernstein, D. I., and Stanberry, L. R., 1986, Zosteriform spread of herpes simplex virus type 2 genital infection in the guinea-pig, *J. Gen. Virol.* **67**:1851–1857.
25. Martin, J. R., and Suzuki, S., 1987, Inflammatory sensory polyradiculopathy and reactivated peripheral nervous system infection in a genital herpes model, *J. Neurol. Sci.* **79**:155–171.
26. Portnoy, R. K., Duma, C., and Foley, K. M., 1986, Acute herpetic and postherpetic neuralgia: Clinical review and current management, *Ann. Neurol.* **20**:651–664.
27. Rapp, F., 1983, The biology of cytomegalovirus, in: *The Herpes Virus* (B. Roizman, ed.), Plenum Press, New York, pp. 1–66.
28. David, G. L., Krawczyk, K. W., and Hawrisiak, M. M., 1979, Age-related neurocytotropism of mouse cytomegalovirus in explanted trigeminal ganglions, *Am. J. Pathol.* **97**:261–276.
29. Schneider, J. F., Carp, R. I., Belisle, E. H., and Rue, C. E., 1972, The response of murine nerve tissue culture to murine cytomegalic virus, *Acta Neuropathol. (Berl.)* **21**:204–212.
30. Willson, J. N., Schneider, J. F., Rosen, M., and Belisle, E. H., 1974, Ultrastructural pathology of murine cytomegalovirus infection in cultured mouse nervous system tissue, *Am. J. Pathol.* **74**:467–480.
31. Plotkin, S. A., Stein, S., Snyder, M., and Immesoete, P., 1977, Attempts to recover varicella virus from ganglia, *Ann. Neurol.* **2**:249.
32. Storey, J. R., Greco, C., Bolan, R., Calanchini, P., Denys, E., Parry, G., and Miller, R., 1988, Electrodiagnostic findings in cytomegalovirus, radiculomyelopathy related to acquired immunodeficiency syndrome, *Muscle Nerve* **11**:971.
33. Ho, H. W., Bailey, R., Rhee, J. M., and Vinters, H. V., 1989, Neuromuscular pathology in patients with acquired immune deficiency syndrome (AIDS): An autopsy study, *J. Neuropathol. Exp. Neurol.* **48**:382.
34. Jordan, M. C., 1983, Latent infection and the elusive cytomegalovirus, *Rev. Infect. Dis.* **5**:205–215.
35. Dutko, F. J., and Oldstone, M. B. A., 1981, Cytomegalovirus causes a latent infection in undifferentiated cells and is activated by induction of cell differentiation, *J. Exp. Med.* **154**:1636–1651.

36. Shanley, J. D., Pesanti, E. L., and Nugent, K. M., 1982, The pathogenesis of pneumonitis due to murine cytomegalovirus, *J. Infect. Dis.* **146:**388–396.

37. Duchowny, M., Caplan, L., and Siber, G., 1979, Cytomegalovirus infection of the adult nervous system, *Ann. Neurol.* **5:**458–461.

38. Ironside, A. G., and Tobin, J. O'H., 1967, Cytomegalovirus infection in the adult, *Lancet* **2:** 615–616.

39. Klemola, E., Weckman, N., Haltia, K., and Kääriäinen, L., 1967, Guillain–Barré syndrome associated with acquired cytomegalovirus infection, *Acta. Med. Scand.* **181:**603–607.

40. Leonard, J. C., and Tobin, J. O'H., 1971, Polyneuritis associated with cytomegalovirus infections, *Q. J. Med.* **40:**435–442.

41. Dowling, P. C., and Cook, S. D., 1981, Role of infection in Guillain–Barré syndrome: Laboratory confirmation of herpesviruses in 41 cases, *Ann. Neurol.* **9:**44–55.

42. Dowling, P., Menonna, J., and Cook, S., 1977, Cytomegalovirus complement fixation antibody in Guillain–Barré syndrome, *Neurology* **27:**1153–1156.

43. Vinters, H. V., Kwok, M. K., Ho, H. W., Anders, K. H., Tomiyasu, U., Wolfson, W. L., and Robert, F., 1989, Cytomegalovirus in the nervous system of patients with the acquired immune deficiency syndrome, *Brain* **112:**245–268.

44. Seneviratne, K. N., and Peiris, O. A., 1970, Peripheral nerve function in chronic liver disease, *J. Neurol. Neurosurg. Psychiatry* **33:**609–614.

45. Chari, V. R., Katiyar, B. C., Rastogi, B. L., and Bhattacharya, S. K., 1977, Neuropathy in hepatic disorders: A clinical, electrophysiological, and histopathological appraisal, *J. Neurol. Sci.* **31:**93–111.

46. Tsukada, N., Koh, C.-S., Owa, M., and Yanagisawa, N., 1983, Chronic neuropathy associated with immune complexes of hepatitis B virus, *J. Neurol. Sci.* **61:**193–211.

47. Farivar, M., Wands, J. R., Benson, G. D., Dienstag, J. L., and Isselbacher, K. J., 1976, Cryoprotein complexes and peripheral neuropathy in a patient with chronic active hepatitis, *Gastroenterology* **71:**490–493.

48. Anders, K. H., Guerra, W. F., Tomiyasu, U., Verity, M. A., and Vinters, H. V., 1986, The neuropathology of AIDS. UCLA experience and review, *Am. J. Pathol.* **124:**537–558.

49. Vinters, H. V., Guerra, W. F., Eppolito, L., and Keith, P. E. III, 1988, Necrotizing vasculitis of the nervous system in a patient with AIDS-related complex, *Neuropathol. Appl. Neurobiol.* **14:** 417–424.

50. Elder, G., Dalakas, M., Pezeshkpour, G., and Sever, J., 1986, Ataxic neuropathy due to ganglioneuronitis after probable acute human immunodeficiency virus infection, *Lancet* **2:** 1275–1276.

51. Bailey, R. O., Baltch, A. L., Venkatesh, R., Singh, J. K., and Bishop, M. B., 1988, Sensory motor neuropathy associated with AIDS, *Neurology* **38:**886–891.

52. Cornblath, D. R., and McArthur, J. C., 1988, Predominantly sensory neuropathy in patients with AIDS and AIDS-related complex, *Neurology* **38:**794–796.

53. Rance, N. E., McArthur, J. C., Cornblath, D. R., Landstrom, D. L., Griffin, J. W., and Price, D. L., 1988, Gracile tract degeneration in patients with sensory neuropathy and AIDS, *Neurology* **38:**265–271.

54. Behar, R., Wiley, C., and McCutchan, J. A., 1987, Cytomegalovirus polyradiculoneuropathy in acquired immune deficiency syndrome, *Neurology* **37:**557–561.

55. Bishopric, G., Bruner, J., and Butler, J., 1985, Guillain–Barré syndrome with cytomegalovirus infection of peripheral nerves, *Arch. Pathol. Lab. Med.* **101:**1106–1108.

56. de la Monte, S. M., Gabuzda, D. H., Ho, D. D., Brown, R. H., Jr., Hedley-Whyte, E. T., Schooley, R. T., Hirsch, M. S., and Bhan, A. K., 1988, Peripheral neuropathy in the acquired immunodeficiency syndrome, *Ann. Neurol.* **23:**485–492.

57. Mah, V., Vartavarian, L. M., Akers, M.-A., and Vinters, H. V., 1988, Abnormalities of peripheral nerve in patients with human immunodeficiency virus infection, *Ann. Neurol.* **24:** 713–717.

58. Lin-Greenberg, A., and Taneja-Uppal, N., 1987, Dysautonomia and infection with the human immunodeficiency virus, *Ann. Intern. Med.* **106**:167.

59. Craddock, C., Bull, R., Pasvol, G., Protheroe, A., and Hopkin, J., 1987, Cardiorespiratory arrest and autonomic neuropathy in AIDS, *Lancet* **2**:16–18.

60. Miller, R. F., and Semple, S. J. G., 1987, Autonomic neuropathy in AIDS, *Lancet* **2**:343–344.

61. Hoffman, P. M., Festoff, B. W., Giron, L. T., Jr., Hollenbeck, L. C., Garruto, R. M., and Ruscetti, F. W., 1985, Isolation of LAV/HTLV-III from a patient with amyotrophic lateral sclerosis, *N. Engl. J. Med.* **313**:324–325.

62. Sher, J. H., Wrzolek, M. A., and Shmuter, Z. B., 1988, Motor neuron disease associated with AIDS, *J. Neuropathol. Exp. Neurol.* **47**:303.

63. Robert, M. E., Geraghty, J. J., III, Miles, S. A., Cornford, M. E., and Vinters, H. V., 1989, Severe neuropathy and a motor neuron disease-like syndrome in a patient with acquired immune deficiency syndrome (AIDS). Evidence for widespread cytomegalovirus infection of peripheral nerve and HIV infection of anterior horn cells, *Acta Neuropathol.* **79**:255–261.

64. Dalakas, M. C., Pezeshkpour, G. H., Gravell, M., and Sever, J. L., 1986, Polymyositis associated with AIDS retrovirus, *J.A.M.A.* **256**:2381–2383.

65. Bailey, R. O., Turok, D. I., Jaufmann, B. P., and Singh, J. K., 1987, Myositis and acquired immunodeficiency syndrome, *Hum. Pathol.* **18**:749–751.

66. Dalakas, M. C., London, W. T., Gravell, M., and Sever, J. L., 1986, Polymyositis in an immunodeficiency disease in monkeys induced by a type D retrovirus, *Neurology* **36**:569–572.

67. Dalakas, M. C., Gravell, M., London, W. T., Cunningham, G., and Sever, J. L., 1987, Morphological changes of an inflammatory myopathy in rhesus monkeys with simian acquired immunodeficiency syndrome, *Proc. Soc. Exp. Biol. Med.* **185**:368–376.

68. Stern, R., Gold, J., and DiCarlo, E. F., 1987, Myopathy complicating the acquired immune deficiency syndrome, *Muscle Nerve* **10**:318–322.

69. Gonzales, M. F., Olney, R. K., So, Y. T., Greco, C. M., McQuinn, B. A., Miller, R. G., and DeArmond, S. J., 1988, Subacute structural myopathy associated with human immunodeficiency virus infection, *Arch. Neurol.* **45**:585–587.

70. Dalakas, M. C., Pezeshkpour, G. H., and Flaherty, M., 1987, Progressive nemaline (rod) myopathy associated with HIV infection, *N. Engl. J. Med.* **317**:1602–1603.

71. Wolfe, D. E., and Simpson, D. M., 1988, Cytoplasmic bodies in AIDS myopathy, *J. Neuropathol. Exp. Neurol.* **47**:388.

72. Simpson, D. M., and Bender, A. N., 1988, Human immunodeficiency virus-associated myopathy: Analysis of 11 patients, *Ann. Neurol.* **24**:79–84.

73. Anders, K., Steinsapir, K. D., Iverson, D. J., Glasgow, B. J., Layfield, L. J., Brown, W. J., Cancilla, P. A., Verity, M. A., and Vinters, H. V., 1986, Neuropathologic findings in the acquired immunodeficiency syndrome (AIDS), *Clin. Neuropathol.* **5**:1–20.

II

Animal Models

4

Visna, a Lentiviral Disease of Sheep

GUDMUNDUR PÉTURSSON,
ÓLAFUR S. ANDRÉSSON,
and GUDMUNDUR GEORGSSON

1. VISNA VIRUS: BIOLOGY AND STRUCTURE

1.1. Introduction

Visna (meaning wasting), an encephalomyelitis, and maedi (meaning dyspnea), an interstitial pneumonia of sheep, were brought to Iceland with imported Karakul sheep in 1933.[1] Almost three decades elapsed until the causative agent of visna, a virus, was isolated from the central nervous system (CNS) of visna-affected sheep.[2] A few years later a virus was isolated from the lungs of a sheep with maedi.[3] The viruses were shown to be serologically related, and early transmission experiments indicated that visna and maedi were but different organ manifestations of infection with the same virus,[4] which is thus frequently referred to as maedi–visna virus (MVV).

It has been shown that MVV is a nononcogenic retrovirus,[5] and based on this property, genome organization, and nucleic acid sequence homologies,[6] it is classified with several other viruses in a separate group called lentiviruses, a term derived from Sigurdsson's concept of slow infections.[7]

GUDMUNDUR PÉTURSSON, ÓLAFUR S. ANDRÉSSON, and GUDMUNDUR GEORGSSON
• Institute for Experimental Pathology, University of Iceland, Keldur, IS-128 Reykjavík, Iceland.

Neuropathogenic Viruses and Immunity, edited by Steven Specter *et al.* Plenum Press, New York, 1992.

1.2. Neurotropism

Visna virus was the first of the lentiviruses to be isolated, but lentiviruses have now been isolated from several mammalian species including humans.[8] As shown in Table 4-1, the lentiviruses vary in their organ tropism and cause a wide variety of disease manifestations. In addition to visna virus, the caprine arthritis–encephalitis virus is also neurotropic, and there is a growing body of evidence that the nervous system may be one of the primary target organs of infection with the human immunodeficiency virus (HIV).[9]

During the epidemic in Iceland the pulmonary affection, maedi, was the predominant disease manifestation. In some sheep flocks, however, visna has been the main cause of morbidity and mortality.[1] This experience seems to be unique for Iceland. In other countries the pulmonary disease dominates the clinical picture, and the CNS infection is usually mild and subclinical.[1] This difference is at least in part a result of an unusual susceptibility of the Icelandic breed of sheep. Thus, in transmission experiments in American sheep using an Icelandic strain of virus, K1514, which regularly causes an encephalitis in Icelandic sheep,[10] a tenfold higher dose was needed to induce an encephalitis, which, in contrast to the progressive encephalitis observed in Icelandic sheep, was self-limiting.[11]

In addition, it seems likely that during the epidemic in Iceland a neuro-virulent strain emerged. Thus, recent results in our laboratory applying restriction enzyme analysis on viral DNA from various visna and maedi strains indicate that visna and maedi viruses may differ approximately 7% in nucleotide sequences (V. Andrésdóttir, unpublished results), a variation comparable to the greatest difference reported for different HIV-1 isolates.[12] In spite of this genetic difference in maedi and visna strains, there have been no reports on *in vitro* differences in any host cells, but such an *in vitro* system to distinguish strains with different organ tropisms would greatly facilitate analysis of genetic determinants of neurovirulence in conjunction with functional molecular clones of visna and maedi proviruses.

TABLE 4-1
Lentiviruses[a]

Virus	Host	Pathology
Maedi–visna virus	Sheep	Encephalitis, interstitial pneumonia
Caprine arthritis–encephalitis virus	Goats	Arthritis, encephalitis
Bovine immunodeficiency-like virus	Cattle	Lymphocytosis, lymphadenopathy
Equine infectious anemia virus	Horses	Hemolytic anemia
Feline immunodeficiency virus	Cats	Immunodeficiency
Simian immunodeficiency virus	Monkeys	Immunodeficiency
Human immunodeficiency virus #1	Humans	Immunodeficiency
Human immunodeficiency virus #2	Humans	Immunodeficiency

[a]Modified from Pétursson *et al.*[8]

1.3. Virus Structure and Genome Organization

The visna virion has a structure similar to other retroviruses, containing two nearly identical polyadenylated RNA molecules of approximately 9200 nucleotides each and several molecules of lysine tRNAs, which act as primers for the enzyme reverse transcriptase, which also is contained in the virion.[13,14] The viral RNA molecules are scaffolded in a characteristic bullet- or wedge-shaped dense core containing three different proteins (p25, p16, and p14). The viral core is surrounded by an envelope derived from the cell membrane containing the viral glycoprotein gp135 and its cleavage products, the surface (SU; gp70) and transmembrane (TM; gp45) glycoproteins. After the initial contact (the receptors for the virus are still uncharacterized), the virion fuses with the cell membrane, and the reverse transcriptase makes use of the lysine tRNAs to initiate a DNA copy of the RNA genome, the RNA is degraded, and a second strand of DNA is produced. This viral DNA contains duplications of both the 5' and 3' ends of the genome, forming long terminal repeats (LTRs, 412 bp in visna virus) characteristic of retroviral DNA replication intermediates.

In permissive cells from sheep choroid plexus (SCP), hundreds[15] of DNA copies are formed, although only one to four copies are found integrated in the cellular chromosomes as proviruses. (Ó. Andrésson, unpublished data). The provirus (and perhaps also the unintegrated DNA) acts as a template for transcription that is regulated by several proteins coded by the virus and cellular genes producing a complex set of mRNAs.[16,17] The action of a positive regulator of visna virus expression has been demonstrated and is probably analogous to the *tat* protein of HIV,[18] but much remains to be elucidated about the factors regulating visna expression.

From DNA sequencing and protein characterization[14,20–22] it has been deduced that the visna virus genome contains three large genes; *gag*, coding for the core proteins p16, p25, and p14; *pol*, coding for the reverse transcriptase and, judging from the nucleic acid sequence, an endonuclease activity (for integration of the provirus) as well as a protease activity for processing the polyprotein products of the *gag* and *pol* genes; and *env*, coding for the glycoproteins of the virion envelope. In between *pol* and *env* there is a gene, termed Q, whose function has not been defined, but is probably analogous to *vif* in the HIV genome. Right after Q there is another short open reading frame, now referred to as *tat*.[18] The third regulatory gene, *rev*, is bipartite, the first part being identical to the start of the *env* gene, and the second part overlapping the terminus of *env*, but out of phase.[18,19]

1.4. Virus Replication *in Vitro*

Most *in vitro* studies of visna virus have been made with monolayers of fibroblastoid cells derived from SCP in which the characteristic cytopathic effect (CPE) is cell fusion resulting in syncytia of multinucleated giant cells. Such fusion is not absolutely dependent on viral infectivity and can readily be observed within an hour after inoculation with a high multiplicity of inactivated virus (fusion from without). When infection is with a lower dose of virus, the CPE correlates with virus replication and may take 4–5 days. In addition to multinucleated syncytia, the formation of mononucleated refractile spindle-shaped cells with dendritic processes and giant multinucleated stellate cells is characteristic, and finally the cells disintegrate. In electron micrographs viral particles can be seen assembling and maturing by budding from the cell surface, where the *env* protein spans the lipid bilayer and can be seen as protruding knobs.[23]

Visna and related viruses have the capacity to infect cells of the monocyte/ macrophage lineage *in vitro*, usually producing a prolonged infection with minimal CPE yielding low titers of virus.[24-26] When macrophages are infected *in vitro*, viral buds are infrequently seen at the cell surface, whereas viral particles are prominent in internal vesicles (G. Georgsson, unpublished data), as has also been observed with HIV-1.[27] Interestingly, the level of transcription from the viral LTRs is higher in macrophages than in SCP cells, but the level of *trans*-activation is much higher in the SCP cells.[18]

Visna virus does not readily grow in cells of other species than the natural hosts (except bovine cells),[28] although Macintyre was able to propagate the virus on a permanent line of human astrocytes.[29]

2. CLINICAL FEATURES AND PATHOLOGY

2.1. Transmission and Incubation Period

The Icelandic sheep that were in contact with the imported Karakul sheep did not show any evidence of clinical disease until 6 years later.[1] Further experience in the field substantiated this extraordinary feature of this viral infection, i.e., the long incubation period. Thus, clinical symptoms were rarely observed in sheep before the age of 3 to 4 years. In transmission experiments with visna virus, using an intracerebral route of inoculation, incubation periods, up to 7 to 8 years have been observed.[30]

According to experience during the epidemic in Iceland, natural transmission was apparently mainly respiratory, and transmission experiments by the respiratory route supported this view.[1] In open pastures the communicability seems to be very low even in the clinical stage of the disease. In Iceland the infection spread mainly during the winter, when the sheep were housed. Later studies in The Netherlands showed that another important mode of spread in endemic areas is from ewe to lamb via colostrum and milk.[31,32]

2.2. Clinical Symptoms

The initial symptoms noticed are that the sheep lag behind when the flock is driven and may fall for no evident reason. An ataxia and weakness of the hindlegs may develop at an early stage, and the sheep lose weight. At this stage the sheep frequently rest on the distal ends of the metatarsals. The head is sometimes tilted to one side, and a fine trembling of the lips and facial muscles is sometimes observed. The symptoms may progress slowly but steadily and lead to a paraplegia or total paralysis within a few months to 1 year. Sometimes a remitting course is observed. There is a gradual loss of weight, but the sheep remain alert to the end and no difficulties are observed in feeding, defecation, or micturition.[1]

2.3. Pathology

2.3.1. Introduction

Visna has now been observed in several breeds of sheep in many countries.[1] The pathological lesions show a similar pattern in the various breeds of sheep and are, as reported by Sigurdsson and co-workers,[33,34] comparable in natural and experimentally transmitted cases regardless of the route of infection.[4,35-37] The following description is mainly based on studies on sheep experimentally infected by an intracerebral route of inoculation. Approximately 150 sheep have been studied. In each case nine standard planes of sections from the brain, three different levels of the spinal cord (and occasionally the entire spinal cord), the sciatic and optic nerves, and the retina were examined. In addition, the cellular exudate in the spinal fluid was studied. The lesions evolving after experimental transmission were analyzed from 2 weeks to 11 years after infection, and the age spectrum varied from fetal to adult sheep. The description of the pathological lesions has been a subject of several reports.[10,38-43]

2.3.2. Macroscopic Changes

The brain and spinal cord usually appear normal on macroscopic examination. The leptomeninges over the brain and spinal cord may show focal grayish thickenings. The choroid plexus is sometimes granular. In severe cases a gray-yellowish softening of the white matter of the cerebrum, brainstem, and cerebellum is present, and rarely relatively sharply demarcated grayish plaques have been observed in the white columns of the spinal cord.

2.3.3. Microscopic Changes

A leptomeningitis over the brain and spinal cord is a very common feature (Fig. 4-1). It is sometimes diffuse, but frequently it is accentuated over the superior frontal gyrus, hippocampic fissure, pyriform and occipital lobe, and the cerebellar lingula. In the spinal cord it is usually most marked over the anterior median fissure, the posterior median sulcus, and around nerve roots, especially the posterior ones. The meningitis has been detected 1 to 2 weeks after infection.[10,34,37,44] The severity of the meningitis varies. Shortly after infection it is sometimes very marked but usually wanes with time, although it may still be present several years after infection.[43] Lymphocytes are usually most numerous in the inflammatory infiltrates, followed by macrophages and plasma cells.

A pleocytosis of the cerebrospinal fluid (CSF), with maximal levels approx-

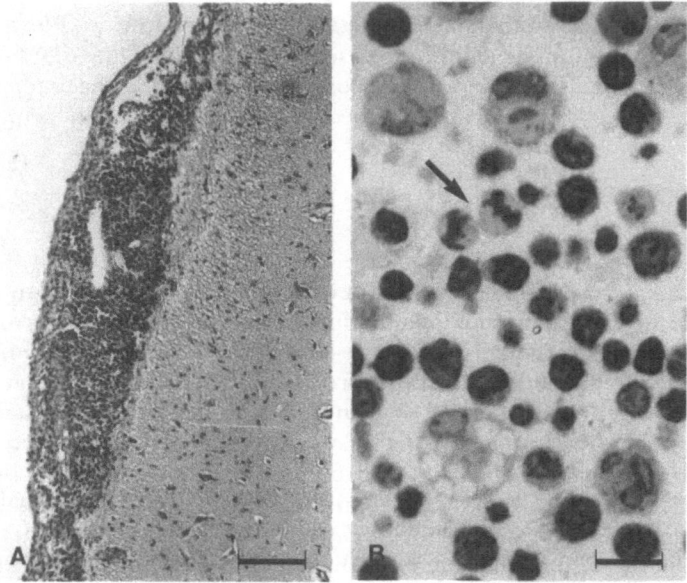

FIGURE 4-1. (A) Heavy infiltration of mononuclear cells in the meninges over the temporal lobe. Bar, 100 μm (hematoxylin and eosin). (B) Cellular exudate in the CSF consisting of lymphocytes and macrophages. Mitotic figure (arrow). Bar, 10 μm (Epon, toluidine blue).

imately 1 month after infection, is observed.[10,37,42,45,46] The number of cells decreases generally within a period of a few months but may stay at slightly increased levels for years.[10,45,46] Occasionally a pleocytosis is not observed until several months after infection, and in long-term studies irregular fluctuations in the number of cells have been observed with peaks of pleocytosis occurring 7 or 8 years after infection.[30,45] The fluctuation in the pleocytosis indicates a remitting lesion activity in the CNS.

The composition of the cellular exudate in the CSF (Fig. 4-1) differs from the inflammatory infiltrates in the meninges such that macrophages are in general more numerous than lymphocytes in the CSF, and plasma cells are very rare.[42,45] This may reflect differences in the migratory potential of these cells.

In an ultrastructural analysis of the CSF, myelin fragments were found,[42] and testing for myelin basic protein in the CSF revealed a transient elevation.[46] This is probably an indication of active myelin breakdown and may have implications for the pathogenesis of lesions. In human demyelinating diseases similar findings have been reported,[47,48] and it has been suggested that myelin entering the CSF may lead to autosensitization to myelin proteins.[49]

In the brain the earliest lesions observed are subependymal inflammation and an inflammatory infiltration of the choroid plexus, which are often present 2 weeks after infection.[10,38,41] The subependymal inflammation begins as small perivascular sleeves scattered underneath the ependymal lining of the ventricles. With increasing severity it becomes more diffuse, and sometimes confluent subependymal inflammation borders the entire ventricular system (Fig. 4-2) of the brain and frequently extends into the spinal cord along the central canal. Sometimes the ependymal lining sloughs off the surface. With increasing severity of the inflammation, the white matter becomes involved and to a lesser degree adjacent nuclei. The white matter involvement at first is in the form of discrete perivascular infiltrates (Fig. 4-2), but with progression of the lesions they become confluent and may in extreme cases involve almost the entire white matter. The cerebral cortex is in general spared, although an occasional glial nodule or discrete perivascular inflammatory cuffs may be present. Glial nodules also occur in the white matter.

In the spinal cord, inflammation usually radiates from the central canal into the adjacent gray matter, but sometimes inflammation is found in the white columns with no apparent relation to that surrounding the central canal.

The myelin is sometimes well preserved even in areas with dense inflammatory infiltration, which pushes the myelinated fibers apart, but eventually the myelin is broken down. The myelin breakdown is sometimes in the form of multiple small foci, but frequently large areas of liquefaction necrosis with

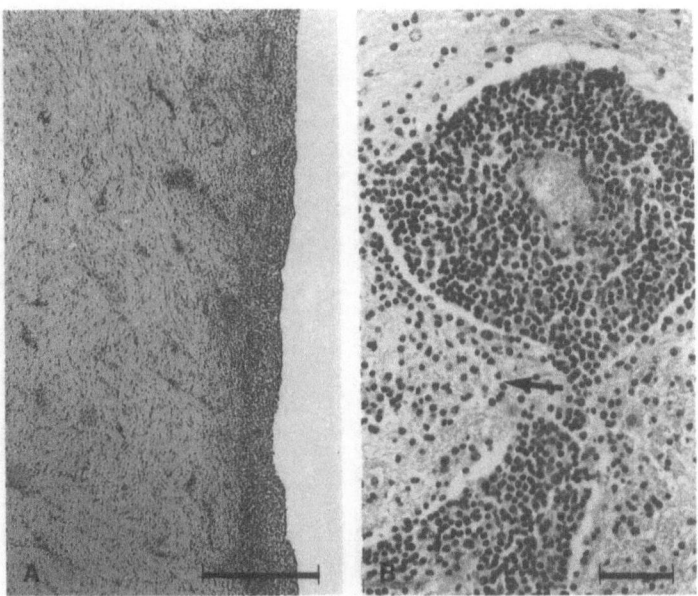

FIGURE 4-2. (A) Lateral ventricle. Confluent subependymal inflammation extending mainly as perivascular cuffs into the white matter. Bar, 500 μm (hematoxylin and eosin). (B) A thick perivascular sleeve of lymphocytes and macrophages in the white matter. Beginning infiltration into adjacent neuroparenchyma (arrow). Bar, 50 μm (hematoxylin and eosin).

destruction of myelin and axons, i.e., secondary demyelination, are found, with massive infiltration of macrophages filled with phagocytosed material ("gitter cells"). Foci of coagulative necrosis are occasionally present.

In addition to foci of secondary demyelination, rather sharply demarcated foci of primary demyelination (Fig. 4-3) resembling chronic active or chronic silent plaques of multiples sclerosis (MS) occur in sheep that are developing clinical signs several years after infection.[43] The demyelinated plaques are mainly found in the spinal cord and may show signs of remyelination, frequently with peripheral-type myelin.

Inflammation of the choroid plexus in the lateral, third, and fourth ventricles is an early and common feature of the pathological lesions observed in visna. The inflammation varies in degree from discrete infiltration with lymphocytes, some macrophages, and plasma cells to very pronounced lymphoid proliferation with formation of lymph follicles with active germinal centers (Fig. 4-3).

Neurons are spared except in areas with frank necrosis and are often well preserved where there is a pronounced inflammation. This is in accord with results of *in situ* hybridization[50] and immunohistochemical studies of the brain.[30,51] Neither the viral genome nor expression of viral proteins has been detected in neurons by these methods, whereas astrocytes and oligodendrocytes apparently harbor the viral genome.[50] Expression of viral proteins has been

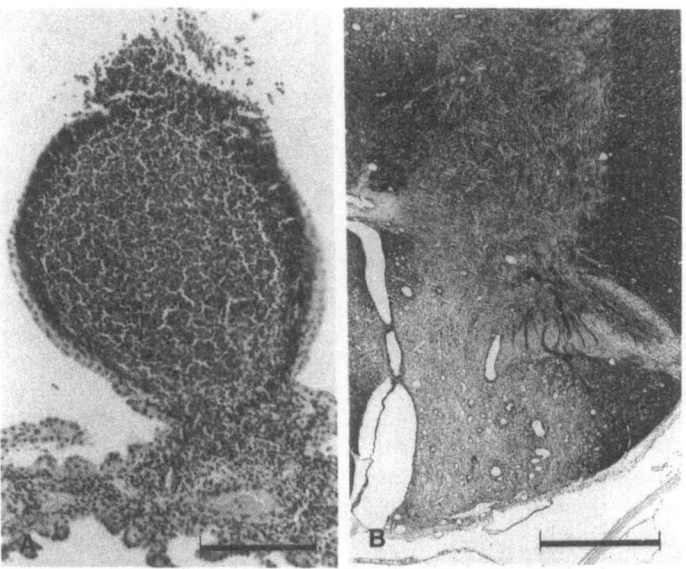

FIGURE 4-3. (A) Lateral ventricle. Inflammation of the choroid plexus with a lymph follicle with an active germinal center. Bar, 200 μm (hematoxylin and eosin). (B) Lumbar cord. Plaque of primary demyelination in the posterior column. Bar, 100 μm (Klüver–Barrera).

detected in a wide variety of cells, i.e., lymphocytes, plasma cells, macrophages, endothelial cells, pericytes, fibroblasts, and choroidal epithelial cells.[51]

Astrocytes surrounding perivascular infiltrates and bordering necrotic foci show a reactive response. In the vessels some swelling of endothelial cells of capillaries and venules in inflammatory foci is commonly observed, and occasionally slight intimal thickening in arteries and veins and some perivascular fibrosis are present.

Multinucleated giant cells, the hallmark of the CPE of visna virus in tissue culture, have only been detected in experiments done with a highly neurovirulent strain, selected by serial passage of virus through sheep.[51]

The peripheral nervous system is rarely affected. In our series nerve roots, the sciatic and optic nerves and retina were normal. But Sigurdsson et al.[33] occasionally found an extension of the inflammation of the meninges into adjacent spinal ganglia and nerve roots as well as isolated inflammatory foci in peripheral nerves at some distance from nerve roots.

Except for the plaques of primary demyelination, which are apparently a rather late manifestation, the character of the pathological lesions does not change with time. Thus, the composition of the inflammatory infiltrates in sheep sacrificed 10 years after infection was similar to those observed 2 weeks after infection. The results of a long-term study indicate that lesion activity may be remitting, and inflammatory lesions that are not accompanied by breakdown of tissue may resolve and reappear later. If sheep survive long enough after necrosis has occurred, glial scars or cystic transformation is present.[33] There are, however, breed differences. Thus, in American sheep a self-limiting encephalitis healing with scar tissue is observed after intracerebral infection with visna virus.[11]

3. IMMUNE RESPONSE

3.1. Humoral Immune Response

Antibodies to viral antigens are induced by natural and experimental infection with maedi–visna virus. They can be demonstrated by various techniques: virus neutralization,[2] complement fixation,[52] immunofluorescence,[53,54] gel immunodiffusion,[55–57] passive hemagglutination,[58] and the ELISA method.[59] Neutralizing antibodies appear relatively slowly following experimental infection and are first detected after 1½ to 3 months or even later.[10,54,60] They may be quite strain specific, and thus some virus isolates are poorly neutralized by antisera that react strongly with other viral strains.[61] Complement-fixing antibodies first appear 3–4 weeks after experimental infection.[10] They are relatively nonspecific and do not distinguish between strains of maedi–visna virus, probably because they are directed at least in part to the p25 group-specific core antigen, whereas the neutralizing antibodies are directed against the envelope glycoprotein.[62]

Precipitating antibodies against both p25 core antigen and the virus envelope glycoprotein are not strain specific. Almost all sheep experimentally infected with visna virus will develop precipitating antibodies to the envelope glycoprotein, whereas only about 40% show a second line in immunodiffusion tests corresponding to the p25 core antigen.[57]

The ELISA method appears to be more sensitive than both complement fixation and immunodiffusion and therefore is the method of choice for general purposes. The immunoglobulin class distribution of viral antibodies in visna is only partially worked out. Both complement-fixing and neutralizing antibodies belong to the IgG_1 subclass but can be separated on the basis of different electrical charge by ion-exchange chromatography. The neutralizing antibodies are more highly charged.[63] Viral antibodies of the IgM class have not been convincingly demonstrated.[63]

Different strains of maedi–visna virus vary in their ability to elicit neutralizing antibodies. Less cytolytic strains, which grow to rather low titers in tissue culture, seem to induce neutralizing antibodies less readily than highly cytolytic strains of visna.[11,57]

Neutralizing antibodies have been found in the CSF of experimentally infected sheep and have been shown to be produced locally in the CNS.[10,45,46] In some sheep with longstanding visna, oligoclonal bands in the γ-globulin region have been demonstrated by electrophoresis.[64] It is not known whether they are directed against virus antigens. An increase in the IgM level in CSF during visna has been reported.[64] The appearance of neutralizing antibodies in the CSF seems to coincide with the disappearance of free infectious virus from the CSF about 3–4 months after experimental infection.[10]

3.2. Cell-Mediated Immune Response

Several reports on lymphocyte blast transformation in response to virus antigens have appeared.[37,65–68] These responses seem to be rather irregular and often transient. No studies on cytotoxic T cells have appeared, and on the whole our knowledge of cell-mediated immune response in maedi–visna infection is still fragmentary.

3.3. Interferon

Earlier work has indicated that visna virus is a poor inducer of interferon[28,69] and that replication of the virus was completely unaffected by high concentrations of sheep interferon induced by polyriboinosinic-polyribocytidylic acid.[69,70] It has been reported recently, however, that lentiviruses of sheep and goats induce a unique type of interferon produced by T lymphocytes during interaction with infected macrophages. This interferon is a nonglycosylated protein, 54–60 kDa, and stable to heat and acid treatment.[71] The role of this interferon in controlling virus replication and in the pathogenesis of lesions is still unknown.

3.4. Vaccination Trials

Attempts to vaccinate sheep against maedi–visna virus have failed to induce protective immunity.[72,73] Either whole virus inactivated by various methods or relatively crude preparations of the envelope glycoprotein have been used in combination with complete Freund's adjuvant. The inoculated sheep have developed antibodies detectable by complement fixation, gel immunodiffusion, and ELISA but no virus-neutralizing antibodies and no protection against challenge with live virus. There are indications that immunization with lentiviruses of sheep and goats may enhance the severity of disease.[74,75]

4. PATHOGENESIS

Inoculation with maedi–visna virus invariably results in a systemic infection, and virus can be recovered from most organs, even those that are apparently free of lesions. Viremia is found in both natural and experimental infection within 1–2 weeks.[10,76] The virus is always cell associated, never found free in the plasma.[10] Since a very low proportion of the peripheral blood leukocytes carry the virus, it has been difficult to identify the target cells of infection. Some of them at least are monocytes[24] in which virus replication seems highly restricted, but it is enhanced when they mature into macrophages.[25,26] There are also reports that lymphocytes of blood and lymph may be infected.[10,23,32]

It has been suggested that visna virus is carried into the CNS by infected monocytes, from which it then may spread to other target cells. This has been referred to as the Trojan horse mechanism of virus spread.[77] Presumably circulating lymphocytes may also carry the virus to the CNS, since there is evidence that there is some lymphocyte traffic into the CNS through an intact blood–brain barrier.[78] By immunohistochemical methods the following cell types in the CNS have been found positive for viral antigen: macrophages, lymphocytes, plasma cells, choroid epithelial cells, pericytes, endothelial cells, and fibroblasts.[51] By in situ hybridization virus nucleic acid has been found in glial cells, astrocytes, and oligodendrocytes.[50] There are no reports of viral infection of neurons.

There is evidence that the early inflammatory CNS lesions of visna are immune mediated. By immunosuppressive treatment of infected sheep with antithymocyte serum and cyclophosphamide, development of lesions could be practically abolished.[79]

Stimulating the immune response by hyperimmunization of already infected sheep seemed to enhance the severity of lesions.[74] The immunopathogenetic mechanism is apparently directed against virus-induced antigens, since the severity of lesions increases with virus dose[23] and shows a correlation with frequency of virus isolations.[10] On the other hand there was no indication of an autoimmune humoral or cell-mediated response in visna-infected sheep to basic protein or galactocerebroside myelin antigens in short-term infection.[80]

It is considered likely that a cell-mediated immune response to virus antigens is responsible for the development of the early inflammatory CNS lesions in visna, since the lesions start to develop before an antiviral antibody response is observed but at a time when virus-specific cell-mediated immune responses have been demonstrated in the CNS.[37]

The pathogenesis of the primary demyelination developing at later stages of visna is incompletely understood. Although the virus genome has been reported in oligodendrocytes by *in situ* hybridization,[50] virus antigen was not found in these cells by immunocytochemical methods.[51] This makes it less likely that an immune-mediated attack on these cells could explain the demyelination. A functional disturbance of these cells by the presence of the viral genome cannot be excluded, nor can a nonspecific destruction by lymphokines or other substances released from inflammatory cells, referred to as "bystander" demyelination.[81]

It has been proposed that antigenic variation of visna virus could explain the persistence of virus in the host as well as the progression of lesions.[24,82,83] It has been amply documented that antigenic variants arise frequently in individual sheep, variants that are poorly neutralized by early antisera from the same sheep. Other studies have not shown any correlation of the appearance of antigenic variants with disease progression, and persistence of virus was apparently not explained by antigenic variation, since variants did not replace the infecting virus type.[61,84]

Persistence of the virus seems to be explained by the provirus theory, that is, the presence of the viral genome in many cells of the body without virus replication or synthesis of viral proteins. This host cell restriction in transcription of the proviral DNA into messenger RNA and also in the translation of messenger RNA into viral proteins[85,86] seems to offer adequate explanation of virus persistence in the face of an active immune response. This restriction may depend on both the cell type involved and the physiological or developmental state of the infected cell.[87] Thus, lentiviral genomes seem to be highly controlled by the cells of the host, which probably explains the slow development of lentiviral diseases.

REFERENCES

1. Pálsson, P. A., 1976, Maedi and visna in sheep, in: *Slow Virus Diseases of Animals and Man* (R. H. Kimberlin, ed.), North Holland, Amsterdam, pp. 17–43.
2. Sigurdsson, B., Thormar, H., and Pálsson, P. A., 1960, Cultivation of visna virus in tissue culture, *Arch. Ges. Virusforsch.* **10**:368–381.
3. Sigurdardóttir, B., and Thormar, H., 1964, Isolation of a viral agent from the lungs of sheep affected with maedi, *J. Infect. Dis.* **114**:55–60.
4. Gudnadóttir, M., and Pálsson, P. A., 1965, Successful transmission of visna by intrapulmonary inoculation, *J. Infect. Dis.* **115**:217–225.
5. Lin, F. H., and Thormar, H., 1970, Ribonucleic acid-dependent deoxyribonucleic acid polymerase in visna virus, *J. Virol.* **6**:702–704.
6. Stephens, R. M., Casey, J. W., and Rice, N. R., 1986, Equine infectious anemia virus *gag* and *pol* genes: Relatedness to visna and AIDS virus, *Science* **231**:589–594.
7. Sigurdsson, B., 1954, Observations on three slow infections of sheep, *Br. Vet. J.* **110**:255–270.
8. Pétursson, G., Pálsson, P. A., and Georgsson, G., 1989, Maedi–visna in sheep: Host–virus interactions and utilization as a model, *Intervirology* **30**:36–44.
9. Price, R. W., Brew, B., Sidtis, J., Rosenblum, M., Scheck, A. C., and Cleary, P., 1988, Central nervous system HIV-1 infection and AIDS dementia complex, *Science* **239**:586–592.
10. Pétursson, G., Nathanson, N., Georgsson, G., Panitch, H., and Pálsson, P. A., 1976, Pathogenesis of visna. I. Sequential virologic, serologic and pathologic studies, *Lab. Invest.* **35**:402–412.

11. Narayan, O., Strandberg, J. D., Griffin, D. E., Clements, J. E., and Adams, R. J., 1983, Aspects of the pathogenesis of visna in sheep, in: *Viruses and Demyelinating Diseases* (C. A. Mims, M. L. Cuzner, and R. E. Kelly, eds.), Academic Press, London, pp. 125–140.

12. Saag, M. S., Hahn, B. H., Gibbons, J., Li, Y., Parks, E. S., Parks, W.P., and Shaw, G. M., 1988, Extensive variation of human immunodeficiency virus type-1 *in vivo*, *Nature* **334:**440–444.

13. Haase, A. T., Garapin, A. C., Faras, A. J., Varmus, H. E., and Bishop, J. M., 1974, Characterization of the nucleic acid product of the visna virus RNA dependent DNA polymerase, *Virology* **57:**251–258.

14. Sonigo, P., Alizon, M., Staskus, K., Klatzmann, D., Cole, S., Danos, O., Retzel, E., Tiollais, P., Haase, A., and Wain-Hobson, S., 1985, Nucleotide sequence of the visna lentivirus: Relationship to the AIDS virus, *Cell* **42:**369–382.

15. Haase, A. T., Stowring, L., Harris, J. D., Traynor, B., Ventura, P., Peluso, R., and Brahic, M., 1982, Visna DNA synthesis and the tempo of infection *in vitro*, *Virology* **119:**399–410.

16. Vigne, R., Barban, V., Quérat, G., Mazarin, V., Gourdou, I., and Sauze, N., 1987, Transcription of visna virus during its lytic cycle: Evidence for a sequential early and late gene expression, *Virology* **161:**218–227.

17. Davis, J. L., and Clements, J. E., 1988, Complex gene expression of lentiviruses, *Microb. Pathogen.* **4:**239–245.

18. Davis, J. L., and Clements, J. E., 1989, Characterization of a cDNA clone encoding the visna virus transactivating protein, *Proc. Natl. Acad. Sci. USA* **86:**414–418.

19. Mazarin, V., Gourdou, I., Quérat, G., Sauze, N., and Vigne, R., 1988, Genetic structure and function of an early transcript of visna virus, *J. Virol.* **62:**4813–4818.

20. Braun, M. J., Clements, J. E., and Gonda, M. A., 1987, The visna virus genome: Evidence for a hypervariable site in the *env* gene and sequence homology among lentivirus envelope proteins, *J. Virol.* **61:**4046–4054.

21. Haase, A. T., and Baringer, J. R., 1974, The structural polypeptides of RNA slow viruses, *Virology* **57:**238–250.

22. Vigne, R., Filippi, P., Quérat, G., Sauze, N., Vitu, C., Russo, P., and Delori, P., 1982, Precursor polypeptides to structural proteins of visna virus, *J. Virol.* **42:**1046–1056.

23. Pétursson, G., Martin, J. R., Georgsson, G., Nathanson, N., and Pálsson, P. A., 1979, Visna, the biology of the agent and the disease, in: *New Perspectives in Clinical Microbiology, Aspects of Slow and Persistent Virus Infections* (D. A. J. Tyrrell, ed.), Martinus Nijhoff, The Hague, pp. 198–220.

24. Narayan, O., Wolinsky, J. S., Clements, J. E., Strandberg, J. D., Griffin, D. E., and Cork, L. C., 1982, Slow virus replication: The role of macrophages in the persistence and expression of visna viruses of sheep and goats, *J. Gen. Virol.* **59:**345–356.

25. Gendelman, H. E., Narayan, O., Molineux, S., Clements, J. E., and Ghotbi, Z., 1985, Slow persistent replication of lentiviruses: Role of tissue macrophages and macrophage precursors in bone marrow, *Proc. Natl. Acad. Sci. U.S.A.* **82:**7086–7090.

26. Gendelman, H. E., Narayan, O., Kennedy-Stoskopf, S., Kennedy, P. G. E., Ghotbi, Z., Clements, J. E., Stanley, J., and Pezeshkpour, G., 1986, Tropism of sheep lentiviruses for monocytes: Susceptibility to infection and virus gene expression increase during maturation of monocytes to macrophages, *J. Virol.* **58:**67–74.

27. Orenstein, J. M., Meltzer, M. S., Phipps, T., and Gendelman, H. E., 1988, Cytoplasmic assembly and accumulation of human immunodeficiency virus types 1 and 2 in recombinant human colony-stimulating factor-1-treated human monocytes: An ultrastructural study, *J. Virol.* **62:**2578–2586.

28. Thormar, H., 1976, Visna–maedi infection in cell cultures and in laboratory animals, in: *Slow Virus Diseases of Animals and Man* (R. H. Kimberlin, ed.), North-Holland, Amsterdam, pp. 97–114.

29. Macintyre, E. H., Wintersgill, C. J., and Vatter, A. E., 1974, A modification in the response of human astrocytes to visna virus, *Am. J. Vet. Res.* **35:**1161–1163.

30. Georgsson, G., Pálsson, P. A., and Pétursson, G., 1987, Pathogenesis of visna, in: *A Multidisciplinary Approach to Myelin Diseases* (G. Serlupi Crescenzi, ed.), Plenum Press, New York, pp. 303–318.

31. de Boer, G. F., 1970, *Zwoegerziekte, a Persistent Infection in Sheep*, Thesis, Utrecht.

32. de Boer, G. F., Terpstra, C., and Houwers, D. J., 1979, Studies in epidemiology of maedi/visna in sheep, *Res. Vet. Sci.* **26:**202–208.

33. Sigurdsson, B., Pálsson, P. A., and Grímsson, H., 1957, Visna, a demyelinating transmissible disease of sheep, *J. Neuropathol. Exp. Neurol.* **16:**389–403.

34. Sigurdsson, B., Pálsson, P. A., and van Bogaert, L., 1962, Pathology of visna. Transmissible demyelinating disease in sheep in Iceland, *Acta Neuropathol. (Berl.)* **1:**343–362.

35. Narayan, O., Silverstein, A. M., Price, D., and Johnson, R. T., 1974, Visna virus infection of American lambs, *Science* **183:**1202–1203.

36. de Boer, G. F., 1975, Zwoergerziekte virus, the causative agent for progressive interstitial pneumonia (maedi) and meningo-leucoencephalitis (visna) in sheep, *Res. Vet. Sci.* **18:** 15–25.

37. Griffin, D. E., Narayan, O., and Adams, R. J., 1978, Early immune responses in visna, a slow viral disease of sheep, *J. Infect. Dis.* **138:**340–350.

38. Georgsson, G., Nathanson, N., Pálsson, P. A., and Pétursson, G., 1976, The pathology of visna and maedi in sheep, in: *Slow Virus Diseases of Animals and Man* (R. Kimberlin, ed.), North Holland, Amsterdam, pp. 61–96.

39. Georgsson, G., Pálsson, P. A., Panitch, H., Nathanson, N. and Pétursson, G., 1977, Ultrastructure of early visna lesions, *Acta Neuropathol. (Berl.)* **37:**127–135.

40. Pálsson, P. A., Georgsson, G., Pétursson, G., and Nathanson, N., 1977, Experimental visna in Icelandic lambs, *Acta Vet. Scand.* **18:**122–128.

41. Georgsson, G., Pétursson, G., Miller, A., Nathanson, N., and Pálsson, P. A., 1978, Experimental visna in foetal Icelandic sheep, *J. Comp. Pathol.* **88:**597–605.

42. Georgsson, G., Martin, J. R., Pálsson, P. A., Nathanson, N., Benediktsdóttir, E., and Pétursson, G., 1979, An ultrastructural study of the cerebrospinal fluid in visna, *Acta Neuropathol. (Berl.)* **48:**39–43.

43. Georgsson, G., Martin, J. R., Klein, J., Pálsson, P. A., Nathanson, N., and Pétursson, G., 1982, Primary demyelination in visna: An ultrastructural study of Icelandic sheep with clinical signs following experimental infection, *Acta Neuropathol. (Berl.)* **57:**171–178.

44. Oliver, R. E., Gorham, J. R., Parish, S. F., Hadlow, W. J., and Narayan, O., 1981, Ovine progressive pneumonia. Pathologic and virologic studies on the naturally occurring disease, *Am. J. Vet. Res.* **42:**1554–1559.

45. Nathanson, N., Pétursson, G., Georgsson, G., Pálsson, P. A., Martin, J. R., and Miller, A., 1979, Pathogenesis of visna. IV. Spinal fluid studies, *J. Neuropathol. Exp. Neurol.* **38:**197–208.

46. Griffin, D. E., Narayan, O., Bukowski, J. F., Adams, R. J., and Cohen, S. R., 1978, The cerebrospinal fluid in visna, a slow viral disease of sheep, *Ann. Neurol.* **4:**212–218.

47. Herndon, R. M., and Johnson, M., 1970, A method for the electron microscopic study of cerebrospinal fluid sediment, *J. Neuropathol. Exp. Neurol.* **29:**320–330.

48. Herndon, R. M., and Kasckow, J., 1978, Electron microscopic studies of cerebrospinal fluid sediment in demyelinating disease, *Ann Neurol.* **4:**515–623.

49. Whitaker, J. N., 1977, Myelin encephalitogenic protein fragments in cerebrospinal fluid of persons with multiple sclerosis, *Neurology* **27:**911–920.

50. Stowring, L., Haase, A. T., Pétursson, G., Georgsson, G., Pálsson, P. A., Lutley, R., Roos, R., and Szuchet, S., 1985, Detection of visna virus antigens and RNA in glial cells in foci of demyelination, *Virology* **141:**311–318.

51. Georgsson, G., Houwers, D. J., Pálsson, P. A., and Pétursson, G., 1989, Expression of viral antigens in the central nervous system of visna-infected sheep: An immunohistochemical study on experimental visna induced by virus strains of increased neurovirulence, *Acta Neuropathol. (Berl.)* **77:**299–306.

52. Gudnadóttir, M., and Kristinsdóttir, K., 1967, Complement-fixing antibodies in sera of sheep affected with visna and maedi, *J. Immunol.* **98**:663–667.
53. Thormar, H., 1969, Visna and maedi virus antigen in infected cell cultures studied by the fluorescent antibody technique, *Acta Pathol. Microbiol. Scand.* **75**:296–302.
54. de Boer, G. F., 1970, Antibody formation in zwoegerziekte, a slow infection in sheep, *J. Immunol.* **104**:414–422.
55. Terpstra, C., and De Boer, G. F., 1973, Precipitating antibodies against maedi–visna virus in experimentally infected sheep, *Arch. Ges. Virusforsch.* **43**:53–62.
56. Cutlip, R. C., Jackson, T. A., and Laird, G. A., 1977, Immunodiffusion test for ovine progressive pneumonia, *Am. J. Vet. Res.* **38**:1081–1084.
57. Klein, J. R., Martin, J. R., Griffing, G., Nathanson, N., Gorham, J., Shen, D. T., Pétursson, G., Georgsson, G., Pálsson, P. A., and Lutley, R., 1985, Precipitating antibodies in experimental visna and natural progressive pneumonia of sheep, *Res. Vet. Sci.* **38**:129–133.
58. Karl, S. C., and Thormar, H., 1971, Antibodies produced by rabbits immunized with visna virus, *Infect. Immun.* **4**:715–719.
59. Houwers, D. J., Gielkens, A. L. J., and Schaake, J. Jr., 1982, An indirect enzyme-linked immunosorbent assay (ELISA) for the detection of antibodies to maedi-visna virus, *Vet. Microbiol.* **7**:209–210.
60. Gudnadóttir, M., and Pálsson, P. A., 1965, Host–virus interaction in visna infected sheep, *J. Immunol.* **95**:1116–1120.
61. Lutley, R., Pétursson, G., Pálsson, P. A., Georgsson, G., Klein, J., and Nathanson, N., 1983, Antigenic drift in visna: Virus variation during long-term infection of Icelandic sheep, *J. Gen. Virol.* **64**:1433–1440.
62. Scott, J. V., Stowring, L., Haase, A. T., Narayan, O., and Vigne, R., 1979, Antigenic variation in visna virus, *Cell* **18**:321–327.
63. Pétursson, G., Douglas, B. M., and Lutley, R., 1974, Immunoglobulin subclass distribution and restriction of antibody response in visna, in *Slow viruses in Sheep, Goats and Cattle* (J. M. Sharp and R. Hoff-Jørgensen, eds.), Commission of the European Communities, Luxembourg, pp. 211–216.
64. Martin, J. R., Goudswaard, J., Pálsson, P. A., Georgsson, G., Pétursson, G., Klein, J., and Nathanson, N., 1982, Cerebrospinal fluid immunoglobulins in sheep with visna, a slow virus infection of the central nervous system, *J. Neuroimmunol.* **3**:139–148.
65. Pétursson, G., Nathanson, N., Pálsson, P. A., Martin, J. R., and Georgsson, G., 1978, Immunopathogenesis of visna, a slow virus disease of the central nervous system, *Acta Neurol. Scand.* **57**:205–219.
66. Sihvonen, L., 1981, Early immune responses in experimental maedi, *Res. Vet. Sci.* **30**:217–222.
67. Larsen, H. J., Hyllseth, B., and Krogsrud, J., 1982, Experimental maedi virus infection in sheep: Early cellular and humoral immune response following parenteral inoculation, *Am. J. Vet. Res.* **43**:379–383.
68. Larsen, H. J., Hyllseth, B., and Krogsrud, J., 1982, Experimental maedi virus infection in sheep: Cellular and humoral immune response during three years following intranasal inoculation, *Am. J. Vet. Res.* **43**:384–389.
69. Trowbridge, R. S., 1975, Long-term visna virus infection of sheep choroid plexus cells: Initiation and preliminary characterization of the carrier cultures, *Infect. Immun.* **11**:862–868.
70. Carroll, D., Ventura, P., Haase, A., Rinaldo, C. R., Jr., Overall, J. C., Jr., and Glasgow, L. A., 1978, Resistance of visna virus to interferon, *J. Infect. Dis.* **138**:614–617.
71. Narayan, O., Sheffer, D., Clements, J. E., and Tennekoon, G., 1985, Restricted replication of lentiviruses. Visna viruses induce a unique interferon during interaction between lymphocytes and infected macrophages, *J. Exp. Med.* **162**:1954–1969.

72. Pétursson, G., 1986, Maedi–visna and scrapie in sheep, recent developments, in: *Proceedings XVth Nordic Veterinary Congress*, Sveriges Veterinärförbund (Swedish Veterinary Society), Stockholm, pp. 273–276.

73. Cutlip, R. C., Lehmkuhl, H. D., Brogden, K. A., and Schmerr, M. J. F., 1987, Failure of experimental vaccines to protect against infection with ovine progressive pneumonia (maedi virus) virus, *Vet. Microbiol.* 13:201–204.

74. Nathanson, N., Martin, J. R., Georgsson, G., Pálsson, P. A., Lutley, R. E., and Pétursson, G., 1981, The effect of post-infection immunization on the severity of experimental visna, *J. Comp. Pathol.* 91:185–191.

75. McGuire, T. C., Adams, S., Johnson, G. C., Klevjer-Anderson, P., Barbee D. D., and Gorham, J. R., 1986, Acute arthritis in caprine arthritis-encephalitis virus challenge exposure of vaccinated or persistently infected goats, *Am. J. Vet. Res.* 47:537–540.

76. Gudnadóttir, M., and Pálsson, P. A., 1967, Transmission of maedi by inoculation of a virus grown in tissue culture from maedi-affected lungs, *J. Infect. Dis.* 117:1–6.

77. Peluso, R., Haase, A., Stowring, L., Edwards, M., and Ventura, P., 1985, A Trojan horse mechanism for the spread of visna in monocytes, *Virology* 147:231–236.

78. Wekerle, H., Linington, C., Lassman, H., and Meyermann, R., 1986, Cellular immune reactivity within the CNS, *Trends Neurosci.* 9:271–277.

79. Nathanson, N., Panitch, H., Pálsson, P. A., Pétursson, G., and Georgsson, G., 1976, Pathogenesis of visna, II. Effect of immunosuppression upon early central nervous system lesions, *Lab. Invest.* 35:444–451.

80. Panitch, H., Pétursson, G., Georgsson, G., Pálsson, P. A., and Nathanson, N., 1976, Pathogenesis of visna, III. Immune response to central nervous system antigens in experimental allergic encephalomyelitis and visna, *Lab. Invest.* 35:452–460.

81. Wisniewski, H. M., and Bloom, B. R., 1975, Primary demyelination as a nonspecific consequence of a cell-mediated immune reaction, *J. Exp. Med.* 141:346–359.

82. Gudnadóttir, M., 1974, Visna–maedi in sheep, *Prog. Med. Virol.* 18:336–349.

83. Narayan, O., Griffin, D. E., and Chase, J., 1977, Antigenic shift of visna virus in persistently infected sheep, *Science* 197:376–378.

84. Thormar, H., Barshatzky, M. R., Arnesen, K., and Kozlowski, P. B., 1983, The emergence of antigenic variants is a rare event in long-term visna virus infection *in vivo*, *J. Gen. Virol.* 64: 1427–1432.

85. Haase, A. T., 1986, The AIDS lentivirus connection, *Microb. Pathogen.* 1:1–4.

86. Haase, A. T., 1986, Pathogenesis of lentivirus infections, *Nature* 322:130–136.

87. Narayan, O., Kennedy-Stoskopf, S., Zink, M. C., 1988, Lentivirus–host interactions: Lessons from visna and caprine arthritis–encephalitis viruses, *Ann. Neurol.* 23:95–100.

5

Theiler's Virus-Induced Demyelinating Disease in Mice

Picornavirus Animal Model

HOWARD L. LIPTON
and MAURO C. DAL CANTO

1. INTRODUCTION

The mouse encephalomyelitis viruses are naturally occurring enteric pathogens of mice. Discovered by Max Theiler in the early 1930s, these viruses are frequently referred to as Theiler's murine encephalomyelitis viruses (TMEV).[1] Their host range appears to be quite narrow, and serological evidence indicates that *Mus musculus* is the natural host. The TMEV are present in virtually all nonbarrier mouse colonies throughout the world, where they cause asymptomatic intestinal infections. Rarely, TMEV spreads to the central nervous system (CNS), producing encephalitis or, more commonly, spontaneous paralysis, i.e., poliomyelitis.[2,3] The incidence of spontaneous paralysis is low, on the order of one paralyzed animal per 1000–5000 mice in a colony. Since TMEV may go undetected unless appropriate serological testing is performed, these agents are a potential hazard for

HOWARD L. LIPTON • Department of Neurology, Northwestern University Medical School, Chicago, Illinois 60611.   MAURO C. DAL CANTO • Departments of Neurology and Pathology (Neuropathology), Northwestern University Medical School, Chicago, Illinois 60611. *Present address of H. L. L.*: Department of Neurology, University of Colorado School of Medicine, Denver, Colorado 80262.

Neuropathogenic Viruses and Immunity, edited by Steven Specter *et al.* Plenum Press, New York, 1992.

investigators using mice in biomedical research. In recent years, this group of viruses has assumed additional importance because TMEV infection in mice provides one of the few available experimental animal models for multiple sclerosis.[4-6] TMEV-induced demyelinating disease is perhaps the most relevant animal model for multiple sclerosis because (1) chronic pathological involvement is essentially limited to the CNS white matter; (2) myelin breakdown is accompanied by mononuclear cell inflammation; (3) pathological changes show demyelinating lesions of different ages, suggesting that disease may be recurrent; (4) demyelination results in clinical disease (spasticity, extensor spasms, and neurogenic bladder) involving upper motor neurons; (5) myelin breakdown is immune mediated; and (6) the disease is under multigenic control with a strong linkage to certain major histocompatibility complex genotypes.

Based on the complete nucleotide sequence and genome organization, TMEV have recently been unofficially classified as cardioviruses in the family *Picornaviridae* along with encephalomyocarditis virus (EMCV) and Mengo virus (Table 5-1).[7,8] The TMEV constitute a separate serological group of cardioviruses, since polyclonal antisera show no cross-neutralization between TMEV and EMCV or Mengo virus.[9] Because the coat proteins share a high degree of identity with the other cardioviruses, cross-reactions are seen on ELISA when disrupted virions are used as antigen.

The single-stranded TMEV RNA genome is of message sense, is approximately 8100 nucleotides in size, and consists of 5' and 3' untranslated regions flanking a large open reading frame. A 20-amino-acid protein, VPg, is covalently linked to the 5' end of the genome.[10] The 5' untranslated region is 1064–1069 nucleotides long but lacks a poly(C) tract.[11] The large size of the 5' noncoding region and the absence of a poly(C) tract distinguish TMEV from the other cardioviruses. The 3' untranslated region is 125 nucleotides in length, similar in size to EMCV and Mengo virus, and a poly(A) tail of indeterminate length is present at the 3' end. As with other picornaviruses, the final gene products of TMEV are the result of posttranslational cleavages of the polyprotein. Thus, the long open reading frame encodes a polyprotein of 2303 amino acids that begins with a short leader protein followed by 11 other gene products in the standard L-4-3-4 picornavirus arrangement.[12] By analogy with other picornaviruses,[13] the functions of many of the TMEV proteins are known (Table 5-2).

TABLE 5-1
Classification of TMEV in the Family Picornaviridae

Human enteroviruses: Polioviruses, Coxsackieviruses, Echoviruses
Human rhinoviruses
Hepatitis A viruses
Aphthoviruses: Foot-and-mouth disease viruses
Cardioviruses
    Group A: EMCV, Mengo, MM, Columbia-SK, Maus–Elberfeld
    Group B: TMEV

TABLE 5-2
Theiler's Murine Encephalomyelitis Virus-Specific Proteins

Protein[a]	Molecular weight[b]	Function
Leader	8,493	Unknown
1A (VP4)	7,102	
1B (VP2)	29,433	
1C (VP3)	25,463	Coat proteins—encapsidate RNA genome
1D (VP1)	30,586	
2A	16,380	?Protease for 1D/2A cleavage
2B	13,836	Unknown
2C	36,845	?Second RNA polymerase
3A	9,934	Unknown
3B	2,169	VPg—attached to RNA 5' end
3C	23,612	Principal viral protease
3D	52,235	Viral RNA polymerase

[a]Arranged in order from 5' to 3' on the RNA genome; standard nomenclature used.[12]
[b]Molecular weights are for BeAn virus.[7]

2. TWO NEUROVIRULENCE GROUPS

All TMEV are transmitted by the fecal–oral route but can be separated into two biological groups based on neurovirulence (Table 5-3). The first group consists of only two isolates, GDVII and FA viruses, which are highly virulent and cause a rapidly fatal encephalitis.[14,15] All of the other TMEV isolates, including viruses recovered from the CNS of spontaneously paralyzed mice and from the feces of asymptomatic mice, form a second, less virulent group. Experimentally, the less virulent viruses produce poliomyelitis (early disease) followed by demyelinating disease (late disease);[16,17] however, the poliomyelitis phase becomes subclinical when tissue-culture-adapted viruses are used.

The following sections of this chapter focus on the pathogenesis of the biphasic disease produced by the less virulent TMEV.

TABLE 5-3
Two Biological Groups of
Theiler's Murine Encephalomyelitis Viruses

Phenotype	Highly virulent	Less virulent
Disease	Encephalitis	Polio/demyelination
Incubation time	1–10 days	7–21 days/50 days
PFU/LD$_{50}$	1–10	>1,000,000
Plaque size	Large	Small
Temp. sensitive	−	+

3. CLINICAL SIGNS

Although the TMEV are enterically transmitted viruses, the pathogenesis of the CNS infection has been primarily studied using the intracerebral (IC) route of inoculation to maximize the incidence of neurological disease. Following IC inoculation, the less virulent strains produce a distinct biphasic CNS disease in susceptible strains of mice, characterized by poliomyelitis during the first few weeks post-inoculation (PI), followed by a chronic, inflammatory demyelinating process that begins during the second or third week PI and becomes manifest clinically between 1 and 3 months PI. Mice with poliomyelitis develop flaccid paralysis, usually of the hindlimbs; only one limb may be affected, or paralysis may spread rapidly to involve all limbs and lead to death. In contrast to the fatal outcome of paralysis produced by the Lansing strain of human poliovirus type 2,[18] complete recovery from TMEV-induced poliomyelitis is usual. Occasionally, residual limb deformities may occur as the result of severe paralysis.

Gait spasticity is the clinical hallmark of the demyelinating or late disease. Late disease is first manifest by slightly unkempt fur and decreased activity, followed by an unstable, waddling gait. Subsequently, generalized tremulousness and ataxia develop, and the waddling gait evolves into overt paralysis. Incontinence of urine and priapism are commonly seen. As the disease advances, prolonged extensor spasms of the limbs (>5–10 sec in duration) followed by difficulty in righting can be induced by abruptly turning a diseased mouse onto its side. The clinical manifestations of late disease are progressive and lead to the animal's demise in 6–14 months.

4. PATHOLOGICAL FEATURES

Motor neurons in the brainstem and spinal cord are the main targets of infection during poliomyelitis (early disease), but sensory neurons and astrocytes are also infected.[19] TMEV does not replicate in endothelial and ependymal cells or initially in oligodendrocytes. A brisk microglial reaction is elicited with the appearance of numerous microglial nodules, particularly in the anterior gray matter of the spinal cord. Examples of neuronophagia are quite frequent at this time, but very little lymphocytic response is seen. the poliomyelitis phase lasts 1–4 weeks, after which time little residual gray matter involvement is apparent other than resolving astrocytosis.

Beginning as early as 2 weeks PI, inflammation of the spinal leptomeninges begins to appear, followed by involvement of the white matter.[19,20] Initially, the inflammatory infiltrates are almost exclusively composed of lymphocytes, but at later times plasma cells and macrophages are numerous. The influx of macrophages is in close temporal and anatomic relationship with myelin breakdown.

Both light and ultrastructural studies of the demyelinating process show that myelin destruction is strictly related to the presence of mononuclear cells,[19,21] which either actively strip myelin lamellae from otherwise normal-appearing axons or are found in contact with myelin sheaths undergoing vesicular disruption.[22] Foci of inflammation and myelin destruction extend from the perivascular spaces into the surrounding white matter, leading to sharply demarcated plaques of demyelination. The ultrastructure of oligodendrocytes during the initial phase of myelin breakdown has not shown alterations in oligodendroglial loops, which are in close apposition with naked but otherwise normal axons, suggesting that myelin injury is not directly related to oligodendroglial cytopathology. However, later in the infection vesicular changes have been seen in the inner cytoplasmic tongues of oligodendrocytes.[23] In addition, immunohistochemical analysis of the demyelinating process has not shown the expected pattern of loss of myelin basic protein (MBP) and myelin-associated glycoprotein (MAG) if oligodendrocytes are degenerating.[24] The MAG is preferentially detected in the inner oligodendroglial component of the myelin sheath, whereas P_o, a major glycoprotein of peripheral myelin, and MBP are found throughout the entire thickness of lamellae of compacted myelin.[25] In acute lesions, MBP disappears before MAG, whereas in recurrent lesions where Schwann cell remyelination has taken place, P_o is lost before MAG. These observations suggest that myelin destruction reflects a direct attack on the myelin sheath rather than on myelinating cells.

One of the most prominent features of TMEV infection is the presence of active inflammatory, demyelinating lesions for many months with old, inactive plaques observed along with the active lesions. The simultaneous presence and close proximity of fresh and inactive demyelinating lesions provides indirect evidence that demyelination is recurrent. Additional evidence comes from observations in DA virus-infected C3H/He mice and in WW virus-infected outbred Swiss mice (DA and WW are less virulent strains), where extensive Schwann cell remyelination occurs in the outer margins of the spinal cord white matter and extensive oligodendroglial remyelination is present along the inner aspects of white matter columns.[26,27] In most of these animals examined several months PI, fresh inflammatory, demyelinating lesions were present in areas previously remyelinated by Schwann cells. Schwann cells in contact with naked axons are normal in appearance and devoid of viral antigen, although surrounded by numerous inflammatory cells.[28] These observations clearly demonstrate the occurrence of more than one episode of demyelination.

5. VIRUS-SPECIFIC IMMUNITY

During the first week, TMEV-infected mice mount a virus-specific humoral immune response that reaches a peak by 1 to 2 months PI and is sustained for the life of the host.[6,17] Both plaque-reduction neutralization[6] and immunoassay[29] have been used to measure TMEV-specific antibody responses. Using a solid-phase particle concentration fluorescence immunoassay (PCFIA), Peterson *et al.*[30] found that the majority of the antiviral IgG in persistently infected mice is of the IgG_{2a} and IgG_{2b} classes, with very little antiviral IgM present by day 21 PI. The antiviral antibody response is substantial and constitutes approximately 10% of the total serum IgG in these mice.[30] Although four neutralizing epitopes have been mapped on the virion of several picornaviruses, TMEV-specific neutralizing epitopes have not yet been determined. Thus, nothing is known about epitope-specific humoral immunity in TMEV infection.

When infected, susceptible strains of mice also develop significant levels of virus-specific cellular immunity. T-cell proliferation and delayed-type hypersensitivity (DTH) appear by approximately 2 weeks PI and remain at high levels for at least 6 months, possibly for the lifetime of the host.[31] Both DTH and T-cell proliferation have been shown to be specific for TMEV and mediated by L3T4+, Lyt-1+2- (CD4+), class-II-restricted T cells.[31]

Although mice mount virus-specific humoral and cellular immune responses on early virus exposure and peak virus titers fall by 2–3 log units, TMEV somehow evades immune clearance to persist at low levels indefinitely in the CNS of the host. Extraneural persistence has not been observed. Current dogma holds that humoral immunity is more important than cellular immunity in clearing infections by nonenveloped viruses such as picornaviruses, but this has not been established for TMEV. The precise mechanism by which TMEV evades immune clearance is presently unknown but does not involve antigenic variation.[32] Although complement and virus–antibody deposition in the CNS parenchyma (e.g., myelin sheaths) has not been found,[33,34] extracellular transport of virus as infectious virus–antibody complexes, in aggregates, or enveloped in cell membranes[35,36] could provide protection against the TMEV-specific immune responses and enable persistence of the infection.

6. SITES OF VIRUS PERSISTENCE

TMEV persistence involves ongoing virus replication, since infectious virus can be readily isolated from the CNS.[16,21] TMEV replication during the persistent phase, like that of other persistent viruses, is known to be restricted. Cash et al.[37] found that the majority of infected cells ($\geq 95\%$) in the spinal cord contain 100 to 500 copies of TMEV RNA. A second but minor population of white matter cells ($\leq 5\%$) contain >1500 copies of virus RNA. Only 10–20% of the cells containing virus RNA produced virus coat proteins.[37] Recently, the restriction in virus RNA replication was shown to be caused by a block at the level of minus-strand synthesis.[38]

Although TMEV-induced demyelinating disease has been studied for more than 15 years, it is still uncertain whether oligodendrocytes, the myelin-maintaining cell, or macrophages are the primary target for virus persistence. Virus antigen,[19,39] virions,[40] and virus genomic RNA[41] have been detected in oligodendrocytes and macrophages in demyelinating lesions in adult mice. Persistence in oligodendrocytes could lead to demyelination by a direct virus cytopathic effect or by triggering an immune-mediated response to virally altered oligodendrocyte membranes as proposed by Rodriguez et al.[42] However, virus antigen has been detected in oligodendrocytes only at day 45 PI, well after the onset of demyelination (20–30 days PI), and apparently in a limited number of these cells.[20] Thus, infection of oligodendrocytes has not as yet been shown to coincide with the onset of myelin breakdown, nor is it clear whether the infection of the myelin-maintaining cell is extensive enough to account for the large size of demyelinating lesions in this infection. In another study in which cell types could be determined for only 50% of the CNS cells positive by *in situ* hybridization for TMEV genomes, 25–40% of these cells were identified as oligodendrocytes by staining for carbonic anhydrase II (CAII).[42] These authors reported finding no overlap in double immunostaining for carbonic anhydrase II (CAII) and the astrocyte indicator, glial fibrillary acidic protein (GFAP). However, anti-GFAP antibodies do not always stain protoplasmic astrocytes, and anti-CAII antibodies can stain protoplasmic astrocytes in addition to oligodendrocytes,[43] so the identity of the TMEV-positive cell remains somewhat unclear. In contrast to the above studies, virus antigen has been readily demonstrated as early as 11 days PI in macrophages located in demyelinating lesions[39] and at 28 days PI[20] as well as throughout the chronic phase of infection,[39] suggesting that macrophages may in fact be the principal target for TMEV persistence. On the other hand, the detection of virus antigen alone in macrophages does not prove that these cells are productively infected.

Recently, mononuclear cells (MNC) isolated directly from CNS inflammatory infiltrates of TMEV-infected mice on discontinuous Percoll gradients were found to contain infectious TMEV.[44] Macrophages appeared to be the principal MNC infected. Infectious center assay and double immunostaining together indicated the presence and possible synthesis of TMEV in approximately one in 225 to one in 1000 CNS macrophages, with one to seven PFU produced per macrophage. Based on these findings, limited replication in macrophages is consistent with the total CNS virus content detected at any time during the persistent phase of the infection as well as the slow pace of the infection.

7. IMMUNE-MEDIATED MECHANISM OF DEMYELINATION

Appropriately timed immunosuppression can prevent and reverse the clinical signs and pathological changes caused by TMEV-induced demyelinating disease, strongly suggesting that myelin breakdown is immune mediated. A number of different immunosuppressive modalities have proven to be effective, including cyclophosphamide, antilymphocyte serum, and monoclonal anti-I-A, L3T4+, and Lyt-2+ antibodies[45–49] (Table 5-4). If given too early, these agents may potentiate the initial neuronal phase of the infection and cause a high mortality from encephalitis.[45–48] Recently, S. D. Miller (unpublished data) adoptively immunized infected SJL mice with a TMEV VP2-specific T-cell line and increased the incidence of demyelinating disease. The recipient mice were inoculated IC with TMEV at a dose that produced a low incidence of demyelinating disease. These results further support an immune-mediated mechanism of demyelination and indicate that the T-cell response is directed at TMEV.

The effector mechanism by which a nonbudding virus, such as TMEV, might lead to immune-mediated tissue injury is unknown. Because TMEV antigens have been found in macrophages,[20,39,44] it has been proposed that myelin breakdown may result from an interaction between virus-specific sensitized lympho-

TABLE 5-4
Immunosuppressive Agents Reported to Prevent TMEV-Induced Demyelinating Disease

Agent	References	Virus strain	Day PI	Endpoints[a]
Cytoxan plus ATS[b]	44	DA	0	H
Cytoxan plus ATS	45	DA	0, 35	H
MAB[c] anti-I-A	46	WW	0, 14, at onset	C, H
MAB anti-L3T4+ plus thymectomy	47	BeAn	0, 65	C, H
MAB anti-Lyt2+	48	DA	−1, 15	H
MAB anti-L3T4+	Unpublished (see text)	BeAn	0, 14	C, H

[a]Endpoints for demyelinating disease; C, clinical signs; H, histology.
[b]Cytoxan, cyclosphosphamide; ATS, antithymocyte serum.
[c]MAB, monoclonal antibody.

cytes trafficking into infected areas in the CNS and the virus. Thus, myelinated axons may be nonspecifically damaged as a consequence of a virus-specific immune response, i.e., an "innocent-bystander" response. Clatch et al.[31,50,51] showed that high levels of TMEV-specific DTH but not TMEV-specific antibody responses correlate with the temporal onset of demyelinating disease as well as with the disease incidence among susceptible and resistant congenic recombinant mice. Thus, in this system, lymphokines produced by MHC class II (I-A)-restricted, TMEV-specific T_{DTH} cells primed by interaction with infected macrophages would lead to the recruitment and activation of additional macrophages in the CNS, resulting in nonspecific macrophage-mediated demyelination. This hypothesis is consistent with the CNS pathological changes observed in mice exhibiting TMEV-induced demyelinating disease and with the classic observations of MacKaness,[52] who demonstrated nonspecific resistance to Mycobacterium in naive recipients of Listeria-immune T cells infected with both viable Mycobacterium and Listeria. Regarding demyelination, antigen-specific T cells and T-cell lines have been shown to cause bystander CNS damage via macrophage activation. Wisniewski and Bloom[53] showed that CNS and peripheral nervous system myelin can be damaged as a nonspecific consequence of a specific DTH reaction directed at non-nervous-tissue antigens, namely, purified protein derivative (PPD) of tuberculin. More recently, Holoshitz et al.[54] showed that encephalitis can be produced in mice by intravenous transfer of PPD-specific T-cell lines following IC inoculation of PPD. However, another study designed to evaluate bystander demyelination in peripheral nervous tissue revealed no evidence for bystander demyelination.[55]

Alternatively, if there is extensive infection of oligodendrocytes, demyelination may result from immune injury to these cells, since they may express TMEV antigens in conjunction with H-2 class I determinants. Because H-2 class I determinants (e.g., H-2D) restrict the development and expression of lyt-2+ (CD8+) cytotoxic T cells to allogeneic and virus-infected syngeneic cells, these T cells might kill infected oligodendrocytes.[42] However, the pathogenic role of

TMEV-specific cytotoxic T-cell responses remains unclear, since such responses have not been reported in susceptible mice, and widespread degeneration of oligodendrocytes is not apparent histologically.[19,20]

Finally, TMEV infection might trigger an autoimmune reaction comparable to that occurring in experimental allergic encephalomyelitis (EAE) and thereby may contribute to the demyelinating process. Such an autoimmune response against CNS antigens could be triggered by one or several mechanisms: (1) direct damage to CNS cells resulting from cytopathic effects of TMEV, (2) damage to CNS constituents as a result of TMEV-specific cell-mediated immune responses, and (3) cross-reactive immune responses between virus and CNS antigens (molecular mimicry). Although it has long been postulated that viruses may share antigenic sites with normal host-cell components, identity between virus and nervous system myelin antigens has been demonstrated only recently.[56,57] Fujinami and Oldstone,[57] who found amino acid sequence identity between the encephalitogenic site of rabbit myelin basic protein and the hepatitis B virus polymerase, showed not only that immune responses were generated in rabbits by the virus peptide that cross-reacted with the self protein but also that mononuclear cell infiltration was present in the CNS of animals immunized with the peptide.

In support of a molecular mimicry pathogenesis, Fujinami et al.[58] have identified a monoclonal antibody from TMEV-infected animals that both neutralizes TMEV and reacts with galactocerebroside, a surface component on myelin; the antibody, on inoculation, causes demyelination. However, using the chronic, relapsing EAE model in SJL mice as a positive control for neuroantigen reactivity, Miller et al.[59] found no evidence of T-cell proliferative or DTH responses to the major neuroantigens in mice with TMEV-induced demyelinating disease. The neuroantigens included mouse whole spinal cord homogenate and purified MBP and PLP.[59] In addition, infected SJL mice also failed to make significant T-cell proliferative responses to peptides representing the immunodominant T-cell epitopes of mouse MBP (amino acids 84–104) and PLP (amino acids 139–151). Finally, the course of demyelinating disease in SJL mice was not altered by a tolerization regimen to the neuroantigens in whole spinal cord homogenate coupled to spleen cells that prevents development of chronic relapsing EAE.[60] Taken together, these observations argue against a role for autoimmunity during TMEV infection.

8. MULTIGENIC CONTROL OF DEMYELINATING DISEASE

Susceptibility to TMEV-induced demyelinating disease differs among inbred mouse strains. In studies with tissue-culture-adapted virus, SJL, SWR, DBA/2, and PL represent susceptible strains, whereas C57BL/6, C57BL/10, BALB/c, and C57L are resistant.[51,61] The availability of such inbred strains and a variety of defined variants has permitted analysis of the genetic basis for virus susceptibility to a much finer degree than is possible in humans, where many genetic variables cannot be controlled. Comparisons of the resistant C57BL/6 and susceptible SJL strains indicate that multiple genes are involved in determination of susceptibility, at least one in the H-2 complex and at least one that segregates independently of H-2.[61] The H-2 gene involved has been localized to the class I locus H-2D,[50,62,63] but the non-H-2 gene(s) remain unidentified.

Differences at H-2 genes, however, do not always appear to be crucial to determination of susceptibility. Recently, Melvold *et al.*[64] noted that in some strain combinations, such as the susceptible SJL and the resistant BALB/c, H-2 genotypes of segregating backcross animals do not correlate well with susceptibility, which appears instead to be primarily determined by multiple non-H-2 loci. In comparisons of the susceptible DBA/2 and the resistant BALB/c, the entire genetic basis for susceptibility must rely on non-H-2 genes because both strains carry the H-2^d haplotype. Comparisons of the susceptible DBA/2 and resistant C57BL/6 strains have indicated an important role for the H-2D locus and for a non-H-2 gene (not involving the β chain of the T-cell receptor) in differential susceptibility.[65,66] Analysis of recombinant-inbred strains (BXD) between the DBA/2 and C57BL/6 strains indicated that this non-H-2 locus is located at the centromeric end of chromosome 3 near the carbonic anhydrase-2 enzyme locus.

The predominant role of different loci in different mouse strains probably reflects the involvement of many genes, and the disease process might be best considered as resembling a metabolic pathway with several stages that can be influenced by different gene products. In comparisons of particular strains, analysis is affected by the facts that (1) only loci that are functionally different in the two strains being compared can be identified (loci at which the strains are functionally identical will have no detectable effects) and (2) the activity of some genes may vary according to the "genetic environment" in which they exist being influenced by the presence or absence of other genes. Thus, a satisfactory description of the genetic control of susceptibility probably requires a composite of numerous strain comparisons to identify a significant portion of the loci involved. This also provides an analogous situation to human studies, where particular HLA genes (e.g., DR2 and Drw2) have positive associations with multiple sclerosis, but the relative risks are so low (ranging between 2 and 3) that genetic factors other than HLA phenotype must also be involved.

REFERENCES

1. Theiler, M., 1937, Spontaneous encephalomyelitis of mice, a new virus disease, *J. Exp. Med.* **65**:705–719.
2. Thompson, R., Harrison, V. M., and Myers, F. P., 1956, A spontaneous epizootic of mouse encephalomyelitis virus, *J. Infect. Dis.* **98**:98–102.
3. Theiler, M., and Gard, S., 1940, Encephalomyelitis of mice. III. Epidemiology, *J. Exp. Med.* **72**:79–90.
4. Lipton, H. L., Miller, S. D., Melvold, R. M., and Fujinami, R. S., 1986, Theiler's virus infection as a model for multiple sclerosis, in: *Concepts in Viral Pathogenesis*, Vol. II (A. L. Notkins and M. B. A. Oldstone, eds.), Academic Press, New York, pp. 248–253.
5. Ohara, Y., and Roos, R., 1987, The antibody response in Theiler's virus infection: New perspectives on multiple sclerosis, *Prog. Med. Virol.* **34**:156–179.
6. Rodriguez, M., Oleszak, E., and Leibowitz, J., 1987, Theiler's murine encephalomyelitis: A model of demyelination and persistence of virus, *CRC Crit. Rev. Immunol.* **7**:325–365.

7. Pevear, D. C., Calenoff, M., Rozhon, E., and Lipton, H. L., 1987, Analysis of the complete nucleotide sequence of the picornavirus Theiler's murine encephalomyelitis virus indicates that it is closely related to cardioviruses, *J. Virol.* **61**:1507–1516.

8. Ohara, Y., Stein, S., Fu, J., Stillman, L., Klaman, L., and Roos, R. P., 1988, Molecular cloning and sequence determination of DA strain of Theiler's murine encephalomyelitis viruses, *Virology* **164**:245–255.

9. Lipton, H. L., 1978, Characterization of the TO strains of Theiler's mouse encephalomyelitis viruses, *Infect. Immun.* **20**:869–872.

10. Lorch, Y., Kotler, M., and Friedmann, A., 1983, GDVII and DA isolates of Theiler's virus: Proteins attached to the 5' end of the RNA are bound covalently to the same nucleotides, *J. Virol.* **45**:1150–1154.

11. Rozhon, E. J., Brown, F. J., and Lipton, H. L., 1982, Characterization of the RNA of Theiler's murine encephalomyelitis virus, *J. Gen. Virol.* **61**:157–161.

12. Rueckert, R. R., and Wimmer, E., 1984, Systematic nomenclature of picornavirus proteins, *J. Virol.* **50**:957–959.

13. Rueckert, R. R., 1986, Picornaviruses and their replication, in: *Fundamental Virology* (B. N. Fields and D. M. Knipe, eds.), Raven Press, New York, pp. 357–390.

14. Theiler, M., and Gard, S., 1940, Encephalomyelitis of mice. I. Characteristics and pathogenesis of the virus, *J. Exp. Med.* **72**:49–67.

15. Liu, C., Collins, J., and Sharp, E., 1967, The pathogenesis of Theiler's GD VII encephalomyelitis virus infection in mice as studied by immunofluorescent technique and infectivity titrations, *J. Immun.* **98**:46–55.

16. Lipton, H. L., 1975, Theiler's virus infection in mice: An unusual biphasic disease process leading to demyelination, *Infect. Immun.* **11**:1147–1155.

17. Lehrich, J. R., Arnason, B. G. W., and Hochberg, F. H., 1976, Demyelinating myelopathy in mice induced by the DA virus, *J. Neurol. Sci.* **29**:149–160.

18. Jubelt, B., Gallez-Hawkins, G., Narayan, O., and Johnson, R. T., 1980, Pathogenesis of human poliovirus infection in mice. I. Clinical and pathological studies, *J. Neuropathol. Exp. Neurol.* **39**:138–148.

19. Dal Canto, M. C., and Lipton, H. L., 1975, Primary demyelination in Theiler's virus infection. An ultrastructural study, *Lab. Invest.* **33**:626–637.

20. Rodriguez, M., Leibowitz, J. L., and Lampert, P. W., 1983, Persistent infection of oligodendrocytes in Theiler's virus-induced encephalomyelitis, *Ann. Neurol.* **13**:426–433.

21. Chamorro, M., Aubert, C., and Brahic, M., 1986, Demyelinating lesions due to Theiler's virus are associated with ongoing central nervous system infection, *J. Virol.* **57**:992–997.

22. Dal Canto, M. C., Wisniewski, H. M., Johnson, A. B., Brostoff, S., and Raine, C. S., 1975, Vesicular disruption of myelin in autoimmune demyelination, *J. Neurol. Sci.* **24**:313–319.

23. Rodriguez, M., 1985, Virus-induced demyelination in mice: "Dying back" of oligodendrocytes, *Mayo Clin. Proc.* **60**:433–438.

24. Dal Canto, M. C., and Barbano, R. L., 1985, Immunocytochemical localization of MAG, MBP, P_o and P_o protein in acute and relapsing demyelinating lesions of Theiler's virus infection, *J. Neuroimmunol.* **10**:129–140.

25. Sternberger, N. H., Quarles, R. H., Itoyama, Y., and Webster, H. deF., 1979, Myelin-associated glycoprotein demonstrated immunocytochemically in myelin and myelin forming cells of developing rat, *Proc. Natl. Acad. Sci. U.S.A.* **76**:1510–1514.

26. Dal Canto, M. C., and Lipton, H. L., 1980, Schwann cell remyelination and recurrent demyelination in the central nervous system of mice infected with attenuated Theiler's virus, *Am. J. Pathol.* **98**:101–110.

27. Dal Canto, M. C., and Barbano, R. L., 1984, Remyelination during remission in Theiler's virus infection, *Am. J. Pathol.* **116**:30–45.

28. Dal Canto, M. C., 1982, Uncoupled relationship between demyelination and primary infection of myelinating cells in Theiler's virus encephalomyelitis, *Infect. Immun.* **35**:1133–1138.

29. Peterson, J. D., Kim, J. Y., Melvold, R. W., Miller, S. D., and Waltenbaugh, C., 1989, A rapid method for quantitation of antiviral antibodies, *J. Virol. Methods* **119**:83–94.

30. Peterson, J. D., Miller, S. D., and Waltenbaugh, C., 1989, Rapid biotin–avidin method for quantitation of antiviral antibody isotypes, *J. Virol. Methods* **27**:189–202.

31. Clatch, R. J., Lipton, H. L., and Miller, S. D., 1986, Characterization of Theiler's murine encephalomyelitis virus (TMEV)-specific delayed-type hypersensitivity responses in TMEV-induced demyelinating disease: Correlation with clinical signs, *J. Immunol.* **136**:920–927.

32. Rozhon, E. J., Kratochvil, J. D., and Lipton, H. L., 1983, Analysis of genetic variation in Theiler's virus during persistent infection in the mouse central nervous system, *Virology* **128**:16–32.

33. Gonzalez-Scariano, F., and Lipton, H. L., 1978, Central nervous system immunity in mice infected with Theiler's virus. I. Local neutralizing antibody response, *J. Infect. Dis.* **137**:145–151.

34. Rodriguez, M., Lucchinetti, C. F., Clark, R. J., Yaksh, T. L., Markowitz, H., and Lennon, V. A., 1988, Immunoglobulins and complement in demyelination induced in mice by Theiler's virus, *J. Immunol.* **140**:800–806.

35. Friedmann, A., and Lipton, H. L., 1980, Replication of GDVII and DA strains of Theiler's murine encephalomyelitis virus in BHK 21 cells: An electron microscopic study, *Virology* **101**:389–398.

36. Frankel, G., Lorch, Y., Karlik, P., and Friedmann, A., 1987, Fractionation of Theiler's virus-infected BHK21 cell homogenates: Isolation of virus-induced membranes, *Virology* **158**:452–455.

37. Cash, E., Chamorro, M., and Brahic, M., 1985, Theiler's virus RNA and protein synthesis in the central nervous system of demyelinating mice, *Virology* **144**:290–294.

38. Cash, E., Chamorro, M., and Brahic, M., 1988, Minus-strand RNA synthesis in the spinal cords of mice persistently infected with Theiler's virus, *J. Virol.* **62**:1824–1826.

39. Dal Canto, M. C., and Lipton, H. L., 1982, Ultrastructural immunohistochemical localization of virus in acute and chronic demyelinating Theiler's virus infection, *Am. J. Pathol.* **106**:20–29.

40. Blakemore, W. F., Welsh, C. J. R., Tonks, P., and Nash, A. A., 1988, Observations on demyelinating lesions induced by Theiler's virus in CBA mice, *Acta Neuropathol. (Berl.)* **76**:581–589.

41. Aubert, C., Chamorro, M., and Brahic, M. (1987). Identification of Theiler's virus infected cells in the central nervous system of the mouse during demyelinating disease, *Microb. Pathogen.* **3**:319–326.

42. Rodriguez, M., Pease, L. R., and David, C. S., 1986, Immune-mediated injury of virus-infected oligodendrocytes. A model of multiple sclerosis, *Immunol. Today* **7**:359–363.

43. Cammer, W., and Tansey, F. A., 1988, Carbonic anhydrase immunostaining in astrocytes in the rat cerebral cortex, *J. Neurochem.* **50**:319–322.

44. Clatch, R. C., Miller, S. D., Metzner, R., Dal Canto, M. C., and Lipton, H. L., 1990, Monocytes/macrophages isolated from the mouse central nervous system contain infectious Theiler's murine encephalomyelitis virus (TMEV), *Virology* **176**:244–254.

45. Lipton, H. L., and Dal Canto, M. C., 1976, Theiler's virus-induced demyelination: Prevention by immunosuppression, *Science* **192**:62–64.

46. Roos, R. P., Firestone, S., Wollmann, R., Variakojis, D., and Arnason, B. G. W., 1982, The effect of short-term and chronic immunosuppression on Theiler's virus demyelination, *J. Neuroimmunol.* **2**:223–234.

47. Friedmann, A., Frankel, G., Lorch, Y., and Steinman, L., 1987, Monoclonal anti-I-A antibody reverses chronic paralysis and demyelination in Theiler's virus-infected mice: Critical importance of timing in treatment, *J. Virol.* **61**:898–903.

48. Rodriguez, M., and Sriram, S., 1988, Successful therapy of Theiler's virus-induced demyelination (DA strain) with monoclonal anti-Lyt-2 antibody, *J. Immunol.* **140**:2950–2955.

49. Welsh, C. J. R., Tonks, P., Nash, A. A., and Blakemore, W. F., 1987, The effect of L3T4 cell

depletion on the pathogenesis of Theiler's murine encephalomyelitis virus infection in CBA mice, *J. Gen. Virol.* **68:**1659–1667.

50. Clatch, R. J., Melvold, R. W., Miller, S. D., and Lipton, H. L., 1985, Theiler's murine encephalomyelitis virus (TMEV)-induced demyelinating disease in mice is influenced by the H-2D region: Correlation with TMEV specific delayed type hypersensitivity, *J. Immunol.* **135:** 1408–1414.

51. Clatch, R. J., Lipton, H. L., and Miller, S. D., 1987, Further support for the importance of Theiler's murine encephalomyelitis virus (TMEV) specific delayed-type hypersensitivity causing demyelination in different inbred mouse strains, *Microb. Pathogen.* **3:**327–337.

52. MacKaness, G. B., 1964, The immunologic basis of acquired cellular resistance, *J. Exp. Med.* **120:**105–120.

53. Wisniewski, H. M., and Bloom, B. R., 1975, Primary demyelination as a nonspecific consequence of a cell-mediated immune reaction, *J. Exp. Med.* **141:**346–359.

54. Holoshitz, J., Naparstek, Y., Ben-Nun, A., Marquardt, P., and Cohen, I. R., 1984, T lymphocyte lines induce autoimmune encephalomyeltis, delayed hypersensitivity, and bystander encephalitis or arthritis, *Eur. J. Immunol.* **14:**729–734.

55. Powell, H. C., Braheny, S. L., Hughes, R. A. C., and Lampert, P. W., 1984, Antigen-specific demyelination and significance of the bystander effect in peripheral nerves, *Am. J. Pathol.* **114:**443–453.

56. Jahnke, U., Fischer, E. H., and Alvord, E. C., Jr., 1985, Sequence homology between certain viral proteins and proteins related to encephalomyelitis and neuritis, *Science* **229:**282–284.

57. Fujanimi, R. S., and Oldstone, M. B. A., 1985, Amino acid homology between the encephalogenic site of myelin basic protein and virus: A mechanism for autoimmunity, *Science* **230:** 1043–1045.

58. Fujanimi, R. S., Zurbriggen, A., and Powell, H. H., 1988, Monoclonal antibody defined determinant between Theiler's virus and lipid-like structures, *J. Neuroimmunol.* **20:**25–30.

59. Miller, S. D., Clatch, R., Pevear, D., and Lipton, H. L., 1987, Specificity of T cell responses in Theiler's virus-induced demyelinating disease. I. Lack of neuroantigen specificity and cross-reactivity to intratypic virus variants, *J. Immunol.* **138:**3776–3784.

60. Miller, S. D., Gerety, S. J., Kennedy, M. K., Peterson, J. D., Trotter, J. L., Waltenbaugh, C., Dal Canto, M. C., and Lipton, H. L., 1989, Class II-restricted T cell responses in Theiler's murine encephalomyelitis virus (TMEV)-induced demyelinating disease. III. Failure of neuroantigen-specific immune tolerance to affect the clinical course of demyelination, *J. Neuroimmunol.* **26:**9–23.

61. Lipton, H. L., and Melvold, R., 1984, Genetic analysis of susceptibility to Theiler's virus-induced demyelinating disease in mice, *J. Immunol.* **132:**1821–1825.

62. Rodriguez, M., and David, C. S., 1985, Demyelination induced by Theiler's virus: Influence of the H-2 haplotype, *J. Immunol.* **135:**2145–2148.

63. Rodriguez, M., Leibowitz, J., and David, C. S., 1986, Susceptibility to Theiler's virus-induced demyelination. Mapping of the gene within the H-2D region, *J. Exp. Med.* **163:**620–631.

64. Melvold, R. M., Jokenen, D., Knobler, R., and Lipton, H. L., 1987, Differences between BALB/c and SJL/J mice in susceptibility to Theiler's murine encephalomyelitis virus (TMEV)-induced demyelinating disease are primarily controlled by non-H-2 genes, *J. Immunol.* **138:**1429–1433.

65. Melvold, R. W., Jokinen, D. K., Miller, S. D., Dal Canto, M. C., and Lipton, H. L., 1988, H-2 genes in TMEV-induced demyelinating disease, a model for multiple sclerosis, in: *Major Histocompatibility Genes and Their Role in Immune Function* (C. S. David, ed.), Plenum Press, New York, pp. 735–745.

66. Melvold, R. W., Jokinen, D. M., Miller, S. D., Dal Canto, M. C., and Lipton, H. L., 1990, Identification of a locus on mouse chromosome 3 involved in differential susceptibility to Theiler's murine encephalomyelitis virus (TMEV)-induced demyelinating disease, *J. Virol.* **64:**686–690.

6

Neurotropic Retroviruses of Mice, Cats, Macaques, and Humans

MURRAY GARDNER, ANDREW LACKNER, and LINDA LOWENSTINE

1. INTRODUCTION

The prominent neurological manifestations associated with infection by the human immunodeficiency virus (HIV) and the human T-lymphotropic retrovirus (HTLV) have focused attention on the neurotropic properties of the family Retroviridae. This family is divided into three subfamilies, the oncoviruses, the lentiviruses, and the spumaviruses, each of which is known to infect the central nervous system (CNS) of their natural animal hosts.[1] The only known naturally occurring neurological disease caused by an oncovirus in animals is the spongiform polioencephalomyelopathy caused by murine leukemia virus (MuLV) discovered in the early 1970s in a population of wild mice in southern California (for review, see Gardner[2]). Since then, several strains of laboratory-derived, temperature-sensitive, and mutant MuLV have produced a similar disease experimentally in mice and rats.[3,4]

MURRAY GARDNER • Department of Pathology, School of Medicine, University of California, Davis, California 95616.   ANDREW LACKNER and LINDA LOWENSTINE • California Primate Research Center, University of California, Davis, California 95616.   *Present address of A.L.*: New Mexico Regional Primate Research Laboratory, New Mexico State University, Holloman Air Force Base, New Mexico 88330-1027.

Neuropathogenic Viruses and Immunity, edited by Steven Specter *et al.* Plenum Press, New York, 1992.

The prototype lentivirus disease of sheep, caused by visna virus, was described in the mid-1950s.[5] The related caprine arthritis encephalitis virus (CAEV) in goats was described about 1980.[6] In retrospect, these animal lentiviruses served notice of pathogenic mechanisms[7] to be observed later with the discovery of similar lentivirus infections in monkeys, cats, and humans. Table 6-1 lists the biological properties shared by the lentivirus infections of animals and man. Visna is the subject of another chapter in this book (Chapter 4). We summarize the MuLV-induced neurological disease in mice, briefly describe a new example of latent, innocuous CNS infection caused by an immunosuppressive type D retrovirus in macaques, and cover the neuropathology associated with infection of macaques and cats with the simian and feline immunodeficiency lentiviruses (SIV and FIV, respectively). These animal models of CNS retroviral infection are briefly compared with the CNS infections caused by the HTLV and HIV. We also call attention to the common recovery of latent, apparently harmless, spumaviruses in the CNS of animals and humans.

2. NEUROTROPIC MURINE LEUKEMIA VIRUS

The best-characterized model of a naturally occurring neurotropic retrovirus is the hind leg paralysis caused by an infectious ecotropic strain of MuLV indigenous in wild mice (*Mus musculus*).[2] This disease occurs in nature in a population of MuLV high-expressor, lymphoma-prone wild mice inhabiting squab farm near Lake Casitas (LC) in southern California.[8,9] About 10% of aging LC mice develop this neurological disease, with or without accompanying lymphoma, between 8 and 18 months of age. Ecotropic MuLV in high titer is uniquely present in the serum and CNS of these paralyzed mice. Affected animals acquire the ecotropic MuLV at birth, primarily from their mothers' milk, and remain viremic and immune tolerant to this virus throughout their lifetime.[10] General immunity, however, remains intact.

The main virus "factory" resides in the B-cell areas of the spleen; from there virus spreads by cell-free viremia to the CNS. Splenectomy and passive immunization early in life lower the amount of virus and prevent development of paralysis.[11]

TABLE 6-1
Properties of Lentiviruses

1. Long incubation period: Healthy carrier state
2. Wide disease spectrum
3. Persistence in face of vigorous immunne response
4. Viral regulatory genes control latency and activation
5. Restricted virus expression
6. Cytopathic effect: Syncytia, nononcogenic
7. Neurotropism
8. Macrophage tropism, "Trojan horse"
9. High rate of envelope antigenic variation
10. Transmission by close physical contact

In the CNS, ecotropic virus replication occurs in endothelial and perithelial cells and eventually reaches a sufficient level to bind to and enter anterior horn neurons and oligodendroglial cells in the lower spinal cord.[12-14] Abortive intracytoplasmic virus replication, cell swelling (vacuolation), and death of affected cell types follow. Endothelial, neuronal, and glial cell membranes may also be damaged by the extracellular accumulation of ecotropic viral proteins, particularly the envelope glycoprotein (gp70). The loss of anterior horn neurons in the lumbosacral spinal cord results in a lower-motor-neuron type of flaccid paralysis of both hind legs with fasciculations and secondary demyelination. Loss of oligodendroglia leads to primary demyelination. The accumulation of intracellular and extracellular edema (spongiosis), the absence of any inflammation, and the loss of anterior horn neurons with reactive gliosis in the lower spinal cord account for the major histopathological features and prompt the naming of this disease, "spongiform polioencephalomyelopathy." Except for the presence of type C particles, the vacuolar changes in the spinal cord of affected mice are remarkably similar to the vacuoles seen in the spongiform encephalopathies caused by atypical agents such as scrapie, kuru, and Creutzfeldt–Jakob disease.

Lake Casitas wild mice also harbor another class of infectious MuLV called "amphotropic" because of its wide *in vitro* host range for murine and nonmurine cells.[15,16] Like ecotropic virus, amphotropic MuLV is also acquired by congenital maternal infection and is accompanied by specific immune tolerance. Amphotropic MuLV is far more prevalent in LC mice than ecotropic virus, being present in about 85% of the mice from birth. Although amphotropic viremia exists in paralyzed LC mice, this class of MuLV is not uniquely associated with CNS infection and paralysis, as seen with the ecotropic virus. Nor is amphotropic virus required for replication or pathogenicity of the ecotropic virus.

Experimental transmission of biologically and molecularly cloned LC MuLVs to laboratory mice has shown that the ecotropic MuLV alone is both paralytogenic and lymphomagenic, whereas the amphotropic MuLV is only lymphomagenic.[17-19] The experimentally induced neurological disease closely mimics the natural disease in histopathology and clinical symptoms except that the lesions often extend more rostrally into the upper spinal cord, brainstem, and cerebellar peduncles. Virtually 100% of FV-1^n (see below) laboratory mice are susceptible to the experimental induction of the neurological disease with a latent period of only several weeks to several months after inoculation of newborns with concentrated virus. Mice that live longer may also develop lymphoma. Amphotropic MuLV induces lymphoma in $\leq$20% recipient newborn mice after a latent period of $\geq$10 months. The lymphomas induced by both viruses in wild and laboratory mice arise in the spleen and are of B- or pre-B-cell origin.[20] Inoculation of laboratory mice in the newborn period is critical for transmission of paralysis because development of immune competence after several days of age prevents the ecotropic virus from replicating to sufficiently high levels in the spleen and CNS.[21]

Recombination of ecotropic MuLV with endogenous MuLV-related sequences, a feature of MuLV-induced lymphomas in laboratory mice, is not involved in the pathogenesis of the neurological disease.[22,23] Nor, apparently, is derivation of mink cell focus (MCF)-forming recombinants, as seen in AKR inbred mice, a feature of natural lymphoma development in LC mice.[17] However, experimental passage of LC amphotropic or ecotropic MuLV in laboratory mice is associated with MCF-like recombination events in the spleen and results in viruses with enhanced lymphogenicity or altered cell tropism.[22,24-26] MuLV recombination probably occurs less frequently in LC wild mice than laboratory mice because of the stronger restriction in wild mice of endogenous xenotropic MuLV-related proviral gene expression.[17]

Sequence analysis of viral genomes of different pathogenicity and induction of disease with viral chimeras have shown that the molecular determinants of paralysis are a number of scattered amino acid alterations confined to the envelope gene (gp70),[27,28] whereas the determinants of lymphomogenesis are distributed throughout the viral genome and include the LTR region.[19] The LC wild mouse viruses are more closely related to the Friend and Moloney strains of exogenous MuLV than to the endogenous AKR-related MuLVs of laboratory mice.[28,29] It seems likely that both Friend–Moloney and wild mouse ecotropic viruses emerged from amphotropic MuLV of wild mice and have existed in mice for many years as a separate group of totally exogenous retroviruses.[10,28-30] This close similarity probably explains why experimental induction of the spongiform polioencephalomyelopathy has also been described with temperature-sensitive mutants of Moloney MuLV and rat-passaged mutant strains of Friend MuLV.[3,4] The similarities in natural history and neurotropism between the wild mouse exogenous MuLVs and the exogenous leukemia virus of humans (HTLV) are far more striking than those exhibited by the prototype AKR-related endogenous MuLVs of inbred laboratory mice.[31]

In laboratory mice the FV-1 locus is the principal dominant gene determining resistance or susceptibility to the experimental induction of paralysis or lymphomas with the LC MuLV.[32] The LC MuLV are all N-tropic, so only laboratory mice of FV-1^{nn} genotype are susceptible. Introduction of the FV-1^{b} allele into LC mice by selective breeding with laboratory mice completely blocks infection with LC MuLV and prevents lymphoma and paralysis.[33] However, segregation of this gene in LC wild mice cannot account for individual resistance or susceptibility to their indigenous MuLV diseases because LC mice are monomorphic for FV-1^{n}.

A dominant gene, different from the FV-1 locus, was discovered segregating in LC mice. This newly recognized gene powerfully blocked or restricted infection with all ecotropic MuLV including that of LC mice, which explains why only 10–20% of LC mice are infected with this neurotropic virus and consequently at risk for paralysis.[34] The MuLV restriction gene was initially called Akvr-1^{R},[34,35] but it was later shown to be allelic with[36] and phenotypically and sequence identical to the FV-4^{R} restriction gene found on chromosome 12 in Japanese wild mice (*Mus molossinus*).[37-39] Presence of this identical gene in wild mice from Japan and California is probably the result of interbreeding in recent times. The FV-4 gene represents an endogenous defective MuLV provirus encoding an ecotropic MuLV-related envelope gp70, which occupies cell surface receptors and interferes with entrance into the cell of ecotropic MuLV.[35,38-40] All ecotropic MuLV use the same receptor and thus are blocked by this gene.

About 75% of LC mice carry at least one of these dominant alleles[34] and are thus resistant to infection with ecotropic MuLV and development of paralysis. These mice are not resistant, however, to late-occurring lymphomas because the FV-4 gene does not block infection with amphotropic MuLV. This type of endogenous provirus interference gene was first described for avian leukosis virus infection of chickens.[41] Although FV-4^R apparently does not occur in North American laboratory mice, a similar interference gene for MCF MuLV has been described in DBA inbred mice.[42] Whether or not this type of leukemia virus restriction gene exists in other animal retrovirus models or in humans remains to be determined.

In summary, the neurotropic oncovirus, an exogenous ecotropic MuLV of LC wild mice, and the resultant spongiform polioencephalomyelopathy exemplify the result of a direct, "slow viral," nonimmunogenic, noninflammatory injury primarily to anterior horn neurons in the lower spinal cord. Although this is clearly caused by maternally transmitted infectious MuLV, the major determinant of this naturally occurring retroviral neurological disease is a dominant host cell gene (FV-4^R) segregating in LC feral mice that represents a defective provirus and blocks infection by interference at the cell surface receptor level.

3. NEUROTROPIC TYPE D RETROVIRUS OF MACAQUES

Type D retroviruses, related to the prototype Mason Pfizer monkey virus, are highly prevalent in macaques, which are their natural host, and are an important cause of a potentially fatal acquired immunodeficiency syndrome (SAIDS) in rhesus and other species of macaques in many primate facilities (for review, see Gardner and Marx[43]). Although apparently not associated with neurological disease, this virus does cause a latent parenchymal infection of the monkey CNS and a productive infection of choroid plexus epithelial cells *in vivo*.[44,45] The type D viruses are nononcogenic, exogenous, and horizontally transmitted, mainly by percutaneous inoculation of saliva via biting and scratching.[46] The virus can be readily transmitted experimentally to juvenile macaques,[47] and fatal SAIDS has been induced with molecularly cloned virus.[48] Three serotypes of type D virus have been molecularly cloned and totally sequenced. Two additional serotypes have recently been discovered. Serotype 1 (SRV-1) causes SAIDS in macaques at the California and New England Primate Centers. Serotype 2 (SRV-2) causes SAIDS and retroperitoneal fibrosis in macaques at the Oregon and Washington primate centers.[49] Natural disease resistance correlates with presence of SRV humoral antibody, including neutralizing antibody,[50] and solid protection against the experimental induction of SAIDS is provided by a formalin-killed SRV-1 vaccine.[51]

In SRV-1-infected macaques the virus is widespread in tissues with an affinity for germinal centers of lymphoid organs, secretory epithelial cells, and epithelial cells in the germinative cell layers of the upper and lower digestive tract.[52] SRV-1 also has a broad cell tropism for rhesus B cells, T helper and suppressor cells, macrophages, and fibroblasts as well as human B and T cell lines.[53] The cellular receptor for type D viruses is as yet unknown, but the gene for this receptor on human cells has been mapped to chromosome 19q13[54]; the same receptor is used by the endogenous type C retroviruses of cats and baboons (RD114 and BaEV, respectively). Except for fusion of human B cells (Raji), there is no direct cytopathology associated with *in vitro* infection with these type D viruses. Syncytial giant cells are not noted *in vivo* in SRV-induced SAIDS. The mechanism of lymphoid depletion and fatal immunosuppression induced by these viruses remains undetermined, but it must involve indirect mechanisms, some of which may be shared with the pathogenesis of HIV and SIV infection.[55]

Of interest is the evidence for latent CNS infection with SRV-1.[44,45] Viral DNA and RNA are readily detected by Southern blot and *in situ* hybridization, respectively, in the brain parenchyma of rhesus monkeys with SAIDS in the complete absence of viral antigen, infectious virus, neurological symptoms, or neuropathology, including a total lack of inflammatory cells. The neural cell types harboring the viral nucleic acid remain unidentified. In these same animals, virus DNA, RNA, antigen, particles, and infectious virus are readily demonstrable in lymphoid organs and salivary glands, and cell-free infectious virus can usually be isolated from the cerebrospinal fluid (CSF). The probable source of virus in the CSF is the choroid plexus, where approximately one in 1000 surface epithelial cells contain viral antigen.[45] Antibodies against SRV-1 are not detected in the CSF even when present in serum, and the CSF contains no cells or alteration in IgG and albumin levels.

The absence of infectious SRV-1, viral antigen, and lesions in the brain parenchyma, despite the detection of SRV-1 nucleic acid in occasional parenchymal cells, suggests that the infection is truly latent and innocuous in the CNS. The virus probably enters the brain parenchyma from the CSF, where it arises from productively infected choroid plexus epithelial cells, which, in turn, are probably infected from the bloodstream. It would be interesting to understand better the molecular mechanisms accounting for the profound restriction of SRV-1 expression in these neural cells. Whether or not this latent CNS virus may become activated later in life and lead to neurological disease in older monkeys remains to be determined. These type D retroviruses may yet find relevance in relation to neurological disease in monkeys and humans.

4. SIMIAN IMMUNODEFICIENCY VIRUS

Simian immunodeficiency virus (SIV) represents a group of African monkey lentiviruses that are the closest known animal relatives to HIV (for review, see Gardner *et al.*[68]). In their natural hosts, which include sooty mangabeys, African green monkeys, mandrills, talapoins, DeBrazza's monkeys, and probably other *Cercopithicus* species,[63,69] the virus apparently exists as an exogenous nonpathogenic infection. The origins of SIV and HIV remain uncertain because closely related DNA sequences have not as yet been identified in primate or nonprimate species.[56] Remarkably, an AIDS-like disease results when the SIV is introduced, inadvertently or purposefully, from an African monkey into an

Asian macaque species, e.g., rhesus monkeys, as appears to have occasionally occurred in several U.S. primate centers.[57–59,64] Macaques are very susceptible to infection with SIV, which uses the same CD4 receptor as HIV.[61] Depending on their ability to mount an immune response to the virus, the infected macaques experience either a relatively short clinical course of several months or a more prolonged chronic infection lasting 1–3 years.[60] The median period of survival appears to be about 9–12 months. Death inevitably occurs, associated with wasting, diarrhea, lymphoid depletion, loss of CD4 helper cells, opportunistic infections including activated cytomegalovirus (CMV) and adenovirus, and lymphomas. Table 6-2 summarizes the major features of SIV infection of macaques.

Of particular interest are the syncytial giant cells noted in the lungs, lymph nodes, spleen, brain, and other organs of SIV-infected macaques. In the CNS, perivascular infiltrates of foamy macrophages and multinucleated giant cells are present throughout the white and gray matter of the brain and spinal cord. These giant cells in brain and spleen contain SIV antigen and lentivirus particles.[58,65,66] Although formation of syncytial giant cells throughout the body is not a feature of HIV infection of humans, identical giant-cell-containing lesions do characterize the neuropathology of HIV encephalitis.[67] In addition to SIV-induced giant-cell encephalitis, neuritis associated with activated CMV is also observed in affected rhesus monkeys. Opportunistic infections such as toxoplasmosis and progressive multifocal leukodystrophy (simian virus 40) have also been seen in stumptailed macaques seropositive for SIV.

In summary, SIV infection of macaques represents a new experimental

TABLE 6-2
SIV Infection of Macaques: Major Features

Lentivirus: same genetic organization and 50–75% sequence homology with HIV-1 and HIV-2; nonpathogenic in natural host—certain species of African monkeys (*Cercopithecus*); exogenous; origin unknown

Persistent infection of all recipients: dose independent

Experimentally transmissible infection via genital mucosa

Long incubation period: months to years

100% mortality: 3 months to 3 years

Vigorous humoral immune response to core and envelope antigens: immune response affects survival period

Decrease in core antibody precedes clinical decline

Cytopathic *in vitro*: syncytia

Restricted cell tropism: T4 cells, macrophages

Restricted virus expression *in vitro*

Tissue distribution of virus similar to HIV

Pathology similar to HIV including neuropathology: syncytia

Opportunistic infections and B-cell lymphomas as in AIDS

Strain variation, mostly in envelope

model system with many similarities to AIDS, especially in respect to the neuro-pathology. The CNS lesions of SIV infection in macaques are virtually identical to HIV encephalitis. Although natural history features, cofactors, and the like remain to be better defined, this experimental system offers, at this time, the most parallels with HIV infection of the human CNS.[68] This represents a very valuable nonhuman primate model system for developing new therapeutic and preventive approaches to lentivirus-induced immunosuppression and neurological disease. Understanding how these viruses remain nonpathogenic in their natural host, African monkeys, while producing fatal immunosuppression in the experimental host, macaques, might provide the key to the riddle of HIV pathogenesis and AIDS.[55] Many new SIV isolates have been recently obtained from healthy African green monkeys from Africa,[69,70] and one of these isolates has been molecularly cloned and totally sequenced.[71] Sequence comparison of various SIV, HIV-1, and HIV-2 isolates and search for related endogenous sequences in various primate and nonprimate animals will also help determine the possible origin and evolution of HIV.

5. COMPARISON OF ANIMAL MODELS OF CNS RETROVIRAL INFECTION WITH CENTRAL NERVOUS SYSTEM INFECTIONS CAUSED BY HUMAN RETROVIRUSES

5.1. Human T-Lymphotropic Virus I

HTLV-I appears to be the etiologic agent of the endemic tropical myelo-neuropathies, which include several clinical syndromes with overlapping features: tropical spastic paraparesis (TSP), tropical ataxic neuropathy (TAN), and HTLV-I-associated myelopathy (HAM).[72] The neuropathology of these disorders is manifested mainly at the spinal cord level and features chronic inflammation with perivascular cuffing, demyelination with reactive gliosis, and secondary spongiform changes in the white matter of the posterior columns and pyramidal tracts.[73] The pathogenesis of the HTLV-I myeloneuropathies remains to be determined.

The location of these spongiform changes in the myelin sheaths rather than neurons, the lack of virus in motor neurons, and the presence of intense inflammation indicate that the pathogenesis of HTLV-I-associated neurological disease is clearly different from the spongiform polioencephalomyelopathy of MuLV-infected wild mice. Possibly, the mechanism of the HTLV-1-associated myeloneuropathies may resemble that of visna or CAEV in sheep and goats in that it reflects, in part, an immunogenic inflammatory response to virus antigens in association with Ia (class II MHC) antigens on CSF lymphocytes or monocyte/macrophages in the CNS, which is consistent with the detection of HTLV-I antigen and nucleic acid in CSF as well as peripheral blood lymphocytes and anti-HTLV-I oligoclonal antibodies (IgG) in serum and CSF.[74] Direct infection of neural cells with HTLV-I has not, as yet, been demonstrated in the limited number of autopsies done on affected individuals. Further study may show, however, that HTLV-I does indeed infect neural tissues.

5.2. Human Immunodeficiency Virus

Central nervous system diseases specifically related to HIV infection and dementia include aseptic meningitis, subacute encephalitis, and vacuolar myelopathy.[75] Grossly, the brain shows mild to moderate atrophy with diffuse myelin pallor. The characteristic histopathological features of HIV encephalitis include diffuse gliosis with focal necrosis of gray and white matter, perivascular mononuclear cells, formation of microglial nodules and multinucleated giant cells, and demyelination of white matter. T4 receptors may be expressed on neural and glial cells,[76,77] and glioma cell lines are susceptible to HIV infection *in vitro*.[78,79]

The human immunodeficiency virus has been demonstrated in the CNS by electron microscopy, immunohistochemistry, and *in situ* hybridization in monocytes, macrophages, and giant cells, and the intrathecal production of HIV-specific immunoglobulin has been detected.[80,81] Similar evidence also suggests the presence of HIV in capillary endothelial cells in the CNS and, less commonly, in glial and neural cells. However, productive HIV infection of neural and glial cells does not appear to be a major feature in the pathogenesis of AIDS dementia. Apparently, the vacuolar myelopathy, which is most prominent in the white matter of the posterior and lateral columns of the thoracic spinal cord, also does not involve neurons. The monocyte/macrophage is, thus, the predominant cell type in the brain infected by HIV. In this respect, HIV neuropathology is similar to other lentiviruses (visna virus, CAEV, SIV) and distinctly different from the gray matter spongiosis and neuronal infection seen in MuLV-infected wild mice and different from the prominent chronic inflammatory response and demyelination seen in HTLV-I-associated myeloneuropathies. Replication of virus in brain endothelial cells is, however, a prominent feature of the MuLV wild mouse neurological disease.

The pathogenesis of HIV CNS damage is not well understood, but it probably reflects primarily the indirect effects of HIV infection of blood monocytes, macrophages, and endothelial cells in the CNS rather than neuronal or glial infection (for summary, see Ho *et al.*[82] and Price *et al.*[83]). After activation or terminal differentiation, infected blood monocytes and macrophages in the CNS may release monokines or proteolytic enzymes that are toxic to neural cells and induce further inflammation or increase capillary permeability. The HIV *env* gp120 may block the neuronal binding of neurotropic factors such as neuroleukins[84] or neuropeptides such as the vasoactive intestinal peptide[77] and thereby lead to neurological dysfunction. Genetic variation of HIV *in vivo* may result in an increased tropism for monocytes/macrophages and glial cells and an increased neurovirulence.[85,86] The HIV-associated lymphoid depletion probably accounts for the less intense chronic inflammatory response in the CNS than seen in the brain of lentivirus-infected sheep or goats.

6. SPUMAVIRUSES

The spumaviruses (foamy viruses) have been isolated from healthy and diseased animals including chickens, cattle, sheep, cats, monkeys, apes, and humans.[87] Of particular note is their regular presence in brains of chimpanzees and humans.[88–90] Although they are not linked to any disease, there is evidence of associated immunosuppression in experimental animals[87] and possibly malignant transformation of human cells *in vitro*.[91] Sequence analysis of a human spumavirus shows an overall genomic organization more similar to lentivirus than oncovirus.[92] In cell culture, spumaviruses cause syncytia and vacuolar degeneration in a wide range of cell types. It is important to be aware of these agents so as not to confuse them with other more pathogenic retroviruses. Indeed, simian foamy virus has been isolated together with HIV-1 from lymphocytes of HIV-1-infected chimpanzees, and the foamy virus has identical reverse transcriptase activity and causes similar cytopathic effects in H9 cells.[84] Further investigation may yet link the spumaviruses to disease in animals and man.

7. FELINE IMMUNODEFICIENCY VIRUS

A lentivirus belonging to the same subfamily as HIV and SIV was isolated in 1987 from a group of domestic house cats suffering from an AIDS-like syndrome.[93] The virus was initially called feline T-lymphotropic virus (FTLV), but, in keeping with the new international nomenclature, it is now designated feline immunodeficiency virus (FIV). The FIV closely resembles HIV and SIV in its *in vitro* T-lymphotropism, Mg^{2+}-dependent reverse transcriptase activity, protein composition, and ultrastructural morphology. However, it is not antigenically related to HIV-1, HIV-2, of SIV. It is also unrelated to feline leukemia virus. In nature, the virus is transmitted mainly by contact during fighting. Infection and disease are, therefore, most common in stray male cats. Experimental infection is readily transmitted by parenteral inoculation of blood, plasma, or infectious tissue culture fluids.

Pathological changes, mostly caused by opportunistic infections, in naturally infected and experimentally infected cats are primarily noted in the oral cavity, nasal passages, intestinal tract, and skin.[94] Clinical neurological signs consisting primarily of behavioral changes such as rage or dementia are observed in fewer than 20% of FIV infected cats, but CNS involvement does not appear to represent a major disease manifestation. Only minor histopathological lesions are noted in the CNS, although FIV can readily be isolated from the CSF. The CNS lesions are apparently not similar to those observed in association with MuLV or the other animal lentiviruses (visna virus, CAEV, SIV) or HIV. The most common finding is choroid plexus fibrosis, occasionally accompanied by mild lymphoplasmocytic inflammation. Although this is a common finding in old cats, many FIV-infected cats showing this feature were less than middle-aged. Another feature, seen in about 70% of FIV-infected cats, is eosinophilic or amphophilic circular to ovoid hyalin bodies 6–30 μm in diameter in the superficial layer of the cerebral cortex. The relationship of these lesions to FIV infection is undetermined. More observations are required to fully describe and evaluate the CNS lesions associated with FIV infection of cats.

TABLE 6-3
Retrovirus Infection of the Central Nervous System in Animals and Humans

Animal	Virus	CNS pathology: Main features	Mechanism of CNS injury
1. Wild mice	Ecotropic MuLV	Spongiosis, no inflammation, gliosis, loss of anterior horn neurons; aberrant type C virus particles in anterior horn neurons	Direct, slow viral, nonimmunogenic injury to anterior horn neurons in spinal cord
2. Macaques	Type D SAIDS retrovirus (SRV-1)	Latent CNS infection with no pathology; viral DNA and RNA present in unidentified neural cells, but no antigen expression	Innocuous CNS infection; virus released into CSF by infected choroid plexus epithelial cells
3. Sheep, goats	Visna virus, CAEV	Chronic inflammation, demyelination	Inflammatory immune response to virus in blood macrophages in the CNS
4. Macaques	SIV	Perivascular monocytes, macrophages, and multinucleated giant cells containing SIV; opportunistic infections, e.g., cytomegalovirus, toxoplasmosis	Direct and indirect injury from virus infection in blood monocytes and macrophages in the CNS; virus infection of neurons undetermined; identical lesions as seen in HIV encephalitis
5. Domestic cats	FIV	Choroid plexus fibrosis; hyaline bodies in cerebral cortex	Undetermined
6. Chickens, cattle, sheep, cats, monkeys, apes, humans	Spumavirus (foamy virus)	Latent CNS infection with no pathology	Innocuous CNS infection; virus causes in vitro cytopathology and has a wide cell and tissue tropism
7. Humans	HTLV-I	Chronic inflammation, gliosis, demyelination; virus antigen in CSF lymphocytes, and virus antibody in CSF	Inflammatory immune response, possibly caused by virus in lymphocytes in CNS; viral infection of neurons undetermined
8. Humans	HIV	Perivascular monocytes, macrophages, and multinucleated giant cells containing HIV; diffuse gliosis, demyelination, vacuolar myelopathy of thoracic spinal cord	Direct and indirect injury from virus infection in monocytes, macrophages, glial cells, and possibly endothelial cells; virus infection of neurons undetermined

8. SUMMARY

This chapter has briefly reviewed six animal models of retrovirus CNS infection and compared them to infection of the human CNS by HTLV-I and HIV. The main neuropathological features and mechanisms of injury associated with retroviral CNS infection in these animal models and in humans are summarized in Table 6-3. Damage to the CNS from retrovirus infection apparently can result from several mechanisms, including (1) direct noninflammatory viral damage to motor neurons, as seen in MuLV-infected feral mice, (2) chronic inflammation with destruction of parenchymal tissue in response to virus infection in the CNS, as seen in HTLV-I-associated myeloneuropathies, (3) a combination of direct viral injury and, perhaps more importantly, indirect effects mediated by virus infection of blood-derived mononuclear (macrophage) cells and possibly capillary endothelial, glial, and neural cells, as seen in HIV dementia and the lentivirus infections of sheep, goats, and monkeys. Type D retrovirus and spumavirus latent infections of the CNS have yet to be associated with neurological disease in animals or humans. It is likely that yet more retroviruses will be found in association with neurological disease of animals and man. The existing animal models of retroviral CNS infection, especially the SIV macaque system, will serve a very useful function in understanding pathogenic mechanisms and developing better therapeutic and preventive measures that are applicable to human AIDS.

REFERENCES

1. Weiss, R., Teich, N., Varmus, H., and Coffin, J., 1985, *RNA Tumor Viruses*, Cold Spring Harbor Laboratory, Cold Spring Harbor, NY, p. 1396.
2. Gardner, M. B., 1985, retroviral spongiform polioencephalopathy, *Rev. Infect. Dis.* **7**:99–110.
3. Zachary, J. F., Knupp, C. J., and Wong, P. K. Y., 1986, Noninflammatory spongiform polioencephalomyelopathy caused by a neurotropic temperature-sensitive mutant of Moloney murine leukemia virus TB, *Am. J. Pathol.* **124**:457–468.
4. Kai, K., and Furuta, T., 1984, Isolation of paralysis-inducing murine leukemia viruses from Friend passaged in rats, *J. Virol.* **50**:970–973.
5. Nathanson, N., Georgsson, G., Palsson, P. A., Najjar, J. A., Lutley, R., and Petursson, G., 1985, Experimental visna in Icelandic sheep: The prototype lentiviral infection, *Rev. Infect. Dis.* **7**:75–82.
6. Narayan, O., and Cork, L. C., 1985, Lentiviral diseases of sheep and goats: Chronic pneumonia leukoencephalomyelitis and arthritis, *Rev. Infect. Dis.* **7**:89–98.
7. Haase, A. T., 1986, Pathogenesis of lentivirus infections, *Nature* **322**:130–136.
8. Gardner, M. B., Henderson, B. E., Officer, J. E., Rongey, R. W., Parker, J. C., Oliver, C., Estes, J. D., and Huebner, R. J., 1973, A spontaneous lower motor neuron disease apparently caused by indigenous type-C RNA virus in wild mice, *J. Nat. Cancer Inst.* **51**:1243–1254.
9. Gardner, M. B., Henderson, B. E., Estes, J. D., Rongey, R. W., Casagrande, J., Pike, M., and Huebner, R. J., 1976, The epidemiology and virology of C-type virus-associated hematological cancers and related diseases in wild mice, *Cancer Res.* **36**:574–581.
10. Gardner, M. B., Chiri, A., Dougherty, M. F., Casagrande, J., and Estes, J. D., 1979, Congenital transmission of murine leukemia virus from wild mice prone to the development of lymphoma and paralysis, *J. Nat. Cancer. Inst.* **62**:63–70.

11. Gardner, M. B., Estes, J. D., Casagrande, J., and Rasheed, S., 1980, Prevention of paralysis and suppression of lymphoma in wild mice by passive immunization to congenitally transmitted murine leukemia virus, *J. Nat. Cancer Inst.* **64**:359–364.
12. Andrews, J. M., and Gardner, M. B., 1974, Lower motor neuron degeneration associated with type C RNA virus infection in mice: Neuropathological features, *J. Neuropathol. Exp. Neurol.* **33**:285–307.
13. Brooks, B. R., Swarz, J. R., and Johnson, R. T., 1980, Spongiform polioencephalomyelopathy caused by a murine retrovirus. 1. Pathogenesis of infection in newborn mice, *Lab. Invest.* **43**:480–486.
14. Oldstone, M. B. A., Jensen, F., Dixon, F. J., and Lampert, P. W., 1980, Pathogenesis of the slow disease of the central nervous system associated with wild mouse virus. II. Role of virus and host gene products, *Virology* **107**:180–193.
15. Rasheed, S., Gardner, M. B., and Chan, E., 1976, Amphotropic host range of naturally occurring wild mouse leukemia viruses, *J. Virol.* **19**:13–18.
16. Gardner, M. B., 1978, Type C viruses of wild mice: Characterization and natural history of amphotropic, ecotropic, and xenotropic MuLV, *Curr. Top. Microbiol. Immunol.* **79**:215–259.
17. Gardner, M. B., and Rasheed, S., 1982, Retroviruses in feral mice, *Int. Rev. Exp. Pathol.* **23**:209–267.
18. Jolicoeur, P., Nicolaiew, N., DesGroseillers, L., and Rassart, E., 1982, Molecular cloning of infectious viral DNA from ecotropic neurotropic wild mouse retrovirus, *J. Virol.* **45**:1159–1163.
19. DesGroseillers, L., and Jolicoeur, P., 1984, Mapping the viral sequences conferring leukemogenicity and disease specificity in Moloney and amphotropic murine leukemia viruses, *J. Virol.* **52**:448–456.
20. Bryant, M. L., Scott, J. L., Pak, B. K., Estes, J. D., and Gardner, M. B., 1981, Immunopathology of natural and experimental lymphomas induced by wild mouse leukemia virus, *Am. J. Pathol.* **104**:272–282.
21. Hoffman, P. M., Ruscetti, S. K., and Morse, H. C. III, 1981, Pathogenesis of paralysis and lymphoma associated with a wild mouse retrovirus infection. Part 1. Age- and dose-related effects in susceptible laboratory mice, *J. Neuroimmunol.* **1**:275–285.
22. Rasheed, S., Gardner, M. B., and Lai, M. M. C., 1983, Isolation and characterization of new ecotropic murine leukemia viruses after passage of an amphotropic virus in NIH Swiss mice, *Virology* **130**:439–451.
23. Oldstone, M. B. A., Jensen, F., Elder, J., Dixon, F. J., and Lampert, P. W., 1983, Pathogenesis of the slow disease of the central nervous system associated with wild mouse virus. III. Role of input virus and MCF recombinants in disease, *Virology* **128**:154–165.
24. Rasheed, S., Pal, B. K., and Gardner, M. B., 1982, Characterization of a highly oncogenic murine leukemia virus from wild mice, *Int. J. Cancer* **29**:345–350.
25. Hoffman, P. M., Davidson, W. F., Ruscetti, S. K., Chused, T. M., and Morse, H. C. III, 1981, Wild mouse ecotropic murine leukemia virus infection of inbred mice: Dual-tropic virus expression precedes the onset of paralysis and lymphoma, *J. Virol.* **39**:597–602.
26. Langdon, W. Y., Hoffman, P. M., Silver, J. E., Buckler, C. E., Hartley, J. W., Ruscetti, S. K., and Morse, H. C. III, 1983, Identification of a spleen focus-forming virus in erythroleukemic mice infected with a wild-mouse ecotropic murine leukemia virus, *J. Virol.* **46**:230–238.
27. DesGroseillers, L., Barrette, M., and Jolicoeur, P., 1984, Physical mapping of paralysis-inducing determinant of a wild mouse ecotropic neurotropic retrovirus, *J. Virol.* **52**:356–366.
28. Rassart, E., Nelbach, L., and Jolicoeur, P., 1986, Cas-Br-E murine leukemia virus: Sequencing of the paralytogenic region of its genome and derivation of specific probes to study its origin and the structure of its recombinant genomes in leukemic tissues, *J. Virol.* **60**:910–919.
29. Barbacid, M., Robbins, K. C., and Aaronson, S. A., 1979, Wild mouse RNA tumor viruses: A nongenetically transmitted virus group closely related to exogenous leukemia viruses of laboratory mouse strains, *J. Exp. Med.* **149**:154–166.

30. O'Neill, R. R., Hartley, J. W., Repaske, R., and Kozak, C. A., 1987, Amphotropic proviral envelope sequences are absent from the *Mus* germline, *J. Virol.* **61**:2225–2231.

31. Gardner, M. B., 1987, Naturally occurring leukaemia viruses in wild mice: How good a model for humans? *Cancer Surv.* **6**:55–71.

32. Oldstone, M. B. A., Lampert, P. W., Lee, S., and Dixon, F. J., 1977, Pathogenesis of the slow disease of the central nervous system associated with WM 1504 E virus. 1. Relationship of strain susceptibility and replication to disease, *Am. J. Pathol.* **88**:193–212.

33. Gardner, M. B., Klement, V., Henderson, B. E., Meier, H., Estes, J. D., and Huebner, R. J., 1976, Genetic control of type C virus of wild mice, *Nature* **259**:143–145.

34. Gardner, M. B., Rasheed, S., Pal, B. K., Estes, J. D., and O'Brien, S. J., 1980, *Akvr-1*, a dominant murine leukemia virus restriction gene, is polymorphic in leukemia-prone wild mice, *Proc. Natl. Acad. Sci. U.S.A.* **77**:531–535.

35. Rasheed, S., and Gardner, M. B., 1983, Resistance of fibroblasts and hematopoietic cells to ecotropic murine leukemia virus infection: An *Akvr-1^R* gene effect, *Int. J. Cancer* **31**:491–496.

36. O'Brien, S. J., Berman, E. J., Estes, J. D., and Gardner, M. B., 1983, Murine retroviral restriction genes *Fv-4* and *Akvr-1* are alleles of a single locus, *J. Virol.* **47**:649–651.

37. Odaka, T., Ikeda, H., Hoshikura, H., Moriwaki, K., and Suzuki, S., 1981, *FV-4*: Gene controlling resistance to NB-tropic Friend murine leukemia virus. Distribution in wild mice, introduction into genetic background of BALB/c mice, and mapping of chromosomes, *J. Nat. Cancer Inst.* **67**:1123–1127.

38. Kozak, C. A., Gromet, N. J., Ikeda, H., and Buckler, C. E., 1984, A unique sequence related to the ecotropic murine leukemia virus is associated with the *Fv-4* resistance gene, *Proc. Natl. Acad. Sci. U.S.A.* **81**:834–837.

39. Dandekar, S., Rossitto, P., Pickett, S., Mockli, L., Bradshaw, H., Cardiff, R., and Gardner, M., 1987, Molecular characterization of the *Akvr-1* restriction gene: A defective endogenous retrovirus-borne gene identical to *Fv-4r*, *J. Virol.* **61**:308–314.

40. Ikeda, H., and Odaka, T., 1984, A cell membrane "gp70" associated with *FV-4* gene: Immunological characterization, and tissue and strain distribution, *Virology* **133**:65–76.

41. Payne, L. N., Poni, P. K., and Weiss, R. A., 1971, A dominant epistatic gene which inhibits cellular susceptibility to RSV (RAV-0), *J. Gen. Virol.* **13**:455–462.

42. Bassin, R. H., Ruscetti, S., Iqbal, A., Haapala, D. K., and Rein, A., 1981, Normal DBA mouse cells synthesize a glycoprotein which interferes with MCF virus infection, *Virology* **23**:139–151.

43. Gardner, M. B., and Marx, P., 1985, Simian Acquired Immunodeficiency Syndrome. *Adv. Viral Oncol.* **2**:57–81.

44. Lackner, A. A., Rodriguez, M. H., Bush, C. E., Munn, R. J., Kwang, H. S., Moore, P. F., Osborn, K. G., Marx, P. A., Gardner, M. B., and Lowenstine, L. J., 1988, Distribution of a macaque immunosuppressive type D retrovirus in neural, lymphoid, and salivary tissues, *J. Virol.* **62**:2134–2142.

45. Lackner, A. A., Marx, P. A., Lerche, N. W., Gardner, M. B., Kluge, J. D., Spinner, A., Kwang, H. S., and Lowenstine, L. J., 1989, Asymptomatic infection of the central nervous system by the macaque immunosuppressive Type D retrovirus, SRV-1, *J. Gen. Virol.* **70**:1641–1651.

46. Lerche, N. W., Osborn, K. G., Marx, P. A., Prahalada, S., Maul, D. H., Henrickson, R. V., Arthur, L. O., Gilden, R. V., Munn, R. J., Bryant, M. L., Heidecker-Fanning, G., and Gardner, M. B., 1986, Inapparent carriers of simian AIDS type D retrovirus and disease transmission with saliva, *J. Natl. Cancer Inst.* **77**:489–496.

47. Maul, D. H., Lerche, N. W., Osborn, K. G., Marx, P. A., Zaiss, C., Spinner, A., Kluge, J. D., MacKenzie, M. R., Lowenstine, L. J., Bryant, M. L., Blakeslee, J. R., Henrickson, R. V., and Gardner, M. B., 1986, Pathogenesis of simian aids in rhesus macaques inoculated with type D retroviruses, *Am. J. Vet. Res.* **47**:863–868.

48. Heidecker, G., Lerche, N. W., Lowenstine, L. J., Lackner, A. A., Osborn, K. G., Gardner, M. B., and Marx, P. A., 1987, Induction of simian acquired immune deficiency syndrome (SAIDS) with a molecular clone of a type D SAIDS retrovirus, *J. Virol.* **61**:3066–3071.

49. Bryant, M. L., Marx, P. A., Shiigi, S. N., Wilson, B. J., McNulty, W. P., and Gardner, M. B.,

1986, Distribution of type D retrovirus sequences in tissues of macaques with simian acquired immune deficiency and retroperitoneal fibromatosis, *Virology* 150:149–160.

50. Kwang, H. S., Pedersen, N. C., Lerche, N. W., Osborn, K. G., Marx, P. A., and Gardner, M. B., 1987, Viremia, antigenemia and serum antibodies in rhesus macaques infected with simian retrovirus, type I and their relationship to disease course, *Lab. Invest.* 56:591–597.

51. Marx, P. S., Pedersen, N. C., Lerche, N. W., Osborn, K. G., Lowenstine, L. J., Lackner, A. A., Maul, D. H., Kwang, H.-S., Kluge, J. D., Zaiss, C. P., Sharpe, V., Spinner, A. P., Allison, A. C., and Gardner, M. B., 1986, Prevention of simian acquired immune deficiency syndrome with a formalin-inactivated type D retrovirus vaccine, *J. Virol.* 60:431–435.

52. Lackner, A. A., Moore, P. F., Marx, P. A., Munn, R. J., Gardner, M. B., and Lowenstine, L. J., 1990, Immunohistochemical localization of Type D retrovirus serotype 1 in the digestive tract of rhesus monkeys with simian AIDS, *J. Med. Primatol.* 19:339–349.

53. Maul, D. H., Zaiss, C. P., MacKenzie, M. R., Shiigi, S. M., Marx, P. A., and Gardner, M. B., 1988, Simian retrovirus D serogroup has a broad cellular tropism for lymphoid and nonlymphoid cells, *J. Virol.* 62:1768–1773.

54. Sommerfelt, M. A., Williams B. P., McKnight, A., Goodfellow, P. N., and Weiss, R. A., 1990, Localization of the receptor gene for type D simian retroviruses on human chromosome 19, *J. Virol.* 64:6214–6220.

55. Gardner, M. B., and Luciw, P. A., 1989, Animal models of AIDS, *FASEB* 3:2593–2606.

56. Gardner, M. B., and Luciw, P. A., 1988, Simian immunodeficiency viruses and their relationship to the human immunodeficiency viruses, *AIDS '88* 2 (Suppl. 1):S3–S10.

57. Daniel, M. D., Letvin, N. L., King, N. W., Kannagi, M., Sehgal, P. K., Hunt, D., Kanki, P. J., Essex, M., and Desrosiers, R. C., 1985, Isolation of T-cell tropic HTLV-III-like retrovirus from macaques, *Science* 228:1201–1204.

58. Letvin, N. L., Daniel, M. D., Sehgal, P. K., Desrosiers, R. C., Hunt, R. D., Waldron, L. M., MacKey, J. J., Schmidt, D. K., Chalifoux, L. V., and King, N. W., 1985, Induction of AIDS-like disease in macaque monkeys with T-cell tropic retrovirus STLV-III, *Science* 230:71–73.

59. Murphey-Corb, M., Martin, L. N., Rangan, S. R. S., Baskin, G. B., Gormus, B. J., Wolf, R. H., Andes, W. A., West, M., and Montelaro, R. C., 1986, Isolation of an HTLV-III-related retrovirus from macaques with simian AIDS and possible origin in asymptomatic mangabeys, *Nature* 321:435–437.

60. Kannagi, M., Kiyotaki, M., Desrosiers, R. C., Reimann, K. A., King, N. W., Waldron, L. M., and Letvin, N. L., 1986, Humoral immune responses to T cell tropic retrovirus simian T lymphotropic virus type III in monkeys with experimentally induced acquired immune deficiency-like syndrome, *J. Clin. Invest.* 78:1229–1236.

61. Kannagi, M., Yetz, J. M., and Letvin, N. L., 1985, *In vitro* growth characteristics of simian T-lymphotropic virus type III, *Proc. Natl. Acad. Sci. U.S.A.* 82:7053–7057.

62. Fultz, P. N., McClure, H. M., Anderson, D. C., Swenson, R. B., Anand, R., and Srinivasan, A., 1986, Isolation of a T-lymphotropic retrovirus from naturally infected sooty mangabey monkeys (*Cercocebus atys*), *Proc. Natl. Acad. Sci. U.S.A.* 83:5286–5290.

63. Lowenstine, L. J., Pedersen, N. C., Higgins, J., Pallis, K. C., Uyeda, A., Marx, P., Lerche, N. W., Munn, R. J., and Gardner, M. B., 1986, Seroepidemiologic survey of captive old world nonhuman primates from North American zoos and vivaria for antibodies to human and simian retroviruses and isolation of a lentivirus from sooty mangabeys (*Cercocebus atys*), *Int. J. Cancer* 38:563–574.

64. Benveniste, R. E., Arthur, L. O., Tsai, C.-C., Sowder, R., Copeland, T. D., Henderson, L. E., and Oroszlan, S., 1986, Isolation of a lentivirus from a macaque with lymphoma: Comparison with HTLV-III/LAV and other lentiviruses, *J. Virol.* 60:483–490.

65. Benveniste, R. E., Morton, W. R., Clark, E. A., Tsai, C.-C., Ochs, H. D., Ward, J. M., Kuller, L., Knott, W. B., Hill, R. W., Gale, M. J., and Thouless, M. E., 1988, Inoculation of baboons and macaques with simian immunodeficiency virus/Mne, a primate lentivirus closely related to human immunodeficiency virus type 2, *J. Virol.* 62:2091–2101.

66. Ward, J. M., O'Leary, T. J., Baskin, G. B., Benveniste, R., Harris, C. A., Nara, P. L., and

Rhodes, R. H., 1987, Immunohistochemical localization of human and simian immunodeficiency viral antigens in fixed tissue sections, *Am. J. Pathol.* **127**:199–205.

67. Sharer, L. R., Epstein, L. G., Cho, E. S., Joshi, V. V., Meyenhofer, M. M., Rankin, L. F., and Petito, C. K., 1986, Pathologic features of AIDS encephalopathy in children: Evidence for LAV/HTLV-III infection of brain, *Hum. Pathol.* **17**:271–284.

68. Gardner, M. B., Luciw, P., Lerche, N., and Marx, P. E., 1988, Non-human primate retrovirus isolates and AIDS, in: *Immunodeficiency Disorders and Retroviruses* (K. Perk, ed.), Academic Press, New York, pp. 171–226.

69. Ohta, Y., Masuda, T., Tsujimoto, H., Ishikawa, K., Kodama, T., Morikawa, S., Naki, M., Honjo, S., and Hayami, M., 1988, Isolation of simian immunodeficiency virus from African green monkeys and seroepidemiologic survey of the virus in various non-human primates, *Int. J. Cancer* **41**:15–122.

70. Daniel M. D., Li, Y., Naidu, Y. M., Durda, P. J., Schmidt, D. K., Finger, C. D., Silva, D. P., MacKey, J. J., Kestler, H. W. III, Sehgal, P. K., King, N. W., Hayami, M., and Desrosiers, R. C., 1988, Simian immunodeficiency virus from African green monkeys, *J. Virol.* **62**:4123–4128.

71. Fukasawa, M., Miura, T., Hasegawa, A., Morikawa, S., Tsujimoto, H., Miki, K., Kitamura, T., and Hayami, M., 1988, Sequence of simian immunodeficiency virus from African green monkey, a new member of the HIV/SIV group, *Nature* **333**:457–461.

72. Romain, G. C., 1988, The neuroepidemiology of tropical spastic paraparesis, *Ann. Neurol.* **23**:S120–S133.

73. Piccardo, P., Ceroni, M., Rodgers-Johnson, P., Mora, C., Asher, D. M., Char, G., Gibbs, C. J., and Gajdusek, D. C., 1988, Pathological and immunological observations on tropical spastic paraparesis in patients from Jamaica, *Ann. Neurol.* **23**:S156–160.

74. Koprowski, H., and DeFreitas, E., 1988, HTLV-1 and chronic nervous diseases: Present status and a look into the future, *Ann. Neurol.* **23**:S166–S170.

75. Petito, C. K., 1988, Review of central nervous system pathology in human immunodeficiency virus infection, *Ann. Neurol.* **23**:S54–S57.

76. Maddon, P. J., Dalgleish, A. G., McDougal, J. S., Clapham, P. R., Weiss, R. A., and Axel, R., 1986, The T4 gene encodes the AIDS virus receptor and is expressed in the immune system and the brain, *Cell* **47**:333–398.

77. Pert, C. B., Hill, J. G., Ruff, M. R., Berman, R. M., Robey, W. G., Arthur, L. O., Ruscetti, F. W., and Farr, W. L., 1986, Octapeptides deduced from the neuropeptide receptor-like pattern of antigen T4 in brain potently inhibit human immunodeficiency virus receptor binding and T cell infectivity, *Proc. Natl. Acad. Sci. U.S.A.* **83**:9254–9258.

78. Koyanagi, Y., Miles, S., Mitsuyasu, R. T., Merrill, J. E., Vinters, H. V., and Chen, I. S. Y., 1987, Dual infection of the CNS by AIDS viruses with distinct cellular tropisms, *Science* **236**:819–822.

79. Cheng-Mayer, C., Rutka, J. T., Rosenblum, N. L., McHugh, T., Stites, D. P., and Levy, J. A., 1987, Human immunodeficiency virus can productively infect cultured human glial cells, *Proc. Natl. Acad. Sci. U.S.A.* **84**:3526–3530.

80. Koenig, S., Gendelman, H. E., Orenstein, J. M., Dal Canto, M. C., Pezeshkpour, G. H., Yungbluth, M., Janotta, F., Aksamit, A., Martin, M. A., and Fauci, A. S., 1986, Detection of AIDS virus in macrophages in brain tissue from AIDS patients with encephalopathy, *Science* **233**:1089–1093.

81. Wiley, C. A., Schier, R. D., Nelson, J. A., Lampert, P. W., and Oldstone, M. B. A., 1986, Cellular localization of human immunodeficiency virus infection within the brains of acquired immune deficiency syndrome patients, *Proc. Natl. Acad. Sci. U.S.A.* **83**:7089–7093.

82. Ho, D. D., Pomerantz, R. J., and Kaplan, J. C., 1987, Pathogenesis of infection with human immunodeficiency virus, *New Engl. J. Med.* **317**:278–286.

83. Price, R. W., Brew, B., Sidtis, J., Rosenblum, M., Scheck, A. C., and Cleary, P., 1988, The brain in AIDS: Central nervous system HIV-1 infection and AIDS dementia complex, *Science* **239**:586–592.

84. Gurney, M. E., Heinrich, S. P., Lee, M. R., and Yin, H. S., 1986, Molecular cloning and expression of neuroleukin, a neurotrophic factor for spinal and sensory neurons, *Science* **234**:566–574.
85. Koyanagi, Y., Miles, S., Mitsuyosu, R. T., Merrill, J. E., Vinters, H. V., and Chen, I. S. Y., 1987, Dual infection of the central nervous system by AIDS viruses with distinct cellular tropisms, *Science* **236**:819–822.
86. Cheng-Mayer, C., Seto, D., Tateno, M., and Levy, J. A., 1988, Biologic features of HIV-1 that correlate with virulence in the host, *Science* **240**:80–82
87. Hooks, J. J., and Detrick-Hooks, B., 1981, Spumaviriniae: Foamy virus group infections: Comparative aspects and diagnoses, *Comp. Diag. Viral Dis.* **4**:599–618.
88. Nara, P. L., Robey, W. G., Arthur, L. O., Gonda, M. A., Asher, D. M., Yanagihara, R., Gibbs, C. J., Jr., Gajudsek, D. C., and Fischinger, P. J., 1987, Simultaneous isolation of simian foamy virus and HTLV-III/LAV from chimpanzee lymphocytes following HTLV-III or LAV inoculation, *Arch. Virol.* **92**:183–186.
89. Rogers, N. G., Basnight, M., Gibbs, C. J., Jr., and Gajdusek, D. C., 1967, Latent viruses in chimpanzees with experimental kuru, *Nature* **216**:446–449.
90. Hooks, J., Gibbs, C. J., Jr., Chopra, H., Lewis, M., and Gajdusek, D. C., 1972, Spontaneous transformation of human brain cells grown *in vitro* and description of associated virus particles, *Science* **176**:1420–1422.
91. Rhodes-Feuillette, A., Mahony, G., Lasneret, J., Flandren, G., and Peries, J., 1987, Characterization of human lymphoblastoid cell line permanently modified by simian foamy virus type 10, *J. Med. Primatol.* **16**:277–289.
92. Maurer, B., Bonnert, H., Dorai, G., and Flugel, R. M., 1988, Analysis of the primary structure of the long terminal repeat and the *gag* and *pol* genes of the human spumaretrovirus, *J. Virol.* **62**:1590–1597.
93. Pedersen, N. C., Ho, E., Brown, M. J., and Yamamoto, K., 1987, Isolation of a T-lymphotropic virus from domestic cats with an immunodeficiency-like syndrome, *Science* **235**: 790–793.
94. Yamamoto, J. K., Sparger, E., Ho, E. W., Andersen, P. R., O'Connor, T. P., Mandell, C. P., Lowenstine, L., Munn, R., and Pedersen, N. C., 1988, Pathogenesis of experimentally induced feline immunodeficiency virus infection in cats, *Am. J. Vet. Res.* **49**:1246–1258.

7

Scrapie

Unconventional Infectious Agent

RICHARD I. CARP

1. SCRAPIE BIOLOGY

In 1954, Bjorn Sigurdsson, an Icelandic virologist, summed up a series of experiments that he had been conducting on a number of diseases of sheep, including visna and scrapie.[1] In comparing his results with those obtained in the burgeoning field of virology, Sigurdsson proposed a new category of infectious diseases, slow infections. He proposed three criteria for slow infections: (1) a very long incubation period lasting from several months to many years; (2) a regular, progressive, and protracted course after the appearance of clinical signs that almost invariably ends in death; and (3) limitation of infection to a single host species and histopathological changes to a single organ or tissue system. Sigurdsson correctly predicted that the third criterion would not stand the test of time, and results on this are detailed later.

The characteristics of these diseases that differentiated them from other types of virus infections were also detailed: for acute versus slow infections, the distinguishing characteristic was simple, the length of the incubation period; for chronic versus slow infections Sigurdsson stressed the precision of the events leading to disease and death in the latter. We quote his comments on slow infections:

> They are chronic mainly in the sense that they are slow. On the other hand, these diseases follow a course which is just as regular as the course of the acute infections, only the time factor is different. In the first place the so-called incubation period,

RICHARD I. CARP • New York State Institute for Basic Research in Developmental Disabilities, Staten Island, New York 10314.

Neuropathogenic Viruses and Immunity, edited by Steven Specter *et al.* Plenum Press, New York, 1992.

although extremely long, apparently does not vary within very wide limits. The appearance and the progression of the clinical signs follow a set pattern.

Certainly, subsequent work on experimental scrapie has documented the precision of events during the course of the interaction between specific strains of scrapie and the specific host.[2-5] Dickinson and colleagues have referred to a "clockwork" predictability in the disease process. Part of the reason for the predictability is the fact that the host does not appear to mount an effective defensive response to the infection; this is discussed in detail below. Key differences between unconventional slow infections and those that would be categorized as acute, chronic, persistent, or latent are noted in Table 7-1.

Two sheep diseases studied by Sigurdsson became the archetypes for the two types of slow infections of the central nervous system (CNS). Visna (see Chapter 4) is the archetype for the conventional slow infections, whereas scrapie is the archetype of the unconventional diseases. The latter group includes three diseases that affect humans—kuru,[6] Creutzfeldt–Jakob disease (CJD),[7] and Gerstmann–Straussler syndrome[8]—and four that are natural diseases of animals—scrapie, transmissible mink encephalopathy,[9] chronic wasting disease of mule deer and elk,[10,11] and a newly discovered disease of cattle termed bovine spongiform encephatholopathy.[12] Among the unconventional group, most experimental work has been done with scrapie, and in those instances in which comparable experiments have been done with other unconventional diseases and agents, the results have been similar to those obtained with scrapie. This chapter focuses on the archetype of these diseases, scrapie, unless otherwise noted.

Scrapie is a natural disease of sheep and goats. The causative agent can also produce disease when introduced experimentally in a variety of small laboratory animals such as mice, hamsters, and rats. It has been demonstrated that under field conditions, scrapie is readily transmissible.[13,14] An example of this is shown in Table 7-2, reproduced from results obtained by Dickinson.[14] In this experiment 75 Scottish blackface sheep were obtained from several farms in the Edinburgh area in which scrapie had not been seen for a number of years. The total number of sheep in these flocks was >18,000. Several years after these 75 animals were placed in fields containing Suffolk sheep in which scrapie was endemic, 21 of the Scottish blackface sheep developed scrapie.

The mechanism(s) involved in transmission of scrapie are not known. A likely

TABLE 7-1
Distinguishing Characteristics of Unconventional Infections

Type of infection	Main characteristic that distinguishes this type of infection from unconventional slow infections
Acute	Short incubation period
Chronic	Neither clinical course nor incubation period is predictable
Persistent	Continuous production of virus in the absence of clinical manifestations
Latent	Incubation period in unpredictable; environmental factors play a role

TABLE 7-2
Incidence of Scrapie Disease in
Scottish Blackface Sheep
with and without Exposure
to Scrapie-Positive Suffolk Sheep

History	Incidence
Exposure	21/75
No exposure	0/>18,000

mechanism can be proposed from several facts that are known: (1) experimentally, scrapie can be transmitted by the oral route,[15–17] as can CJD[18]; (2) for scrapie, we know that the placenta contains high levels of infectivity[16]; and (3) sheep have a tendency to eat placenta.[16] These findings suggest that the oral route is a likely means of natural transmission. Maternal transmission also plays a role in the spread of scrapie. In reciprocal crosses, lambs obtained from scrapie-positive ewes were more likely to develop disease than were lambs from scrapie-positive rams.[14,19] Furthermore, lambs removed from scrapie-positive ewes at birth and reared in an area away from postnatal contact with scrapie agent developed disease.[19] A remarkable example of our lack of understanding of the mechanisms involved in the epidemiology of scrapie can be derived from quarantine efforts done in Iceland.[20] Fields that had contained infected sheep were depopulated and allowed to remain free of sheep and goats for 1–3 years. The fields were then repopulated with sheep from flocks that did not then or subsequently show scrapie disease. Some of the sheep placed in those "contaminated" fields developed scrapie disease.[20] Do these data imply that there is a vector or an alternate host? Does the remarkable resistance of scrapie infectivity to inactivation (see Section 7) play a role? Could infectivity have persisted on contaminated fence posts, etc., throughout the time when fields were empty?

2. DISEASES ASSOCIATED WITH NERVOUS SYSTEM INFECTION

There is only a single disease associated with the scrapie agent, and that disease appears to be based on effects on the CNS. The clinical manifestations of the disease, with an exception noted later, always include incoordination and ataxia that progress to paresis, paralysis, and eventual death. There are recognizable differences in clinical symptoms that are dependent on the strain of agent and on the host affected. For example, there is a scratchy form of goat scrapie and a drowsy form.[19] The early finding of genetic differences in clinical manifestations in the natural disease preceded findings of genetic differences in experimental scrapie, which are discussed in Section 3.5. In sheep there is the tendency of animals to scratch against fence posts and other hard objects; this phenomenon prompted farmers to give the disease its name. In contrast, in experimental animals such as mice and hamsters, scratching is not seen in most scrapie strain–host combinations. In mice, we provide a test requiring good coordination, traversing a series of narrow parallel bars, as a sensitive indication of incoordination and thus an early monitor of clinical scrapie disease.[21] Hamsters injected with the most commonly used hamster-adapted agent (263K) develop characteristic head bobbing and erratic movements as early signs.

Recent data from our laboratory combined with a variety of observations noted in the literature suggest that there may be an entire panoply of nonmotor signs that have thus far received little attention. In certain scrapie strain–mouse strain combinations obesity develops during the preclinical phase of disease.[21] The increased weight is caused by an accumulation of fat and not by an overall enlargement of the animal[21] or of visceral organs. The induction of obesity is augmented by direct injection of the agent in the hypothalamus[22] and can be countered by removal of the adrenal glands.[23,24] There is adrenal cortical hypertrophy. Thus, it appears that scrapie-induced obesity is a function of an effect on the hypothalamus–pituitary–adrenal axis. The obese animals showed an inability to process a glucose overload effectively. The level of glucose after an overload was consistently high in obese mice and in some instances was sufficiently high (greater than 5 S.D. above normal) to qualify as diabetic.[24]

Obesity is also seen in some scrapie strain–hamster combinations. However, there are marked differences from the situation in mice.[24a] With regard to glucose metabolism, hamsters show hypoglycemia and marked hyperinsulinemia. In mice there are no changes in the pancreas, whereas hamsters show an increase in the number of islets of Langerhans, with marked hypertrophy and hyperplasia of islet cells. In mice the late period of clinical disease results in a loss of weight, whereas in hamsters animals continue to gain weight. Finally, hamsters injected with those strains that induce obesity (139H, 22CH, ME7H) show very little incoordination but rather become listless and slow-moving. It appears that clinical manifestations of the disease induced by some scrapie strains in hamsters are different from those seen in mice and different from those seen in other scrapie strain–hamster combinations. In those combinations that show obesity and very little incoordination, an endocrine or neuroendocrine process rather than a motor process seems to be the primary manifestation of disease.

There are additional changes seen in scrapie-injected animals that do not appear to be related to motor function (Table 7-3). Behavioral changes have been noted early in the incubation period; these include changes in eating and drinking,[25] in emergence times,[26] in defecation scores,[27] and in open-field and Y-maze tests.[28] In a little-noted study, luteinizing changes in ovaries and alteration of coat color were seen.[29] The ovarian changes occurred late in the disease and were accompanied by very low scrapie infectivity titers in the ovaries. The author postulated that the ovarian changes were caused by an effect on a neuroendocrine system.

The wide spectrum of nonmotor changes seen in different scrapie strain–host combinations is not surprising. The concept that there is specific targeting to cells (almost certainly neurons) with differing capacities and/or different loca-

TABLE 7-3
Changes in Scrapie-Injected Animals that Are Not Based on Motor Function

Change	Characterization	Reference
Obesity	Clinical	21
Hyperglycemia	Clinical	24
Hypoglycemia–hyperinsulinemia	Clinical	24a
Emergence times	Clinical	26
Defecation scores	Clinical	27
Open field and Y-maze	Clinical	28
Alteration of coat color	Clinical	29
Ovarian dystrophy	Histopathological	29
Adrenal cortical hypertrophy	Histopathological	22
Increase in number and size of islets of Langerhans	Histopathological	24a
Hypertrophy and hyperplasia of the islets of Langerhans cells	Histopathological	24a

tions within the CNS comes from a number of studies. Using several scrapie strains and inbred mouse strains, Fraser and Dickinson and their colleagues established that the pattern of vacuolation in the brain was a function of both scrapie strain and mouse strain.[4,30] For example, whereas some combinations yielded extreme vacuolation in the cerebellum, others did not; some combinations showed extensive white matter vacuolation, but others showed none.[4] By the intraocular route,[31] stereotactic injection of the cerebellum, or stereotactic injection of the nigrostriatum,[32,33] it was possible to get regional targeting of vacuolation[31,32] and/or the induction of a specific clinical manifestation.[32,33]

Kimberlin and Walker[34,35] have proposed that there are clinical target areas, i.e., areas within the brain that when affected by scrapie lead to clinical disease. This theory was confirmed by stereotactic injection into different brain areas.[32] Kim et al.[32] were also able to show that the area of the brain infected that resulted in the shortest incubation period differed with different scrapie strains, further supporting the concept that there is scrapie-strain-specific targeting to different cells, most probably different types of neurons.[36] Finally, Beck et al.[37] noted that vacuolar changes in scrapie-affected sheep were seen primarily either in the cerebellar motor system or the hypothalamoneurohypophyseal system. Effects on the latter system correlated, in part, with clinical manifestations of obesity. The authors suggested that scrapie in sheep could yield motor or "metabolic and autonomic disturbances" or, in some instances, both. All of these findings point to some form of specific targeting for the scrapie agent within the brain and to the fact that this targeting can differ depending on the scrapie strain–host combination. In turn, the difference in targeting can lead to different clinical manifestations. This concept has implications with regard to the potential of unconventional slow infections of animals to serve as models for a variety of human diseases.

3. PATHOGENESIS

3.1. Pathogenesis and Route of Injection

The pathogenesis of experimental scrapie follows two distinct patterns depending on the route of injection: one pattern is seen with intracerebral (i.c.) injection, and a different pattern with all other (non-CNS) routes. In both patterns there is extensive replication at the site that leads to disease, the brain, as well as early replication in the organs of the lymphoreticular system (LRS), such as the spleen and lymph nodes.[38,39] For the CNS route, replication in the LRS is irrelevant as far as disease progression is concerned, whereas for non-CNS routes spleen and lymph nodes play a role in the time required for the initiation of clinical disease. Early experiments in mice established that the infectivity of scrapie after subcutaneous injection reached high titers in spleen and lymph nodes long before infectivity was detectable in the brain.[38] The role of the LRS organs in pathogenesis was explored by a series of splenectomy studies.[40–42] A sample of our data in this area is shown in Table 7-4. The basic finding was that splenectomy prior to peripheral (non-CNS) injection significantly lengthened the incubation period. In contrast, splenectomy prior to CNS injection had no effect on incubation period. Finally *dh/dh* mice, which are genetically devoid of a spleen, have longer incubation periods after injection by a peripheral route than do +/*dh* or +/+ mice, which contain spleens.[43]

3.2. The Cell Type Involved in Early Steps of Pathogenesis

By use of a number of immunomodulators it was shown that administration of drugs at the time of or just prior to peripheral injection alters the scrapie incubation period. With the typical perversity of scrapie, the results are opposite to those seen in most standard virus diseases in that drugs that primarily cause immunostimulation such as phytohemagglutinin,[44] methanol extraction residue of BCG,[45] and human lymphokines,[46] which usually ameliorate virus diseases, shorten scrapie incubation periods. In contrast, prednisone, an immunosuppressant, extends the incubation period.[47] The effect of these drugs is not seen after i.c. injection of scrapie. In another study using a mouse-adapted agent obtained

TABLE 7-4
Effect of Splenectomy on the Incubation Period
of Intraperitoneally Injected ME7 Scrapie Strain

Experiment no.	Treatment	Number of mice	Incubation period (days) (mean ± S.E.)	Difference from control, P value
1	Control	8	198 ± 10	
	Splenectomy	15	250 ± 10	<0.01
2	Control	12	271 ± 7	
	Splenectomy	21	334 ± 8	<0.001

from a CJD patient, high-infectivity titers of agent are found in low-density lymphoblastoid cell fractions from density-gradient-separated splenocytes.[48]

All of the above data suggest that the cell of the LRS that is important is one of the immunocompetent cell types. Compelling counterarguments can be raised using several lines of investigation. In a long series of irradiation studies, Fraser et al.[49,50] show that the key LRS cell in scrapie pathogenesis has to be relatively radiation resistant and long-lived, clearly not characteristics of T and B cells or their progenitors. Secondly, in a study of fractionated spleens from scrapie-infected mice, most of the infectivity is associated with the stroma fraction rather than the pulp fraction.[51] Depletion of T cells has no effect on the scrapie incubation period, clinical course, or pathology.[52] Finally, in recent unpublished results (R. I. Carp and S. M. Callahan, in preparation) from in vivo experiments, the level of infectivity on a per cell basis was much higher in adherent than in nonadherent splenocytes. In a second set of experiments, unfractionated thymocytes, unfractionated splenocytes, and T-enriched and T-depleted fractions of splenocytes from normal mice were exposed to agent in vitro. In each preparation scrapie infectivity decreased at a significantly greater rate in the presence of cells than in scrapie aliquots incubated in the absence of cells. The effect was not reversed by addition of phytohemagglutinin or lipopolysaccharide. The rate of loss was greater in culture medium than in cell pellets from scrapie-exposed lymphoid cultures (R. I. Carp and S. M. Callahan, in preparation).

In reviewing the data on drug treatments that modulate scrapie pathogenesis, Outram postulates a "Trojan horse" concept in which a cell that ordinarily acts to protect the host acts to further the progression of the scrapie disease process after peripheral injection.[53] The concept that this cell type is not present during the first few days of life in the mouse is used to explain results after neonatal injection of scrapie.[54,55] In these studies, intraperitoneal (i.p.) injection of neonates leads to an extremely large variation in incubation period times, with most mice having a much longer incubation period than mice injected as weanlings and a few mice having a shorter incubation period than weanlings. It also results in some neonatally injected mice surviving what would be lethal doses for weanling mice.

Investigation of the role of the macrophage in scrapie pathogenesis has led to conflicting conclusions. On one hand, in vitro evidence suggests that peritoneal macrophages inactivate scrapie.[56,57] These data are supported by the finding that the incubation period of mice injected i.p. with scrapie 5 days after thioglycollate is significantly longer than if scrapie is preceded at 5 days by injection of phosphate-buffered saline.[58] Five days post-injection is the time when thioglycollate stimulates the maximum influx of macrophages into the peritoneum. There is no effect of prior thioglycollate treatment on the incubation period after i.c. injection of scrapie. These studies corroborate the in vitro studies and suggest that peritoneal macrophages affect pathogenesis by inactivation of some of the input infectivity.

In contrast, several other studies in which stimulation of peritoneal macrophages is followed by i.p. injection of scrapie failed to show an increase in the incubation period.[59,60] In addition, the findings with dextran sulfate administration show a marked lengthening of incubation period even when the drug is given as long as 1 month prior to or as much as 2 weeks after scrapie injection.[61] The drug persists over this time span in macrophage-like cells in the spleen and lymph nodes. The studies with dextran sulfate suggest that macrophage-type cells are important in an early step of scrapie pathogenesis by directly abetting either replication of the agent or its dispersal to sites where replication occurs. Interpretation of the findings noted above is complicated by the fact that other compounds (e.g., silica, trypan blue) that destroy the phagocytic activity of macrophages have no effect on scrapie incubation periods or mouse survival rates.[61] It is known that macrophages from different organs can function differently in their interactions with viruses,[62] so the contradictory results obtained for splenic[61] and peritoneal[56-58] macrophages may be a function of the location of the cells. This potential difference could be related to the fact that peritoneal macrophages are circulating, whereas many of the macrophage-type cells in the spleen would be stationary.

In summary, the key cell in the LRS that is important in scrapie pathogenesis remains a mystery. Clearly, the LRS organs play a role. It appears that cells that are radiation resistant and long-lived are important. Therefore, the effect of immunomodulators on the scrapie incubation period must be mediated through cells other than T and B cells. It is possible that the macrophage/histiocyte group of cells is important and that different types within this group play different roles.

3.3. The Spread of Infectivity in the Infected Host

In natural disease the site of entry of the scrapie agent is unknown. As stated previously (Section 1), there is strong evidence that the oral route of entry is a possibility.[15-18]

There must be two phases of spread within the naturally infected host: in the first, the agent must get from the port of entry to the spleen and lymph nodes, and in the second phase, infectivity must move from the LRS organs to the brain. Since the site of entry in natural disease is uncertain, there is no information about the first phase. In an experimental situation it is known that after intragastric administration in the mouse, infectivity is first found in Peyer's patches.[62a] With this route, access to cells of the LRS is direct. With regard to the second phase, there is evidence of neural spread, which is discussed shortly. The role that the circulatory system plays in either phase is unclear. Early studies fail to show infectivity in blood except within the first few hours after infection.[63,64] However, several recent studies showed continuous low titers of these agents in blood.[65] In one, low levels of infectivity are found throughout the incubation period in the 263K–hamster model when a method for concentrating infectivity is applied to cardiac blood. In another study, viremia of the CJD agent is seen in experimentally infected guinea pigs at various times throughout the incubation period.[66] Infectivity is found only in buffy coat fractions and is absent from blood cell and plasma fractions. In another study, viremia is seen in CJD-infected mice,[48] and infectivity is detected in the blood of humans with CJD by transmission studies in animals.[67] The circulatory system appears to be important in the spread of the agent to peripheral sties that are known to become infected.[38]

The concept that neural spread plays a role in the second phase of pathogenesis is supported by a series of experiments done by Kimberlin and colleagues. In these experiments, after peripheral injection and the appearance of infectivity in spleen, the first area of the CNS that shows infectivity is the thoracic region of the spinal cord between thoracic vertebrae 4 and 9, which is where the splanchnic nerve enters the cord in mice.[34,68–70] From there, infectivity spreads to the lumbar and cervical regions of the cord. Infectivity then spreads to the brain and is found first in posterior regions and subsequently appears in anterior regions.[69,70] Evidence for neural spread also comes from analysis of incubation periods following intrasciatic injection versus deposition of equivalent quantities of inoculum just adjacent to the sciatic nerve.[71]

Further support for the concept of neural spread comes from the study of intraocular injections in the mouse, in which both the sequential occurrence of infectivity and the development of scrapie-induced vacuolation occur as predicted from the pathway followed by the optic nerve.[31,72,73] Thus, following injections of the right eye, vacuolation is seen first in the contralateral superior colliculus and lateral geniculate.[31] Further, analysis of infectivity shows that it also follows the route dictated by the crossover of nerve fibers at the optic chiasma.[31,72]

3.4. The Concept of Clinical Target Areas

One further aspect of pathogenesis concerns the concept of clinical target areas.[34,35] It has long been known that different routes of injection yield different incubation periods. For example, with the same dilution of homogenate, the incubation periods from shortest to longest are produced by i.c., intravenous (i.v.), i.p., and subcutaneous injection. If one divides the incubation period into two phases, the time from injection to the initiation of replication in the brain and the time from initiation of replication in the brain to the start of clinical disease, the comparison of i.c. with the non-CNS routes yields a surprising result[34,70]: the time from initiation of replication in the brain to the start of clinical disease is shorter for the non-CNS routes than for i.c. injection. From these findings Kimberlin and Walker[34,35] put forward the concept of clinical target areas, i.e., the areas in which scrapie replication leads to clinical manifestations of disease. Thus, the replication of scrapie in the brain after peripheral injection is shorter because the agent more quickly reaches the clinical target area(s) than does the agent injected i.c.

One logical consequence of this hypothesis is that intraspinal injection might yield a shorter incubation period than i.c. injection, and that has been shown to be the case.[74] The corollary of the hypothesis is that much of the replication of scrapie in the brain is irrelevant as far as inducing the clinical changes that we traditionally measure. In experiments with stereotactic injection of different areas within the brain, it has been possible to show that injection of some areas more quickly leads to clinical disease than injection of others.[32] For example, after stereotactic injection of the 22L strain into the cerebral cortex, thalamus, caudate nucleus, substantia nigra, and cerebellum, the shortest incubation period is seen in mice injected in the cerebellum.[32]

3.5. Genetic Aspects of Pathogenesis

A discussion of scrapie pathogenesis cannot be complete without a description of the influence of the genetics of both host and agent. A variety of parameters are affected (see Table 7-5). We discuss two of the genetic markers in detail—length of incubation period and differences in clinical target areas.

The length of the incubation period is clearly a fundamental characteristic of the pathogenesis of any infectious disease process. Scrapie strains can differ in incubation periods in the same inbred mouse strain by fourfold or more.[80] The mouse strain also plays a role through a gene called *Sinc*.[2,3,75,81] The allelic designation is determined on the basis of the incubation period for the ME7 scrapie stain: mouse strains with a short incubation period for ME7 are designated s7s7, whereas those mouse strains with a prolonged incubation period are termed p7p7. As an example, the incubation period of ME7 in C57BL mice, an s7s7 strain, is 140 days, whereas its incubation period in a p7p7 mouse strain such as VM is 300 days. The influence of scrapie strain can be seen by noting the incubation periods of the 22A scrapie strain in C57BL and VM mice, which are 380 days and 180 days, respectively. The cause of the difference in incubation periods is not clear; it does appear that in longer-incubation models both the time from injection to initiation of replication in brain and the interval from that time to clinical disease onset are both longer.[81a]

There is a gene in sheep, termed *Sip*, that plays a similar role to that of *Sinc* in mice.[3]

Genetic control with regard to clinical target areas is evidenced by the fact that different scrapie strains differ in the area that yields the shortest incubation

TABLE 7-5
Characteristics of Scrapie that Are under
Genetic Control of Both Host and Agent

Characteristic	Reference
Incubation period	75
Vacuolation pattern (lesion profile)	30
Presence of amyloid plaques	76
Resistance to inactivation by heat	77[a]
Induction of obesity and altered glucose tolerance	21, 24
Clinical target areas	32
Behavior	25, 28
Species specificity	78
Characteristics of scrapie-associated fibrils	79

[a]Differences in heat inactivation are under control of agent alone.

period following stereotactic injection.[32] As stated previously, for the 22L strain the shortest incubation period is seen after injection of the cerebellum, whereas for the ME7 strain both thalamus and cerebellum injection yield short incubation periods.[32] For 139A, injections of the cerebellum, thalamus, caudate nucleus, and substantia nigra yield equivalent incubation periods, all significantly shorter than cortex injection. This does not mean that the areas mentioned are the clinical target areas for the particular scrapie strain but that injection of these areas more quickly leads to access of infectivity to the scrapie-strain-specific clinical target areas.

The concept that clinical target areas can differ depending on the scrapie strain and the host provides a rationale for the development of either motor or nonmotor changes in scrapie as described previously.[21,24–29,37] For example, if the targeted neurons[36] in a particular scrapie strain–host combination are in the hypothalamic–pituitary axis, the induction of obesity and changes in glucose tolerance would be possible early clinical manifestations[21–24] of the infectious process, whereas cerebellar neurons as targets would lead to motor changes predominantly.

4. THE ABSENCE OF AN IMMUNOPATHOLOGICAL EFFECT IN SCRAPIE

Classical immunologic responses in scrapie infections have never been demonstrated to play a role in pathological changes. There is no evidence of any kind of antibody response to the agent.[82–85] A variety of studies have failed to show positive responses in any antibody test of serum obtained from infected animals. Immune competence is not compromised, since the antibody response of scrapie-infected mice to antigens such as sheep red blood cells is similar to that seen in normal mice.[86,87] There are reports of an autoimmune reactivity to neurofilament proteins in some cases of sheep with natural scrapie as well as in some cases of kuru and CJD.[88] The importance of this in the pathogenesis of scrapie is unclear, since a similar response is seen in a variety of other types of infections.[88] Furthermore, many individuals with these diseases fail to show reactivity to neurofilament proteins.

There does not appear to be a cellular immune response, nor is there an effect of scrapie infection on the responsiveness of T cells. Mixed lymphocyte reactions between splenocytes from scrapie-infected and from normal mice fail to show any scrapie-specific reactivity.[89] Thymectomy has no effect on incubation period, pathology, or the characteristics of the clinical course.[52] Analysis of the *in vitro* reactivity of splenocytes from scrapie-infected mice to mitogens that affect T and/or B cells failed to reveal any changes in Swiss, C57BL, and BALB/c mice.[89] The mitogens used were phytohemagglutinin, concanavalin A, bacterial lipopolysaccharide, and pokeweed mitogen. In one study scrapie-infected C3H/HeJ mice showed a decreased responsiveness to the B-cell mitogen bacterial lipopolysaccharide.[90] This change is seen only between 20 and 40 days postinjection, and it coincides with a period in which mice show transient splenomegaly. The responsiveness to phytohemagglutinin and concanavalin A are unaffected in these mice throughout the incubation period. This group also reports transient splenomegaly in scrapie-infected Swiss, C57BL, and BALB/c mice, although the responsiveness to mitogens in these mouse strains is normal.[90] An attempt to repeat this work failed to show splenomegaly in C3H/HeJ, BALB/c or C57BL strains (Swiss mice were not tested), nor did the study show a decrease in response to bacterial lipopolysaccharide in C3H/HeJ mice.[89] The cause of the discrepancy in the results is deemed to relate either to the difference in the strain of scrapie used or to the possible contamination of the inoculum used by one group[90] with a virus that induces splenomegaly.[89]

In another series of studies the macrophage electrophoresis mobility test was used to diagnose scrapie.[91,92] In these tests sensitized lymphocytes from scrapie-infected animals are exposed to "scrapie-specific antigen" derived from brain or spleen from scrapie-infected mice. This mixture is then placed in contact with normal guinea pig macrophages, and the mobility of these cells is measured. The lymphocytes are derived either from guinea pigs injected 8 days previously with homogenates from scrapie or normal mice or from natural or experimental sheep or goat scrapie. Migration is inhibited in instances in which "scrapie antigen" is mixed with scrapie-sensitized lymphocytes.[92] However, in another study no effect on macrophage mobility was seen in comparing groups of scrapie-infected and normal sheep.[93]

Certainly an examination of the histopathological changes in the brain fails to reveal any evidence of a host immune response: there is no inflammation and no cellular infiltration of any kind.[4,6,80] There is gliosis and astrocytosis,[4,6] but these are standard responses within the brain to a variety of injuries, and there is no evidence for an immune component in these responses.

Scrapie does not affect the capacity of the host to mount an immune response to a number of antigens, although in some instances it does lead to a change in the quantities of immunoglobulin (IgG) subclasses. In one of these studies an effect was seen in a proportion of infected sheep during the clinical phase of disease.[94] Results of the second study showed changes in the subclasses of IgG in p7p7 mouse strains injected with the 87V scrapie strain.[94a] These scrapie strain–mouse combinations are characterized by very long incubation periods and the presence of numerous amyloid plaques in the CNS. Most scrapie strain–mouse combinations, including some that yielded CNS plaques, fail to show IgG changes.

Infection with scrapie does not lead to the induction of interferon (IFN), nor does treatment with IFN or with IFN inducers affect the incubation period or incidence of scrapie in injected animals.[96–98] Scrapie-infected animals can mount an IFN response after infection with viruses that induce IFN in a normal host.[98]

In summary, the immune system does not play a role in the development of disease in scrapie. In fact, no consistent immunologic responses occur in scrapie disease. The few changes noted are either limited with respect to the scrapie strain–host combinations showing the change (e.g., increase in IgG[94]) or have not been confirmed by subsequent experimentation (e.g., decreased responsiveness to bacterial lipopolysaccharide in C3H/HeJ mice[89,90]).

5. AGENT PERSISTENCE

Scrapie agent replicates over an extended period of time before any clinical changes are seen, and, as stated before, infectivity is present in a variety of organs before the onset of pathological or clinical changes. However, in all instances except those noted below, the replication and persistence of the agent proceeds irrevocably toward clinical disease and death. As noted previously, if experimental conditions such as dose and strain of agent, route of injection, and the strain, age, and sex of host are all kept constant, then disease progression is remarkably precise and predictable. This is the salient feature of the pattern of slow infections with unconventional agents, but it is not the pattern seen in persistent infections with conventional viruses.[1]

There are several experimental situations in which a phenomenon similar to persistence occurs. In one, it has been shown that i.p. injection of the 87V scrapie strain into p7p7 mice rarely results in clinical disease; however, the agent replicates in lymphoid organs, and infectivity persists in the spleen for more than 400 days.[99,102] Infectivity is not found in the brain, and there are no apparent clinical changes in the mice.

In another example of persistence a large series of mice were injected with brain homogenates from Icelandic sheep that had clinical scrapie disease. A few of these mice became sick with clinical scrapie; however, most mice lived a full life span or close to it.[100] Brains removed from animals that had lived a full life span without clinical disease were then assessed for histopathological changes typical of scrapie, and in some instances brain preparations were assayed for infectivity. Many of the brains from clinically normal animals had lesions typical of scrapie. On blind passage into additional mice, many of the preparations caused clinical scrapie within a time frame consistent with second passage in a new host species.

It should be emphasized that the situation in these two examples may be consistent with the slow infection pattern. The failure to demonstrate clinical disease was probably a function of the fact that with the scrapie inoculum, host, and route used, the incubation period for scrapie was longer than the life span of the host.[101] In fact, with i.p. injection of the 87V strain in IM mice, one study showed that a small proportion of mice injected with a high concentration of brain homogenate proceeded to clinical disease after a very long incubation period.[102] The data with the sheep isolates are consistent with the idea that infectivity was present in the brain of mice injected at first passage but had not had enough time either to attain sufficient titer or to reach clinical target areas by the end of the natural life span of the mice.

6. TREATMENT

The search for an effective treatment of scrapie and related diseases has failed to reveal a compound and regimen that is very promising. In Section 3, we detailed data concerning administration of drugs at or very close to the time of experimental infection that could affect the length of scrapie incubation. Even though some of these drugs cause a lengthening of the incubation period,[44–46] they can not be considered treatments in that they are effective only if given prior to or at the time of infection. Three drugs have shown limited efficacy: HPA-23, dextran sulfate 500, and amphotericin B.

Administration of HPA-23, a tungstoantimoniate that has sodium at its center, can extend the incubation period and reduce effective titer by 1 to 2.6 $\log_{10}$ units.[103,104] The greatest effect is seen if the interval between infection (by the i.v. route) and the first HPA-23 treatment (by the i.p. route) is 4 hr. As the interval is increased there is a diminishing effect, with virtually no effect if treatment is initiated 48 hr after infection. In instances in which the time interval between infection and the first treatment is 4 hr, there is no difference in effect between 12 and 3 daily doses. HPA-23 is known to suppress the replication of a number of viruses, but the mechanism of action is not known.[105]

Using dextran sulfate 500 (DS 500) in a variety of regimens, investigators have produced either prolongation of incubation period or survival of animals infected with lethal doses of scrapie.[61,103,106,107] Thus, the efficiency of infection is reduced by DS 500. In perhaps the most striking result, i.p. administration of DS 500 as much as 2 weeks after i.p. injection of scrapie causes a significant lengthening of the incubation period.[107] There is an effect on some mice even if the time interval between scrapie and a single dose of the drug is as much as 2 months. With this interval there is a bimodal distribution, with some mice having an extended incubation period and others showing absolutely no effect. DS 500 has an effect on either i.p. or i.v. scrapie when given as a single dose i.p. as much as 3 days after scrapie injection.[61] Various regimens of treatment have no effect on the incubation period of scrapie after i.c. injection.[106] There is a reduction in spleen titer and an extension of the incubation period in mice given a single dose of DS 500 i.p. 3 days after i.v. injection of scrapie.[106] The effects on both parameters are much more pronounced if DS 500 is given three times at 3, 10, and 17 days after scrapie.[106]

Questions concerning the mode of action of DS 500 on scrapie replication remain. The suggestion has been made that the drug affects the aggregation of scrapie infectious agents in the blood when administered shortly before or after scrapie but that it must also have a different mode of action at a later time during scrapie replication and spread.[103] As noted earlier (Section 3), there is the contention that the continued presence of the drug in tissue-bound mononuclear phagocytes in spleen and lymph nodes for as much as 7 months after administration would provide the opportunity for a direct effect on the scrapie agent or an effect on the replication of the agent within those cells.[61] Of course we do not know if the scrapie agent enters or replicates in this cell type. Furthermore, there is evidence that other compounds that depress the phagocytic activity of mononuclear phagocytes, e.g., silica and trypan blue, fail to affect the scrapie incubation period or endpoint titers.[61,103] In a study in which the effects of high dose of DS 500 were used at a series of intervals prior to scrapie infection that extended from the time of maximum depression of phagocytosis to complete recovery of this capacity, there were no differences in the reduction of effective scrapie titers by the drug.[103]

The final drug that shows potential as a treatment is amphotericin B.[108] Again the mode of action of this antifungal drug is unknown. The incubation period of i.c. scrapie (strain 263K) in hamsters is extended by as much as 45 days. Treatment is initiated on the day of injection and continued 6 days per week for 50, 75, or 100 days. The incubation periods for the three treatment times were 83, 96, and 102 days, respectively, and this compared to 55 days in hamsters administered saline rather than drug throughout the incubation period.[108] The drug is also effective after i.p. injection of a scrapie-containing homogenate. If drug is mixed with the inoculum for 2 hr prior to infection by the i.p. route and animals are subsequently treated with saline for 50 days (as are controls), there is no effect on the incubation period, suggesting that there is no direct action of drug on the agent. The drug has no effect on the length of the clinical course if administration of drug is started when the animals first exhibit clinical signs. It is interesting that this is the only drug that is effective in treatment after i.c. injection of the agent.

It appears that the Amphotericin B effect is highly specific for the 263K-hamster combination. In studies with short incubation mouse models (139-SJL mice; ME7-SJL mice), a long incubation mouse model (87V-IM mice), and a long incubation hamster model (139H-hamster), there were no effects of Amphotericin B treatment on the characteristics of agent–host interactions (Y. S. Kim, S. M. Callahan, and R. I. Carp, unpublished).

7. MISCELLANEOUS

No review of scrapie would be complete without reference to the unusual characteristics of the infectious agent and the theories concerning its nature that have been proposed. We do not go into great detail about this because there have been a number of recent informative reviews.[109–115] Rather, we outline the unusual characteristics of the agent, describe the three theories that are currently in vogue, and relate these theories to some of the information presented in this chapter.

The unusual characteristics of the agent include resistance to a variety of physical and chemical treatments such as ultraviolet irradiation, x rays, boiling, exposure to 10% neutral formalin, and exposure to β-propriolactone.[109–116] In interactions with living systems the unusual findings include the lack of cytopathic effects in tissue culture, the absence of any immune response to the agent, the absence of inflammatory changes in the primary organ affected (brain), and the close association of infectivity with cellular membrane components.[80,111,116] It is really the combination of characteristics that marks scrapie and the other agents of the group as unconventional, since there are viruses or virus groups that have characteristics with regard to single parameters that do not differ markedly from scrapie.

In recent years, findings in three areas have been the subject of intense study. In one, fibrillar structures, termed scrapie-associated fibrils (SAF), are found consistently in preparations of brain and spleen from slow infection diseases caused by unconventional agents but never in normal material or in preparations from other diseases.[117–120] There is a correlation between SAF and infectivity, both in relation to kinetics of their occurrence in the infected animal and in their occurrence in different partially purified (for infectivity) preparations of brain.[119,121,122]

Another area of great interest revolves around the finding of a protein that is found in brain and spleen preparations that are partially purified for scrapie infectivity.[123–126] Subsequent studies show that this protein, termed PrP for protease-resistant protein, is coded for by a cellular gene and that the level of mRNA is similar in organs obtained from scrapie and normal animals.[127–129] In fact, the protein is found in normal preparations, but its characteristics are different from the protein found in scrapie in several respects: (1) the normal protein is completely degraded by proteases, whereas the scrapie protein is only partially degraded[130–132]; (2) the scrapie protein is easily sedimentable, whereas the normal is not[130–132]; (3) the scrapie protein is part of SAF, which are found in all scrapie strain–host combinations, and the scrapie protein constitutes at least part of the amyloid plaques found in some combinations.[133,134] In contrast, fibrils and plaques are not found in normal material.

The final area of intense effort is related to studies on the genetics of scrapie–host interactions. The importance of these studies is that they all enforce the idea that scrapie contains an independent genome. In Section 3 we noted a series of genetic markers that can be used to distinguish different scrapie strains[21,25,28,30,32,75–79] (see Table 7-5). In another study, progeny of the scrapie agent from mice injected with an agent that has a long incubation period and yields extensive plaque formation (in that particular mouse strain) was analyzed for these two parameters.[135,136] In some instances, which occur at random, agent is found that has entirely different characteristics in that mouse strain (short incubation period, no plaque formation). Detailed analysis of the histopathology of the brains indicates that these "new" agents arise in localized areas. The best explanation of these data is that mutation occurs. In another study, several "cloned" mouse-adapted scrapie strains were passaged in hamsters several times and then injected into mice.[137,137a] For some mouse-adapted scrapie strains the agent produced after hamster passage has the same characteristics as the agent put into hamsters. For other strains, however, the hamster-passaged material has characteristics (incubation period and vacuolation pattern) that are markedly different from the starting strain. This suggests that in the latter instances a mutational event(s) occurs that is selected for during passage in the hamster.

There are three theories concerning the nature of the scrapie agent that have evolved from the data noted above (Table 7-6) and from other information in the reviews that are referenced. The first is a virus model. In this theory the scrapie agent is a virus with standard virus molecular components (a nucleic acid genome that codes for its protective protein coat) but with unusual characteristics that confer resistance to a variety of physical and chemical inactivating procedures and provide the basis for its unusual biological characteristics.[138–140] The second is a prion or modified host protein theory. In this theory the scrapie agent is composed of protein only, and there are no exogenous (nonhost) components required for infection.[109,110,112,115] The third is the virino theory in which the agent

TABLE 7-6
The Relationship of Recent Findings to the Three Theories
about the Nature of the Scrapie Agent

| | Theory | | |
	Virus	Modified host protein (prion)	Virino
Finding			
SAF	Byproduct of infection; not part of infectious agent	Aggregates of PrP; structure is not required for infectivity	Structures that contain the small scrapie-strain-specific nucleic acid
PrP	Byproduct of infection; not part of infectious agent	The only macromolecule required for infectivity	A host-derived protein that protects the scrapie-strain-specific nucleic acid
Genetics of scrapie	The infectious agents contain nucleic acid; sequence differences yield strains	Differences in host coding sequences for PrP define a limited number of strains	The infectious agents contain nucleic acid; sequence differences yield strains

is composed of a small, scrapie-strain-specific nucleic acid (perhaps noncoding and regulatory, similar to a viroid) surrounded by host protein.[110–113,141]

In these three theories the role of SAF and scrapie protein are very different (Table 7-6). In the virus theory, SAF and the scrapie protein are pathological products that are a host response to the infectious agent. Proponents of this theory contend that we have found neither the protein nor the nucleic acid of the putative virus. In the modified host protein theory, the scrapie protein described above is thought to constitute the sum and substance of the infectious agent, and SAF are aggregates of that protein. In the virino theory, the scrapie protein described above is a candidate for the host protein that protects the scrapie-strain-specific nucleic acid, and SAF are structures that could contain the scrapie-specific nucleic acid plus the protective host protein.

The proponents of these theories attempt to explain some of the unusual characteristics of the scrapie agent and of the disease process on the basis of their proposed theories. For example, the remarkable resistance of scrapie to UV irradiation at 256 nm and to x rays could be explained by either the protein-only theory, since proteins are more resistant to irradiation (at the listed wavelength) than nucleic acids, or by an informational molecule that is very small (virino theory) and therefore presents a small target for inactivation. It is very easy to explain the genetics of scrapie strains in those theories in which the informational molecule is a nucleic acid (virus and virino) but is more difficult to explain if the agent is composed exclusively of modified host protein. The difficulty in isolating a nucleic acid from partially purified preparations of infectious material could be explained because there is none (modified host protein), because it is present in small amounts and is small and/or tightly bound to protective material (virino), or because preparations are not sufficiently purified to distinguish agent-specific nucleic acid from contaminating host nucleic acid. The failure of the host to mount an immune response can be explained if a host protein is the only protein that is an integral part of the infectious moiety (virino and modified host protein).

Since preparation of the above summary of theories on the nature of the agent, there have been significant findings in a number of areas, particularly in relation to the role of PrP in the infectious process. It has been shown that the gene coding for PrP is closely linked to the *Sinc* gene in mice that controls incubation period.[142] Furthermore, using transgenic technology it was shown that mice with the hamster-specific PrP transgene behaved like hamsters with regard to sensitivity to hamster passaged scrapie strains.[143,144] Thus, the PrP gene appears to play a key role in the species barrier phenomenon.

Despite extensive studies on the chemical characteristics of the normal and scrapie isoforms of this protein, there are no known differences in their chemical composition that would explain the differences noted in protease sensitivity and sedimentability.[145,146] Analysis of primary sequence and examination of post-translational modifications have failed to reveal any differences between the normal and scrapie isoforms nor between different scrapie strains.[145–147] An additional difference in the two isoforms is that although both appear to have a phosphotidylinositol glycolipid linkage,[148] the normal isoform can be released from the cell surface by the enzyme phosphotidylinositol specific phospholipase C, whereas the scrapie isoform is not released under similar conditions.[149] There are no definitive explanations of this finding although in tissue culture systems the scrapie isoform appears to be located primarily intracellularly, whereas the normal isoform is on the cell surface.[150]

Additional genetic studies have shown that characteristic differences among scrapie strains are maintained after repeated passages of strains in a single inbred host, further establishing that differences in PrP coding sequences cannot be the sole delineator of scrapie strains.[145,151]

Despite all of these recent findings the relationship of PrP to the infectious agent and its role in the replication cycle of the agent are still a matter of contention. The three theories on the nature of the agent are still au courant and each theory has its dedicated proponents. Further studies will be required to establish the nature of these important and unusual infectious agents.

ACKNOWLEDGMENTS.   The author wishes to thank Ms. Patricia Merz and Drs. Richard Kimberlin, Richard Kascsak, Richard Rubenstein, and Paul Behdheim for their careful criticism of the manuscript. The author greatly appreciates the opportunity granted by Dr. Kimberlin to describe results that he and his colleagues have not yet published. Excellent assistance in preparing the manuscript was provided by Ms. Jennifer Parese. Dr. George Merz provided invaluable assistance in generating the reference list. This work was supported in part by the Office of Mental Retardation and Developmental Disabilities and by grant No. NS21349.

REFERENCES

1. Sigurdsson, S. B., 1954, Rida, a chronic encephalitis of sheep with general remarks on infections which develop slowly and some of their special characteristics, *Br. Vet. J.* **110**:341–354.
2. Dickinson, A. G., and Fraser, H., 1977, Scrapie pathogenesis in inbred mice: An assessment of host control and response involving many strains of agent, in: *Slow Virus Infections of the Central Nervous System* (V. ter Meulen and M. Katz, eds.), Springer-Verlag, New York, pp. 3–14.
3. Dickinson, A. G., and Fraser, H., 1979, An assessment of the genetics of scrapie in sheep and mice, in: *Slow Transmissible Diseases of the Nervous System*, Vol. 1 (S. B. Prusiner and W. J. Hadlow, eds.), Academic Press, New York, pp. 367–385.
4. Fraser, H., 1979, Neuropathology of scrapie: The precision of the lesions and their diversity, in: *Slow Transmissible Diseases of the Nervous System*, Vol. 1 (S. B. Prusiner and W. J. Hadlow, eds.), Academic Press, New York, pp. 387–406.
5. Outram, G. W., 1976, The pathogenesis of scrapie in mice, in: *Slow Virus Diseases of Animals and Man* (R. H. Kimberlin, ed.), North Holland, Amsterdam, pp. 325–357.
6. Gajdusek, D. C., 1977, Unconventional viruses and the origin and disappearance of kuru, *Science* **197**:943–960.
7. Gibbs, C. J., Jr., Gajdusek, D. C., Asher, D. M., Alpers, M. P., Beck, E., Daniel, P. M., and Matthews, W. B., 1968, Creutzfeldt–Jakob disease (spongiform encephalopathy): Transmission to the chimpanzee, *Science* **161**:388–389.
8. Masters, C. L., Gajdusek, D. C., and Gibbs, C. J., Jr., 1981, Creutzfeldt–Jakob disease virus isolations from the Gerstmann–Straussler syndrome, *Brain* **104**:559–588.
9. Marsh, R. F., and Hanson, R. P., 1979, On the origin of transmissible mink encephalopathy, in: *Slow Transmissible Diseases of the Nervous System*, Vol. 1 (S. B. Prusiner and W. J. Hadlow, eds.), Academic Press, New York, pp. 451–460.
10. Williams, E. S., and Young, S., 1980, Chronic wasting disease of captive mule deer: A spongiform encephalopathy, *J. Wildl. Dis.* **16**:89–98.
11. Williams, E. S., and Young, S., 1982, Spongiform encephalopathy of Rocky Mountain elk, *J. Wildl. Dis.* **18**:465–471.
12. Wells, G. A. H., Scott, A. C., Johnson, C. T., Gunning, R. G., Hancock, R. D., Jeffrey, M., Dawson, M., and Bradley, R., 1987, A novel progressive spongiform encephalopathy in cattle, *Vet. Rec.* **121**:419–420.
13. Brotherson, J. G., Renwick, C. C., Stamp, J. T., Zlotnick, I., and Patterson, I. H., 1968, Spread of scrapie by contact to goats and sheep, *J. Comp. Pathol.* **68**:9–17.
14. Dickinson, A. G., Stamp, J. T., and Renwick, C. C., 1974, Maternal and lateral transmission of scrapie in sheep, *J. Comp. Pathol.* **84**:19–25.
15. Pattison, I. H., and Millson, G. C., 1961, Experimental transmission of scrapie to goats and sheep by the oral route, *J. Comp. Pathol.* **71**:171–176.
16. Pattison, I. H., Hoare, M. N., Jebbett, J. N., and Watson, W. A., 1972, Spread of scrapie to sheep and goats by oral dosing with fetal membranes from scrapie-infected sheep, *Vet. Res.* **9**:465–468.
17. Carp, R. I., 1982, Transmission of scrapie by the oral route: Effect of gingival scarification, *Lancet* **1**:170–171.
18. Gibbs, C. J., Jr., Amyx, H., Bacote, A., Masters, C. L., and Gajdusek, D. C., 1980, Oral transmission of kuru, Creutzfeldt–Jakob disease and scrapie to non-human primates, *J. Infect. Dis.* **142**:205–208.
19. Dickinson, A. G., 1976, Scrapie in sheep and goats, in: *Slow Virus Diseases of Animals and Man* (R. H. Kimberlin, ed.), North Holland, Amsterdam, pp. 209–241.
20. Palsson, P. A., 1979, Rida (scrapie) in Iceland and its epidemiology, in: *Slow Transmissible Diseases of the Nervous System*, Vol. 1 (S. B. Prusiner and W. J. Hadlow, eds.), Academic Press, New York, pp. 357–366.

21. Carp, R. I., Callahan, S. M., Sersen, E. A., and Moretz, R. C., 1984, Preclinical changes in weight of scrapie-infected mice as a function of scrapie agent–mouse strain combination, *Intervirology* **21**:61–69.

22. Kim, Y. S., Carp, R. I., Callahan, S. M., and Wisniewski, H. M., 1987, Scrapie-induced obesity in mice, *J. Infect. Dis.* **156**:402–405.

23. Kim, Y. S., Carp, R. I., Callahan, S. M., and Wisniewski, H. M., 1988, Adrenal involvement in scrapie-induced obesity, *Proc. Soc. Exp. Biol. Med.* **189**:21–27.

24. Carp, R. I., Kim, Y. S., Callahan, S. M., and Wisniewski, H. M., 1989, Development of preclinical weight increases and aberrant glucose tolerance tests in scrapie injected mice, in: *Unconventional Virus Diseases of the Central Nervous System* (L. A. Court, D. Dormont, P. Brown, and D. T. Kingsbury, eds.), Commissariat a l'Energie Atomique, Departement de Protection Sanitaire, Service de Documentation, Fontenay-aux-Roses Cedex, Paris, pp. 587–601.

24a. Carp, R .I., Kim, Y. S., and Callahan, S. M., 1990, Pancreatic lesions and hypoglycemia-hyperinsulinemia in scrapie-injected hamsters, *J. Infect. Dis.* **161**:462–466.

25. Outram, G. W., 1972, Changes in drinking and feeding habits of mice with experimental scrapie, *J. Comp. Pathol.* **82**:415–427.

26. Heitzman, R. S., and Corp, C. R., 1968, Behavior in emergence and open-field tests of normal and scrapie mice, *Res. Vet. Sci.* **9**:600–601.

27. Savage, R. D., and Field, E. J., 1965, Brain damage and emotional behavior: The effects of scrapie on the emotional response of mice, *Anim. Behav.* **13**:443–446.

28. McFarland, D. J., Baker, F. D., and Hotchin, J., 1980, Host and viral genetic determinants of the behavioral effects of scrapie encephalopathy, *Physiol. Behav.* **24**:911–914.

29. Sturman, L. S., 1972, Endocrinopathy in mice affected with scrapie, *Nature* **238**:208–209.

30. Fraser, H., and Dickinson, A. G., 1973, Scrapie in mice: Agent–strain differences in the distribution and intensity of grey matter vacuolation, *J. Comp. Pathol.* **83**:29–40.

31. Fraser, H., and Dickinson, A. G., 1985, Targeting of scrapie lesions and spread of agent via the retino-tectal projection, *Brain Res.* **346**:32–41.

32. Kim, Y. S., Carp, R. I., Callahan, S. M., and Wisniewski, H. M., 1987, Clinical course of three scrapie strains in mice injected stereotaxically in different brain regions, *J. Gen. Virol.* **68**:695–702.

33. Gorde, J. M., Tamalet, J., Toga, M., and Bert, J., 1982, Changes in the nigrostriatal system following microinjection of an unconventional agent, *Brain Res.* **240**:87–93.

34. Kimberlin, R. H., and Walker, C. A., 1983, Invasion of the CNS by scrapie agent and its spread to different parts of the brain, in: *Virus Non Conventionnels et Affections du Systeme Nerveux Central* (L. A. Court and F. Cathala, eds.), Masson, Paris, pp. 17–33.

35. Kimberlin, R. H., and Walker, C. A., 1986, Pathogenesis of scrapie (strain 263K) in hamsters infected intracerebrally, intraperitoneally or intraocularly, *J. Gen. Virol.* **67**:255–263.

36. Fraser, H., McBride, P. A., Scott, J. R., and Bruce, M. E., 1986, Infectious degeneration of the nervous system, in: *Advanced Medicine* (D. R. Triger, ed.), Bailliere Tindall, London, pp. 371–383.

37. Beck, E., Daniel, P. M., and Parry, H. B., 1964, Degeneration of the cerebellar and hypothalamic neurohypophysial systems in sheep with scrapie and its relationship to human system degenerations, *Brain* **87**:153–176.

38. Eklund, C. M., Hadlow, W. J., and Kennedy, R. C., 1963, Some properties of the scrapie agent and its behavior in mice, *Proc. Soc. Exp. Biol. Med.* **112**:974–979.

39. Dickinson, A. G., and Fraser, H., 1969, Genetical control of the concentration of ME7 scrapie agent in mouse spleen, *J. Comp. Pathol.* **79**:363–366.

40. Fraser, H., and Dickinson, A. G., 1970, Pathogenesis of scrapie in the mouse: The role of the spleen, *Nature* **226**:462–463.

41. Clarke, M. C., and Haig, D. A., 1971, Multiplication of scrapie agent in mouse spleen, *Res. Vet. Sci.* **12**:195–197.

42. Fraser, H., and Dickinson, A. G., 1978, Studies of the lymphoreticular system in the pathogenesis of scrapie: Role of spleen and thymus, *J. Comp. Pathol.* **88:**563–573.
43. Dickinson, A. G., and Fraser, H., 1972, Scrapie: Effect of Dh gene on the incubation period of extraneurally injected agent, *Heredity* **29:**91–93.
44. Dickinson, A. G., Fraser, H., McConnell, I., and Outram, G. W., 1978, Mitogenic stimulation of the host enhances susceptibility to scrapie, *Nature* **272:**54–55.
45. Kimberlin, R. H., and Cunnington, P. G., 1978, Reduction of scrapie incubation time in mice and hamsters by a single injection of methanol extraction residue of BCG, *FEMS Microbiol. Lett.* **3:**169–172.
46. Carp, R. I., and Warner, H. B., 1980, Effect of human lymphokines on scrapie incubation, in: *80th Annual Meeting American Society for Microbiology*, American Society for Microbiology, Washington, D.C., p. 262.
47. Outram, G. W., Dickinson, A. G., and Fraser, H., 1974, Reduced susceptibility to scrapie in mice after steroid administration, *Nature* **249:**855–856.
48. Kuroda, Y., Gibbs, C. J., Jr., Amyx, H., and Gajdusek, D. C., 1983, Creutzfeldt–Jakob disease in mice: Persistent viremia and preferential replication of virus in low-density lymphocytes, *Infect. Immun.* **41:**159–161.
49. Fraser, H., and Farquhar, C. F., 1987, Ionising radiation has no influence on scrapie incubation period in mice, *Vet. Microbiol.* **13:**211–223.
50. Fraser, H., Davies, D., McConnell, I., and Farquhar, C. F., 1989, Are radiation-resistant, post-mitotic, long-lived (RRPMLL) cells involved in scrapie replication? in: *Unconventional Virus Diseases of the Central Nervous System* (L. A. Court, D. Dormont, P. Brown, and D. T. Kingsbury, eds.), Commissariat a l'Energie Atomique, Departement de Protection Sanitaire, Service de Documentation, Fontenay-aux-Roses Cedex, Paris, pp. 563–574.
51. Clarke, M. C., and Kimberlin, R. H., 1984, Pathogenesis of mouse scrapie: Distribution of agent in the pulp and stroma of infected spleens, *Vet. Microbiol.* **9:**215–225.
52. McFarlin, D. E., Raff, M. C., Simpson, E., and Nehlsen, S. H., 1971, Scrapie in immunologically deficient mice, *Nature* **233:**336.
53. Outram, G. W., Dickinson, A. G., and Fraser, H., 1975, Slow encephalopathies, inflammatory responses, and arachis oil, *Lancet* **1:**198–203.
54. Outram, G. W., Dickinson, A. G., and Fraser, H., 1973, Developmental maturation of susceptibility to scrapie in mice, *Nature* **241:**103–104.
55. Hotchin, J., and Buckley, S., 1977, Latent form of scrapie virus: A new factor in slow-virus disease, *Science* **196:**668–671.
56. Carp, R. I., and Callahan, S. M., 1981, *In vitro* interaction of scrapie agent and mouse peritoneal macrophages, *Intervirology* **16:**8–13.
57. Carp, R. I., and Callahan, S. M., 1982, Effect of mouse peritoneal macrophages on scrapie infectivity during extended *in vitro* incubation, *Intervirology* **17:**201–207.
58. Carp, R. I., and Callahan, S. M., 1985, Effect of prior treatment with thioglycolate on the incubation period of intraperitoneally injected scrapie, *Intervirology* **24:**170–173.
59. Marsh, R. F., 1981, Effect of vaccinia-activated macrophages on scrapie infection in hamsters, *Adv. Exp. Med. Biol.* **134:**359–363.
60. Michel, B., Tamalet, J., Bongrand, P., Gambarelli, D., and Gastaut, J. L., 1987, Role des cellules phagocytaires dans la scrapie experimentale du hamster, *Rev. Neurol.* **143:**526–531.
61. Ehlers, B., Rudolph, R., and Diringer, H., 1984, The reticuloendothelial system in scrapie pathogenesis, *J. Gen. Virol.* **65:**423–428.
62. Rodgers, B., and Mims, C. A., 1981, Interaction of influenza virus with mouse macrophages, *Infect. Immun.* **31:**751–757.
62a. Kimberlin, R. H., and Walker, C. A., 1989, Pathogenesis of scrapie in mice after intragastric infection, *Virus Res.* **12:**213–230.
63. Field, E. J., Caspary, E. A., and Joyce, G., 1968, Scrapie agent in blood, *Vet. Rec.* **83:**109–110.

64. Clarke, M. C., and Haig, D. A., 1967, Presence of transmissible agent of scrapie in the serum of affected mice and rats, *Vet. Rec.* **80**:504.

65. Diringer, H., 1984, Sustained viremia in experimental hamster scrapie, *Arch. Virol.* **82**: 105–109.

66. Manuelidis, E. E., Gorgacz, E. J., and Manuelidis, L., 1978, Viremia in experimental Creutzfeldt–Jakob disease, *Science* **200**:1069–1071.

67. Manuelidis, E. E., Kim, J. H., Mericangas, J. R., and Manuelidis, L., 1985, Transmission to animals of Creutzfeldt–Jakob disease from human blood, *Lancet* **2**:896–897.

68. Kimberlin, R. H., and Walker, C. A., 1979, Pathogenesis of mouse scrapie: Dynamics of agent replication in spleen, spinal cord and brain after infection by different routes, *J. Comp. Pathol.* **89**:551–562.

69. Kimberlin, R. H., and Walker, C. A., 1980, Pathogenesis of mouse scrapie: Evidence for neural spread of infection to the CNS, *J. Gen. Virol.* **51**:183–187.

70. Kimberlin, R. H., and Walker, C. A., 1982, Pathogenesis of mouse scrapie: Patterns of agent replication in different parts of the CNS following intraperitoneal infection, *J. R. Soc. Med.* **75**:618–624.

71. Kimberlin, R. H., Hall, S. M., and Walker, C. A., 1983, Pathogenesis of mouse scrapie: Evidence for direct neural spread of infection to the CNS after injection of sciatic nerve, *J. Neurol. Sci.* **61**:315–325.

72. Fraser, H., 1981, Neuronal spread of scrapie agent and targeting of lesions within the retino-tectal pathway, *Nature* **294**:149–150.

73. Buyukmihci, N., Goehring-Harmon, F., and Marsh, R. F., 1983, Neural pathogenesis of experimental scrapie after intraocular inoculation of hamsters, *Exp. Neurol.* **81**: 396–406.

74. Kimberlin, R. H., Cole, S., and Walker, C. A., 1987, Pathogenesis of scrapie is faster when infection is intraspinal instead of intracerebral, *Microb. Pathogen.* **2**:405–415.

75. Dickinson, A. G., and Meikle, V. M. H., 1969, A comparison of some biological characteristics of the mouse-passaged scrapie agents, 22A and ME7, *Genet. Res.* **13**:213–225.

76. Fraser, H., and Bruce, M. E., 1973, Argyrophilic plaques in mice inoculated with scrapie from particular sources, *Lancet* **1**:617–618.

77. Dickinson, A. G., and Taylor, D. M., 1978, Resistance of scrapie agent to decontamination, *N. Engl. J. Med.* **299**:1413–1414.

78. Kimberlin, R. H., and Walker, C. A., 1978, Evidence that the transmission of one source of scrapie agent to hamsters involves separation of agent strains from a mixture, *J. Gen. Virol.* **39**:487–496.

79. Kascsak, R. J., Rubenstein, R., Merz, P. A., Carp, R. I., Robakis, N. K., Wisniewski, H. M., and Diringer, H., 1986, Immunological comparison of scrapie associated fibrils isolated from animals infected with four different scrapie strains, *J. Virol.* **59**:676–683.

80. Carp, R. I., Merz, P. A., Moretz, R. C., Somerville, R. A., Callahan, S. M., and Wisniewski, H. M., 1985, Biological properties of scrapie: An unconventional slow virus, in: *Subviral Pathogens of Plants and Animals: Viroids and Prions* (K. Maramorosch and J. J. McKelvey, eds.), Academic Press, New York, pp. 425–463.

81. Dickinson, A. G., and Meikle, V. M. H., 1971, Host–genotype and agent effects in scrapie incubation: Change in allelic interaction with different strains of agent, *Mol. Gen. Genet.* **112**:73–79.

81a. Kimberlin, R. H., and Walker, C. A., 1988, Incubation periods in six models of intra-peritoneally injected scrapie depend mainly on the dynamics of agent replication within the nervous system and not the lymphoreticular system, *J. Gen. Virol.* **69**:2953–2960.

82. Clarke, M. C., and Haig, D. A., 1966, Attempts to demonstrate neutralizing antibodies in the sera of scrapie-affected animals, *Vet. Rec.* **78**:647–649.

83. Pattison, I. H., Millson, G. C., and Smith, K., 1964, An examination of the action of whole blood, blood cells or serum on goat scrapie agent, *Res. Vet. Sci.* **5**:116–121.

84. Gardiner, A. C., 1965, Gel diffusion reactions of tissues and sera from scrapie-infected animals, *Res. Vet. Sci.* **7**:190–195.

85. Porter, D. D., Porter, H. G., and Cox, N. A., 1973, Failure to demonstrate a humoral immune response to scrapie infection in mice, *J. Immunol.* **111**:1407–1410.

86. Clarke, M. C., 1968, The antibody response of scrapie-affected mice to immunisation with sheep red blood cells, *Res. Vet. Sci.* **9**:595–597.

87. Gardiner, A. C., and Marucci, A. A., 1969, Immunological responsiveness of scrapie infected mice, *J. Comp. Pathol.* **79**:233–235.

88. Aoki, T., Gibbs, C. J., Jr., Sotelo, J., and Gadjusek, D. C., 1982, Heterogeneic auto antibody against neurofilament protein in the sera of animals with experimental kuru and Creutzfeldt–Jakob disease and natural scrapie infection, *Infect. Immun.* **38**:316–324.

89. Kingsbury, D. T., Smeltzer, D. A., Gibbs, C. J., Jr., and Gajdusek, D. C., 1981, Evidence for normal cell-mediated immunity in scrapie-infected mice, *Infect. Immun.* **32**:1176–1180.

90. Garfin, D. E., Stites, D. P., Zitnik, L. A., and Prusiner, S. B., 1978, Suppression of polyclonal B cell activation in scrapie-infected C3H/HeJ mice, *J. Immunol.* **120**:1986–1990.

91. Field, E. J., and Shenton, B. K., 1973, Rapid immunological method for diagnosis of natural scrapie in sheep, *Nature* **244**:96–97.

92. Field, E. J., and Shenton, B. K., 1974, A rapid immunologic test for scrapie in sheep, *Am. J. Vet. Res.* **35**:393–395.

93. Fraser, H., and Hancock, P. M., 1977, An investigation of the macrophage electrophoretic mobility test in the diagnosis of scrapie in sheep, *J. Comp. Pathol.* **87**:267–274.

94. Collis, S. C., Kimberlin, R. H., and Millson, G. C., 1979, Immunoglobulin G concentration in the sera of Herdwick sheep with natural scrapie, *J. Comp. Pathol.* **89**:389–396.

95. Collis, S. C., and Kimberlin, R. H., 1989, Polyclonal increase in certain IgG subclasses in mice persistently infected with the 87V strain of scrapie, *J. Comp. Path.* **101**:131–141.

96. Katz, M., and Koprowski, H., 1968, Failure to demonstrate a relationship between scrapie and the production of interferon, *Nature* **219**:639–640.

97. Field, E. J., Joyce, G., and Keith, A., 1969, Failure of interferon to modify scrapie in the mouse, *J. Gen. Virol.* **5**:149–150.

98. Worthington, M., 1972, Interferon system in mice infected with the scrapie agent, *Infect. Immun.* **6**:643–645.

99. Collis, S. C., and Kimberlin, R. H., 1985, Long term persistence of scrapie infection in mouse spleens in the absence of clinical disease, *FEMS Microbiol. Lett.* **29**:111–114.

100. Fraser, H., 1983, A survey of primary transmission of Icelandic scrapie (rida) to mice, in: *Virus Non Conventionnels et Affections du Systeme Nerveux Central* (L. A. Court and F. Cathala, eds.), Masson, Paris, pp. 34–46.

101. Dickinson, A. G., Fraser, H., and Outram, G. W., 1975, Scrapie incubation period can exceed natural lifespan, *Nature* **256**:732–733.

102. Bruce, M. E., 1985, Agent replication dynamics in a long incubation period model of mouse scrapie, *J. Gen. Virol.* **66**:2517–2522.

103. Kimberlin, R. H., and Walker, C. A., 1986, Suppression of scrapie infection in mice by heteropolyanion 23, dextran sulfate and some other polyanions, *Antimicrob. Agents Chemother.* **30**:409–413.

104. Kimberlin, R. H., and Walker, C. A., 1979, Antiviral compound effective against experimental scrapie, *Lancet* **2**:591–592.

105. Jasmin, C., Chermann, J. C., Raynard, M., Werner, G., Rayboud, N., Simoussi, F., and Boy-Loustau, C., 1974, *In vivo* and *in vitro* antiviral activity of the mineral condensed heteropolyanion 5-tungsto-2-antimoniate, in: *Proceedings of the 8th International Congress of Chemotherapy, Progress in Chemotherapy*, Vol. 2 (G. K. Daikos, ed.), Hellenic Society for Chemotherapy, Athens, pp. 956–962.

106. Ehlers, B., and Diringer, H., 1984, Dextran sulfate 500 delays and prevents mouse scrapie by impairment of agent replication in spleen, *J. Gen. Virol.* **65**:1325–1330.

107. Farquhar, C. F., and Dickinson, A. G., 1986, Prolongation of scrapie incubation period by an injection of dextran sulfate 500 within the month before or after infection, *J. Gen. Virol.* **67**:463–473.

108. Pocchiari, M., Schmittinger, S., and Masullo, C., 1987, Amphotericin B delays the incubation period of scrapie in intracerebrally inoculated hamsters, *J. Gen. Virol.* **68**:219–223.

109. Prusiner, S. B., 1982, Novel proteinaceous infectious particles cause scrapie, *Science* **216**: 136–144.

110. Kimberlin, R. H., 1982, Scrapie agent: Prions or virinos? *Nature* **297**:107–108.

111. Carp, R. I., Merz, P. A., Kascsak, R. J., Merz, G. S., and Wisniewski, H. M., 1985, Nature of the scrapie agent: Current status of facts and hypotheses, *J. Gen. Virol.* **66**:1357–1368.

112. Robertson, H. D., Branch, A. D., and Dahlberg, J. E., 1985, Focusing on the nature of the scrapie agent, *Cell* **40**:725–727.

113. Hope, J., and Kimberlin, R. H., 1987, The molecular biology of scrapie: The last two years, *Trends Neurosci.* **10**:149–151.

114. Prusiner, S. B., 1987, Prions and neurodegenerative diseases, *N. Engl. J. Med.* **317**:1571–1598.

115. Bolton, D. C., and Bendheim, P. E., 1988, A modified host protein model of scrapie, in: *Novel Infectious Agents and the Central Nervous System (Ciba Foundation Symposium 135)* (G. Bock and J. Marsh, eds.), John Wiley & Sons, Chichester, pp. 164–181.

116. Hunter, G. D., 1972, Scrapie: A prototype slow infection, *J. Infect. Dis.* **125**:427–438.

117. Merz, P. A., Somerville, R. A., Wisniewski, H. M., and Iqbal, K., 1981, Abnormal fibrils from scrapie-infected brains, *Acta. Neuropathol. (Berl.)* **54**:63–74.

118. Merz, P. A., Somerville, R. A., Wisniewski, H. M., Manuelidis, L., and Manuelidis, E. E., 1983, Scrapie-associated fibrils in Creutzfeldt–Jakob disease, *Nature* **306**:474–476.

119. Diringer, H., Gelderbloom, H., Hilmert, H., Ozel, M., Edelbluth, C., and Kimberlin, R. H., 1983, Scrapie infectivity, fibrils and low molecular weight protein, *Nature* **306**:476–478.

120. Merz, P. A., Rohwer, R. G., Kascsak, R., Wisniewski, H. M., Somerville, R. A., Gibbs, C. J., Jr., and Gajdusek, D. C., 1984, Infection-specific particle from the unconventional slow virus diseases, *Science* **225**:437–440.

121. Merz, P. A., Somerville, R. A., and Wisniewski, H. M., 1988, Abnormal fibrils in scrapie and senile dementia of the Alzheimer type, in: *Virus Non Conventionnels et Affections du Systeme Nerveux Central* (L. A. Court and F. Cathala, eds.), Masson, Paris, pp. 259–281.

122. Somerville, R. A., Merz, P. A., and Carp, R. I., 1986, Partial copurification of scrapie-associated fibrils and scrapie infectivity, *Intervirology* **25**:48–55.

123. Bolton, D. C., McKinley, M. P., and Prusiner, S. B., 1982, Identification of a protein that purifies with the scrapie prion, *Science* **218**:1309–1311.

124. McKinley, M. P., Bolton, D. C., and Prusiner, S. B., 1983, A protease resistant protein is a structural component of the scrapie prion, *Cell* **35**:57–62.

125. Bolton, D. C., McKinley, M. P., and Prusiner, S. B., 1984, Molecular characteristics of the major scrapie prion protein, *Biochemistry* **23**:5898–5906.

126. Bolton, D. C., Meyer, R. K., and Prusiner, S. B., 1985, Scrapie PrP 27–30 is a sialoglycoprotein, *J. Virol.* **53**:596–606.

127. Oesch, B., Westaway, D., Walchli, M., McKinley, M. P., Kent, S. B. H., Aebersold, R., Barry, R. A., Tempst, P., Teplow, D. B., Hood, L. E., Prusiner, S. B., and Weissmann, C., 1985, A cellular gene encodes scrapie PrP 27–30 protein, *Cell* **40**:735–746.

128. Chesebro, B., Race, R., Wehrly, K., Nishio, J., Bloom, M., Lechner, D., Bergstrom, S., Robbins, K., Mayer, L., Keith, J. M., Garon, C., and Haase, A., 1985, Identification of scrapie prion protein-specific mRNA in scrapie-infected and uninfected brain, *Nature* **315**: 331–333.

129. Robakis, N. K., Sawh, P. R., Wolfe, G. C., Rubenstein, R., Carp, R. I., and Innis, M. A., 1986, Isolation of a cDNA clone encoding the leader peptide of prion protein and expression of the homologous gene in various tissues, *Proc. Natl. Acad. Sci. U.S.A.* **83**:6377–6381.

130. Meyer, R. K., McKinley, M. P., Bowman, K. A., Braunfeld, M. B., Barry, R. A., and Prusiner, S. B., 1986, Separation and properties of cellular and scrapie prion proteins, *Proc. Natl. Acad. Sci. U.S.A.* **83:**2310–2314.

131. Bolton, D. C., Bendheim, P. E., Marmorstein, A. D., and Potempksa, A., 1987, Isolation and structural studies of the intact scrapie agent protein, *Arch. Biochem. Biophys.* **258:**579–590.

132. Hope, J., Multhaup, G., Reekie, L. J. D., Kimberlin, R. H., and Beyreuther, K., 1988, Molecular pathology of scrapie associated fibril protein (PrP) in mouse brain affected by the ME7 strain of scrapie, *Eur. J. Biochem.* **172:**271–277.

133. Merz, P. A., Kascsak, R. J., Rubenstein, R., Carp, R. I., and Wisniewski, H. M., 1987, Antisera to scrapie-associated fibril protein and prion protein decorate scrapie-associated fibrils, *J. Virol.* **61:**42–49.

134. DeArmond, S. J., McKinley, M. P., Barry, R. A., Braunfeld, M. B., McColloch, J. R., and Prusiner, S. B., 1985, Identification of prion amyloid filaments in scrapie-infected brain, *Cell* **41:**221–235.

135. Bruce, M. E., and Dickinson, A. G., 1979, Biological stability of different classes of scrapie agent, in: *Slow Transmissible Diseases of the Nervous System*, Vol. 2 (S. B. Prusiner and W. J. Hadlow, eds.), Academic Press, New York, pp. 71–86.

136. Bruce, M. E., and Dickinson, A. G., 1987, Biological evidence that scrapie agent has an independent genome, *J. Gen. Virol.* **68:**79–89.

137. Kimberlin, R. H., Cole, S., and Walker, C. A., 1987, Temporary and permanent modifications to a single strain of mouse scrapie on transmission to rats and hamsters, *J. Gen. Virol.* **68:**1875–1881.

137a. Kimberlin, R. H., Walker, C. A., and Fraser, H., 1989, The genomic identity of different strains of mouse scrapie is expressed in hamsters and preserved on reisolation in mice, *J. Gen. Virol.* **70:**2017–2025.

138. Rohwer, R. G., 1984, Scrapie infectious agent is virus-like in size and susceptibility to inactivation, *Nature* **308:**658–662.

139. Manuelidis, L., and Manuelidis, E. E., 1986, Recent developments in scrapie and Creutzfeldt–Jakob disease, in: *Progress in Medical Virology*, Vol. 33 (J. L. Melnick, ed.), S. Karger, Basel, pp. 78–98.

140. Braig, H. R., and Diringer, H., 1985, Scrapie: Concept of a virus-induced amyloidosis of the brain, *EMBO J.* **4:**2309–2312.

141. Dickinson, A. G., and Outram, G. W., 1979, The scrapie replication-site hypothesis and its implication for pathogenesis, in: *Slow Transmissible Diseases of the Nervous System*, Vol. 2 (S. B. Prusiner and W. J. Hadlow, eds.), Academic Press, New York, pp. 13–31.

142. Westaway, D., Goodman, P. A., Mirenda, C. A., McKinley, M. P., Carlson, G. A., and Prusiner, S. B., 1987, Distinct prion proteins in short and long scrapie incubation period mice, *Cell* **51:**651–662.

143. Scott, M., Foster, D., Mirenda, C. A., Serban, D., Coufal, F., Walchli, M., Torchia, M., Groth, D., Carlson, G. A., DeArmond, S. J., Westaway, D., and Prusiner, S. B., 1989, Transgenic mice expressing hamster prion protein produce species-specific scrapie infectivity and amyloid plaques, *Cell* **59:**847–857.

144. Prusiner, S. B., Scott, M., Foster, D., Pan, K.-M., Groth, D. F., Mirenda, C. A., Torchia, M., Yang, S.-L., Serban, D., Carlson, G. A., Hoppe, P. C., Westaway, D., and DeArmond, S. J., 1990, Transgenic studies implicate interactions between homologous PrP isoforms in scrapie prion replication, *Cell* **63:**673–686.

145. Carp, R. I., Kascsak, R. J., Wisniewski, H. M., Merz, P. A., Rubenstein, R., Bendheim, P. E., and Bolton, D. C., 1989, The nature of the unconventional slow infection agents remains a puzzle, *Alzheimer Dis. Assoc. Disord.* **3:**79–99.

146. Turk, E., Teplow, D. B., Hood, L. E., and Prusiner, S. B., 1988, Purification and properties of the cellular and scrapie hamster prion proteins, *Eur. J. Biochem.* **176:**21–30.

147. Hope, J., Morton, L. J., Farquhar, C. F., Multhaup, G., Beyreuther, K., and Kimberlin, R. H.,

1986, The major polypeptide of scrapie-associated fibrils (SAF) has the same size, charge distribution and N-terminal protein sequence as predicted for the normal brain protein (PrP), *EMBO J.* **5**:2591–2597.

148. Stahl, N., Borchelt, D. R., Hsiao, K., and Prusiner, S. B., 1987, Scrapie prion protein contains a phosphatidylinositol glycolipid, *Cell* **51**:229–240.
149. Borchelt, D. R., Scott, M., Taraboulos, A., Stahl, N., and Prusiner, S. B., 1990, Scrapie and cellular prion proteins differ in the kinetics of synthesis and topology in cultured cells, *J. Cell Biol.* **110**:743–752.
150. Taraboulos, A., Serban, D., and Prusiner, S. B., 1990, Scrapie prion proteins accumulate in the cytoplasm of persistently infected cultured cells, *J. Cell Biol.* **110**:2117–2132.
151. Carp, R. I., Kim, Y. S., Kascsak, R. J., Merz, P. A., and Rubenstein, R., 1989, Classic genetics of scrapie, in: *Alzheimer's Disease and Related Disorders* (K. Iqbal, H. M. Wisniewski, and B. Winblad, eds.), Alan R. Liss, Inc., New York, pp. 567–582.

III

Human Infections
of the CNS

8

Enteroviruses

JOSEPH L. MELNICK

1. INTRODUCTION

The polioviruses are the only enteroviruses for which there is sufficient information to devote a chapter to their neurotropism and prevention by immunologic
methods. They are thus the main feature of this chapter. However, it is important
to bear in mind that other enteroviruses may also invade the central nervous
system (CNS). Especially prominent in this regard are coxsackie B viruses and
enterovirus type 71, but other nonpolio enteroviruses also have been associated
with CNS disease.

The history of poliomyelitis and of the development of knowledge concerning the polioviruses and the other enteroviruses has been described elsewhere.[1-3]

2. DESCRIPTION AND CLASSIFICATION

The enteroviruses are classified as a genus, *Enterovirus*, within the family
Picornaviridae. In addition to polioviruses, the genus includes coxsackieviruses of
the A and the B groups, echoviruses, and the high-numbered (68–72) enteroviruses. The picornavirus family has many other members, including genus
Rhinovirus, whose members also infect humans. In addition, there are enteroviruses and rhinoviruses that infect different groups of lower animals; and within
the Picornaviridae there are two other genera—*Aphthovirus*, agents that cause
foot-and-mouth disease of cattle, and *Cardiovirus*, viruses that infect rodents. The
enteroviruses of humans are listed in Table 8-1.

JOSEPH L. MELNICK • Division of Molecular Virology, Baylor College of Medicine, Houston,
Texas 77030.

Neuropathogenic Viruses and Immunity, edited by Steven Specter *et al.* Plenum Press, New York,
1992.

TABLE 8-1
Enteroviruses of Humans

Subgroup	Number of Serotypes
Polioviruses	3 (types 1–3)
Coxsackieviruses, group A	23 (types A1–A22, A-24)[a]
Coxsackieviruses, group B	6 (types B1–B6)
Echoviruses	31 (types 1–9, 11–27, 29–33)[b]
Enteroviruses	5 (types 68-72)[c]

[a]Coxsackievirus A23 turned out to be the same virus as that previously identified as echovirus 9.
[b]Echovirus 10 has been reclassified as reovirus, echovirus 28 as rhinovirus type 1A, and echovirus 34 as coxsackievirus A24.
[c]Enterovirus 72—hepatitis A virus—now seems to warrant placement as a separate genus within the picornavirus family. The name *Heparnavirus* has been proposed.

Enteroviruses are transient inhabitants of the human alimentary tract; they are isolated most frequently from stool specimens but may also be recovered from the throat. Enterovirus 70, the agent of acute hemorrhagic conjunctivitis (AHC),[4] has been found almost exclusively in conjunctival and throat specimens, but a few fecal isolations have been reported. Wild polioviruses vary in their virulence and neurotropism. Although their most severe effect on the infected individual is produced by invasion of the CNS, poliovirus, unlike some other enteroviruses, is seldom isolated from the cerebrospinal fluid (CSF). Most enteroviruses can be cultivated in cell cultures, where they can be detected by their characteristic cytopathic effects.

Enteroviruses and the other picornaviruses are among the smallest viruses of animals, with virions about 28 nm in diameter. The icosahedral capsid shell, composed of 60 subunits, has no envelope; it surrounds a genome made up of a single strand of positive-sense infectious RNA of relatively small molecular weight (2.5×10^6). For poliovirus, the complete sequence of RNA has been determined, and the molecular biology of poliovirus continues to be intensively studied.[5] In addition, by means of x-ray diffraction studies the three-dimensional structure of poliovirus has been determined.[6] The details of structure that have been revealed have shed much light on the antigenic sites. Recent information has greatly advanced research on the neutralization of the polioviruses.[7,8]

3. EPIDEMIOLOGY

3.1. General Epidemiology of Enteroviruses

Enteroviruses can cause a variety of illnesses. Different viruses may produce the same syndrome; on the other hand, the same enterovirus may cause more than a single syndrome. But by far the most common form of enterovirus infection is inapparent or is accompanied only by minor malaise. This characteristic makes it difficult to trace the course of transmission; poliomyelitis remained an epidemiologic enigma until the predominance of inapparent infections and mild illnesses was recognized. Although overt illnesses represent only a "very small tip of a very large iceberg," wide dissemination has been repeatedly documented for a number of the enteroviruses.[9,10]

Humans are the only known reservoir for members of the human enterovirus group, and close human contact is the primary avenue of spread of the enteroviruses. From infected individuals, whether or not they develop clinical illness, the oropharynx and intestine can yield virus—in stools for as long as a month or two, and in oropharyngeal secretions for a shorter period. Fecal contamination is the usual source of transmission. Enteroviruses are most readily spread within a household. Commonly, by the time an infection is recognized in one individual within a family, all susceptible members have already been infected. The extent of intrafamilial spread appears to be closely related to duration of virus shedding, particularly by young children.

The close correlation between living under low socioeconomic conditions and the acquisition of infection with the enteroviruses early in life has been emphasized repeatedly in both tropical and temperate environments and reflects the general level of hygiene of the population group.[1,11]

3.2. Epidemiology of Poliomyelitis

The epidemiology of poliomyelitis, as the most severe and the most studied of the diseases caused by enteroviruses, serves to illustrate patterns typical of other members of the group as well.[3]

Under the historical endemic conditions, polioviruses circulated widely and constantly under conditions of poor community sanitation and family hygiene, infecting new susceptible individuals early in life. Since almost all women of childbearing age had antibody to all three poliovirus types, passive immunity was transferred from mother to offspring. Most infants experienced their first poliovirus infections, which provided active immunization, in the first few months of life while maternal antibodies still provided some protection. Because so large a proportion of poliovirus infections are subclinical, such rare paralytic cases as did occur often went unrecorded in populations faced with very high infant and child mortality rates from many other causes, known and unknown.

This endemic phase is not merely of historical interest, for these conditions still exist in some parts of the world. Also, some areas are currently experiencing a transition to an epidemic phase, while others have fully entered the "vaccine era" of polio epidemiology in which paralytic poliomyelitis has been brought under virtually complete control.

The transition from the endemic phase to the severe epidemic phase in industrialized countries is explained as follows. With increased economic development, resources for community and household hygiene were enhanced. With the polioviruses, this meant that opportunities for early immunizing infections acquired from household and environmental contamination were reduced among infants, so that many persons were infected for the first time in later childhood or in adult life—ages at which poliovirus infections are more likely to take the paralytic form. In temperate climates, initial exposures were even further delayed, allowing the pool of susceptible persons to increase. Thus, when virulent polioviruses did enter the population, they spread rapidly and explosively, in contrast to the steady endemic transmission of the preceding phase.

Large epidemics of paralytic polio appeared within the past century, first in northern Europe and soon thereafter in the northeastern United States. By 1955, just before the inactivated polio vaccine (IPV) became generally available, combined totals of more than 76,000 cases of paralytic polio were being reported annually from the USSR, 23 other European countries, the United States, Canada, Australia, and New Zealand. The United States alone was experiencing 10,000 to 20,000 paralytic cases annually. After IPV came into use, cases in the United States were reduced markedly, yet about 2500 cases continued to occur each year, some even among the fully vaccinated. Live, attenuated, orally administered polio vaccine (OPV), licensed in 1961–1962, has become widely used in the United States, and the number of cases has been reduced to fewer than ten annually. This excellent result is being seen in most of the other industrialized, developed parts of the world, where paralytic poliomyelitis is now a rare disease.[12]

Unfortunately, this is not yet true for many developing countries in tropical and subtropical areas. In recent years many of these countries—some of them with very limited health care resources in facilities, personnel, and funds—have been facing the epidemic phase of polio epidemiology. In a number of areas where diagnostic and surveillance activities are severely limited, few cases are being reported, yet special "lameness" surveys conducted with the assistance of the World Health Organization (WHO) Expanded Programme on Immunization (EPI)[13–15] indicate that the actual incidence of polio in the years prior to the surveys has been at least as high as the rates seen in more privileged countries just before vaccines were introduced. The WHO has estimated that worldwide, about 250,000 cases of paralytic polio are still occurring each year.[12]

4. CLINICAL DISEASES CAUSED BY ENTEROVIRUSES

4.1. Poliomyelitis

When a susceptible individual is exposed to poliovirus, responses may range from inapparent infection to paralytic poliomyelitis. The disease may progress from a minor illness to the major severe illness, in some instances after an intervening few days without symptoms.

The minor illness is characterized by fever, malaise, drowsiness, headache, nausea, vomiting, constipation, or sore throat in various combinations. The patient recovers in a few days. In addition to the symptoms and signs mentioned above, the patient may present with stiffness and pain in the back and neck (aseptic meningitis). The disease may last 2 to 10 days, and recovery can be rapid and complete. In rare instances, this form may advance to paralysis. Poliovirus is only one of many viruses that produce aseptic meningitis.

In the absence of virological laboratory diagnosis, paralytic poliomyelitis must be suspected if disease occurs in persons associated with paralytic patients, since paralysis is rare in other enterovirus infections. In poliomyelitis, the major paralytic illness may follow the minor illness described above, particularly in young children, but it usually develops without an antecedent first phase. The predominating sign is flaccid paralysis resulting from lower motor neuron damage. However, incoordination secondary to brainstem invasion may occur, and there may be painful spasms of nonparalyzed muscles. The amount of damage and destruction varies from case to case. Muscle involvement is usually maximal within a few days after the paralytic phase begins. Maximal recovery of function usually has been reached within 6 months, but it may take longer.

At times, nonpolio enteroviruses have been associated with cases of polio-like paralytic disease, but this has been uncommon.[16] Enterovirus 71 also has been involved in several outbreaks of CNS disease, including polio-like paralysis, with some fatal cases.[17] Coxsackievirus A7 has been associated with outbreaks of paralytic disease.[16,18]

4.2. Meningitis and Mild Paresis Caused by Nonpolio Enteroviruses

Common early symptoms are fever, malaise, headache, nausea, and abdominal pain. One to two days later there may be signs of meningeal irritation with stiffness of the neck or back; vomiting may also appear at this time. The disease sometimes progresses to mild muscle weakness that is often confused clinically with paralytic poliomyelitis.

Enteroviruses that have been associated, to at least some degree, with meningitis or transient mild paresis and on very rare occasions with paralytic CNS disease include the polioviruses, almost all coxsackieviruses of both A and B groups, and most echoviruses.[3] The chief types repeatedly associated with meningitis are coxsackieviruses B1–B6, A7 and A9 and echoviruses 4, 6, 9, 11, 14, 16, 25, 30, 31, and 33; types 3, 18, and 29 also have been responsible for some outbreaks. Muscle weakness and mild, transient paralysis have been observed with echoviruses 6 and 9. Type 9 also has been recovered from the medulla of a fatal case.

Among the newer, high-numbered enteroviruses, type 70—the agent of AHC—in rare instances has been involved in neurological complications including poliomyelitis-like illnesses.[19] Infections with enterovirus 71, which exhibit a variety of clinical manifestations, have been associated with meningitis and with some cases of more severe polio-like CNS disease; some of the latter cases have been fatal.[17]

Patients almost always recover completely from paresis caused by nonpolio enteroviruses. However, among infants infected during their first year of life there is a risk of serious neurological sequelae.[20]

4.3. Other Diseases Caused by Enteroviruses

In addition to the CNS diseases indicated above, illnesses caused by enteroviruses include pleurodynia, myocarditis, hepatitis, vesicular and exanthematous skin lesions, mucocutaneous lesions, respiratory and intestinal illnesses, undifferentiated febrile illness, and conjunctivitis. There is also considerable evidence indicating that enteroviruses may have a role in some cardiovascular diseases and perhaps in the development of diabetes.[3]

5. PATHOGENESIS AND IMMUNITY

As the virus travels from the portal of entry (the mouth), implantation and multiplication take place in the oropharynx and the small intestine (Fig. 8-1). The incubation period (defined as the time from exposure to onset of disease) is usually between 7 and 14 days. By 3 to 5 days after exposure, virus can be recovered from blood, throat, and feces. At this time symptoms of the "minor illness" may appear, or the infection may remain asymptomatic, but viremia begins several days before the onset of CNS signs in those who develop either "nonparalytic polio" (aseptic meningitis) or the paralytic disease. Antibodies develop early, usually before paralysis appears. After free virus can no longer be found in the blood, virus bound to antibody may be detected for a few additional days.[21]

After initial multiplication in the tonsils, the lymph nodes of the neck, Peyer's patches, and the small intestine, the virus then spreads by way of the bloodstream to other susceptible tissues (other lymph nodes, brown fat, and the CNS).

Poliovirus can also spread along axons of peripheral nerves to the CNS; there it continues to progress along the fibers of the lower motor neurons, increasingly involving the spinal cord and/or parts of the brain. Tonsillectomy or other surgery in the oropharynx increases the risk of CNS involvement at times when polioviruses are prevalent. This may result from virus in the pharynx gaining direct access to cut nerve fibers or may be a secondary consequence of the removal of immunologically active lymphoid tissue.

Poliovirus invades only certain types of nerve cells; in the process of its intracellular multiplication, it may damage or completely destroy these cells. The anterior horn cells of the spinal cord are most prominently involved, but in severe cases the intermediate gray ganglia and even the posterior horn and dorsal root ganglia are often affected. Lesions are found as far forward as the hypothalamus and thalamus. In the brain, the reticular formation, the vestibular nuclei, the cerebellar vermis, and the deep cerebellar nuclei are most often affected. The cortex is virtually spared, with the exception of the motor cortex along the precentral gyrus.

Although flaccid paralysis is the hallmark of poliomyelitis, the virus does not multiply in muscle *in vivo*. The changes that occur in peripheral nerves and voluntary muscles are secondary to destruction of nerve cells within the CNS. Cells that are not killed but lose function temporarily as a result of edema may recover completely within 3 to 4 weeks after onset. Inflammation occurs secondary to the attack on nerve cells.

The development of immunity to the polioviruses is typical of that for the enteroviruses generally. Virus-neutralizing antibody develops within a few days after exposure to the virus and may persist for life.[22]

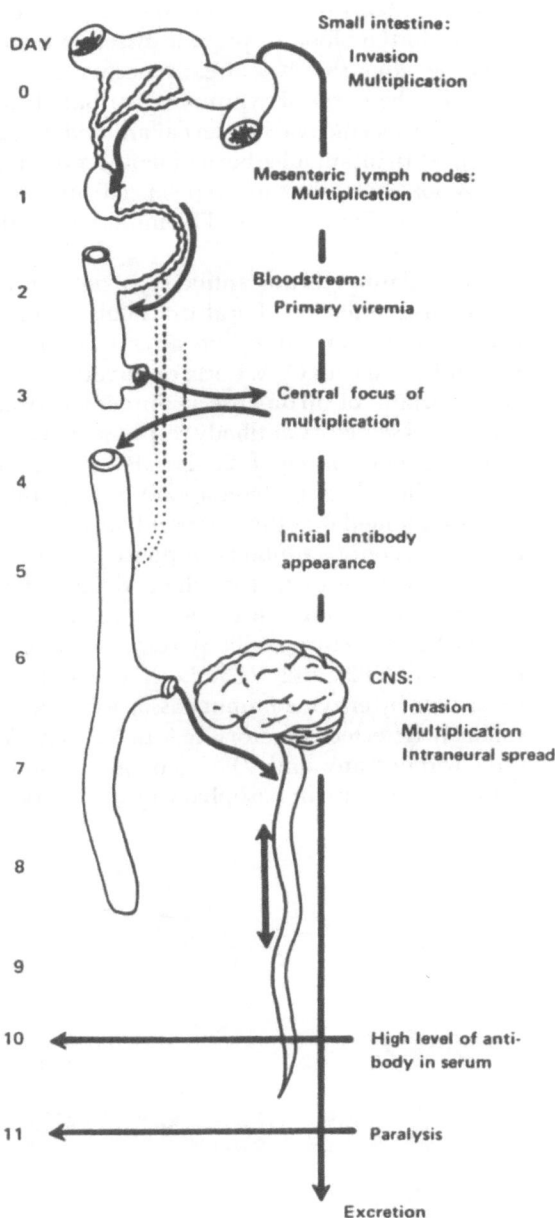

FIGURE 8-1. Schematic illustration of the pathogenesis of poliomyelitis (modified from Fenner). Virus enters by way of the alimentary tract and multiplies locally at the initial sites of virus implantation (tonsils, Peyer's patches) or the lymph nodes that drain these tissues, and virus begins to appear in the throat and in the feces. Secondary virus spread occurs by way of the bloodstream to other susceptible tissues, namely, other lymph nodes, brown fat, and the CNS. Within the CNS the virus spreads along nerve fibers. If a high level of multiplication occurs as the virus spreads through the CNS, motor neurons are destroyed, and paralysis occurs. The shedding of virus into the environment does not depend on secondary virus spread to the CNS.

Passive immunity is transferred from mother to offspring. The maternal antibodies gradually disappear during the first 6 months of life. Passively administered antibody (i.e., γ-globulin) lasts only 3 to 5 weeks. Since antibodies must be present in the blood to prevent dissemination of virus to the brain and are not effective after this has already occurred, immunization is of value only if it precedes the onset of symptoms attributed to CNS infection.

Type specificity of maternal antibodies has been studied in a model animal system.[23] In infant mice born of mothers immunized with coxsackieviruses, cross-protection shows the same type-specificity as that observed in neutralization and complement-fixation tests. The immunity conferred by the mother's milk is also type specific.

Circulating serum antibody is not the only source of protection against enterovirus infection. Local or cellular immunity is manifested by protection against intestinal reinfection after recovery from a natural infection or after immunization with OPV. Local or secretory IgA is generally recognized as having an important role in defense against enteroviral infections.[24] The development of serum and secretory antibody responses to OPV and to intramuscular inoculation of IPV is shown in Fig. 8-2. The IPV used at that time was found to be not very effective in inducing secretory antibody in the respiratory or intestinal tracts. It had been hoped that the newer enhanced-potency IPV[25] would stimulate a more effective secretory antibody response. In a recent study[26] on the development of antibody responses to the whole virus and to the subunit virion proteins in humans, infants immunized with enhanced IPV or with OPV were studied for serum and secretory antibody responses to the poliovirus itself and to polypeptides VP1, VP2, and VP3. Both vaccines induce neutralizing IgG and IgG detectable by enzyme immunoassay to the whole virus and to VP1 and VP3 and similarly detected secretory IgA to VP1 and VP2 in the nasopharyngeal secretions without any anti-VP3 response. However, in regard to the neutralizing antibody response in nasopharyngeal secretions OPV was markedly more effec-

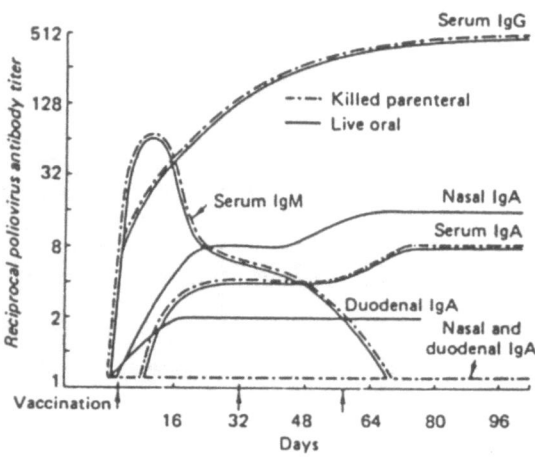

FIGURE 8-2. Serum and secretory antibody response to oral administration of live attenuated polio vaccine (OPV) and to intramuscular inoculation of killed polio vaccine (IPV). (From Ogra *et al.*,[24] with permission.)

tive than the enhanced IPV; 70% of the infants developed this response after OPV, as compared with only 27% of those who had received enhanced-potency IPV.[26]

6. PERSISTENCE

The usual events in an enterovirus infection, whether subclinical or clinically apparent, follow a relatively brief course: except in the rare instance of a fatal outcome, antibodies develop, the infection is resolved, and the virus is completely cleared from the host. Recent studies indicate that enteroviruses may produce a persistent infection particularly associated with myalgic encephalomyelitis, also known as postviral fatigue syndrome.[27]

In persons with deficiencies in either humoral or cell-mediated immunity, enterovirus infections (including those with OPV) present a considerably increased risk and may take a different course that may include persistence of the virus for long periods. In such persons poliovirus infection (either by wild virus or by live vaccine strains) may develop in an atypical manner, with an incubation period longer than 28 days and unusual lesions in the CNS. A number of persistent or fatal infections of immunodeficient persons by echoviruses, particularly types 9, 11, 19, 30, and 33, have been reported.[28–30] A prominent feature of the infections was the patients' inability to eradicate the virus from the CSF; some continued to yield virus from CSF for up to 3 years.[30]

7. CONTROL OF ENTEROVIRAL DISEASES

7.1. Control of Paralytic Poliomyelitis

Both OPV[31] and IPV[32] are available and are excellent vaccines.[33] The IPV, administered intramuscularly, was used in a number of countries for several years after it was licensed, and its use has continued since the mid-1950s in several countries or provinces with small populations served by excellent health care systems; these areas are located in cooler climates.

The IPV formulations used until recently were of low immunogenicity, and continuing booster inoculations were needed to maintain satisfactory antibody levels; furthermore although humoral antibodies induced by IPV can protect the vaccinated individual from developing paralytic poliomyelitis, IPV does not induce local secretory antibodies that could regularly block intestinal carriage by vaccinees, who thus can still serve to transmit the virus. Therefore, when wild virulent strains of poliovirus have entered countries that depend on IPV, they have been able to spread, even through such well-vaccinated populations as those of the Netherlands[34] and Finland.[35,36] In the Netherlands, cases occurred only among persons who had refused vaccination on religious grounds, but a disturbingly high percentage of children fully vaccinated with enhanced IPV (eIPV) in the affected communities also excreted the epidemic virus. In Finland, cases occurred even in some fully vaccinated persons, and subsequent studies indicated that at least 100,000 persons throughout the country had been infected with the epidemic strain. The outbreak was attributed to a combination of low immunogenicity of the type 3 component of the IPV that had been used for many years and the introduction of a wild, virulent type 3 strain that showed considerable antigenic divergence from the vaccine strain.[35,36]

New formulations of eIPV have stimulated new investigations.[37] Results have been promising in trials in the United States[38] and also in some areas within developing countries in the tropics.[39] However, reliance on IPV or eIPV alone has proven dangerous.[40] The IPV protects the vaccinated person from paralysis but not from gut infection by wild poliovirus. Consequently IPV-vaccinated persons may circulate wild virus and expose susceptible contacts. In one such area where young children received only eIPV, an outbreak of paralytic polio occurred among the susceptible adults living in the community.

Most countries have continued to rely primarily on OPV, and its use has virtually eliminated poliomyelitis from most of the industrialized regions of the world.[12] By 1988, OPV, the vaccine recommended by WHO's EPI, was preventing at least 350,000 cases of paralytic poliomyelitis each year in developing countries.

The wider use of OPV has been related to its greater ease of administration by the oral route, lower cost, ability to induce both serum antibodies and intestinal resistance, and the rapidity with which vaccinees develop long-lasting immunity. A number of serological studies have indicated that the proportion of individuals found to possess antibodies is considerably greater than would appear to be explainable either by their vaccination histories or by the circulation of wild polioviruses in their communities. The spread of vaccine-derived virus through the community, thus immunizing nonvaccinated persons, is viewed by some as an advantage despite the fact that such individuals may be infected with modified virus excreted by the vaccinees—a virus that obviously has not been tested for safety, as was the original vaccine administered to the vaccinees.

Problems associated with live poliovirus vaccine relate chiefly to the fact that the vaccine consists of living viruses, which can mutate. Some poliovirus strains excreted by vaccinees, although still attenuated, are indeed less attenuated than the vaccine viruses administered. There have been rare cases of paralytic poliomyelitis in vaccine recipients and in close contacts of vaccinees that are temporally and epidemiologically associated with vaccine administration. Among such vaccine-associated cases are those in persons with immune deficiencies, either long-term or induced in the course of chemotherapy or organ transplantation. If compromised immunity has been recognized, neither OPV nor any other living infectious agent should be given to the immune-deficient person or to his or her close contacts.

The risks of paralytic polio associated with OPV are exceedingly small, and by the 1980s such cases have decreased to an almost vanishing number. In a long-term WHO study among 12–15 nations and covering three sequential 5-year periods, live poliovirus vaccine has been judged repeatedly to be an extraordinarily safe vaccine, with less than one reported case for every million babies vaccinated.[41]

In the recent evaluation of cases in the United States for the period 1973–1984, it has been estimated that, overall, there was one vaccine-associated case per 2.6 million vaccine doses distributed. When first and subsequent doses were considered separately, however, the rate was estimated to be one case per 520,000 first doses versus one case per 12.3 million subsequent doses.[42] Of 105 vaccine-associated cases during the 12-year period, 35 were in recipients, and 50 were in contacts. In many tropical countries at present, there may be even less risk to contacts because almost all parents and older siblings of vaccinees would have been naturally immunized previously by infections with wild polioviruses.

The problems of controlling paralytic polio in developing countries by use of vaccine are often exacerbated by the fact that in some areas newborn infants have only a short time in which to acquire protective immunity because exposure comes so early, in their first months of life.[43]

The recommended primary schedule for routine administration of OPV to infants in the United States starts at 6–12 weeks of age; two subsequent doses are given at intervals of 6–8 weeks, and a booster is recommended at 4–6 years of age.[44]

In developing countries in tropical areas, the schedule must be accelerated: not only should primary immunization begin very early—even at birth—but in particular should be completed early in infancy. In an evaluation of the effectiveness of vaccinating newborns, the Global Advisory Group for WHO's EPI[13] has stated that although the serological response to trivalent OPV administered in the first week of life is less than that observed in older infants, 30–50% of the infants develop serum antibodies to one or more poliovirus types. Furthermore, 70–100% of neonates benefit by developing local immunity in the intestinal tract. Many of the remaining infants have been immunologically primed, and they respond promptly and to higher antibody levels when additional doses are given later in life.

For those infants in many countries whose only encounter with preventive services is at the time of birth, this single dose of vaccine will offer some protection, and the vaccinated infants will be less likely to be a source of transmission of wild polioviruses.[13] The EPI schedule designed to provide protection at the earliest possible age is at birth and then at 6, 10, and 14 weeks of age.

In some tropical countries live vaccines have not induced antibody production in a satisfactorily high percentage of vaccinees.[45] This lower rate of vaccine "takes" has been ascribed to various factors: interference from other enteroviruses in the intestinal tract, antibody in breast milk, cellular resistance in the intestinal tract because of previous exposure to wild polioviruses (or perhaps to related viruses), and an inhibitor in the alimentary tract (saliva) of infants. This low response, regardless of the reason, may be overcome by proper and repeated use of live vaccine.[46,47] The vaccine should be protected against thermal inactivation by the use of a stabilizer[48–50] and constant maintenance at low (4°C) temperature.[13]

A combined schedule utilizing both OPV and IPV may be necessary in some situations. In one such program,[51,52] the schedule included administration of OPV (type 1 monovalent) during the first month of an infant's life; then at 2½ months and again at 4 months of age trivalent OPV is given, and at the same time the infant is inoculated with a quadruple vaccine consisting of diphtheria/tetanus/pertussis plus IPV; trivalent OPV is given at 5½ months and again at 12 months. The rationale for this schedule is that, under conditions of regular and heavy importation of virus resulting in frequent challenge from virulent wild polioviruses early in infancy, features of both types of vaccine are needed.[51,52] The OPV acts by inducing protective immunity both in the form of circulating humoral antibodies and in the form of intestinal immunity; furthermore, the immunity that ensues is long-lasting. The IPV, on the other hand, provides an immediate immunogenic stimulus that is not subject to the interfering or inhibiting factors that may prevent live vaccine "takes" in some young infants. With the schedule of combined vaccination, immediate protection can be provided in the critical first weeks or months of life, and long-lasting protection—both humoral and intestinal—also is provided.

Studies of the results of this combined vaccine program indicate the following. (1) Protection provided by OPV alone was about 90% effective; that is, the case rate in those who received OPV alone was one-tenth the rate in those who were not vaccinated or who received only part of the series of live vaccine feedings. (2) In the children who received two doses of IPV plus three doses of OPV, virtually 100% protection was achieved. Two serological surveys in children aged 9 to 36 months have substantiated their protection.[52]

7.2. Control of Other Enteroviral Diseases

For the nonpolio enteroviruses, no specific control measures are known. In view of the numerous reports of serious or even fatal enterovirus infections of newborns, hospital personnel need to be especially alert to even "minor illnesses" compatible with enterovirus infections in mothers delivering babies who enter newborn nurseries or special-care units. The staff members of these units also need to be constantly aware of possible hazards from their own "minor illnesses."[53,54]

Particularly for infants threatened by severe nursery outbreaks of infection with group B coxsackieviruses and other serious enteroviral disease, as in an outbreak of echovirus 11 infections in particularly vulnerable infants, providing passive protection with γ-globulin may be considered.[55]

In outbreaks of infection by enterovirus 70, precautions may be taken on the basis of the ability of this virus to be transmitted by fomites. Special precautions are indicated for eye clinics, and infected children may be excluded from schools. In the household, such hygienic precautions as care to avoid touching an infected eye or sharing towels have been judged to be helpful in limiting outbreaks.[56]

Development of vaccines for selected nonpolio enteroviruses is technically possible, and it may be prudent to consider seriously having such vaccines available for some of these viruses, for example, for coxsackieviruses of the B group that not only can cause serious CNS diseases but also myocarditis in infants and cardiovascular disorders in adults. Such vaccines may not be required for use in the general population but could be important for specifically targeted groups at high risk.

REFERENCES

1. Paul, J. R., 1971, *A History of Poliomyelitis*, Yale University Press, New Haven.
2. Melnick, J. L., 1983, Portraits of viruses: The picornaviruses, *Intervirology* **20**:61–100.
3. Melnick, J. L., 1990, Enteroviruses: Polioviruses, coxsackieviruses, echoviruses, and newer enteroviruses, in: *Virology*, 2nd ed. (B. N. Fields, D. M. Knipe, R. M. Chanock, M. Hirsch, J. L. Melnick, T. Monath, and B. Roizman, eds.), Raven Press, New York, pp. 549–605.
4. Mirkovic, R. R., Kono, R., Yin-Murphy, M., Sohier, R., Schmidt, N. J., and Melnick, J. L., 1973, Enterovirus type 70: The etiologic agent of pandemic acute haemorrhagic conjunctivitis, *Bull. WHO* **49**:341–346.
5. Nomoto, A., and Wimmer, E., 1987, Genetic studies of the antigenicity and the attenuation phenotype of poliovirus, in: *Molecular Basis of Virus Disease, Symposium 40, Soc. Gen. Microbiol.* (W. D. Russell and J. W. Almond, eds.), Cambridge University Press, Cambridge, pp. 107–134.
6. Hogle, J. M., Chow, M., and Filman, D. J., 1985, Three-dimensional structure of poliovirus at 2.9 Å resolution, *Science* **229**:1358–1365.
7. Burke, K. L., Dunn, G., Ferguson, M., Minor, P. D., and Almond, J. W., 1988, Antigen chimaeras of poliovirus as potential new vaccines, *Nature* **332**:81–82.
8. Murray, M. G., Kuhn, R. J., Arita, M., Kawamura, N., Nomoto, A., and Wimmer, E., 1988, Poliovirus type 1/type 3 antigenic hybrid virus constructed *in vitro* that elicits type 1 and type 3 neutralizing antibodies in rabbits and monkeys, *Proc. Natl. Acad. Sci. U.S.A.* **85**:3203–3207.
9. Sabin, A. B., Krumbeigel, E. R., and Wigand, R., 1958, ECHO type 9 virus disease: Virologically controlled clinical and epidemiologic observations during 1957 epidemic in Milwaukee with notes on concurrent similar diseases associated with Coxsackie and other ECHO viruses, *Am. J. Dis. Child.* **96**:197–219.
10. Hall, C. E., Cooney, M. K., and Fox, J. P., 1970, The Seattle Virus Watch Program. I. Infection and illness experience of Virus Watch families during a community-wide epidemic of echovirus type 30 aseptic meningitis, *Am. J. Publ. Health* **60**:1456–1465.
11. Grist, N. R., Bell, E. J., and Assaad, F., 1978, Enteroviruses in human disease, *Prog. Med. Virol.* **24**:114–157.
12. World Health Organization, 1987–1989, Poliomyelitis in 1985; Poliomyelitis in 1986, 1987, and 1988, *Weekly Epidemiol. Rec.* **62**:273–280; **64**:273–279.
13. WHO EPI Global Advisory Group, 1985, Summary of conclusions and recommendations, *Weekly Epidemiol. Rec.* **60**:13–16.
14. LaForce, F. M., Lichnevski, M. S., Keja, J., and Henderson, R. H., 1980, Clinical survey techniques to estimate prevalence and annual incidence of poliomyelitis in developing countries, *Bull. WHO* **58**:609–620.
15. Foster, S. O., Kesseng-Maben, G., N'jie, H., and Coffi, E., 1984, Control of poliomyelitis in Africa, *Rev. Infect. Dis.* **6**:S433–S437.
16. Grist, N. R., and Bell, E. J., 1984, Paralytic poliomyelitis and nonpolio enteroviruses: Studies in Scotland, *Rev. Infect. Dis.* **6**:S385–S386.
17. Melnick, J. L., 1984, Enterovirus type 71 infections: A varied clinical pattern sometimes mimicking paralytic poliomyelitis, *Rev. Infect. Dis.* **6**:S387–S390.
18. Voroshilova, M. K., and Chumakov, M. P., 1959, Poliomyelitis-like properties of AB-IV-Coxsackie A7 group of viruses, *Prog. Med. Virol.* **2**:106–170.
19. Kono, R., Miyamura, K., Tajiri, E., Shiga, S., Sasagawa, A., Irani, P. F., Katrak, S. M., and Wadia, N. H., 1974, Neurologic complications associated with acute hemorrhagic conjunctivitis virus infection and its serological confirmation, *J. Infect. Dis.* **129**:590–593.
20. Sells, C. J., Carpenter, R. L., and Ray, C. G., 1975, Sequelae of central-nervous-system enterovirus infections, *N. Engl. J. Med.* **293**:1–4.
21. Melnick, J. L., Proctor, R. O., Ocampo, A. R., Diwan, A. R., and Ben-Porath, E., 1966, Free and bound virus in serum after administration of oral poliovirus vaccine, *Am. J. Epidemiol.* **84**: 329–342.

22. Paul, J. R., Riordan, J. T., and Melnick, J. L., 1951, Antibodies to three different antigenic types of poliomyelitis virus in sera from North Alaskan Eskimos, *Am. J. Hyg.* **54:**275–285.
23. Melnick, J. L., Clark, N. A., and Kraft, L. M., 1950, Immunological reactions of the Coxsackie viruses. III. Cross-protection tests in infant mice born of vaccinated mothers. Transfer of immunity through the milk, *J. Exp. Med.* **92:**499–505.
24. Ogra, P. L., Fishaut, M., and Gallagher, M. R., 1980, Viral vaccination via the mucosal routes, *Rev. Infect. Dis.* **2:**352–369.
25. van Wezel, A. L., van Steenis, G., Hannick, C. A., and Cohen, H., 1978, New approach to the production of concentrated and purified inactivated polio and rabies tissue culture vaccines, in: *Developments in Biological Standardization*, Vol. 41, *Vaccinations in the Developing Countries*, (R. H. Regamey, ed.) S. Karger, Basel, pp. 159–168.
26. Zhaori, G., Sun, M., and Ogra, P. L., 1988, Characteristics of the immune response to poliovirus virion polypeptides after immunization with live or inactivated polio vaccines, *J. Infect. Dis.* **158:**160–165.
27. Yousef, G. E., Bell, E. J., Mann, G. F., Murugesan, V., Smith, D. G., McCartney, R. A., and Mowbray, J. F., 1988, Chronic enterovirus infection in patients with postviral fatigue syndrome, *Lancet* **1:**146–150.
28. Wilfert, C. M., Buckley, R. H., Mohanakumar, T., Griffith, J. F., Katz, S. L., Whisnant, J. K., Eggleston, P. A., Moore, M., Treadwell, E., Oxman, M. N., and Rosen, F. S., 1977, Persistent and fatal central-nervous-system echovirus infections in patients with agammaglobulinemia, *N. Engl. J. Med.* **296:**1485–1489.
29. Ziegler, J. B., and Penny, R., 1975, Fatal echo 30 virus infection and amyloidosis in X-linked hypogammaglobulinemia, *Clin. Immunol. Immunopathol.* **3:**347–352.
30. Hodes, D. S., and Espinoza, D. V., 1981, Temperature sensitivity of isolates of echovirus type 11 causing chronic meningoencephalitis in an agammaglobulinemic patient, *J. Infect. Dis.* **144:**377.
31. Melnick, J. L., 1988, Live attenuated poliovaccines, in: *Vaccines* (S. A. Plotkin and E. A. Mortimer, Jr., eds.), W. B. Saunders, Philadelphia, pp. 115–157.
32. Salk, J., and Drucker, J., 1988, Noninfectious poliovirus vaccine, in: *Vaccines* (S. A. Plotkin and E. A. Mortimer, Jr., eds.), W. B. Saunders, Philadelphia, pp. 158–181.
33. Robbins, F. C., 1988, Polio—historical, in: *Vaccines* (S. A. Plotkin and E. A. Mortimer, Jr., eds.), W. B. Saunders, Philadelphia, pp. 98–114.
34. Schaap, G. J. P., Bijkerk, H., Coutinho, R. A., Kapsenberg, J. G., and van Wezel, A. L., 1984, The spread of wild poliovirus in the well-vaccinated Netherlands in connection with the 1978 epidemic, *Prog. Med. Virol.* **29:**124–149.
35. Hovi, T., Huovilainen, A., Kuronen, T., Poyry, T., Salama, N., Cantell, K., Kinnunen, E., Lapinleimu, K., Roivainen, M., Stenvik, M., Silander, A., Thoden, C.-J., Salminen, S., and Weckstrom, P., 1986, Outbreak of paralytic poliomyelitis in Finland: Widespread circulation of antigenically altered poliovirus type 3 in a vaccinated population, *Lancet* **2:**1427–1435.
36. Magrath, D. I., Evans, D. M. A., Ferguson, M., Schild, G. C., Minor, P. D., Horaud, F., Crainic, R., Stenvik, M., and Hovi, T., 1986, Antigenic and molecular properties of type 3 poliovirus responsible for an outbreak of poliomyelitis in a vaccinated population, *J. Gen. Virol.* **76:**899–905.
37. van Wezel, A. L., van Steenis, G., van der Marel, P., and Osterhaus, A. D. M. E., 1984, Inactivated poliovirus vaccine: Current production methods and new developments, *Rev. Infect. Dis.* **6:**S335–S340.
38. McBean, A. M., Thoms, M. L., Albrecht, P., Cuthie, J. C., Bernier, R., and the Field Staff and Coordinating Committee, 1988, Serologic response to oral polio vaccine and enhanced-potency inactivated polio vaccines, *Am. J. Epidemiol.* **128:**615–628.
39. Stoeckel, P., Schlumberger M., Parent, G., Maire, B., van Wezel, A., van Steenis, G., Evans, A., and Salk, D., 1984, Use of killed poliovirus vaccine in a routine immunization program in West Africa, *Rev. Infect. Dis.* **6:**S463–S466.

40. Melnick, J. L., 1989, Poliomyelitis: A vanishing but not vanished disease, in: *Current Topics in Medical Virology* (Y. C. Chan, S. Doraisingham, and A. E. Ling, eds.), World Scientific, Singapore, pp. 226–253.
41. Cockburn, W. C., 1988, The work of the WHO Consultative Group on Poliomyelitis Vaccines, *Bull. WHO* **66:**143–154.
42. Nkowane, B. M., Wassilak, S. G. F., Orenstein, W. A., Bart, K. J., Schonberger, L. B., and Hinman, A. R., 1987, Vaccine-associated paralytic poliomyelitis. United States: 1973 through 1984, *J.A.M.A.* **257:**1335–1340.
43. Lasch, E. E., Abed, Y., Abdulla, K., El Tibbi, A. G., Marcus, O., El Massri, M., Handscher, R., Gerichter, C. B., and Melnick, J. L., 1984, Successful results of a program combining live and inactivated poliovirus vaccines to control poliomyelitis in Gaza, *Rev. Inf. Dis.* **6:**S467–S470.
44. Advisory Committee on Immunization Practices (ACIP) 1986–1987, Poliomyelitis prevention, *Morbid. Mortal. Weekly Rep.* **35:**577–579; **36:**795–798.
45. Domok, I., Balayan, M. S., Fayinka, O. A., Skrtic, N., Soneji, A. D., and Harland, P. S. E. G., 1974, Factors affecting the efficacy of live poliovirus vaccines in warm climates, *Bull. WHO* **51:**333–347.
46. Robinson, D. A., 1982, Polio vaccination—a review of strategies, *Trans. R. Soc. Trop. Med. Hyg.* **76:**575–581.
47. Sabin, A. B., 1985, Oral poliovirus vaccine: History of its development and use and current challenge to eliminate poliomyelitis from the world, *J. Inf. Dis.* **151:**420–436.
48. Melnick, J. L., Ashkenazi, A., Midulla, V. C., Wallis, C., and Bernstein, A., 1963, Immunogenic potency of $MgCl_2$-stabilized oral poliovaccine, *J.A.M.A.* **185:**406–408.
49. Peetermans, J., Colinet, G., and Stephenne, J., 1976, Activity of attenuated poliomyelitis and measles vaccines exposed at different temperatures, in: *Proceedings of the Symposium on Stability and Effectiveness of Measles, Poliomyelitis and Pertussis Vaccines*, Yugoslav Academy of Sciences and Arts, Zagreb, pp. 61–66.
50. Mirchamsy, H., Shafyi, A., Mahinpour, M., and Nazari, P., 1978, Stabilizing effect of magnesium chloride and sucrose on Sabin live polio vaccine, in: *Developments in Biological Standardization*, Vol. 41, *Vaccinations in the Developing Countries* (R. H. Regamey, ed.) S. Karger, Basel, pp. 255–257.
51. Melnick, J. L., 1984, Recent developments in the worldwide control of poliomyelitis, in: *Control of Virus Diseases* (E. Kurstak and E. G. Marusyk, eds.), Marcel Dekker, New York, pp. 3–31.
52. Lasch, E. E., Abed, Y., Marcus, O., Gerichter, C. B., and Melnick, J. L., 1986, Combined live and inactivated poliovirus vaccine to control poliomyelitis in a developing country: five years after, in: *Developments in Biological Standardization*, Vol. 65, *Use and Standardization of Combined Vaccines*, S. Karger, Basel, pp. 137–143.
53. Nagington, J., Wreghitt, T. G., Gandy, G., Roberton, N. R. C., and Berry, P. J., 1978, Fatal echovirus 11 infections in outbreak in special-care baby unit, *Lancet* **2:**725.
54. Modlin, J. F., 1986, Perinatal echovirus infection: Insights from a literature review of 61 cases of serious infection and 16 outbreaks in nurseries, *Rev. Infect. Dis.* **8:**918–926.
55. Nagington, J., Gandy, G., Walker, J., and Gray, J. J., 1978, Use of normal immunoglobulin in an echovirus outbreak in a special-care baby unit, *Lancet* **2:**443–446.
56. Patriarca, P. A., Onorato, I. M., Sklar, V. E. F., Schonberger, L. B., Kaminski, R. M., Hatch, M. H., Morens, D. M., and Forster, R. K., 1983, Acute hemorrhagic conjunctivitis. Investigation of a large-scale community outbreak in Dade County, Florida, *J.A.M.A.* **249:**1283–1289.

9

Pathogenesis and Immunology of Herpesvirus Infections of the Nervous System

ANTHONY A. NASH and J. MATTHIAS LÖHR

1. INTRODUCTION

The herpesviruses are ubiquitous in nature, infecting fish, amphibians, reptiles, birds, and mammals. They are highly successful parasites, requiring only a small host range in which to maintain an infection: typically 10^2–10^3 individuals are sufficient to maintain a varicella–zoster virus (VZV) infection in the population, compared with $>10^5$ individuals for measles virus.[1] The great success of the herpesviruses is attributed to a remarkable strategy for persisting within their host, termed latency. This strategy involves the virus persisting as genetic material, but without expressing any detectable viral proteins and thereby evading host immune defenses. Although the ability to establish a latent infection is a characteristic of all herpesviruses, we consider classical latency to be a property of those viruses involved with infections of the nervous system, notably herpes simplex virus (HSV) types 1 and 2 and VZV.

In this review we consider the properties of HSV, the best-studied virus of the

ANTHONY A. NASH • Department of Pathology, University of Cambridge, Cambridge CB2 1QP, England.   J. MATTHIAS LÖHR • Department of Internal Medicine, University of Erlangen, D-8520 Erlangen, Germany.

Neuropathogenic Viruses and Immunity, edited by Steven Specter *et al.* Plenum Press, New York, 1992.

α-herpesvirinae, and in particular the infection of the nervous system and resultant disease and the role of the host response in controlling the infection or in precipitating pathology.

However, it is worth noting that many herpesviruses are associated with nervous system infection (see Table 9-1), and in addition to the classical pathology associated with HSV and VZV infection, other neuropathies are seen, for example, polyneuritis associated with Marek's disease (herpesvirus of chickens), which is a useful model system for the Guillain–Barré syndrome.[2,3]

2. Properties of Herpes Simplex Virus

HSV 1 and 2 are the prototype viruses of the α-herpesvirinae. The virus has a ds-DNA genome encoding some 70 virus proteins[4,5] including seven major glycoproteins. As with other herpesviruses, replication is initiated by α-gene products (immediate early genes, notably ICP0, ICP4), which in turn activate β-genes (delayed early genes, responsible for viral DNA replication) and finally activation of γ-genes (late genes, encoding the structural proteins). In tissue culture the whole process takes 10–12 hr (see Roizman[6] for a review).

The tropism of the virus is determined by envelope glycoproteins, of which gB, gD, and gH appear to be essential for successful virus propagation.[7] Herpes simplex virus infects a wide variety of tissue culture cells and can cause infection of a variety of tissues in the host. However, with rare exceptions, the virus causes only local mucocutaneous infections and subsequent infection of the peripheral nervous system. A disseminated infection can occur, as seen in the newborn and in patients undergoing immunosuppression. Despite our knowledge of the glycoproteins of this virus, the nature of the cellular receptor(s) involved in virus spread is far from clear.

3. THE RELATIONSHIP BETWEEN HERPES SIMPLEX VIRUS AND THE NERVOUS SYSTEM

The natural history of herpes simplex virus can be subdivided into three distinct phases for convenience: the primary or acute infection, latency, and reactivation/recurrence.

3.1. Primary Infection

Our knowledge of the primary infection has come largely from studying animal models of this human infectious disease. Clearly, experimental animals, although not ideal, provide important information on the progression of the infection and on the host response.

The extent of a primary HSV infection is dependent upon a number of factors: the strain of virus (whether type 1 or type 2 virus or genetic variants of these); the susceptibility of the host to the virus—this is highlighted in mice,

TABLE 9-1
Diversity of Neurological Infections Associated with Herpesviruses

Name	Natural host	Comments	Reference
Viruses of humans			
Herpes simplex virus (types 1 and 2)	Man	Latency in ganglia and CNS Encephalitis, meningitis, radiculitis	122–125
Varicella–zoster virus	Man	Latency in ganglia/recurrence Encephalitis, radiculitis, angiitis, neuritis, myelitis, postherpetic neuralgia Associated with postinfectious encephalomyelitis	126
Epstein–Barr virus	Man	Associated with chronic fatigue syndrome (neuromyasthenia)	127–131
Cytomegalovirus	Man	Encephalitis and other neuropathies arise following immunosuppression	
Viruses of nonhuman primates			
Herpesvirus simiae	Rhesus monkey	Fatal encephalitis in man and various experimental animals	132
Spider monkey herpes virus	Spider monkey	Fatal encephalitis in various experimental animals	133
Herpesvirus tamarinus	Squirrel monkey	Latency in ganglia and recurrences in experimental animals	133, 134
Viruses of other mammals			
Bovine herpesvirus 1	Cattle	Latency in ganglia Meningoencephalitis with widespread lesions in the brain	135
Equine herpesvirus 1	Horse	Latency/recurrence Vasculitis leading to widespread neurological infection	136, 137
Pseudorabies virus	Swine	Latency in ganglia/recurrence Meningoencephalitis and ganglioneuritis in swine and experimental animals Usually fatal disease in cat, dog, cattle, and experimental animals	138–140
Feline herpesvirus 1	Cat	Latency in ganglia/recurrence	141
Viruses of other vertebrates			
Gallid herpesvirus (Marek's virus)	Chicken	Polyneuritis Model for Guillain–Barré syndrome	2, 3, 142
Channel catfish virus	Catfish	Histopathological changes in the brain without apparent disease	143

where inbred strains display a range of susceptibility to CNS infection, e.g., C57BL/10 mice are highly resistant, whereas A/J mice are highly susceptible.[8,9] Resistance in this instance does not equate with the immune system but rather relates to intrinsic or natural resistance properties of the host; the route of infection—mice are more susceptible to an intravenous injection of virus than a subcutaneous injection, and the age at infection—young animals are more susceptible to the virus than adults.

These variables, though determined from experiments in animals, nevertheless apply to the situation in man.[10]

In using animal models it is important to mimic as far as possible the natural disease process in man. To this end experimental HSV infections of the skin, lip, or vagina have been widely used. In our description of a primary HSV infection, the mouse ear model[11] and the zosteriform model[12] are used as examples.

Following an intradermal/subcutaneous route of infection, virus is observed to replicate in epidermal cells. Sensory nerve endings become infected, and the virus travels, intraaxonally, to the neuronal cell body in the sensory ganglion. Here a productive infection of neurons can occur, though clearly some neurons will harbor the virus in a latent form. The virus can travel into the CNS via the root entry zone. Secondary and tertiary neurons can become infected, and the virus may spread into the brain. Glial cells also become infected, notably astrocytes and oligodendrocytes. The latter infection results in local demyelination, which causes some sensory loss. The extent of demyelination may depend on whether autoimmune responses become activated, which could exacerbate the pathology. These events take place during the first week of infection and, as noted above, are dependent on the strain of virus used (for a review of these processes see Wildy et al.[13] and Wildy and Gell[14]). Several investigators have observed that in mice HSV-2 strains are more neurovirulent then type 1 strains.[15] However, in man, most cases of encephalitis arise from type 1 infections[16] (see below).

An important feature of the primary infection is the role of the nervous system in the spread of virus to cutaneous sites. This is highlighted, most dramatically, in the zosteriform model.[12,17,18] Here virus is observed to spread within a dermatome (neurodermatome) to produce a band-like lesion of the skin some 6–7 days after infection. Just how virus spreads within the sensory ganglion to cause this effect is not known. A likely area is on the CNS side of the root entry zone, where naked nerve fibers occur. This dynamic action of the virus may be important in establishing sites of latency in neurons not directly connected to the initial site of infection. An example of this is the trigeminal complex, with ophthalmic, maxillary, and mandibular compartments. Consequently, infection of the lip can produce infection of the ophthalmic and maxillary compartments following spread within the CNS—the so-called "backdoor" route of infection.[19] A summary of the movement of virus between the skin and nervous system is depicted in Fig. 9-1.

The relative ease by which herpes simplex virus spreads within the nervous system has significant implications for disease processes arising at sites not directly involved in the natural infection. An example of this is the observation of type 1 virus associated with peptic ulcers of man.[20] The implication that this

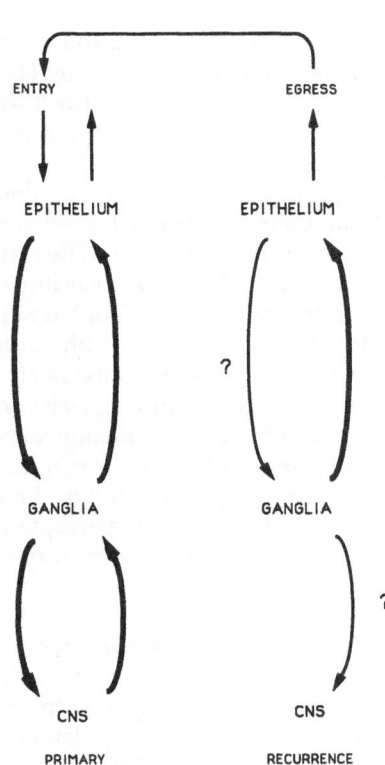

FIGURE 9-1. The spread of herpes simplex virus between epithelial cells and the nervous system during a primary and recurrent infection. In the primary infection virus travels from epithelium to the sensory ganglion by intraaxonal transport. Progression into the central nervous system can occur, and the virus ascends neurons by crossing synaptic junctions. Centrifugal spread of virus can occur during the primary infection, in which other nerves within a dermatome become infected, leading to infection of the epithelium (zosteriform spread). In recurrent infections virus originates in the sensory neuron following reactivation from latency and travels by retrograde axonal transport to infect cells in the epithelium. Transmission of virus to the CNS and reinfection of the same dermatome are possible, although these are probably rare events. The extent of this spread is dependent on the efficacy of the immune system. Egress of virus leads to infection of a new host or to reinfection of the same host at a different anatomic site, e.g., cold sore to eye.

pathology arises from a neurotropic infection (via the vagal nerve) is plausible in light of the biology of this virus. Indeed, transmission of the virus over long tracts of the nervous system, involving numerous neuron connections, has been shown by Ugolini *et al.*[21] in mapping the neuronal connections between the ulnar nerve and the region of the brain controlling motor function of the limb.

Normally, the primary infection lasts 10–12 days in the mouse.[11,14] A similar time course is observed in man, where clinical evidence of a primary infection is known to exist. Usually, primary infections in man are asymptomatic.

3.2. Latency

There are several excellent reviews of latency,[13,22–25] and the reader is referred to these for a fuller picture of this phase of infection. The virus persists in a latent form in the neuron of a sensory ganglion.[26] Other sites have been implicated in experimental animals, notably in the CNS[27,28] and in the skin.[29] Until recently the method for detecting latent virus involved explanting ganglia in tissue culture for a few days and then determining infectious virus by plaque assay.[30] However, efforts to reactivate virus from the CNS by this procedure have proved difficult (reviewed by Hill[22]).

The nature of the virus during this quiescent stage has been the subject of much investigation. Two important observations have emerged. (1) The virus genome lacks the inverted terminal repeats and most likely exists as circular or concatameric DNA.[28] In a productive virus infection, the genome is linear, containing the terminal repeats, and (2) limited virus transcription occurs that is localized to the nucleus. The transcripts, referred to as latency-associated transcripts (LATS), only appear during this stage of the infection and hence provide an important "marker" for latent HSV.[31-34] No translation product has so far been detected, which agrees with *in situ* hybridization data in that the transcripts are specifically localized to the nucleus of the cell. What role LATS play during the latent infection is unclear. Deletion of this region does not appear to interfere with the establishment or maintenance of latency. LATS-defective virus can also be reactivated, although this may be with a reduced efficiency.[35] These are the only virus transcripts detected in the latently infected neuron.

Consequently, our perception of HSV latency is of a static viral genome. In this state the virus is refractory to the immune system.

3.3. Reactivation/Recurrence

The ability to reactivate from the latent state is clearly an essential step for the spread of HSV in the population. The biochemical and molecular events involved in the initiation of this process are not understood. Clearly, a change in the physiological state of the neuron is important. Such changes presumably arise when the host is subjected to bouts of stress, UV irradiation, or trauma, the latter arising as stimuli to the skin causing damage to the epidermal cells.[36]

Whatever the processes involved, the reactivated virus may pass intraaxonally to skin, where reinfection of epidermal cells can arise, resulting in recurrent or recrudescent lesions. The virus egresses and becomes available to infect a new host. The cycle of events is depicted diagrammatically in Fig. 9-1.

The best account of recurrence or recrudescence in man was described by Spruance *et al.*[37] The early clinical signs involved pain, which subsided before the progression of erythema → papule → vesicle → ulcer → healing was observed. Virus could be isolated at the papule and vesicle stage but thereafter rapidly declined and was absent 5 days after the first clinical signs of recurrence. Not surprisingly, recurrent lesions tend to persist in the immunocompromised, with virus isolated for up to 3 weeks in some individuals.[38]

The precipitation of recurrent lesions in animal models has been difficult to achieve. In the mouse, such events occur at a low frequency following skin trauma such as cellophane tape striping of the ear skin.[36] A more reliable species is the guinea pig, with up to 45–90% of animals exhibiting spontaneous recrudescent lesions.[39] Rabbits have also been used to study ocular herpes infections. Virus can be recovered from conjunctival swabs following the use of epinephrine iontophoresis to trigger the reactivation of latent virus.[40]

Reactivation and recurrent infections with HSV occur in the face of an existing immunologic response to the virus. Although the immune response does not inhibit this process from occurring, it may reduce the intensity of recurrent lesions to a subclinical level. Clearly, in some individuals encountering frequent recrudescent lesions, defects in the potency of the immune response may be a principal cause. Wilton et al.[41] noted that in patients with recurrent lesions, virus-specific lymphocyte proliferation was inhibited, suggesting that such patients exhibit high levels of immunosuppression.

4. NEUROLOGICAL DISEASES ARISING FROM HERPES SIMPLEX VIRUS INFECTIONS

Table 9-1 presents a list of various neuropathies associated with members of the Herpesviridae. In man, the two most common neuropathies arising from HSV infection are encephalitis and meningitis.[42] Transient demyelinating disorders are noted in mice and rats infected with the virus,[43] though in man such events are rare.

Encephalitis is a rare and devastating consequence of HSV infection. In the adult, virtually all cases of this disease are caused by HSV 1, whereas in the neonate and newborn the disease results mainly from HSV 2.[44,45]

In the adult, HSV encephalitis is mainly associated with the temporal lobes.[46] This anatomic location is consistent with the virus entering and spreading via the olfactory nerve. Evidence for this is found in postmortem material[47] and from studies on animal models of HSV encephalitis.[48,49] This route of infection could account for all cases of encephalitis arising from (1) primary infection of nasal mucosa, (2) secondary infection with a different virus strain, and (3) reactivation or recurrence of resident virus that has recurred peripherally and subsequently entered the olfactory nerve. These variations would be consistent with the observations of Whitley et al.,[50] who found in half of the patients with encephalitis that the virus strain isolated from the brain was different from isolates from the lip or oral cavity. In the remaining patients virus isolates from the lip and brain were the same. It is possible that virus could reactivate from a peripheral site of latency and enter the CNS. An example would be reactivation of virus in the trigeminal ganglia and spread along the fifth nerve to cause infection of the brainstem. This could account for the very rare cases of brainstem encephalitis.[51] Similarly, since HSV DNA is found in the brainstem of latently infected mice,[27,28,52] it is possible that virus within the CNS could reactivate to cause disease. Stroop[53] proposed that latency in the entorhinal cortex could reactivate to cause encephalitis in the temporal lobe. However, attempts to reactivate virus from the CNS by various approaches have been largely unsuccessful.[28,52] This is in marked contrast to the ease of reactivating and detecting virus in sensory ganglia.

In HSV encephalitis of the newborn, the virus causes a more widespread necrosis in the brain.[54] This is consistent with a blood-borne infection such as occurs in HSV 2 infections of the newborn.

Since the advent of acyclovir (ACV) the prognosis of HSV encephalitis has greatly improved. When treatment is implemented early in the course of infection, complete recovery with minimal residual damage to the brain is usual. However, the later the infection is treated, the greater is the likelihood of neurological or neuropsychological damage occurring, including destruction of the temporal lobes.[55]

Herpes simplex has also been implicated in a number of other neurological disorders, notably multiple sclerosis (reviewed by Hill[43]), psychopathic disorders,[56] and Bell's palsy.[57]

5. THE HOST RESPONSE TO HERPES SIMPLEX VIRUS INFECTIONS

The early host response to HSV is composed of innate immunity (natural resistance) in which macrophages, natural killer cells, α/β interferon (IFN), and other factors become active during the first few hours to days after infection and serve to restrict virus replication and limit the initial spread (this work is reviewed by Lopez[8]). During the first week the immune system becomes fully active and is responsible for recovery from infection. By the second week infective virus is eliminated and effector T-cell activity declines, but neutralizing antibodies are present in serum and are important in preventing reinfection. Antibody and T-cell memory are long-lived and may periodically become stimulated by the virus following reactivation/recurrence.

A more detailed analysis of the immune response has come from studies on animal models. In this respect the mouse ear model and the zosteriform model have proved to be particularly useful. Studies in these models of HSV infection have revealed the importance of T-cell immune responses in the control of a primary infection and in the importance of antibody at interrupting the spread of virus to and within the nervous system (for reviews see references 14, 58, 59).

5.1. Induction and Activity of T Cells during a Herpes Simplex Virus Infection

Following an infection of mouse skin with HSV, Langerhans/dendritic cells are detected in the draining lymph node after 12–24 hr. These cells present viral antigens to T cells and initiate the induction of T-cell immune responses (H. Dickson and A. A. Nash, unpublished observations). Similar observations occur in man, where Langerhans cells treated with virus *in vitro* act as potent stimulators of HSV-specific MHC class II restricted T cells.[60] More recently, dendritic cells have been used successfully to induce *in vitro* primary cytotoxic T lymphocytes (CTL) active against HSV[61] or influenza virus[62] infected target cells.

Four days after infection both MHC class I and class II restricted T cells specific for the virus are detected in the lymph node. In the case of "classical" CTLs, i.e., MHC class I restricted/CD8+, active cell killing is only apparent after a period of culture *in vitro*.[63,64] This absolute requirement suggests that CTL development in the lymph node is inhibited during the infection and may be regulated by suppressor cells.[65] The CTLs are also difficult to detect in man, where their activity in peripheral blood is masked by the dominant NK cell response and virus-specific CD4+ T cells.[66] Equally, it could be argued that the natural route of infection, i.e., mucocutaneous surfaces, is not conducive to the effective generation of classical CTLs but favors the induction of other T-cell subsets. In support of this argument is the observation that active CD4+ T cells are very effective in mediating antiviral immunity in the skin of mice.[67] This protection, which is achieved in the absence of CTLs or antiviral antibodies, suggests a direct anti-HSV role for these T cells. One such function that is detected in the lymph node of infected mice as early as day 4 is the ability of CD4+ T cells to adoptively transfer delayed hypersensitivity (DH) responses to syngeneic recipients.[68] This pronounced T-cell-mediated inflammatory response accompanies cutaneous HSV infections and has been considered to be important in antiviral immunity.[68] However, this response does not always correlate with protection,[69] and CD4+ T cells can presumably function by other mechanisms, e.g., direct cytotoxicity of MHC class II infected cells[66] (keratinocytes can be induced to express these MHC antigens) or via lymphokine-mediated antiviral activity, e.g., IFN-γ or tumor necrosis factor.[70]

The evidence above indicates that CD4+ T cells play a central role in the recovery of the host from HSV infection. In addition to direct antiviral activity, these T cells are also involved in B-cell help and the rapid induction of CTLs. Despite the difficulty in detecting, directly, classical CTL activity *in vitro*, there is clear evidence that these cells contribute to the recovery of the host from the primary infection.[71–74] In experiments on MHC matching of donor antiviral T cells with infected recipients, a requirement for compatibility at the H-2K and H-2 I-A loci was necessary for optimal antivirus response.[73] This observation was highlighted in experiments on CD8+-deficient mice (rendered deficient by injection of anti-CD8 monoclonal antibody *in vivo*[75]). In such mice the rate of clearance of virus from skin was unaltered (implying that other antiviral mechanisms were operating), but clearance from the sensory ganglia and spinal cord was markedly delayed.[74] Depletion of CD4+ T cells delayed clearance of virus from the skin, supporting previous observations on the importance of these cells at this site of infection.

Based on our knowledge of events in the mouse, we propose that CD4 T cells function by similar mechanisms in man. MHC class II restricted CD4+ T cells are readily identified in the peripheral blood of infected individuals.[66] Cloned T-cell lines recognizing gD and gB have been produced, some of which are cytotoxic for infected target cells expressing the MHC class II antigens.[76–78] Similar specificities for gD and gB have been detected with murine CD4+ T cells.[79–82]

The nature of the viral antigens recognized by classical CTLs is less clearly defined. To a large part, this is dependent on the strain of mouse used in particular studies. For example, 40–50% of T cells from CBA and Balb/c mice recognize immediate early antigens of the virus,[83] whereas T cells from C57BL/6 (H-2^b) mice do not recognize these antigens (A. A. Nash and C. R. Bland, unpublished observations). CBA mouse CTLs also recognize gB, although far less efficiently.[79] Other glycoproteins of HSV 1 have also been investigated, namely, gD, gE, gG, gI, and gH, but none of these appear to be recognized by HSV-specific CTLs (A. A. Nash and C. R. Bland, unpublished observations). However, gC is recognized by CTLs from infected C57BL/6 mice (H-2^b) (S. Martin, personal communication), although CBA CTLs do not recognize this antigen. As is evident from the studies reported here, there is diversity of antigen recognition by CTLs, though a preference for recognizing nonstructural proteins is consistent with observations on other CTL–virus systems.[84–86]

5.2. T Cells in Recurrent Infection

In the last section we portray an efficient antiviral T-cell system in the recovery of the host from a primary HSV infection. However, the efficiency of T cells is dependent on identifying infected cells. Clearly, during latency the virus is not expressing any antigens, and also following reactivation the virus remains unchallenged during intraaxonal transport to the skin. From the nerve ending virus enters the epidermis, a tissue poorly patrolled by the immune system. Only after the virus cytocidal effect has caused damage to the epidermal basement membrane is the inflammatory response triggered, and with it the immune system. These events are clearly demonstrated in the zosteriform model, where antiviral T cells given 2 days after infection are unable to prevent a "recurrent" lesion[12] (it should be stressed that similar antiviral T cells given at the time of infection or 1 day later are extremely effective at reducing zosteriform spread). Histological studies of the zosteriform lesion revealed virus replication in epidermal cells 24–48 hr before there were any signs of inflammation and T-cell recruitment to the lesion. These observations highlight the remarkable survival strategy of HSV and the difficulty the immune system faces in eradicating this virus from its parasitic existence.

In other studies immune suppression has been suggested as a mechanism favoring the appearance of recurrent lesions. In man[41,87,88] and guinea pig,[89] recurrences are accompanied by a reduced T-cell proliferative response or lymphokine production. This effect may be attributed in part to the action of suppressor T cells.[90]

In support of these observations is the known affect of UV radiation on the induction of recurrent infections in man and mouse.[91] A consequence of UV-B radiation on the skin of mice is to negate the inductive effects of dendritic cells,[92] with the result that DH responses are inhibited. Interestingly, such animals develop virus-specific suppressor T cells.[93] Clearly, UV radiation can tip the balance between positive and negative immune responses, and this may have profound effects on the severity and frequency of recurrent lesions.

5.3. Significance of Antibody in Herpes Simplex Virus Infections

The importance of antibodies against herpes simplex virus in protection against infection and recurrence has long been questioned. After all, virus recurrence occurs in the face of an existing immune response, notably serum neutralizing antibodies. Furthermore, patients with agammaglobulinemia (Bruton type) do not appear to suffer any increase in number or severity of recurrent lesions.[94]

A similar picture is seen in mice deficient in B-cell activity. Such animals control a primary infection as efficiently as normal, immunocompetent mice. However, Kapoor et al.[95] and Simmons and Nash[96] noted that although clearance of virus from the skin was unaffected in B-cell-deficient animals, there was a more florid infection of the sensory ganglion and adjacent segments of the spinal cord and an increase in the number of latently infected mice. Experiments performed in nude mice support this observation. Here, infection of athymic nude mice leads to an overwhelming infection of the nervous system, with animals dying 14–21 days later. However, the administration of neutralizing polyclonal antisera against the virus or monoclonal antibodies to gD of HSV 1, 2–3 days after infection, led to a reduction in the virus isolated from the nervous system.[97] It was also apparent from this study that T cells were effective at reducing infection in the skin and antibody more effective at controlling the spread to or infection within the nervous system. This division of labor by the immune system would make good sense, insofar as T-cell activity in the nervous system is restricted by a lack of MHC class I antigens and constitutive MHC class II antigen expression.

Simmons and Nash[98] further examined the mechanism of antibody action *in vivo* using the zosteriform model. As discussed earlier, the zosteriform reaction involves the spread of virus in a particular dermatome. The result is a band-like cutaneous lesion similar to shingles (herpes zoster). Neutralizing antibody is highly effective at interrupting this pathway when given up to 68 hr after infection. Beyond this time the lesion develops. The implication of this observation is that the antibody neutralizes the virus as it emerges from the nerve endings to infect cells in the epidermis. In the same study the amount of antibody required to inhibit this response was investigated. Serum antibody titers of 1/256–1/512 were required for maximum effect; at lower titers the number of mice with lesions increased.

This observation has two important implications. First, the zosteriform reaction may be viewed as a model for recurrent infections. As with true recurrence, the virus emerges from nerve endings to infect epidermal cells. Clearly, antibody can interrupt this "recurrent" infection, which implies that similar mechanisms could be operating in man. The second point relates to the quality of the antibody or antiserum. Few studies have really addressed this point; most are concerned with either specificity, neutralizing titer, or ELISA titer. However, the antibody isotype or subtype and the affinity of the antibodies are highly relevant. Clearly the quality of an antibody response will vary among individuals.

There have been a number of studies on the protection of mice by passive administration of antibody. Monoclonal antibodies to gD, gB, gC, and gE have all proved effective in the control of infections when investigated in a variety of animal models.[99–101] In some experiments nonneutralizing monoclonal antibodies were observed to prevent lethal infections.[102] Protection in this instance is most likely mediated via antibody-dependent cell cytotoxicity.

So far we have discussed the role of antibody on the peripheral nervous system and the skin/mucous membranes. However, virus-specific antibodies can also be identified in the CNS of patients recovering from herpes encephalitis. Intrathecal production of IgG, IgA, and IgM has been observed, and in some instances antibody is detected up to 1 year after infection.[103] As with other viral infections of the CNS, the antibodies appear to be oligoclonal. These observations suggest that B cells persist within the CNS for prolonged periods and may provide a barrier to recurrent infections in this vital tissue.

Studies reported in this section indicate the potential use of anti-HSV antibodies in therapy of HSV infection. One important study has shown that a combination of acyclovir and anti-HSV antibody produces greater protection against encephalitis than either antibody or the drug alone.[104] This suggests that antibody may serve as an important adjunct to acyclovir therapy for the treatment of HSV encephalitis.

6. IMMUNOPATHOLOGY ASSOCIATED WITH HERPES INFECTIONS OF THE NERVOUS SYSTEM

Any infection of the nervous system is likely to result in the immune system contributing to the pathogenesis of disease. This is seen in various pathologies associated with HSV, notably stromal keratitis of the eye,[105] encephalitis,[106] and demyelination.[107] The latter is frequently observed in mice and rats following skin infection (though only rarely in man) and is located on the CNS side of the root entry zone of sensory ganglia. T lymphocytes certainly contribute to this process, since the demyelinating disorder is more pronounced and extensive in immuno-competent mice than in athymic nude mice.[108] As discussed earlier, the virus can directly infect oligodendrocytes and result in demyelination.

A major factor in the immunopathology of HSV infections is CD4 T cells able to mediate a DH response to the virus. Such T cells can readily be suppressed either by injecting virus intravenously[69] or by coinjecting virus plus lipopolysaccharide.[109] In both cases suppressor T cells are induced, which actively inhibit DH T cells from developing and functioning. In this situation animals are protected from severe demyelination and death by encephalitis.[109]

A consequence of the immune response to HSV in the CNS is the induction of an autoimmune response to myelin basic protein (MBP) or galactocerebroside. Although weak T-cell responses to MBP have been detected during an acute HSV infection of mice (W. A. Blyth and T. J. Hill, personal communication), such responses are transient, and remyelination is observed by the second or third week after infection.

7. TOWARD THE PREVENTION OF PRIMARY AND RECURRENT HERPES INFECTIONS

The advent of ACV has clearly revolutionized the treatment of HSV and other herpesvirus infections. It is not our intention here to consider any of this work, but we refer the reader to excellent reviews on the subject.[110–112]

An unfortunate drawback with ACV therapy is the failure of this drug to eradicate latent infections.[113] Consequently, recurrences and recrudescences still occur, and herpes infections continue to be transmitted. Another drawback is the presence of acyclovir-resistant mutants in treated patients. The significance of this is reviewed above.

Vaccination against HSV infections is still a real possibility, although it is argued that mass vaccination is unnecessary, since this virus is generally not life threatening.[114] Nevertheless, a recent survey of subjects in the United States shows some 25,000,000 infected with HSV 2.[115] In many cases the recurrent infection leads to painful lesions and can result in spread to uninfected individuals.

If vaccination is to be implemented, two possible patient groups have to be considered: those without prior experience of HSV and those with the infection. Vaccination of the former group should take place around the time of puberty to provide the maximum effect against type 2 herpes infections. However, type 1 infections may occur throughout life, with children under 5 years at particular risk from overaffectionate grandparents and grand aunts sporting orofacial lesions. Whether one should advocate vaccination during this period is indeed questionable.

The latter group (those with the infection) offers more scope for immuno-therapeutic intervention. In individuals undergoing frequent recrudescences, the opportunity to selectively boost the immune system to counter overt lesions is a possibility. As discussed earlier, if the quality and quantity of particular antiviral antibodies are important in reducing virus spread from nerves to epidermal cells, then selecting for such antibody species by vaccination with specific viral proteins could have beneficial affects. Furthermore, passively introducing antiviral anti-bodies (as reported earlier[98]) could serve the same function as vaccination in the short term.

If vaccination is considered an important preventative, then the type of vaccine is important. Whole-live-virus vaccines have the advantage of immunizing for all components of the host response. Attenuated viruses lacking selected virus genes have been developed and have met with rigorous trials in various animal models.[116,117] Such defective viruses must lack the ability to establish latent infections and cause overt neurological infections. Inactivated, DNA-free HSV vaccines have been used to vaccinate individuals at risk from type 2 infections or recurrences. The success of these vaccination trials varies between groups. Whereas one group[118] using a subunit vaccine reported seroconversion, there was no protection against acquiring the infection. Another group[119] reported consid-erable success in reducing the incidence of recurrent disease. The latter study did not include a double-blind trial, and therefore these results must be considered inconclusive.

The identification of important antigenic regions of the virus has enabled various proteins to become expressed via a number of expression vectors.[79,80,82,120] The most immunogenic proteins of HSV are gD and gB. They have the added advantage of being antigenically common between HSV 1 and 2.

Injection of gD into mice[79,82] leads to a level of protection comparable to the whole virus. However, a single injection of gD expressed constitutively on L cells

(H-2^k fibroblast cell line) leads to only a short-lived protection of mice when compared to the whole virus.[121] On the other hand, two injections 14 days apart lead to long-lived protection. Neutralizing antibodies are detected at day 270 post-infection and serve to inhibit the establishment of latent infections in animals challenged with live virus. This study indicates that subunit viral proteins can be used as suitable vaccines, although more than one injection would be required for optimal protection.

In a similar way, such glycoproteins could be used to augment preexisting antibody and T-cell responses in infected individuals. This would fulfill the requirement discussed earlier in this section for increasing antibody affinity and/ or augmenting or selecting antibody subtypes. Such a strategy will not eradicate HSV latency but may reduce the extent of clinical recurrences to an acceptable level for the patient. Trials are now urgently needed to implement theory and practice of vaccination against HSV infections.

ACKNOWLEDGMENTS. A.A.N.'s work reported herein was supported by grants from the Medical Research Council of Great Britain. Gay Schilling is especially thanked for typing the manuscript.

REFERENCES

1. Hope-Simpson, R. E., 1965, The nature of herpes zoster: A long term study and a new hypothesis, *Proc. R. Soc. Med.* **58:**9–20.
2. Lampert, P., Garrett, R., and Powell, H., 1977, Demyelination in allergic and Marek's disease virus induced neuritis: Comparative electron microscopic studies, *Acta Neuropathol. (Berl.)* **40:**103–109.
3. Stevens, J. G., Pepose, J. S., and Cook, M. L., 1981, Marek's disease: A natural model for the Landry–Guillain–Barré syndrome, *Ann. Neurol. [Suppl.]* **9:**102.
4. McGeoch, D. J., Dolan, A., Donald, S., and Rixon, F. J., 1985, Sequence determination and genetic content of the short unique region in the genome of herpes simplex virus type 1, *J. Mol. Biol.* **181:**1–13.
5. McGeoch, D. J., Dalrymple, M. A., Davison, A. J., Dolan, A., Frame, M. C., McNab, D., Perry, L. J., Scott, J. E., and Taylor, P., 1988, The complete DNA sequence of the long unique region in the genome of herpes simplex virus type 1, *J. Gen. Virol.* **69:**1531–1574.
6. Roizman, B., and Batterson, W., 1985, Herpesviruses and their replication, in: *Virology* (B. N. Fields, J. L. Melnick, B. Roizman, and R. E. Shope, eds.), Raven Press, New York, pp. 497–526.
7. Marsden, H. S., 1987, Herpes simplex virus glycoproteins and pathogenesis, in : *Molecular Basis of Virus Disease, S.G.M. Symposium 40* (W. C. Russell and J. W. Almond, eds.), Cambridge University Press, London.
8. Lopez, C., 1985, Natural resistance mechanisms in herpes simplex virus infections, in: *The Herpes Viruses*, Volume 4 (B. Roizman and C. Lopez, eds.), Plenum Press, New York, pp. 37–68.
9. Simmons, A., 1989, H-2 linked genes influence the severity of herpes simplex virus infection of the peripheral nervous system, *J. Exp. Med.* **169:**1503–1507.
10. Rawls, W. E., 1985, Herpes simplex virus, in: *Virology* (B. N. Fields, D. M. Knipe, and R. M. Channock, eds.), Raven Press, New York, pp. 527–561.
11. Hill, T. J., Field, H. J., and Blyth, W. A., 1975, Acute and recurrent infection with herpes simplex virus in the mouse: A model for studying latency and recurrent disease, *J. Gen. Virol.* **28:**341–353.

12. Simmons, A., and Nash, A. A., 1984, Zosteriform spread of herpes simplex virus as a model of recrudescence and its use to investigate the role of immune cells in prevention of recurrent disease, *J. Virol.* **52:**816–821.
13. Wildy, P., Field, H. J., and Nash, A. A., 1982, Classical herpes latency revisited, in: *Virus Persistence* (B. W. J. Mahy, A. C. Minson, and G. K. Darby, eds.), Cambridge University Press, Cambridge, pp. 133–167.
14. Wildy, P., and Gell, P. G. H., 1985, The host response to herpes simplex virus, *Br. Med. Bull.* **41:**86–91.
15. Richards, J. T., Kern, E. R., Overall, J. O., and Glasgow, L. A., 1981, Differences in neurovirulence among isolates of herpes simplex virus types 1 and 2 in mice using four routes of infection, *J. Infect. Dis.* **144:**464–471.
16. Nahmias, A. J., Whitley, R. J., Visintine, A. N., Takei, Y., and Alford, C. A., 1982, Herpes simplex encephalitis: Laboratory evaluations and their diagnostic significance, *J. Infect. Dis.* **145:**829–836.
17. Goodpasture, E. W., and Teague, O., 1923, Transmission of the virus of herpes along nerves in experimentally infected rabbits, *J. Med. Res.* **44:**139–184.
18. Blyth, W. A., Harbour, D. A., and Hill, T. J., 1984, Pathogenesis of zosteriform spread of herpes simplex virus in the mouse, *J. Gen. Virol.* **65:**1477–1486.
19. Shimeld, C., Tullo, A. B., Hill, T. J., Blyth, W. A., and Easty, D. L., 1985, Spread of herpes simplex virus and distribution of latent infection after intraocular infection of the mouse, *Arch. Virol.* **85:**175–187.
20. Löhr, M., Nelson, J. A., and Oldstone, M. B. A., 1990, Is herpes simplex virus associated with peptic ulcer disease, *J. Virol.* **64:**2168–2174.
21. Ugolini, G., Kuypers, H. G. M., and Strick, P. L., 1989, Transneural transfer of herpes simplex virus from peripheral nerves to cortex and brainstem, *Science* **243:**89–91.
22. Hill, T. J., 1982, Herpes simplex virus latency, in: *The Herpes Viruses*, Volume 3 (B. Roizman, ed.), Plenum Press, New York, pp. 175–240.
23. Klein, R. J., 1976, Pathogenetic mechanisms of recurrent herpes simplex viral infections, *Arch. Virol.* **51:**1–13.
24. Stevens, J., 1980, Herpetic latency and reactivation, in: *Oncogenic Viruses*, Volume 2 (F. Rapp, ed.), CRC Press, Boca Raton, Fl, pp. 1–11.
25. Roizman, B., and Sears, A. E., 1987, An inquiry into the mechanisms of herpes simplex virus latency, *Annu. Rev. Microbiol.* **41:**543–571.
26. Cook, M. L., Bastone, V. B., and Stevens, J. G., 1974, Evidence that neurons harbor latent herpes simplex virus, *Infect. Immun.* **9:**946–951.
27. Rock, D. L., and Fraser, N. W., 1983, Detection of HSV-1 genome in the central nervous system of latently infected mice, *Nature* **302:**523–525.
28. Efstathiou, S., Minson, A. C., Field, H. J., Anderson, J. R., and Wildy, P., 1986, Detection of herpes simplex virus-specific DNA sequences in latently infected mice and in humans, *J. Virol.* **57:**446–455.
29. Al-Saadi, S. A., Gross, P., and Wildy, P., 1988, Herpes simplex virus type 2 latency in the footpad of mice: Effect of acycloguanosine on the recovery of virus, *J. Gen. Virol.* **69:**433–438.
30. Stevens, J. G., and Cook, M. L., 1971, Latent herpes simplex virus in spinal ganglia of mice, *Science* **173:**843–845.
31. Stevens, J. G., Wagner, E. K., Devi-Rao, G. B., Cook, M. L., and Feldman, L. T., 1987, RNA complementary to a herpesvirus alpha gene mRNA is prominent in latently infected neurons, *Science* **235:**1056–1059.
32. Stevens, J. G., Haarr, L., Porter, D. D., Cook, M. L., and Wagner, E. K., 1988, Prominence of the herpes simplex virus latency-associated transcript in trigeminal ganglia from seropositive humans, *J. Infect. Dis.* **158:**117–123.
33. Spivack, J. G., and Fraser, N. W., 1987, Detection of herpes simplex virus type 1 transcripts during latent infection in mice, *J. Virol.* **61:**3841–3847.

34. Deatly, A. M., Spivack, J. G., Lavi, E., O'Boyle, D. R. II, and Fraser, N. W., 1988, Latent herpes simplex virus type 1 transcripts in peripheral and central nervous system tissue of mice map to similar regions of the viral genome, *J. Virol.* **62:**749–756.

35. Leib, D. A., Bogard, C. L., Kosz-Vnenchak, M., Hicks, K. A., Coen, D. M., Knipe, D. M., and Schaffer, P. A., 1989, A deletion mutant of the latency-associated transcript of herpes simplex virus type 1 reactivates from the latent state with reduced frequency, *J. Virol.* **63:** 2893–3000.

36. Hill, T. J., Blyth, W. A., and Harbour, D. A., 1978, Trauma to the skin causes recurrence of herpes simplex in the mouse, *J. Gen. Virol.* **39:**21–28.

37. Spruance, S. L., Overall, J. C., Kern, E. R., Keueger, G. G., Pliam, V., and Miller, W., 1977, The natural history of recurrent herpes simplex labialis, *N. Engl. J. Med.* **297:**69–74.

38. Daniels, C. A., LeGoff, S. G., and Notkins, A. L., 1975, Shedding infectious virus/antibody complexes from vesicular lesions of patients with recurrent herpes labialis, *Lancet* **2:** 524–528.

39. Scriba, M., 1975, Herpes simplex infection in guinea pigs: An animal model for studying latent and recurrent herpes simplex infection, *Infect. Immun.* **12:**162–165.

40. Hill, J., Kwon, B. A., Shimomura, Y., Colborn, G. L., Yaghmai, F., and Gangarosa, L., 1983, Herpes simplex recovery in neural tissue after ocular herpes simplex virus shedding induced by epinephrine iontophoresis to the rabbit cornea, *Infect. Ophthalmol. Vis. Sci.* **24:**243.

41. Wilton, J. M. A., Ivanyi, L., and Lehner, T., 1972, Cell-mediated immunity in herpesvirus hominis infections, *Br. Med. J.* **1:**723–726.

42. Baringer, J. R., 1975, Herpes simplex virus infection of nervous tissue in animals and man, *Prog. Med. Virol.* **20:**1–26.

43. Hill, T. J., 1983, Herpesviruses in the central nervous system, in: *Viruses and Demyelinating Diseases* (C. A. Mims, M. L. Cuzner, and R. E. Kelly, eds.), Academic Press, London, pp. 23–45.

44. Whitley, R. J., 1984, Treatment of human herpes virus infections with special reference to encephalitis, *J. Antimicrob. Chemother.* **14**(Suppl. A):57–74.

45. Whitley, R. J., and Alford, C. A., 1982, Herpes infections in childhood: Diagnostic dilemmas and therapy, *Pediatr. Infect. Dis.* **1:**81–84.

46. Anderson, J. R., 1988, Viral encephalitis and its pathology, *Curr. Top. Pathol.* **76:**23–60.

47. Esiri, M. M., 1982, Herpes simplex encephalitis. An immunohistological study of the distribution of viral antigen within the brain, *J. Neurol. Sci.* **54:**209–226.

48. Anderson, J. R., and Field, H. J., 1983, The distribution of herpes simplex type 1 antigen in mouse central nervous system after different routes of inoculation, *J. Neurol. Sci.* **60:**181–195.

49. Tomlinson, A. H., and Esiri, M. M., 1983, Herpes simplex encephalitis: Immunohisto-chemical demonstrations of spread of virus via olfactory pathways in mice, *J. Neurol. Sci.* **60:** 473–484.

50. Whitley, R., Lakeman, A. D., Nahmias, A. J., and Roizman, B., 1982, DNA restriction analysis of herpes simplex virus isolates obtained from patients with encephalitis, *N. Engl. J. Med.* **307:**1060–1062.

51. Dayan, A. D., Gooddy, W., Harrison, M. J. G., and Rudge, P., 1972, Brain stem encephalitis caused by herpes virus hominis, *Br. Med. J.* **4:**405–406.

52. Cabrera, C. V., Wholenberg, C., Openshaw, H., Rey-Mendez, M., Puga, A., and Notkins, A. L., 1980, Herpes simplex virus DNA sequences in the CNS of latently infected mice, *Nature* **288:**648–650.

53. Stroop, W. G., 1986, Herpes simplex virus encephalitis of the human adult: Reactivation of latent brain infection, *Pathol. Immunopathol. Res.* **5:**156–169.

54. Whitley, R. J., and Hutto, C., 1985, Neonatal herpes simplex virus infections, *Pediatr. Rev.* **7:** 119–126.

55. Greenwood, R., Bhalla, A., Gordon, A., and Roberts, J., 1983, Behavior disturbances during recovery from herpes simplex encephalitis, *J. Neurol. Neurosurg. Psychiatry* **46:**809–817.

56. Cleobury, J. R., Skinner, G. R. B., Thouless, M. E., and Wildy, P., 1971, Association between psychopathic disorder and serum antibody to herpes simplex virus (type 1), *Br. Med. J.* **1:** 438–445.

57. Vahlne, A., Edstrom, S., Arstila, P., Beran, M., Ejnall, M., Nylen, O., and Lycke, E., 1981, Bell's palsy and herpes simplex virus, *Arch. Otolaryngol.* **107:**79.

58. Nash, A. A., Leung, K.-N., and Wildy, P., 1985, The T cell mediated immune response of mice to HSV, in: *The Herpesviruses*, Volume 4 (B. Roizman and C. Lopez, eds.), Plenum Press, New York, pp. 87–102.

59. Nash, A. A., and Wildy, P., 1983, Immunity in relation to the pathogenesis of herpes simplex virus, in: *Human Immunity to Viruses* (F. Ennis, ed.), Academic Press, New York, pp. 179–192.

60. Bjercke, S., Elg, J., Braathen, L., and Thorsby, E., 1984, Enriched epidermal Langerhans cells are potent antigen-presenting cells for T cells, *J. Invest. Dermatol.* **83:**286–289.

61. Hengel, H., Lindner, M., Wagner, H., and Heeg, K., 1987, Frequency of herpes simplex virus-specific murine cytotoxic T lymphocyte precursors in mitogen and antigen-driven primary *in vitro* T cell responses, *J. Immunol.* **139:**4196–4202.

62. Macatonia, S. E., Taylor, P. M., Knight, S. C., and Askonas, B. A., 1989, Primary stimulation by dendritic cells induces antiviral proliferative and cytotoxic T cell responses *in vitro*, *J. Exp. Med.* **169:**1255–1264.

63. Pfizenmaier, K., Starzinski-Powitz, A., Rollinghoff, M., Falke, D., and Wagner, H., 1977, T-cell mediated cytotoxicity against herpes simplex virus-infected target cells, *Nature* **265:** 630–632.

64. Nash, A. A., Quartey-Papafio, R., and Wildy, P., 1980, Cell-mediated immunity in herpes simplex virus-infected mice: Functional analysis of lymph node cells during periods of acute and latent infection, with reference to cytotoxic and memory cells, *J. Gen. Virol.* **49:**309–317.

65. Horohov, D. W., Wyckoff, J. H. III, Moore, R. N., and Rouse, B. T., 1986, Regulation of herpes simplex virus-specific cell-mediated immunity by a specific suppressor factor, *J. Virol.* **58:**331–338.

66. Schmid, D. S., 1988, The human MHC-restricted cellular response to herpes simplex virus type 1 is mediated by CD4+, CD8− T cells and is restricted to the DR region of the MHC complex, *J. Immunol.* **140:**3610–3616.

67. Nash, A. A., and Gell, P. G. H., 1983, Membrane phenotype of murine effector and suppressor T cells involved in delayed hypersensitivity and protective immunity to herpes simplex virus, *Cell. Immunol.* **75:**348–355.

68. Nash, A. A., Field, H. J., and Quartey-Papafio, R., 1980, Cell-mediated immunity in herpes simplex virus infected mice: Induction, characterization and antiviral effects of delayed type hypersensitivity, *J. Gen. Virol.* **48:**351–357.

69. Nash, A. A., Gell, P. G. H., and Wildy, P., 1981, Tolerance and immunity in mice infected with herpes simplex virus: Simultaneous induction of protective immunity and tolerance to delayed-type hypersensitivity, *Immunology* **43:**153–159.

70. Seid, J. M., Leung, K. N., Pye, C., Phelan, J., Nash, A. A., and Godfrey, H. P., 1987, Clonal analysis of the T-cell response of mice to herpes simplex virus: Correlation between lymphokine production *in vitro* and induction of DTH and anti-viral activity *in vivo*, *Viral Immun.* **1:**35–44.

71. Larsen, H. S., Russell, R. G., and Rouse, B. T., 1983, Recovery from lethal herpes simplex virus type 1 infection is mediated by cytotoxic T lymphocytes, *Infect. Immun.* **41:**197–204.

72. Howes, E. L., Taylor, W., Mitchison, N. A., and Simpson, E., 1979, MHC matching shows that at least two T-cell subsets determine resistance to HSV, *Nature* **277:**67–68.

73. Nash, A. A., Phelan, J., and Wildy, P., 1981, Cell-mediated immunity in herpes simplex virus-infected mice: H-2 mapping of the delayed-type hypersensitivity response and the antiviral T cell response, *J. Immunol.* **126:**1260–1262.

74. Nash, A. A., Jayasuriya, A., Phelan, J., Cobbold, S. P., Waldmann, H., and Prospero, T., 1987, Different roles for L3T4+ and Lyt2+ T cell subsets in the control of an acute herpes simplex virus infection of the skin and nervous system, *J. Gen. Virol.* **68**:825–833.

75. Cobbold, S. C., Jayasuriya, A., Nash, A., Prospero, T. D., and Waldmann, H., 1984, Therapy with monoclonal antibodies by elimination of T cell subsets *in vivo*, *Nature* **312**:548–551.

76. Yasukawa, M., and Zarling, J. M., 1985, Human cytotoxic T cell clones directed against herpes simplex virus-infected cells. III. Analysis of viral glycoproteins recognized by CTL clones by using recombinant herpes simplex viruses, *J. Immunol.* **134**:2679–2682.

77. Zarling, J. M., Moran, P. A., Burke, R. L., Pachl, C., Berman, P. W., and Lasky, L. A., 1986, Human cytotoxic T cell clones directed against herpes simplex virus-infected cells. IV. Recognition and activation by cloned glycoproteins gB and gD, *J. Immunol.* **136**:4669–4673.

78. Zarling, J. M., Moran, P. A., Lasky, L. A., and Moss, B., 1986, Herpes simplex virus (HSV)-specific human T-cell clones recognize HSV glycoprotein D expressed by a recombinant vaccinia virus, *J. Virol.* **59**:506–509.

79. Blacklaws, B. A., Nash, A. A., and Darby, G., 1987, Specificity of the immune response of mice to herpes simplex virus glycoproteins B and D constitutively expressed on L cell lines, *J. Gen. Virol.* **68**:1103–1114.

80. Chan, W. L., Lukig, M. L., and Liew, F. Y., 1985, Helper T cells induced by an immuno-purified herpes simplex virus type 1 (HSV-1) 115 kilodalton glycoprotein (gB) protect mice against HSV-1 infection, *J. Exp. Med.* **162**:1304–1318.

81. Martin, S., and Rouse, B. T., 1987, The mechanism of antiviral immunity induced by a vaccinia recombinant expressing herpes simplex virus type 1 glycoprotein D. Clearance of local infection, *J. Immunol.* **138**:3431–3437.

82. Krishna, S., Blacklaws, B. A., Overton, H. A., Bishop, D. H. L., and Nash, A. A., 1989, Expression of glycoprotein D of herpes simplex virus type 1 in a recombinant baculovirus: Protective responses and T cell recognition of the recombinant-infected cell extracts, *J. Gen. Virol.* **770**:1805–1814.

83. Martin, S., Courtney, R. J., Fowler, G., and Rouse, B. T., 1988, Herpes simplex virus type 1-specific cytotoxic T lymphocytes recognize virus nonstructural proteins, *J. Virol.* **62**:2265– 2273.

84. Townsend, A. R. M., Gotch, F. M., and Davey, J., 1985, Cytotoxic T cells recognize fragments of influenza nucleoprotein, *Cell* **42**:457–467.

85. Whitton, J. L., Southern, P. J., and Oldstone, M. B. A., 1988, Analyses of the cytotoxic T lymphocyte responses to glycoprotein and nucleoprotein components of lymphocytic choriomeningitis virus, *Virology* **162**:321–327.

86. Del Val, M., Volkmer, H., Rothbard, J. B., Jonjic, S., Messerle, M., Schickendanz, J., Reddehase, M. J., and Koszinowski, U. H., 1988, Molecular basis for cytolytic T-lymphocyte recognition of the murine cytomegalovirus immediate-early protein pp89, *J. Virol.* **62**:3965– 3972.

87. Shillitoe, E. J., Wilton, J. M. A., and Lehner, T., 1977, Sequential changes in cell mediated immune responses to herpes simplex virus after recurrent herpetic infections in humans, *Infect. Immun.* **18**:130–137.

88. O'Reilly, R. J., Chibbaro, A., Anger, E., and Lopez, C., 1977, Cell mediated responses in patients with herpes simplex infections. II. Infection-associated deficiency of lymphokine production in patients with recurrent herpes labialis or herpes progenitalis, *J. Immunol.* **118**: 1095–1102.

89. Donnenberg, A. D., Chaikof, E., and Aurelian, L., 1980, Immunity to herpes simplex virus type 2: Cell-mediated immunity in latently infected guinea pigs, *Infect. Immun.* **30**:99–109.

90. Iwasaka, T., Sheridan, J. F., and Aurelian, L., 1983, Immunity to herpes simplex virus type 2: Recurrent lesions are associated with the induction of suppressor cell and soluble suppressor factors, *Infect. Immun.* **42**:955–964.

91. Blyth, W. A., Hill, T. J., Field, H. J., and Harbour, D. A., 1976, Reactivation of herpes

simplex virus infection by ultraviolet light and possible involvement of prostaglandins, *J. Gen. Virol.* **33:**547–550.

92. Howie, S. E. M., Norval, M., and Maingay, J. P., 1986, Alterations in epidermal handling of HSV-1 antigens *in vitro* induced by *in vivo* exposure to UV-B light, *Immunology* **57:**225–230.

93. Howie, S. E., Norval, M., Maingay, J., and Ross, J. A., 1986, Two phenotypically distinct T cell (Ly1+2- and Ly1-2+) are involved in ultraviolet-B light-induced suppression of the efferent DTH response to HSV-1 *in vivo*, *Immunology* **58:**653–658.

94. Merigan, T. C., and Stevens, D. A., 1971, Viral infections in man associated with acquired immunological deficiency states, *Fed. Proc.* **30:**1858–1864.

95. Kapoor, A. K., Nash, A. A., and Wildy, P., 1982, Pathogenesis of herpes simplex virus in B cell-suppressed mice: The relative roles of cell-mediated and humoral immunity, *J. Gen. Virol.* **61;**127–131.

96. Simmons, A., and Nash, A. A., 1987, Effect of B cell suppression on primary and reinfection of mice with herpes simplex virus, *J. Infect. Dis.* **155:**649–654.

97. Kapoor, A. K., Nash, A. A., Wildy, P., Phelan, J., McClean, C. S., and Field, H. J., 1982, Pathogenesis of herpes simplex virus in congenitally athymic mice: The relative roles of cell mediated and humoral immunity, *J. Gen. Virol.* **60:**225–233.

98. Simmons, A., and Nash, A. A., 1985, Role of antibody in primary and recurrent herpes simplex infection, *J. Virol.* **53:**944–948.

99. Dix, R. D., Pereira, L., and Baringer, J. R., 1981, Use of monoclonal antibody directed against herpes simplex virus glycoproteins to protect mice against acute virus-induced neurological disease, *Infect. Immun.* **34:**192–199.

100. Kino, Y., Eto, T., Ohtomo, Y., Yamamato, M., and Mori, R., 1985, Passive immunization of mice with monoclonal antibodies to glycoprotein gB of herpes simplex virus, *Microbiol. Immunol.* **29:**143–149.

101. Balachandran, N., Bacchetti, S., and Rawls, W. E., 1982, Protection against lethal challenge of BALB/c mice by passive transfer of monoclonal antibodies to five glycoproteins of herpes simplex virus type 2, *Infect. Immun.* **37:**1132–1137.

102. Rector, J. T., Lausch, R. N., and Oakes, J. E., 1984, Identification of infected cell-specific monoclonal antibodies and their role in host resistance to ocular herpes simplex virus type 1 infection, *J. Gen. Virol.* **65:**657–661.

103. Vandrik, B., Vartdal, F., and Norrby, E., 1982, Herpes simplex virus encephalitis: Intrathecal synthesis of oligoclonal virus specific IgG, IgA and IgM antibodies, *J. Neurol.* **228:**25–38.

104. Erlich, K. S., and Mills, J., 1986, Passive immunotherapy for encephalitis caused by herpes simplex virus, *Rev. Infect. Dis.* 8(Suppl. 4):439–445.

105. Metcalf, J. F., and Kaufman, H. E., 1976, Herpetic stromal keratitis: Evidence for cell mediated immunopathogenesis, *Am. J. Ophthalmol.* **82:**827–832.

106. Chan, W. L., Javanovic, T., and Lukig, M. L., 1989, Infiltration of immune T cells in the brain of mice with herpes simplex virus-induced encephalitis, *J. Neuroimmunol.* **23:** 195–201.

107. Kristensson, K., Svennerholm, B., Persson, L., Vahlne, A., and Lycke, E., 1979, Latent herpes simplex virus trigeminal ganglionic infection in mice and demyelination in the central nervous system, *J. Neurol. Sci.* **43:**253–264.

108. Townsend, J. J., 1981, The demyelinating effect of corneal HSV infections in normal and nude (athymic) mice, *J. Neurol. Sci.* **50:**435–441.

109. Altmann, D. M., and Blyth, W. A., 1984, Lipopolysaccharide-induced suppressor cells for delayed type hypersensitivity to herpes simplex virus: Nature of suppressor cell and effect on pathogenesis of herpes simplex, *Immunology* **53:**473–480.

110. O'Brien, J. J., and Campoli-Richards, D. M., 1989, Acyclovir. An updated review of its antiviral activity, pharmacokinetic properties and therapeutic efficacy, *Drugs* **37:**233–309.

111. Collins, P., 1983, The spectrum of antiviral activities of acyclovir *in vitro* and *in vivo*, *J. Antimicrob. Chemother.* **12**(Suppl. B):19–27.

112. Dorsky, D. I., and Crumpacker, C. S., 1987, Drugs five years later: Acyclovir, *Ann. Intern. Med.* **107**:859–874.
113. Field, H. J., Bell, S. E., Elion, G. B., Nash, A. A., and Wildy, P., 1979, Effect of acycloguanosine treatment on acute and latent herpes simplex infections in mice, *Antimicrob. Agents Chemother.* **15**:554–561.
114. Wildy, P., 1984, Prospects for vaccines against herpes simplex types 1 and 2, in: *New Approaches to Vaccine Development* (R. Bell and G. Torrigiani, eds.), Schwabe and Co., Basel, pp. 300–312.
115. Johnson, R. E., Nahmias, A. J., Magder, L. S., Lee, F. K., Brooks, C. A., and Snowden, C. B., 1989, A seroepidemiologic survey of the prevalence of herpes simplex virus type 2 infection in the United States, *N. Engl. J. Med.* **321**:7–12.
116. Meignier, B., 1985, Vaccination against herpes simplex virus infections, in: *The Herpes Viruses*, Volume 4 (B. Roizman and C. Lopez, eds.), Plenum Press, New York, pp. 265–296.
117. Roizman, B., Warren, J., Thuning, C. A., Fanshaw, M. S., Norrild, B., and Meignier, B., 1982, Application of molecular genetics to the design of live herpes simplex virus vaccines, *Dev. Biol. Stand.* **53**:287–304.
118. Ashley, R., Mertz, G. J., and Corey, L., 1987, Detection of asymptomatic herpes simplex virus infections after vaccination, *J. Virol.* **61**:264–268.
119. Skinner, G. R. B., Woodman, G., Hartley, C., Buchan, A., Fuller, A., Wiblin, C., Wilkins, G., and Melling, J., 1982, Early experience with "antigenoid" vaccine AcNFUI(S')MRC towards prevention or modification of herpes genitalis, *Dev. Biol. Stand.* **52**:333–344.
120. Dix, R. D., 1987, Prospects for a vaccine against herpes simplex virus types 1 and 2, *Prog. Med. Virol.* **34**:89–128.
121. Blacklaws, B. A., and Nash, A. A., 1990, Immunological memory to herpes simplex virus type 1 glycoproteins B and D in mice, *J. Gen. Virol.* **71**:863–871.
122. Appelbaum, E., Kreps, S. I., and Sunshine, A., 1962, Herpes zoster encephalitis, *Am. J. Med.* **32**:25–31.
123. Gold, E., and Nankervis, G. A., 1973, Varicella–zoster viruses, in: *The Herpesviruses* (A. S. Kaplan, ed.), Academic Press, New York, pp. 327–351.
124. Esiri, M. M., and Tomlinson, A. H., 1972, Herpes zoster: Demonstration of virus in trigeminal nerve and ganglion by immunofluorescence and electron microscopy, *J. Neurol. Sci.* **15**:35–48.
125. Linneman, C. C., and Alvira, M. M., 1980, Pathogenesis of varicella–zoster angiitis in the CNS, *Arch. Neurol.* **37**:239–240.
126. Straus, S. E., 1988, The chronic mononucleosis syndrome, *J. Infect. Dis.* **157**:405–412.
127. Behar, R., Wiley, C., and McCutchan, J. A., 1987, Cytomegalovirus polyradiculoneuropathy in acquired immune deficiency syndrome, *Neurology* **37**:557–561.
128. Richert, J. R., Potolicchio, S., Garagusu, V. F., Manz, H. J., and Cohan, S. L., 1987, Cytomegalovirus encephalitis associated with episodic neurologic deficits and OKT-8 positive pleocytosis, *Neurology* **37**:149–152.
129. Wiley, C. A., Schrier, R. D., Denaro, F. J., Nelson, J. A., Lampert, P. W., and Oldstone, M. B. A., 1986, Localization of cytomegalovirus proteins and genome during fulminant central nervous system infection in an acquired immune deficiency syndrome patient, *J. Neuropathol. Exp. Neurol.* **45**:127–139.
130. Cordonnier, C., Feuilhade, F., Vernant, J. P., Marsault, C., Rochant, H., and Rodet, H., 1983, Cytomegalovirus encephalitis occurring after bone marrow transplantation, *Scand. J. Haematol.* **31**:248–252.
131. Bishopric, G., Bruner, J., and Butler, J., 1985, Guillain–Barré syndrome with cytomegalovirus infection of peripheral nerves, *Arch. Pathol. Lab. Med.* **109**:1106–1108.
132. Ludwig, H., Pauli, G., Gelderblom, H., Darai, G., Koch, H.-G., Flugel, R. M., Norrild, B., and Daniel, M. D., 1983, B virus (*Herpesvirus simiae*), in: *The Herpesviruses*, Volume 2 (B. Roizman, ed.), Plenum Press, New York, pp. 385–428.

133. Barahona, H., Melendez, L. V., and Melnick, J. L., 1974, A compendium of herpesviruses isolated from non-human primates, *Intervirology* **3:**175–192.

134. McCarthy, K., and Tosolini, F. A., 1975, Hazards from simian herpes viruses: Reactivation of skin lesions with shedding, *Lancet* **1:**649–650.

135. Ludwig, H., 1982, Bovine herpesviruses, in: *The Herpesviruses*, Volume 2 (B. Roizman, ed.), Plenum Press, New York, pp. 135–214.

136. Pursell, A. R., Sangster, L. T., Byars, T. D., Divers, T. J., and Cole, J. R., Jr., 1979, Neurologic disease induced by equine herpesvirus 1, *J. Am. Vet. Med. Assoc.* **175:**473–474.

137. Jackson, T. A., Osburn, B. I., Cordy, D. R., and Kendrick, J. W., 1977, Equine herpes 1 infection of horses: Studies on the experimentally induced neurologic disease, *Am. J. Vet. Res.* **38:**709–719.

138. Sabo, A., and Rajcani, J., 1976, Latent pseudorabies virus infection in pigs, *Acta. Virol. (Praha)* **20:**208–214.

139. McCracken, R. M., McFerran, J. B., and Dow, C., 1973, The neural spread of pseudorabies virus in calves, *J. Gen. Virol.* **20:**17–28.

140. McKercher, D. G., 1973, Viruses of other vertebrates, in: *The Herpesviruses* (A. S. Kaplan, ed.), Academic Press, New York, pp. 427–493.

141. Gaskell, R. M., and Povey, R. C., 1979, Feline rhinotracheitis: Sites of virus replication and persistence in acutely and persistently infected cats, *Res. Vet. Sci.* **27:**107–174.

142. Payne, L. N., 1982, Biology of Marek's disease virus and the herpes virus of turkeys, in: *The Herpesviruses*, Volume 1 (B. Roizman, ed.), Plenum Press, New York, pp. 347–431.

143. Chumnonqsitathum, B. J., Plumb, J. A., and Hilge, V., 1988, Histopathology, electron microscopy and isolation of channel catfish virus in experimentally infected European catfish, *Silurus glanis* L, *J. Fish Dis.* **11:**351–358.

10

Paramyxoviruses

CLAES ÖRVELL

1. INTRODUCTION

Paramyxoviruses may cause an involvement of the central nervous system (CNS) during infection of the natural host as well as in experimental infection of laboratory animals.[1] These viruses may invade the CNS during the clinical or subclinical acute stage of the infection. When the virus is not effectively eliminated by the immune response, it may persist in the CNS in a clinically silent phase for variable periods of time. After the silent phase the virus may cause symptoms attributable to cerebral dysfunction; serious disease and even death of the host may follow.

The interaction between a paramyxovirus and the host depends on a multitude of factors, i.e., the virus, the host, and the competence of the host's immune system. These factors decide whether there will be involvement of the CNS and the outcome of such involvement. In order for an infection of the CNS to occur, there must be an effective spread of virus from the primary site of infection, and the virus must be capable of entering the CNS through the blood–brain barrier or by other routes. Once virus is inside the CNS, the result of the infection is dependent on the ability of the virus to replicate in the different heterogeneous cell populations in the brain. This capacity may vary among different types of the same virus. The outcome of the infection is dependent on the ability of the host to mount an efficient immune response that will clear the virus before or after entry of virus into the CNS. In certain situations the virus may remain latent in the CNS, escaping the immune surveillance of the host. The age of the individual may be important for his or her susceptibility to the paramyxoviruses.

CLAES ÖRVELL • Department of Virology, National Bacteriological Laboratory and Karolinska Institute, School of Medicine, S-105 21 Stockholm, Sweden.

Neuropathogenic Viruses and Immunity, edited by Steven Specter *et al.* Plenum Press, New York, 1992.

2. CLASSIFICATION AND STRUCTURAL PROPERTIES OF PARAMYXOVIRUSES

During recent years there has been a considerable accumulation of new information about the molecular and immunobiological properties of paramyxoviruses. The application of new technologies, mainly recombinant DNA techniques and hybridoma methodology, has led to a better understanding of the complex relationships between these viruses and their hosts, as it has improved our possibilities to study the expression of the virus in terms of viral proteins and nucleic acid *in vivo*.

The family Paramyxoviridae contains three genera, paramyxoviruses, morbilliviruses, and pneumoviruses.[2] The best-characterized members of the paramyxovirus genus include mumps virus, four types of human parainfluenza virus, bovine parainfluenza type 3 virus, canine parainfluenza virus, Sendai virus of mice, simian virus 5, and Newcastle disease virus (NDV). In addition there exist a number of less well-characterized types isolated from birds and mammals. Several of the viruses belonging to the paramyxovirus genus may invade the CNS during natural and experimental infections.

The morbillivirus genus comprises measles virus, canine distemper virus (CDV), phocine distemper virus of seals, rinderpest virus (RV) of cattle, bovine paramyxovirus 107, and peste des petits ruminants virus of sheep and goat. Pronounced immunologic relationships are known to occur between these viruses.[3-5] Acute infection with these viruses may involve the CNS. Persistent infections in the CNS can be established by measles virus and are described as two separate disease entities, measles inclusion body encephalitis (MIBE) and subacute sclerosing panencephalitis (SSPE). Persistent infection with CDV has been described with varying manifestations.

The third genus, pneumoviruses, comprises respiratory syncytial virus (RSV) of human subgroup A and B, bovine RSV, and pneumonia virus of mice (PVM).[6-8] The RSV is an important respiratory tract pathogen in young individuals; these viruses have not been implicated in CNS disease and are thus not within the scope of this review.

The paramyxoviruses are enveloped, usually spherical in shape, with a diameter of 120 to 300 nm[2,9] enclosing a centrally located helical nucleocapsid. The nucleic acid within the nucleocapsid consists of one piece of single-stranded RNA. Five major virus-specific proteins, the nucleoprotein (NP), phospho (P), matrix (M), fusion (F), and hemagglutinin–neuraminidase (HN) protein, which can be defined by radioimmune precipitation assays (RIPA) with monoclonal antibodies, and one protein that is cellularly derived make up the structure of the virus[2,6,10] (Table 10-1, Fig. 10-1). In addition to the five major virus-specific proteins, the virion contains a sixth protein of relative low concentration in the virion. This protein, which has a high molecular weight, is called the large (L) protein.

Four of the proteins, NP, P, L, and actin, are surrounded by the virion envelope. The envelope of the virus is built up by a bimolecular lipid layer derived

TABLE 10-1
Structural Proteins of the Genera *Paramyxovirus* and *Morbillivirus*

Designation	Relative molecular weight		Location	Function
	Paramyxovirus	Morbillivirus		
NP	55–70	58–65	Nucleocapsid	Protecting RNA
P	45–79	72–78	Nucleocapsid	In transcriptive complex
L	~200	~200	Nucleocapsid	Not known; possibly in transcriptive complex
Actin	43	43	Not known (not at the surface of virion)	Not known
M	34–40	34–37	Inner side of virion envelope	Virion assembly
HN/H	70–80	76–85	Spike	Adsorption to cells; neuraminidase activity in paramyxovirus
F	60–70	60	Spike	Fusion of cells, hemolysis, virus entry

from the host cell with the virus-specified M protein located on its inner side. Hemagglutination and neuraminidase activity are associated with a glycosylated protein, the HN protein, and hemolytic and cell-fusion activities are associated with a second glycosylated protein, the F protein. These two glycoproteins form the spikes or peplomers of the virus. They are transmembraneous, traversing the lipid bilayer of the envelope and protruding 10 to 15 nm outside the virion envelope. Neuraminidase activity has only been found in members of the paramyxovirus genus. In morbilliviruses the corresponding surface projection is designated the hemagglutinin (H) protein. The molecular masses of the HN or H, F, NP, P, and M proteins vary from 80 to 34 kDa (Table 10-1). It has been demonstrated by RIPA with radiolabeled dissociated viral proteins and convalescent sera that an antibody response against all the major structural proteins is raised during natural infection. Antibodies against each of the two surface glycoproteins are important for protection against infection with paramyxoviruses.[2,9]

Most genes of at least one type of representative members of paramyxoviruses have been sequenced, and the amino acid sequences of the corresponding proteins have been deduced. The linear arrangement of the different genes from the 3' to the 5' end of the genome of paramyxoviruses in terms of their products is NP, P, M, F, HN or H proteins and, finally, a gene directing the synthesis of the L protein.

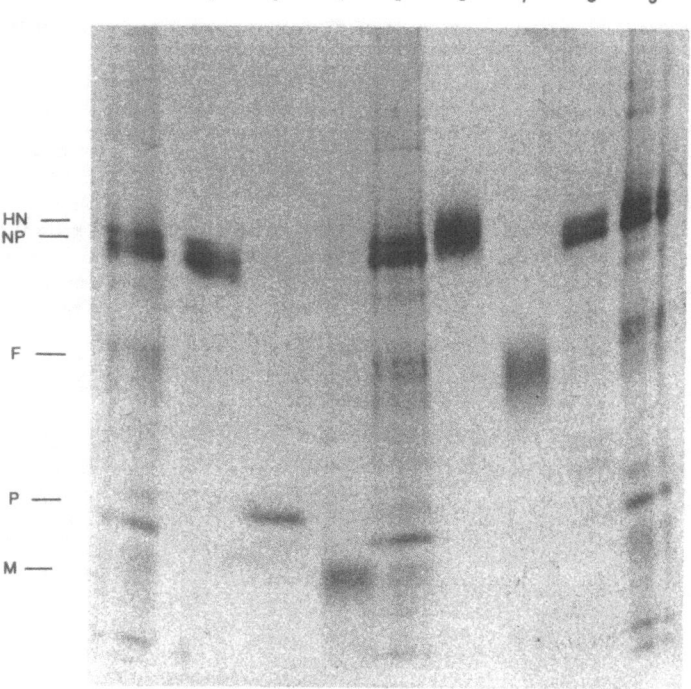

FIGURE 10-1. Radioimmune precipitation assay (RIPA) with monoclonal antibody from mouse ascites reacting with different mumps virus proteins. Materials included (the positions in the gel are given in parentheses) [35S]methionine-labeled purified mumps virions (lanes 1,5); monoclonal antibodies against NP (2,8), P (3), M (4), HN (6), and F (7). (From Örvell[10] with permission of the *Journal of Immunology*.)

3. PATHOGENESIS OF CNS INFECTIONS CAUSED BY PARAMYXOVIRUSES

3.1. Modes of Spread of Infection to the CNS

Viruses can spread to the CNS via the neural route (i.e., along peripheral nerves or along the olfactory nerve from the olfactory mucosa in the nasopharynx) and via the hematogenous route.[11]

Spread via the hematogenous route requires that the virus produce a generalized acute infection of the host with sustained viremia. These requirements are met by several viruses of the paramyxo and morbillivirus genera, i.e., mumps virus, NDV, measles, CDV, and RV. The brain is separated from circulating blood by the blood–brain barrier, which is made up by the relative impermeability of cerebral capillaries, which lack fenestrations, and by the astrocytic footpads densely packed against them. In order to invade the CNS the virus must pass through the endothelial cells of the vessels. The spread of paramyxoviruses to the CNS has been studied in experimental animals. In one study experimental infection of dogs with a neurovirulent strain of CDV resulted in hematogenous dissemination of the virus to the CNS.[12–14] Viral antigen was first detected within CNS capillary and venular endothelia and perivascular astrocytic foot processes (Fig. 10-2) 5 to 7 days post-infection (p.i.). Leukocyte infiltration with antigen-positive cells followed 1 to 2 days later.

In other studies newborn hamsters were inoculated intraperitoneally with neuroadapted mumps virus.[15,16] After local replication in different visceral organs the virus was disseminated via a low-level viremia and appeared to enter the CNS through endothelial cells of the plexus choroideus. The virus was found in choroidal and ependymal epithelial cells 3 days p.i. and on the fifth day had spread to neuronal cells.

Involvement of the brain in connection with measles and mumps infection appears to be frequent. In mumps infection, clinical meningoencephalitis is found in 10% of patients, and there is pleocytosis in the cerebrospinal fluid (CSF) in 50% of all cases. Mumps virus may sometimes be isolated from the CSF of patients with mumps meningoencephalitis, but the extent of virus replication in the brain is not known. Pleocytosis in the CSF is found in all patients with measles,

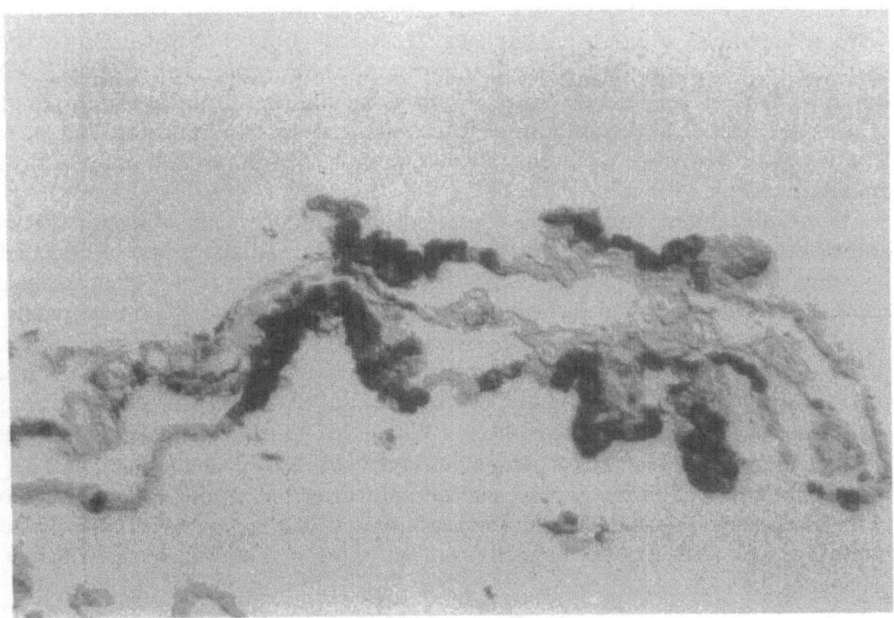

FIGURE 10-2. Appearance of viral antigen in canine-distemper-virus-induced encephalitis in the cells of plexus choriodeus of a dog, traced with a peroxidase-labeled monoclonal antibody directed against the NP. (Courtesy of Dr. W. Baumgärtner.)

and electroencephalographic changes are seen in 50% of infected children. In uncomplicated measles it is not possible to isolate the virus from the CSF.

3.2. Models for Immunobiological Characterization of Paramyxoviruses Involved in Brain Infection

The degree of neurovirulence of a virus has been suggested to reflect the capacity of the virus to invade the brain parenchyma and infect neurons.[17–19] Today virtually nothing is known about paramyxovirus characteristics involved in neuropathogenicity.

Monoclonal antibodies offer a rapid and simple tool for studying the immunobiological characteristics of viral strains. Each monoclonal antibody is specific for a particular epitope, and by using a large number of well-characterized monoclonal antibodies it is possible to obtain an estimation of antigenic differences in different strains of the same virus.

Neutralizing monoclonal antibodies may also be used to select for viral mutants, which may then be studied *in vitro* and *in vivo* with respect to their new immunobiological properties. In one study four mutants of the Kilham neurotropic strain of mumps virus were isolated with the aid of neutralizing monoclonal antibodies directed against a specific site on the HN surface glycoprotein.[20] Two mutants had lost their hemagglutination capacity with human O eythrocytes, and a third one showed a change in the molecular weight of the HN protein. These three mutants showed unaltered capacity to infect tissue cultures and to cause encephalitis in newborn hamsters. A fourth mutant (M13) retained the hemagglutinating activity and the capacity to infect Vero cell cultures but showed significantly decreased neurovirulence in suckling hamster brain. The number of infected neurons and the amount of infectious virus in the brain were reduced. There was no apparent difference in the amount of viral antigen expressed in ependymal cells.

In a subsequent study the HN genes of the Kilham strain of mumps virus and three of the neutralization escape mutants described above were sequenced by using their genomes as template.[21] The predicted amino acid sequences were compared. The biological differences of the different mutants were primarily associated with strain-specific amino acid changes outside the region of the presumed neutralizing epitope. A substitution in position 297 of a leucine for a phenylalanine probably correlated with the reduced neurovirulence of strain M13. Two other strains of mumps virus, RW and SBL-1, which, like M13, exhibit a low neurovirulence, also have a phenylalanine in position 297. Sequence comparisons of the genes of neurovirulent and nonneurovirulent strains with deduction of the amino acid sequences of the corresponding proteins may eventually help to identify amino acids and amino acid positions of importance for neurovirulence.

3.3. Persistent Infections with Paramyxoviruses *in Vitro*

A number of studies have described persistent infections with paramyxoviruses *in vitro*. Such studies may help to elucidate the mechanisms operating in viral persistence *in vivo*.

In most cases persistent infections in tissue culture cells have been established by infecting cells at high multiplicity of infection.[22-27] This procedure has resulted in a fraction of cells surviving the acute lytic infection. The few surviving cells may grow out, and a persistent culture may be formed that continues to divide. Later the virus-producing culture may undergo a lytic cycle of viral replication, and the cells may be lost. The survival of the cells in the culture may be through the presence of defective interfering viral particles, interferon production, or the formation of viral mutants defective in replicative capacity. Investigation of the extracellular virions produced and of different viral proteins expressed by the persistently infected cells with monoclonal antibodies against the major structural components of the virus may help establish whether an imbalance exists in the production of some viral component necessary for virion formation. In one study a HeLa cell line persistently infected with the Edmonston strain of measles virus was studied with monoclonal antibodies directed against the NP, P, M, HN, and F components.[28] Immunofluorescence analysis revealed that all cells synthesized NP and P proteins, but detectable amounts of H, F, and M components were found in only 50, 10, and 30% of cells, respectively. The low percentage of cells with detectable F protein formation may explain the low-grade cell fusion that was observed in the culture.

3.4. Persistent Paramyxovirus Infection in the Brain

Persistence of a virus in the brain is a complex situation because of the multitude of cell types that may be involved and the existence of a functionally active immune system. Because paramyxoviruses are RNA viruses and do not contain a reverse transcriptase enzyme, they cannot become integrated in the cellular DNA to rest there as silent proviruses. Present knowledge dictates that some expression of the virus, either complete or incomplete, must take place to uphold a state of persistence. Under these circumstances there are some restraints on viral replication; i.e., the infection must be nonlytic or cause limited cell death only, and the virus-infected cell must not express viral antigens on the surface that would make it identifiable by immune defense mechanisms. Under normal conditions the HN/H and F proteins are inserted into and change the antigenic structure of the cell membrane. As a consequence of this the cell will be recognized as foreign by the host, and immune mechanisms will come into play.

A number of experimental animals, i.e., mice and hamsters of different ages, ferrets, dogs, and monkeys, have been used to study the involvement of the brain, the distribution of viral antigen, and cellular lesions after infection with various paramyxoviruses by different routes. Ferrets and dogs have been used to study CDV. The outcome of experimental infection in dogs varies with the strain of virus used[29] and the intensity of the immune lymphocyte-mediated cytotoxicity (ILMC) response elicited by that particular strain.[30] Dogs surviving the acute phase of the infection show evidence of viral persistence. In one study 12 4-week-old puppies were inoculated intraperitoneally with the R252 strain of CDV.[31] Nine of 12 dogs died between 4 and 5 weeks p.i., but the other three dogs survived the acute episode and appeared to recover. The three dogs that recovered did not show any clinical signs. Five months p.i. one of the dogs developed convulsions and signs of brainstem dysfunction and died. When the brain of this dog was investigated by immunofluorescence, foci of CDV antigen were found throughout the brain, both coincident with and independent of inflammatory lesions. Viral antigen was found exclusively in neurons, which contained beaded inclusions in their cell bodies and processes. In another study the presence of viral activity in the brain of dogs was revealed by the presence of elevated levels of myelin basic protein (MBP) in the CSF.[32] None of 17 dogs infected via the intranasal route with the Cornell A75-17 strain of CDV had elevated levels of MBP at day 20 p.i.; five had elevated levels at day 40; and seven had elevated levels at the last sampling 50 to 72 days p.i. Dogs with elevated levels of immunoreactive MBP in the CSF had severe demyelination in their brains. The pathological changes were most severe in the white matter adjacent to the fourth ventricle. Myelin loss, manifest as sponginess and pallor, was accompanied by astrocyte hypertrophy and proliferation and by a mixed lymphocytic inflammatory response.

Persistence *in vivo* has been described for measles virus and CDV in the brain of their natural hosts, and other paramyxoviruses are known to persist in the CNS of experimentally infected animals. A classical example of persistent infection in the brain of humans is SSPE.[1,33-35] It is a rare, fatal disease of children and young adults starting on average 6 to 8 years after an uncomplicated measles infection that usually occurred before the age of 2 years. An invasion of the brain at the time of the primary infection with measles virus is likely to have occurred. Different mechanisms have been proposed over the years to explain the establishment of nonproductive, cell-associated CNS infections by measles virus, but experimental support for most of these theories is lacking. Absence of virion budding because of defects in the formation or function of the M protein has been suggested. In an experimental model of SSPE in hamsters studied by Johnson *et al.*,[36,37] measles virus in the brain was shown to evolve from a complete infectious form to a cell-associated form 8 to 12 days after intracerebral inoculation. Rabbit hyperimmune sera containing antibodies directed against NP and M were used to trace these proteins. During the early phase of the infection, when complete infectious virus could be isolated from brain tissue, both proteins were labeled, but in later stages only the NP was demonstrable.

Monoclonal antibodies against NP, P, M, H, and F were used in immunofluorescence to investigate the expression of measles virus proteins in brains from cases of SSPE.[38,39] In one study all of the five structural components were detectable in four brains,[38] but staining with antibodies against NP and P gave a more intense immunofluorescence than with the other antibodies. In the second study, brains from four other patients with SSPE were investigated. The NP protein and P proteins were found in every diseased brain area, whereas the H protein was detected in two brains, the F protein in three, and the M protein in only one.[39] Furthermore, the three envelope proteins were detected in only a few cells and in some diseased areas of the brain. The results from these and other studies[40,41] show that, compared to lytically infected cells, there may exist an underrepresentation of the envelope proteins H, F, and M in comparison to the internal proteins of the virus NP and P in both experimental and natural SSPE.

This proposed imbalance between the expression of internal and envelope proteins does not take into account the possibility of pronounced mutations of the envelope proteins that would render them undetectable by the serological reagents used. An estimation of the quantity of transcripts of the different measles virus genes in the brains of two SSPE cases and one case of MIBE has been reported by Cattaneo et al.[42] By the use of quantitative Northern blots on the brain materials it was demonstrated that in all three cases the mRNA transcripts from the first measles virus genes (NP and P) were relatively abundant, amounting to 10% of that in lytically infected cells. The transcription of successive measles virus genes declined sharply compared to that in the lytically infected cells. These results explain on a molecular basis the reduced expression of viral envelope proteins typical of persistent measles virus brain infections.[37–39]

In a subsequent report the same group studied the alterations of viral gene expression by cloning full-length transcripts of the different genes from three diseased brains.[43] About 2% of all nucleotides were mutated during persistence, and 35% of differences resulted in amino acid changes. One of the nucleotide substitutions and one deletion resulted in alteration of the reading frames of the F gene in two cases, resulting in a reduction of 15 and 24 amino acids at the C-terminal part of the F protein, respectively. In one case the M gene exhibited one exceptional cluster of mutations; 50% of uracil residues were changed to cytosine, resulting in a grossly changed M protein. In another study the M gene from brain cells of a SSPE patient was expressed in IP-3-Ca cells.[44] This resulted in a unstable M protein being produced, and viral particles could not be formed. Transfection of the viral genome into other cell lines did not abrogate the defects. The authors concluded that the mutated M protein was nonfunctional in viral assembly.

The distribution of different viral proteins also has been studied in persistently infected mouse brains inoculated as newborns with Sendai virus.[45] The levels of infectious virus in the brain declined before the delayed appearance of serum antibodies at 12 days p.i. Immunofluorescence analysis showed viral antigen in neurons and their dendritic processes at 12 and 24 days p.i. Staining with fluorescein-labeled monoclonal antibodies to the NP, P, M, F, and HN proteins of Sendai virus demonstrated that all were present in choroid plexus epithelial cells and ependymal cells during acute infection. In contrast, during acute and persistent infections in neurons, the NP, P, and M proteins were found, but the two surface glycoproteins, F and HN, were lacking or found at low levels.

In another study newborn hamsters and newborn mice were infected intracerebrally with the neuroadapted Kilham virus strain of mumps virus.[46] The newborn hamsters that were inoculated succumbed to an acute fatal encephalitis, and high titers of infectious virus were recovered from their brains. By immunofluorescence using monoclonal antibodies, all five structural proteins were demonstrable in the brains of infected animals.

In contrast to hamsters, newborn mice showed no evidence of disease and no infectious mumps virus in brain tissue. Large numbers of neurons showed the presence of the NP and P proteins using immunofluorescence, but no M, F, and HN proteins were detectable. To ascertain whether the defect in virus replication was caused by a cellular restriction of replication in neurons or by the host's defense mechanisms, explant cultures of spinal ganglia and cord from mice and hamsters were infected with mumps or Sendai virus.[47] Expression of the five structural proteins NP, P, M, F, and HN was examined with monoclonal antibodies. In Sendai-virus-infected mouse neurons and mumps-virus-infected hamsters neurons all five viral proteins were detected. In mouse neurons infected with mumps virus, on the other hand, only the NP and P proteins were detected. The authors concluded that there may exist a species-specific cellular restriction in mouse neurons. There was little or no reduction in the absolute number of neurons in cultures of mouse brain tissue infected with mumps virus between days 4 and 20 p.i., but the proportion of infected neurons diminished fourfold during that time.

The distribution and occurrence of the major structural proteins also has been studied in mice and rats inoculated intracerebrally with neurotropic strains of measles virus.[48,49] In one study Lewis and Brown Norway (BN) rats were inoculated.[48] Suckling rats died from acute encephalitis, but with increasing age Lewis rats developed a subacute encephalomyelitis, whereas BN rats developed an encephalitis without clinical signs. In the two mouse strains studied, the NP and P proteins were detected in the infected brain cells of animals with acute, subacute, or symptomless encephalitis, whereas the M, F, and H proteins were reduced or absent. These data indicated that the different diseases of the two rat strains were related to the immunogenetic background of the animals and not to different replication of measles virus in the CNS of the two rat strains. In newborn mice inoculated with a hamster neurotropic measles virus strain, a similar restriction of the expression of viral envelope proteins was observed, whereas the NP and P antigens appeared in clusters of neurons in the cerebral cortex.[49] Only single infected neurons occurred in the striatum, and in the hippocampus only a few groups of infected neurons were observed in the pyramidal cell layer. The antigens had a typical beaded distribution within neurites of individual nerve cells.

3.5. Local Antibody Production Caused by Paramyxovirus Infections in the CNS

Because of the presence of immunocompetent cells, the brain has the capacity to respond to viral infection with antibody production. This fact may be used in the diagnosis of CNS viral infection unless the blood–brain barrier has been damaged with nonspecific leakage of serum proteins into the CSF. Under normal conditions the antibody titer in serum is about 300 times higher than the antibody titer in the CSF. With sensitive serological techniques such as the enzyme-linked immunosorbent assay (ELISA) the antibody content in the CSF can be estimated and compared to the titer in serum. If the serum/CSF antibody ratio for a particular virus is significantly lower than the ratios of antibodies for other viruses or the serum/CSF ratio of albumin, this may be taken as an indication of an ongoing brain infection with that virus. In one recent study the ELISA was used for demonstrating specific viral antibodies in 21 patients with mumps meningitis.[50] Nineteen (91%) patients had mumps-specific IgG antibodies, and 11 (52%) mumps-specific IgM antibodies, demonstrable in their CSF, whereas none of a control group of 21 patients with aseptic meningitis of other etiology did. A decreased ratio of serum/CSF mumps IgG antibodies, less than 125, was demonstrated in 18 (85%) patients, whereas none of the patients had a decreased serum/CSF IgG ratio with measles virus, which was used as a control. In a previous study on 38 mumps meningitis patients similar results were obtained.[51] The CSF contained demonstrable mumps IgG and IgM antibodies in 88% and 46% of cases, respectively.

Using a direct ELISA, Chiodi et al.[52] examined serum and CSF from six patients with SSPE and control subjects for presence of measles-virus-specific IgM antibodies. They were able to demonstrate IgM antibodies in the CSF as a sign of ongoing infection in all six patients. The levels of measles IgM antibodies were higher in CSF diluted 1 : 5 than in serum diluted 1 : 50, reflecting a local production of IgM antibodies in the CNS. The antibody titers in three patients who were followed for 3 to 6 months remained constant. In addition to ELISA, another more cumbersome technique used to study an ongoing infection in the CNS is electrophoresis of CSF and demonstration of virus-specific oligoclonal bands. Measles antibodies in CSF of SSPE patients correlate with the presence of oligoclonal IgG bands.[53–55] Oligoclonal IgG bands are also found in the serum of most patients with SSPE, which may be an indication of antibody synthesis outside the CNS as well. The same technique has also been used for demonstration of mumps-specific oligoclonal bands in the CSF of patients with mumps meningitis.[56–58] From the latter studies some evidence has been presented that a polyclonal activation of cells in the CNS can take place in mumps meningitis.

The antibody response to individual viral proteins of paramyxoviruses has been studied in complement-fixation tests and ELISA with purified NP and M components in the case of measles virus[59,60] and by RIPA with all of the major structural components of measles virus, CDV, and mumps virus.[59,61–65] In serum and CSF from SSPE patients there was a strong antibody response to the NP component, a relatively weaker response to the F and H components, and a very weak or absent antibody response to the M component.[59,61–63] However, this pattern of response was not pathognomonic for SSPE but was also found in measles convalescent sera and in sera from patients with high measles antibody titers in other disease conditions such as multiple sclerosis and chronic active hepatitis.[59,63]

In contrast, an accentuated antibody response to the M component was found in sera from patients with atypical measles,[59,61] a rare form of measles that occurs in individuals who have been vaccinated with inactivated measles virus preparations. The vaccine does not give long-term protection against the infection because of an antigenic defect but instead worsens the clinical signs and symptoms through the occurrence of immune pathological phenomena. In a recent study the response to the M protein of measles virus was examined in patients with SSE and in controls.[60] Antibodies were considerably higher against NP than against M in both serum and CSF. In spite of this, antibodies against M were demonstrable in CSF samples from all eight patients with SSPE but not in CSF from patients with other neurological diseases. In two of the eight patients the level of antibodies against M was higher in CSF than in serum, and in three there was also a suggestion of intrathecal synthesis of such antibodies.

In one study on CDV the humoral immune response in sera and CSF of dogs with various forms of encephalitis was assessed by RIPA with NP, P, M, F, and H proteins.[64] Sera from vaccinated dogs and hyperimmune sera contained antibodies to all five antigens. In two cases of old-dog encephalitis (symptoms of encephalitis develop in old or immunized dogs without clinical signs of a preceding primary infection), the sera and CSF showed a restricted response to the M protein, whereas in three other dogs with old-dog encephalitis, two dogs with chronic distemper meningoencephalitis, and four with experimentally induced encephalitis, the antibody response was directed against the NP, P, and M proteins but not against the F and H proteins. It should be noted that most sera precipitated the M protein only when the antigen had been prepared by *in vitro* translation.

Serum and CSF from a patient with chronic encephalomyelitis caused by mumps virus were investigated by RIPA.[65] Antibodies against the envelope glycoproteins and the NP and M protein could be demonstrated, and the antibody patterns in serum and CSF were similar, but the antibodies to individual proteins were not quantitated. Mumps antibodies were not found in the CSF of 57 control patients.

3.6. Cellular Immune Response in CNS Caused by Paramyxovirus Infections

The cellular immune response has been investigated in brain infections with selected paramyxoviruses.[30,60,66–73] The responses to NP, M, F, and H were examined in four patients with SSPE by lymphoproliferation and were found to be of a similar order of magnitude as in five health control patients.[60] The ILMC was measured in dogs that had been either vaccinated with the Rockborn strain of CDV or exposed to virulent CDV strains.[30,67] A T-cell-mediated ILMC response was measurable for 10 days beginning at 6 days post-vaccination.[67] After intranasal exposure to one of three virulent strains of CDV, a strong correlation was found between ILMC and the outcome of the infection.[30] The dogs that survived the infection generally had the highest activity, whereas dogs that died of encephalitis had low or unmeasurable ILMC. Dogs with reduced ILMC or with a late response developed a subclinical CNS infection that persisted. The results suggested that the ILMC response is an important factor in determining the effect of infection. With certain strains of CDV capable of inducing persistent infection, a delayed ILMC response correlates with the establishment of persistence in the CNS.

During mumps meningitis there are increased numbers of T cells with suppressor/cytotoxic activity in the CSF.[68–73] In one study lymphocytes from venous blood and CSF from ten children with mumps meningitis were tested in a ^{51}Cr-release assay for cytotoxic activity against uninfected and mumps-virus-infected phytohemagglutinin (PHA)-induced blast cells from cryopreserved lymphocytes.[70] Lymphocytes from all patients were cytotoxic to autologous mumps-virus-infected target cells but failed to lyse histoincompatible infected PHA blasts. Cytotoxicity was specific for mumps virus and was mediated by E-rosette-forming lymphocytes. The cytotoxic cells were present 2 to 3 weeks after the start of meningitis. These results demonstrated that specifically sensitized cytotoxic T cells are induced in patients with mumps meningitis. In a subsequent study T lymphocytes from the CSF of one patient with mumps meningitis were analyzed at a clonal level.[72] From the cloning experiments 84 colonies were established and analyzed for the ability to effectuate PHA-dependent cell killing. Forty-one colonies exhibited cytotoxic activity, and of these, 39 were specific for autologous mumps-virus-infected target cells. The results from the latter two studies and from another study[68] suggest that there is a relative concentration of antigen-specific immune-competent cells in the CSF compared to venous blood of patients with mumps meningitis.

4. DISEASES OF THE CNS CAUSED BY PARAMYXOVIRUSES

4.1. Diseases of the CNS Caused by Viruses Belonging to the Paramyxovirus Genus

4.1.1. Mumps Virus

Different strains of mumps virus with unique antigenic characteristics can be defined with monoclonal antibodies.[10,74] Man is the natural host for mumps virus. The incubation period for mumps is usually 18 to 21 days. The virus causes a general systemic infection with swelling of the parotid gland and involvement of the CNS in a high proportion of cases. Rarely, meningoencephalitis may be the only symptom of mumps. When the patient develops meningoencephalitis, the body temperature increases, the patient suffers from headache and vomiting, and the neck and back of the patient become stiff. There is pleocytosis in the CSF, usually with dominance of lymphocytes. The clinical phase of mumps meningitis usually abates within a few days, and there is complete recovery without sequelae. The diagnosis of mumps is mostly made on clinical grounds, but in unclear cases it can be established by demonstration of IgM antibodies against mumps virus in a single serum sample or a significant titer rise in mumps-virus-specific antibodies between an acute and a convalescent serum.

Few reports have addressed the capacity of mumps virus to persist in the brain of man. One case has been described by Finnish workers.[65] The patient, a previously healthy man, developed symptoms of severe encephalomyelitis at 31 years of age. He had a high serum antibody titer to mumps virus associated with a polymorphic cell reaction and an increased protein concentration in the CSF. Fourteen years later the patient became severely handicapped by a chronic encephalomyelitis. By electromyographic investigation the patient showed evidence of neurogenic lesion without active denervation. There was serological evidence of intrathecal production of mumps antibodies and by electrophoresis the locally produced antibodies showed an oligoclonal pattern. In an earlier study, Vandvik et al.[75] reported a prolonged pleocytosis of the CSF 1-year after mumps meningitis, which suggested persistence of mumps virus in the CNS. Another case report described the occurrence of aqueductal stenosis and hydrocephalus 2 years after mumps encephalitis, but mumps-specific antibodies were not recorded in the CSF.[76]

Newborn hamsters and newborn mice have been used to study neurovirulence and persistence of different strains of mumps virus.[15,16,46,77-79] In one investigation the growth and neuropathogenicity of five strains of mumps virus were studied.[78] The ability to invade the brain parenchyma and infect neurons of neonatal hamsters differed in different strains and appeared to correlate with the cytopathology caused by the strain in tissue cultures. There was no correlation between virus growth in neurons and in ependymal cells of the animals.

It has also been shown that a mumps virus strain that is highly neuropathogenic possesses strong cell fusion and weak neuraminidase activity.[80,81] Less neuropathogenic strains, on the other hand, have low cell-fusing activity and strong neuraminidase activity. In addition to the neuropathogenicity of the strain, the route of inoculation is of importance for the infection outcome. When inoculated intracerebrally with a neuropathogenic strain of mumps virus, newborn hamsters developed widespread CNS disease and succumbed to the infection.[46,77,82,83] After intraperitoneal inoculation the same virus caused disseminated disease with moderate mortality, and viral persistence was established in the CNS.[77] Viral antigen could be demonstrated in the brains 50 days p.i.

In those experiments the pathology both at light microscopic and at an ultrastructural level was followed from day 3 through 50.[16] Three to 5 days p.i. viral mucleocapsids were found in ependymal cells, and neurons and virions were formed at the plasma membrane of these cells by budding up to 7 days postinoculation. Accumulation of nucleocapsids without evidence of virion formation was found in the ependymal cells up to 33 days p.i. Several animals had aqueductal occlusion, and at these sites the ependymal cells were swollen and distorted. Also, the ependymal cells lining the lateral ventricles were changed: interstitial edema separated parenchymal elements from the underlying white matter. Hydrocephalus with symmetrical enlargement of the lateral ventricles and stenosis or obliteration of the posterior third ventricle and aqueduct was present in all animals sacrificed 16 pays p.i. or later. In another study, a mumps virus strain of low neurovirulence has been reported to cause developmental disturbances in the hamster retina after intracerebral inoculation into newborn animals.[79]

Two reports of successful immunotherapy of experimental mumps meningoencephalitis have been presented. In these experiments the ability of monoclonal antibodies against the HN and F proteins of mumps virus to protect newborn hamsters against intracerebral inoculation with a lethal dose of a neurovirulent strain of virus was assessed.[82,83] Monoclonal antibodies reactive with epitopes on the HN glycoprotein inhibited hemagglutination and neutralized infectivity in vitro.[82] Such antibodies protected the animals when given intraperitoneally 24 hr p.i. More than 50% of the animals were alive after an observation period of 40 days. Virus antigen and infectivity titers were diminished in the brains of the animals. In the second study a monoclonal antibody against the F protein of mumps virus was administered.[83] The antibody could inhibit hemolysis of the virus but did not neutralize the infectivity *in vitro*. The antibody was found to confer marked protection when given subcutaneously at the same time as the virus. Almost total prevention of extensive brain damage was found. The study indicated that the F protein is directly involved in pathogenesis of brain necrosis. However, all animals eventually died over an observation period of 24 days.

4.1.2. Parainfluenza Viruses

In natural infections the tissue tropism of parainfluenza viruses is limited to the respiratory tract. One report has been published on a generalized infection with parainfluenza type 3 virus with isolation of the virus from the brian, but the patient was an 8-month-old infant with severe combined immunodeficiency.[84]

Newborn, suckling, and adult mice have been inoculated intracerebrally with parainfluenza viruses to study the effects on the brain.[45,85–95] Sendai virus multiplied to higher titers in newborn than in adult mice.[45,85] Titers of infectious virus in the brains peaked on day 1 to 2 and disappeared 5 to 6 days p.i. Newborn mice had a lower and more delayed antibody response to the virus than adults. As measured by immunofluorescence, the viral antigen appeared initially in ependyma, choroid plexus, and leptomeninges 1 to 2 days p.i., and histologically there were signs of meningitis, ependymitis, and choroiditis with periventricular and perivascular mononuclear cell infiltration 2 to 3 days post-inoculation.[45,85,86] Early in infection viral antigen was found in many neurons in the cortex, hippocampus, and basal ganglia. Viral antigen in the neurons gradually disappeared and could not be detected 2 weeks p.i. in adult mice, whereas it persisted in the neurons of the newborn mice for several months.[45,85,87] Intracerebral infection of mice with the closely related 6/94 strain of human parainfluenza type 1 results in a similar process of infection including the establishment of a persistent infection in the neurons of mice that had been inoculated as newborn.[88–91] This strain establishes lifelong persistence in C129 mice. Although no viral antigen could be detected more than 1 year p.i., viral RNA could be demonstrated in the brains by hybridization with virus-specific cDNA cloned in the bacteriophage λ system.[92]

Strains of human parainfluenza type 3 virus are markedly different antigenically from strains of bovine parainfluenza type 3 virus.[96] Studied by Shibuta *et al.*[93–95] have demonstrated that different strains of bovine parainfluenza type 3 virus show different pathogenicity after intracerebral inoculation into mice. The 910N strain of bovine parainfluenza type 3 virus induces hydrocephalus in newborn mice, and most mice can survive, whereas the YN strain causes an acute lethal disease with marked thymic and splenic atrophy.[94,95] The YN strain was not phenotypically uniform but contained three distinct plaque-type variants, which were isolated. Two variants induced hydrocephalus, and the third (M strain) induced a fatal encephalitis in newborn mice. Infection with the 910N virus was limited to the ventricular epithelial cells, whereas the M variant invaded the subependymal parenchyma, thalamus, hypothalamus, and brainstem. The strain and variants that showed low neurovirulence *in vivo* had low cell-fusing activity *in vitro*, whereas the virulent M variant had strong cell-fusing activity and extremely low neuraminidase activity. Similar observations had been previously made on different mumps virus strains by Merz and Wolinsky.[80,81]

Canine parinfluenza virus is antigenically related to simian virus 5 and causes tracheobronchitis in its natural host. A neurotropic strain of the virus has been isolated from the CSF of a dog with temporary posterior paralysis.[97,98] When inoculated into gnotobiotic dogs and ferrets, this strain caused encephalitis with subsequent internal hydrocephalus.[99] The development of hydrocephalus is a common long-term effect of experimental infection of mice with parainfluenza viruses and is probably a result of obliteration of the aqueduct.[85,87,93]

4.2. Diseases of the CNS Caused by Viruses Belonging to the Morbillivirus Genus

4.2.1. Measles Virus

Variations in the amino acid sequence of gene products of different strains of measles virus have been found,[43] and different serotypes of the virus have been defined by monoclonal antibodies.[100]

Acute measles encephalitis in most cases appears during the first week after appearance of rash. There are two clinically indistinguishable forms of the disease. In one form, acute MIBE, there is active replication of virus in the brain, measles-like inclusions and viral antigen can be demonstrated, and measles virus can be isolated from the CSF. There is a mild to moderate inflammatory response with some perivenous myelin loss. Glial cell proliferation and focal necrosis are common.

Demonstration of measles virus in the CNS during acute measles encephalitis is rare when functional immunity to the virus exists.[101–103] This second form of disease is called acute allergic measles encephalitis or postinfectious measles encephalitis. In most instances it occurs a few days to 1 week after appearance of rash. The pathological changes in the brain include perivascular lymphocytic infiltration, vascular damage, and demyelination. The pattern of local demyelination resembles that reported in experimental allergic encephalomyelitis.

The other syndrome, SSPE, described above, is a rare, slowly progressing CNS disease that mostly affects children and young adults between 5 and 14 years of age.[33–35,103–105] Boys are affected twice as often as girls. The disease onset is insidious, with a decline of intellectual functions and mental and behavioral changes. In a period of weeks to months, neurological signs such as myoclonus and incoordination follow. Myoclonic jerks that gradually affect all somatic muscle groups are a typical symptom of the disease. The disease worsens with neurological dysfunction progressing to convulsions, coma, and death usually within 1 to 2 years after onset, but there may exist large variations in the rate of progression. The laboratory findings include abnormal electroencephalographic changes and increased levels of γ-globulin and measles antibodies in CSF with evidence of intrathecal synthesis. Measles virus has been isolated from brains of SSPE patients,[106] but this is a difficult procedure that requires long-term cocultivation and fusion with sensitive cells.

The pathological changes in the brain are found in both gray and white matter with widespread inflammatory changes. There are perivascular cuffing of small vessels by lymphocytes and plasma cells, foci of demyelination, loss of dendrites in the cortex, and proliferation of glia cells and astrocytes. Measles virus antigen and viral inclusions resulting from aggregation of nucleocapsids can be traced in the brain by immunologic techniques such as immunofluorescence and peroxidase–antiperoxidase techniques and by electron microscopy. Measles virus antigen and viral inclusions can be found in the cytoplasm, nuclei, and processes of neurons and in oligodendrocytes. Patients with large amounts of antigen in the brain usually die at an earlier age and after a shorter duration of disease than patients with few antigen-containing cells.[107] Mononuclear cells in leptomeningeal and perivascular infiltrates and astrocytes rarely contained measles virus antigen.[107] In contrast, IgG deposition was found mainly in astrocytes and mononuclear cells.

It is not known if persistence of measles virus in the brain is a regular event after measles. It has been suggested that the immune response may be delayed during the first infection with measles virus in SSPE patients, as they are unable to clear the virus from the brain.[104] These questions may be studied post-mortem by performing hybridization experiments with sensitive nucleic acid probes on brain materials. In one study using dot–blot hybridization, measles-virus-specific RNA sequences were readily detected in brains of SSPE patients but not in 11 brains from patients with multiple sclerosis or eight brains from control patients.[108]

A number of animal models using monkeys, ferrets, mice, rats, and hamsters have been established to study measles virus persistence in the brain.[36,48,49,109–118] When young adult ferrets were inoculated intracerebrally with a cell-associated encephalitogenic SSPE strain, they developed an acute encephalitis and died within 1 to 3 weeks without detectable antibody formation.[109,110] The brains of the animals did not show the characteristic viral nucleocapsids or the pronounced inflammatory response seen in SSPE patients. Unstructured viral antigen was detected by immunologic techniques, especially in postsynaptic regions of all brain areas.

Some ferrets were immunized with measles vaccine 5 weeks before challenge with SSPE virus. About half of the preimmunized animals showed no signs of acute illness but instead developed a subacute encephalitis weeks or months p.i. They had symptoms similar to those seen in SSPE, including tremor, paralysis, and seizures. Perivascular cuffing of inflammatory cells and large cytoplasmic inclusions of nucleocapsids were found in infected cells in the brain and spinal cord. Antibodies were synthesized in the brain in response to the persistent measles virus infection. Protein A conjugated to horseradish peroxidase was used to localize immunoglobulins in the brain of the animals.[111] Immunoglobulins were found in plasma cells in different stages of antibody production both in perivascular inflammatory lesions and scattered throughout the cerebral cortex. Antibody was also demonstrated in glial and neuronal cell bodies and processes and postsynaptic profiles. The distribution of antibodies correlated with the distribution of viral antigen and suggested the possibility of immune-complex formation.

In two reports, newborn, weanling, and adult mice were inoculated with the hamster neurotropic strain of measles virus.[49,112] In newborn mice a severe meningoencephalitis developed with extensive necrosis of the cerebral cortex, and in addition there were focal areas of necrosis in the hippocampus. All the newborn mice died within 8 days p.i. Older mice that were inoculated at 4 or 6 weeks of age developed disease later. In these animals a bilateral selective destruction of the whole pyramidal cell layer of the hippocampus was seen.

In one study newborn hamsters were inoculated intracerebrally with measles virus from carrier cell lines.[113] No animal developed acute encephalitis, but seven animals in a group of 50 developed a neurological disease (unsteady gait, myoclonic jerks) 79 to 212 days p.i. On histological examination the animals showed spongy degeneration in the mesencephalon with perivascular cuffing of mononuclear inflammatory cells. Intranuclear and cytoplasmic eosinophilic inclusion bodies occurred in both neurons and glia cells, and by electron microscopy nucleocapsids were identified in the cells. No budding of virus particles was observed. Some of the animals had hydrocephalus, which was also found in many apparently healthy animals.

In another study of weanling (21-day-old) hamsters inoculated intracerebrally with a hamster-adapted strain of SSPE virus and studied between days 6 and 51 of infection, similar findings were made.[114] Focal concentrations of bound hamster IgG occurred within foci of infected cells. The authors suggested that the focal concentrations of IgG acted as a "blocking factor" to protect infected cells from immune surveillance and destruction.

The influence of antibodies on the outcome of an experimental measles virus infection in the brain has been studied.[109,110,115,116,119] As already mentioned, immunization with measles virus vaccine 5 weeks prior to intracerebral challenge of ferrets with SSPE virus changed the infection from an acute encephalitis to a subacute encephalitis in about 50% of the ferrets.[109] In another study the protective effects of different monoclonal antibodies directed against the H protein of measles virus were evaluated.[119] In the absence of immune protection all newborn mice inoculated intracerebrally with measles virus died of acute encephalitis. All monoclonal antibodies, with a single exception, neutralized virus infectivity *in vitro*. The antibodies could be divided into three groups in the passive protection experiments: protective, inducer of a retarded disease, and nonprotective. Only the antibodies with hemagglutination-inhibiting activity protected against the acute disease. One such antibody lead to a retarded disease in up to 40% of the animals. These mice died 2 to 3 months p.i., and it was only possible to demonstrate viral antigen in 20% of the cases with the retarded disease. There were no histological lesions except a discrete inflammation in the brains of these animals.

4.2.2. Canine Distemper Virus

Three different serotypes of CDV with little antigenic variation have been defined by monoclonal antibodies,[120] and a new strain that closely resembles CDV has been isolated from seals.[5] Clinical or subclinical CNS infection is a regular event in systemic CDV infection. The lesions in the brain of dogs with natural

CDV encephalitis have been described.[121–123] The exact duration of the disease was not always known. In one report the clinical signs had lasted from 2 to 5 months in the majority of the dogs,[121] and in the other report prodromal symptoms of fever, anorexia, cough, vomiting, and diarrhea occurred 1 to 2 months before serious neurological symptoms such as ataxia and convulsions.[122] Shortly after the appearance of ataxia and convulsions, the animals were subjected to euthanasia. In the cerebral cortex, cerebellum, and spinal cord lesions consisted of foci of demyelination and necrosis, perivascular cuffing of inflammatory cells, neuronal loss, and proliferation of astrocytes and microglial cells.

In ten out of 14 dogs, neutralizing anti-CDV antibody titers were determined.[121] In four dogs neutralizing antibodies were found in the serum only, in one in the CSF only, and in five both in serum and CSF. There was a correlation among the presence of inflammation, intrathecal antiviral antibodies, and disappearance of CDV from the lesions. In another study the spread and distribution of viral antigen in nervous canine distemper was followed from 16 to 170 days p.i.[124] Infection of glial cells preceded demyelination, and the degree of myelin destruction correlated with the amount of viral antigen in the tissue. Ependymal infection and spread of virus to the subependymal white matter were common. In two dogs with chronic progressive neurological distemper, viral antigen persisted in the brain for longer periods, whereas in dogs that recovered, viral antigen was no longer demonstrable.

Two phases of demyelination were described in this and in another study.[124,125] The first phase may be caused by early direct cell damage resulting from replication of the virus, and the second, late phase may result from the reaction of antibodies with persisting viral antigen. In dogs experimentally infected with three different strains of CDV, variation was found in the severity, clinical course, and neuropathology of the encephalomyelitis.[29] Infection with one strain was acute, and dogs either died between 14 and 19 days p.i. or recovered. Lesions in the neuraxis were similar to polioencephalomyelitis. The second and third strains produced a subacute to chronic disease with prominent demyelination. Some dogs died, other recovered, and a third group developed persistent CNS infection but remained clinically stable. Neutralizing antibody responses correlated with the clinical course. Dogs that died had no or only low serum antibody levels. High titers and early development of antibodies were found in dogs that recovered. A few dogs with persistent CNS infection had demonstrable titers of antibodies in the CSF.

Intracerebral CDV infection of newborn or weanling mice has also been studied.[126–128] Lethality was found to be mouse strain dependent.[126,127] Mice of resistant strains survived the effects of acute infection, appeared well for several weeks, and then developed signs of a subacute encephalitis. In the acute encephalitis, pathological changes consisted of mononuclear cell meningitis, parenchymal inflammation with prominent perivascular mononuclear cell infiltrates, necrosis, and microglial hypertrophy. In the subacute disease the same lesions were found, but they were less pronounced. Virus was readily isolated from acute encephalitis but could not be isolated from brains of mice with subacute encephalitis. Intracerebral inoculation of newborn mice with either the parent or mouse-adapted

Onderstepoost strain of CDV led to 100% mortality.[128] In weanling mice the parent virus produced less than 10% mortality, whereas the mouse-adapted strain killed 40% of the animals in a dose-independent manner. Viral antigen was less widespread in the brains of weanling mice than in those of newborn mice and could not be detected later than 5 months p.i. At 13 to 17 months p.i. a number of mice became paralyzed and viral antigen could again be detected in the brain and lymph glands, but infectious virus could not be isolated. The CDV cannot infect oligodendrocytes *in vitro*.[129–131] Therefore, degeneration of oligodendrocytes leading to demyelination must be related to other factors, which have not yet been clarified.

4.3. Other Less-Common Viruses Causing Diseases of the CNS of Animals

The natural hosts for NDV are chickens and other birds. There exist several serotypes of the virus which can be defined by monoclonal antibodies.[132] In a large study 300 chickens were inoculated by natural routes, i.e., via the eyes and external nasal openings with the Missouri–(H) Len 1950 strain.[133] Four days p.i. all the chickens developed respiratory symptoms from severe pneumonitis. At this stage virus was isolated from the brain of all chickens, and viral antigens could be demonstrated in Purkinje neurons by immunofluorescence. The most extensive lesions were found in the cerebellum, with perivascular accumulations of round cells in the white matter, focal accumulations of glial cells, and disappearance of Purkinje neurons. More than 50% of the chickens recovered from the first attack of the virus. Several days later, beginning at about 11 or 12 days p.i., about one-third of the remaining animals developed severe signs of encephalitis with ataxia and paralysis of wings and legs. At this stage viral antigen could not be detected in the brains by immunologic techniques. Such aggravation of clinical signs in the absence of viral replication was a unique situation and led to the conclusion that other factors might be involved in pathogenesis.

In another study newborn and adult mice were inoculated intracerebrally with the virus.[134] Newborn mice became ill during the third day p.i. and died by the fourth day, whereas adult mice became ill 4 to 6 days p.i. and were all dead by 7 days p.i. Infectious virus was demonstrated in newborn mouse brain at 24 hr p.i. and increased rapidly until the animals died. Viral antigen was traced by immunofluorescence in neurons, where it presented itself as cytoplasmic granular inclusions. Viral antigen was found throughout gray matter but was predominantly concentrated in the frontal cortex, pyramidal cell layer of the hippocampus, and Purkinje cell layer of the cerebellum. On histological examination there were few or no observable pathological changes in brains from both newborn and adult mice. As a parallel to the situation in chickens described above, the adult mice failed to produce high titers of virus in the brain.

Different strains of NDV show variations in virulence *in vivo*. In one study five virulent and five avirulent strains were analyzed.[135] In the avirulent strains glycoprotein F, which is responsible for hemolysis and cell fusion, was not found because of lack of proteolytic cleavage from the precursor glycoprotein F_0, a cleavage required for infectivity of the virus.[135,136] In contrast, the virulent strains always contain mature F glycoprotein.

There are other less-characterized members of the paramyxovirus family with neuropathogenic capacity such as Nariva virus, LPM virus, and bovine morbillivirus 107.[137–139] Intracranial inoculation of suckling hamsters with Nariva virus produced an acute necrotizing encephalitis with high titers of infectious virus in the brain, whereas weanling hamsters only showed low titers of infectious virus early p.i., when they were clinically well.[137] Weanling hamsters died later than suckling hamsters and with less cerebral necrosis. During the late stages of infection, when clinically ill, they had a nonproductive infection with evidence of viral antigens but no infectious virus in their brains. Besides the maturation of the immune system, a change in susceptibility of the neural cells in older hamsters may also influence the course of the disease.

The LPM virus has been isolated from the brain of a piglet with CNS disorder, pneumonia, and corneal opacity.[138] Experimental transmission of the virus to young pigs resulted in disease with similar characteristics. Following intracerebral inoculation adult mice died in 3 to 5 days, whereas mice that had been inoculated intraperitoneally remained healthy.

Bovine morbillivirus 107 resembles viruses belonging to the morbillivirus genus in nucleocapsid diameter and antigenically.[139] The neuropathological changes resulting from infection of calves were perivascular cuffing, diffuse lymphocytic infiltration, proliferation of astrocytes and microglia, and moderate neuronal loss.

5. SUMMARY AND CONCLUDING REMARKS

During recent years much has been learned about the persistence of paramyxoviruses in the CNS. This is a result of recent techniques that permit study of viral expression in terms of viral protein and nucleic acid. In future years, it seems likely that the newly introduced sensitive technique, the polymerase chain reaction, will advance our knowledge of viral persistence in the CNS.

There are many questions to be answered by future research. Which are the viral characteristics involved in the neuropathogenicity of a virus? Which are the factors that prevent identification and destruction of the viral proteins by immune defense mechanisms in persistent infections? The mechanisms that prevent complete viral maturation during persistence in vivo must also be further studied. For example, which are the reasons behind the low expression of paramyxovirus envelope proteins? The role of cellular and humoral immunity also must be attentively studied. There are many examples of persistent infections that can be established in young individuals but not in older individuals with a more mature immune system. On the other hand, there are several examples of the fact that an immune response or passively transferred antibodies can change the course of an acute encephalitis to a chronic or subacute one. Answering these basic questions will create better possibilities for therapeutic intervention in diseases caused by these neurotropic viruses.

REFERENCES

1. Kristensson, K., and Norrby, E., 1986, Persistence of RNA viruses in the central nervous system, *Annu. Rev. Microbiol.* **40:**159–184.
2. Örvell, C., and Norrby, E., 1985, Antigenic structure of paramyxoviruses, in: *Immunochemistry of Viruses—the Basis for Serodiagnosis and Vaccines* (A. R. Neurath and M. H. V. van Regenmortel, eds.), Elsevier, Amsterdam, pp. 241–264.
3. Örvell, C., and Norrby, E., 1980, Immunological relationships between homologous structural polypeptides of measles and canine distemper virus, *J. Gen. Virol.* **50:**231–245.
4. Sheshberadaran, H., Norrby, E., McCullough, K. C., Carpenter, W. C., and Örvell, C., 1986, The antigenic relationship between measles, canine distemper and rinderpest viruses studied with monoclonal antibodies, *J. Gen. Virol.* **67:**1381–1392.
5. Cosby, S. L., McQuaid, S., Duffu, N., Lyons, C., Rima, B. K., Allan, G. M., McCullough, S. J., Kennedy, S., Smyth, J. A., McNeilly, F., Craig, C., and Örvell, C., 1988, Characterization of a seal morbillivirus, *Nature* **336:**115–116.
6. Mufson, M. A., Örvell, C., Rafnar, B., and Norrby, E., 1985, Two distinct subtypes of human respiratory syncytial (RS) virus of different epidemiological occurrence, *J. Gen. Virol.* **66:**2111–2124.
7. Lerch, R. A., Stott, E. J., and Wertz, G. W., 1989, Characterization of bovine respiratory syncytial virus proteins and mRNAs and generation of cDNA clones to the viral mRNAs, *J. Virol.* **63:**833–840.
8. Ling, R., and Pringle, C. R., 1989, Polypeptides of pneumonia virus of mice. I. Immunological cross-reactions and post-translational modifications, *J. Gen. Virol.* **70:**1427–1440.
9. Grandien, M., Örvell, C., and Norrby, E., 1988, Paramyxoviridae, in: *Laboratory Diagnosis of Infectious Diseases. Principles and Practice*, Volume II, *Viral, Rickettsial and Chlamydial Diseases* (E. H. Lennette, P. Halonen, and F. A. Murphy, eds.), Springer Verlag, New York, pp. 484–539.
10. Örvell, C., 1984, The reactions of monoclonal antibodies with structural proteins of mumps virus, *J. Immunol.* **132:**2622–2629.
11. Johnson, R. T., 1982, *Viral Infections of the Central Nervous System*, Raven Press, New York.
12. Axthelm, M. K., and Krakowa, S., 1987, Canine distemper virus: The early blood–brain lesion, *Acta Neuropathol. (Berl.)* **75:**27–33.
13. Krakowa, S., Cork, L. C., Winkelstein, J. A., and Axthelm, M. K., 1987, Establishment of central nervous system infection by canine distemper virus: breach of the blood–brain barrier and facilitation by antiviral antibody, *Vet. Immunol. Immunopathol.* **17:**471–482.
14. Krakowa, S., Axthelm, M. K., and Gorham, J. R., 1987, Effects of induced thrombocytopenia on viral invasion of the central nervous system in canine distemper virus infection, *J. Comp. Pathol.* **97:**441–450.
15. Wolinsky, J. S., Klassen, T., and Baringer, J. R., 1976, Persistence of neuroadapted mumps virus in brains of newborn hamsters after intraperitoneal inoculation, *J. Infect. Dis.* **133:**260–267.
16. Wolinsky, J. S., 1977, Mumps virus-induced hydrocephalus in hamsters. Ultrastructure of the chronic infection, *Lab. Invest.* **37:**229–235.
17. Duffy, P. E., Wolf, A., Harter, D. H., Gamboa, E. T., and Hsue, K. C., 1973, Murine influenza virus encephalomyelitis. II. Electron microscopic observations, *J. Neuropathol. Exp. Neurol.* **32:**72–91.
18. Margolis, G., Kilham, L., and Baringer, J. R., 1974, A new look at mumps encephalitis: Inclusion bodies and cytopathic effects, *J. Neuropathol. Exp. Neurol.* **33:**13–28.
19. Wolinsky, J. S., Baringer, J. R., Margolis, G., and Kilham, L., 1974, Ultrastructure of mumps virus replication in newborn hamster central nervous system, *Lab. Invest.* **31:**403–412.
20. Löve, A., Rydbeck, R., Kristensson, K., Örvell, C., and Norrby, E., 1985, Hemagglutinin-

neuraminidase glycoprotein as a determinant of pathogenicity in mumps virus hamster encephalitis: Analysis of mutants selected with monoclonal antibodies, *J. Virol.* **53**:67–74.

21. Kövamees, J., Rydbeck, R., Örvell, C., and Norrby, E., 1990, Hemagglutinin–neuraminidase (HN) amino acid alterations in neutralization escape mutants of Kilham mumps virus, *Virus Res.* **17**:119–130.

22. Roux, L., and Waldvogel, F. A., 1981, Establishment of Sendai virus persistent infection: Biochemical analysis of the early phase of standard plus defective interfering virus infection of BHK cells, *Virology* **112**:400–410.

23. Wechsler, S. L., Lambert, D. M., Galinski, M. S., Mink, M. A., Rochovansky, O., and Pons, M. W., 1987, Immediate persistent infection by human parainfluenza virus 3: Unique fusion properties of the persistently infected cells, *J. Gen. Virol.* **68**:1737–1748.

24. Rima, B. K., Davidson, W. B., and Martin, S. J., 1977, The role of defective interfering particles in persistent infection of Vero cells by measles virus, *J. Gen. Virol.* **35**:89–97.

25. Wild, T. F., Bernard, A., and Greenland, T., 1981, Measles virus: Evolution of a persistent infection in BGM cells, *Arch. Virol.* **67**:297–308.

26. Metzler, A. E., Krakowa, S., Axthelm, M. K., and Gorham, J. R., 1983, *In vitro* propagation of canine distemper virus: Establishment of persistent infection in Vero cells, *Am. J. Vet. Res.* **45**:2211–2215.

27. Cosby, S. L., Lyons, C., Rima, B. K., and Martin, S. J., 1985, The generation of small-plaque mutants during undiluted passage of canine distemper virus, *Intervirology* **23**:157–166.

28. Norrby, E., Chen, S. N., Togashi, T., Sheshberadaran, H., and Johnson, K. P., 1982, Five measles virus antigens demonstrated by use of mouse hybridoma antibodies in productively infected tissue culture cells, *Arch. Virol.* **71**:1–11.

29. Summers, B. A., Greissen, H. A., and Appel, M. J. G., 1984, Canine distemper encephalomyelitis: Variation with virus strain, *J. Comp. Pathol.* **94**:65–75.

30. Appel, M. G. J., Shek, W. R., and Summers, B. A., 1982, Lymphocyte-mediated immune cytotoxicity in dogs infected with virulent canine distemper virus, *Infect. Immun.* **37**: 592–600.

31. Johnson, G. C., Krakowa, S., and Axthelm, M. K., 1987, Prolonged viral antigen retention in the brain of a gnotobiotic dog experimentally infected with canine distemper virus, *Vet. Pathol.* **24**:87–89.

32. Summers, B. A., Whitaker, J. N., and Appel, M. G. J., 1987, Demyelinating canine distemper encephalomyelitis: Measurement of myelin basic protein in cerebrospinal fluid, *J. Neuroimmunol.* **14**:227–233.

33. Wechsler, S. L., and Meissner, H. C., 1982, Measles and SSPE viruses: Similarities and differences, *Prog. Med. Virol.* **28**:65–95.

34. ter Meulen, V., and Carter, M. J., 1984, Measles virus persistency and disease, *Prog. Med. Virol.* **30**:44–61.

35. Dyken, P. R., 1985, Subacute sclerosing panencephalitis. Current status, *Neurol. Clin.* **3**: 179–196.

36. Johnson, K. P., and Norrby, E., 1974, Subacute sclerosing panencephalitis (SSPE) agent in hamsters. III. Induction of defective measles in hamster brain, *Exp. Mol. Pathol.* **21**: 166–178.

37. Johnson, K. P., Norrby, E., Swoveland, P., and Carrigan, D. R., 1981, Experimental subacute sclerosing panencephalitis: Selective disappearance of measles virus matrix protein from the central nervous system, *J. Infect. Dis.* **144**:161–169.

38. Norrby, E., Kristensson, K., Brzosko, W. J., and Kaspenberg, J. A., 1985, Measles virus matrix protein detected by immune fluorescence with monoclonal antibodies in the brain of patients with subacute sclerosing panencephalitis, *J. Virol.* **56**:337–340.

39. Liebert, U. G., Baczko, K., Budka, H., and ter Meulen, V., 1986, Restricted expression of measles virus proteins in brains from cases of subacute sclerosing panencephalitis, *J. Gen. Virol.* **67**:2435–2444.

40. Baczko, K., Liebert, U. G., Billeter, M., Cattaneo, R., Budka, H., and ter Meulen, V., 1986, Expression of defective measles virus genes in brain tissues of patients with subacute sclerosing panencephalitis, *J. Virol.* **59**:472–478.

41. Baczko, K., Liebert, U. G., Cattano, R., Billeter, M. A., Roos, R. P., and ter Meulen, V., 1988, Restriction of measles virus gene expression in measles inclusion body encephalitis, *J. Infect. Dis.* **158**:144–150.

42. Cattaneo, R., Rebmann, G., Baczko, K., ter Meulen, V., and Billeter, M. A., 1987, Altered ratios of measles virus transcripts in diseased human brains, *Virology* **160**:523–526.

43. Cattaneo, R., Schmid, A., Eschele, D., Baczko, K., ter Meulen, V., and Billeter, M. A., 1988, Biased hypermutation and other genetic changes in defective measles viruses in human brain infection, *Cell* **55**:255–266.

44. Cattaneo, R., Schmid, A., Billeter, M. A., Sheppard, R. D., and Udem, S. A., 1988, Multiple viral mutations rather than host factors cause defective measles virus gene expression in a subacute sclerosing panencephalitis cell line, *J. Virol.* **62**:1388–1397.

45. Kristensson, K., Örvell, C., Leestma, J., and Norrby, E., 1983, Sendai virus infection in the brains of mice: Distribution of viral antigens studied with monoclonal antibodies, *J. Infect. Dis.* **147**:297–301.

46. Kristensson, K., Örvell, C., Malm, M., and Norrby, E., 1984, Mumps virus infection of the developing mouse brain. Appearance of structural virus proteins demonstrated with monoclonal antibodies, *J. Neuropathol. Exp. Neurol.* **43**:131–140.

47. Löve, A., Andersson, T., Norrby, E., and Kristensson, K., 1987, Mumps virus infection of dissociated rodent spinal ganglia *in vitro*. Expression and disappearance of viral structural proteins from neurons, *J. Gen. Virol.* **68**:1755–1759.

48. Liebert, U. G., and ter Meulen, V., 1987, Virological aspects of measles virus-induced encephalomyelitis in Lewis and BN rats, *J. Gen. Virol.* **68**:1715–1722.

49. Löve, A., Norrby, E., and Kristensson, K., 1986, Measles encephalitis in rodents: Defective expression of viral proteins, *J. Neuropathol. Exp. Neurol.* **45**:258–267.

50. Forsberg, P., Fryden, A., Link, H., and Örvell, C., 1986, Viral IgM and IgG antibody synthesis within the central nervous system in mumps meningitis, *Acta Neurol. Scand.* **73**:372–380.

51. Ukkonen, P., Granström, M.-L., Räsänen, J., Salonen, E.-M., and Penttinen, K., 1981, Local production of mumps IgG and IgM antibodies in the cerebrospinal fluid of meningitis patients, *J. Med. Virol.* **8**:257–265.

52. Chiodi, F., Sundqvist, V.-A., Norrby, E., Mavra, M., and Link, H., 1986, Measles IgM antibodies in cerebrospinal fluid and serum in subacute sclerosing panencephalitis, *J. Med. Virol.* **18**:149–158.

53. Lowenthal, A., Van Sande, M., and Karcher, D., 1960, The differential diagnosis of neurological diseases by fractionating electrophoretically the CSF gamma globulins, *J. Neurochem.* **6**:51–56.

54. Link, H., Panelius, M., and Salmi, A. A., 1972, Measles antibodies and immunoglobulins in serum and cerebrospinal fluid in subacute sclerosing panencephalitis, *Acta Neurol. Scand. [Suppl.]* **51**:385–387.

55. Vandvik, B., and Norrby, E., 1973, Oligoclonal IgG antibody response in the central nervous system to different measles virus antigens in subacute sclerosing panencephalitis, *Proc. Natl. Acad. Sci. U.S.A.* **70**:1060–1063.

56. Fryden, A., Link, H., and Norrby, E., 1978, Cerebrospinal fluid and serum immunoglobulins and antibody titers in mumps meningitis and aseptic meningitis of other etiology, *Infect. Immun.* **21**:852–861.

57. Link, H., Laurenzi, M. A., and Fryden, A., 1981, Viral antibodies in oligoclonal and polyclonal IgG synthesized within the central nervous system over the course of mumps meningitis, *J. Neuroimmunol.* **1**:287–298.

58. Vandvik, B., Nilsen, R. E., Vartdal, F., and Norrby, E., 1982, Mumps meningitis: Specific

and nonspecific antibody responses in the central nervous system, *Acta Neurol. Scand.* **65:** 468–487.

59. Norrby, E., Örvell, C., Vandvik, B., and Cherry, J. D., 1981, Antibodies against measles virus polypeptides in different disease conditions, *Infect. Immun.* **34:**718–724.

60. Dhib-Jalbut, S., McFarland, H. F., Mingioli, E. S., Sever, J. S., and McFarlin, D. E., 1988, Humoral and cellular immune responses to matrix protein of measles virus in subacute sclerosing panancephalitis, *J. Virol.* **62:**2483–2489.

61. Hall, W. W., Lamb, R., and Choppin, P. W., 1979, Measles and subacute sclerosing panencephalitis virus proteins: Lack of antibodies to the M protein in patients with subacute sclerosing panencephalitis, *Proc. Natl. Acad. Sci. U.S.A.* **76:**2047–2051.

62. Wechsler, S. L., Weiner, H. L., and Fields, B. N., 1979, Immune response in subacute sclerosing panencephalitis: Reduced antibody response to the matrix protein of measles virus, *J. Immunol.* **123:**884–889.

63. Stephenson, J. R., and ter Meulen, V., 1979, Antigenic relationship between measles and canine distemper virus: Comparison of immune response in animals and humans to individual virus-specific polypeptides, *Proc. Natl. Acad. Sci. U.S.A.* **76:**6601–6605.

64. Rima, B. K., Baczko, K., Imagawa, D. T., and ter Meulen, V., 1987, Humoral immune response in dogs with old dog encephalitis and chronic distemper meningo-encephalitis, *J. Gen. Virol.* **68:**1723–1735.

65. Vaheri, A., Julkunen, J., and Koskiniemi, M.-L., 1982, Chronic encephalomyelitis with specific increase in intrathecal mumps antibodies, *Lancet* **25:**685–689.

66. Ilonen, J., Reunanen, M., Herva, G., Ziola, B., and Salmi, A., 1980, Stimulation of lymphocytes from subacute sclerosing panencephalitits patients by defined measles antigens, *Cell. Immunol.* **51:**201–214.

67. Shek, W. R., Schultz, R. D., and Appel, M. J. G., 1980, Natural and immune cytolysis of canine distemper virus-infected target cells, *Infect. Immun.* **28:**724–734.

68. Frydén, A., Link, H., and Möller, E., 1978, Demonstration of cerebrospinal fluid lymphocytes sensitized against virus antigens in mumps meningitis, *Acta Neurol. Scand.* **57:** 396–404.

69. Kam-Hansen, S., Frydén, A., and Link, H., 1978, B and T lymphocytes in cerebrospinal fluid and blood in multiple sclerosis, optic neuritis and mumps meningitis, *Acta Neurol. Scand.* **58:**95–103.

70. Kreth, H. W., Kress, L., Kress, H. G., Ott, H. F., and Eckert, G., 1982, Demonstration of primary cytotoxic T cells in venous blood and cerebrospinal fluid of children with mumps meningitis, *J. Immunol.* **128:**2411–2415.

71. Reunanen, M., Salonen, R., and Salmi, A., 1982, Intrathecal immune responses in mumps meningitis patients, *Scand. J. Immunol.* **15:**419–426.

72. Fleischer, B., and Kreth, H. W., 1983, Clonal analysis of HLA-restricted virus specific cytotoxic T-lymphocytes from cerebrospinal fluid in mumps meningitis, *J. Immunol.* **130:** 2187–2190.

73. Nagai, H., Morishima, T., Morishima, Y., Isomura, S., and Suzuki, S., 1983, Local T subsets in mumps meningitis, *Arch. Dis. Child.* **11:**927–928.

74. Rydbeck, R., Löve, A., Örvell, C., and Norrby, E., 1986, Antigenic variations of envelope and internal proteins of mumps virus strains detected with monoclonal antibodies, *J. Gen. Virol.* **67:**281–287.

75. Vandvik, B., Norrby, E., Steen-Johnson, J., and Stensvold, K., 1978, Mumps meningitis: Prolonged pleocytosis and occurrence of mumps virus-specific oligoclonal IgG in the cerebrospinal fluid, *Eur. Neurol.* **17:**13–22.

76. Timmons, G. D., and Johnson, K. P., 1970, Aqueductal stenosis and hydrocephalus after mumps encephalitis, *N. Engl. J. Med.* **283:**1505–1507.

77. Wolinsky, J. S., and Stroop, W. G., 1978, Virulence and persistence of three prototype strains of mumps virus in newborn hamsters, *Arch. Virol.* **57:**355–359.

78. McCarthy, M., Jubelt, B., Fay, D. B., and Johnson, R. T., 1980, Comparative studies of five strains of mumps virus *in vitro* and in neonatal hamsters: Evaluation of growth, cytopathogenicity and neurovirulence, *J. Med. Virol.* **5**:1–15.

79. Löve, A., Malm, G., Rydbeck, R., Norrby, E., and Kristensson, K., 1985, Developmental disturbances in the hamster retina caused by a mutant of mumps virus, *Dev. Neurosci.* **7**: 65–72.

80. Merz, D. C., and Wolinsky, J. S., 1981, Biochemical features of mumps virus neuraminidases and their relationship with pathogenicity, *Virology* **114**:218–227.

81. Merz, D. C., and Wolinsky, J. S., 1983, Conversion of nonfusing mumps virus infections to fusing infections by selective proteolysis of the HN protein, *Virology* **131**:328–340.

82. Wolinsky, J. S., Waxham, M. N., and Server, A. C., 1985, Protective effects of glycoprotein-specific monoclonal antibodies on the course of experimental mumps virus meningoencephalitis, *J. Virol.* **53**:727–734.

83. Löve, A., Rydbeck, R., Utter, G., Örvell, C., Kristensson, K., and Norrby, E., 1986, Monoclonal antibodies against the fusion protein are protective in necrotizing mumps meningoencephalitis, *J. Virol.* **58**:220–222.

84. Frank, J. A., Warren, R. W., Tucker, A., Zeller, J., and Wilfert, C. M., 1983, Disseminated parainfluenza infection in a child with severe combined immunodeficiency, *Am. J. Infect. Dis.* **137**:1172–1174.

85. Shimokata, K., Nishiyama, Y., Ito, Y., Kimura, Y., Nagata, I., Iida, M., and Sobue, I., 1976, Pathogenesis of Sendai virus infection in the central nervous system of mice, *Infect. Immun.* **13**:1497–1502.

86. Schwendemann, G., and Löhler, J., 1979, Pathological alterations of ependyma and choroid plexus after experimental cerebral infection of mice with Sendai virus, *Acta Neuropathol. (Berl.)* **46**:85–94.

87. Kristensson, K., Leestma, J., Lundh, B., and Norrby, E., 1984, Sendai virus infection in the mouse brain: Virus spread and long-term effects, *Acta Neuropathol. (Berl.)* **63**:89–95.

88. Rorke, L. B., Gilden, D. H., Wroblewska, Z., and Wolinsky, J. S., 1976, Experimental panencephalitis induced in suckling mice by parainfluenza type 1 (6/94) virus. I. Clinical and pathological features, *J. Neuropathol. Exp. Neurol.* **35**:247–258.

89. Gilden, D. H., Wroblewska, Z., Chesler, M., Wellish, M. C., Lief, F. S., Wolinsky, J. S., and Rorke, L. B., 1976, Experimental panencephalitis induced in suckling mice by parainfluenza type 1 (6/94) virus, *J. Neuropathol. Exp. Neurol.* **35**:259–270.

90. Wolinsky, J. S., Gilden, D. H., and Rorke, L. B., 1976, Experimental panencephalitis induced in suckling mice by parainfluenza I (6/94) virus, *J. Neuropathol. Exp. Neurol.* **35**: 271–286.

91. Zgorniak-Nowosielska, I., Iwasaki, Y., Tachovsky, T., Tanaka, R., and Koprowski, H., 1976, Experimental parainfluenza type 1 virus-induced encephalopathy in adult mice. Pathogenesis of chronic degenerative changes in the CNS, *Arch. Neurol.* **33**:55–62.

92. Koch, E. M., Neubert, W. J., and Hofschneider, P. H., 1984, Lifelong persistence of paramyxovirus Sendai-6/94 in C129 mice: Detection of a latent viral RNA by hybridization with a cloned genomic cDNA probe, *Virology* **136**:78–88.

93. Shibuta, H., Adachi, A., Kanda, T., and Shimada, H., 1978, Experimental parainfluenza-virus infection. I. Hydrocephalus of mice due to infection with parainfluenza virus type 1 and type 3, *Microbiol. Immunol.* **22**:505–508.

94. Shibuta, H., Adachi, A., Kanda, T., and Matumoto, M., 1982, Experimental parainfluenza-virus infection in mice: Fatal illness with atrophy of thymus and spleen in mice caused by a variant of parainfluenza 3 virus, *Infect. Immun.* **35**:437–441.

95. Shibuta, H., Kanda, T., Nozawa, A., Sato, S., and Kumanishi, T., 1985, Experimental parainfluenza virus infection in mice: Growth and spread of a highly pathogenic variant of parainfluenza 3 virus in the mouse brain, *Arch. Virol.* **83**:43–52.

96. Rydbeck, R., Löve, A., Örvell, C., and Norrby, E., 1987, Antigenic analysis of human and

bovine parainfluenza virus type 3 strains with monoclonal antibodies, *J. Gen. Virol.* **68:** 2153–2160.

97. Baumgärtner, W., Metzler, A. E., Krakowka, S., and Koestner, A., 1981, *In vitro* identification and characterization of a virus isolated from a dog with neurological dysfunction, *Infect. Immun.* **31:**1177–1183.

98. Evermann, J. F., Krakowa, S., McKeirnan, A. J., and Baumgärtner, W., 1981, Properties of an encephalitogenic canine parainfluenza virus, *Arch. Virol.* **68:**165–172.

99. Baumgärtner, W., Krakowa, S., Koestner, A., and Evermann, J., 1982, Acute encephalitis and hydrocephalus in dogs caused by canine parainfluenza virus, *Vet. Pathol.* **19:**79–92.

100. Sheshberadaran, H., Chen, S.-N., and Norrby, E., 1983, Monoclonal antibodies against five structural components of measles virus, *Virology* **128:**341–353.

101. Johnson, R. T., Griffin, D. E., Hirsch, R., and Vaisberg, A., 1983, Measles encephalitis, *Clin. Exp. Neurol.* **19:**13–16.

102. Gendelman, H. E., Wolinsky, J. S., Johnson, R. T., Pressman, N. J., Pezeshkpour, G. H., and Boisset, G. F., 1984, Measles encephalomyelitis: Lack of evidence of viral invasion of the central nervous system and quantitative study of the nature of demyelination, *Ann. Neurol.* **15:**353–360.

103. Kipps, A., Dick, G., and Moodie, J. W., 1983, Measles and the central nervous system, *Lancet* **17:**1406–1410.

104. Sever, J. L., 1983, Persistent measles infection of the central nervous system: subacute sclerosing panencephalitis, *Rev. Infect. Dis.* **5:**467–473.

105. Graves, M. C., 1984, Subacute sclerosing panencephalitis, *Neurol. Clin.* **2:**267–280.

106. Horta-Barbosa, L., Fuccillo, D. A., London, W. J., Jabbour, J. T., Zeman, W., and Sever, J. L., 1969, Isolation of measles virus from brain cell cultures of two patients with subacute sclerosing panencephalitis, *Proc. Soc. Exp. Biol. Med.* **132:**272–277.

107. Budka, H., Lassmann, H., and Popow-Kraupp, T., 1982, Measles virus antigen in panencephalitis. An immunomorphological study stressing dendritic involvement in SSPE, *Acta Neuropathol. (Berl.)* **56:**52–62.

108. Dowling, P. C., Blumberg, B. M., Kolakofsky, D., Cook, P., Jotkowitz, A., Prineas, J. W., and Cook, S. D., 1986, Measles virus nucleic acid sequences in human brain, *Virus Res.* **5:** 97–107.

109. Thormar, H., Metha, P. D., Barshatzky, M. R., and Brown, H. R., 1985, Measles virus encephalitis in ferrets as a model for subacute sclerosing panencephalitis, *Lab. Animal Sci.* **35:**229–232.

110. Thormar, H., Metha, P. D., Barshatzky, M. R., and Brown, H. R., 1985, Localization of measles virus antigens in subacute sclerosing panencephalitis in ferrets, *Lab. Anim. Sci.* **35:** 233–237.

111. Brown, H. R., Pessolano, T. L., Nostro, A. F., and Thormar, H., 1985, Demonstration of immunoglobulin in brains of ferrets inoculated with a SSPE strain of measles virus: Use of protein A conjugated to horseradish peroxidase, *Acta Neuropathol. (Berl.)* **65:**195–201.

112. Roos, R. P., Griffin, D. E., and Johnson, R. T., 1978, Determinants of measles virus (hamster neurotropic strain) replication in mouse brain, *J. Infect. Dis.* **137:**722–727.

113. Norrby, E., and Kristensson, K., 1978, Subacute encephalitis and hydrocephalus in hamsters caused by measles virus from persistently infected cell cultures, *J. Med. Virol.* **2:** 305–317.

114. Johnson, K. P., and Swoveland, P., 1977, Measles antigen distribution in brains of chronically infected hamsters. An immunoperoxidase study of experimental subacute sclerosing panencephalitis, *Lab. Invest.* **37:**459–465.

115. Cremer, N. E., Hagens, S. J., Taylor, D. O. N., and Lennette, E. H., 1977, Complications and immunological studies of measles virus in antithymocyte-treated hamsters, *Infect. Immun.* **16:**155–162.

116. Thormar, H., Arnesen, K., and Mehta, P. D., 1977, Encephalitis in ferrets caused by a

nonproductive strain of measles virus (D.R.) isolated from a patient with subacute sclerosing panencephalitis, *J. Infect. Dis.* **136:**229–238.

117. Albrecht, P., Burnstein, T., Klutch, M. J., Hicks, J. T., and Ennis, F. A., 1977, Subacute sclerosing panencephalitis: Experimental infection in primates, *Science* **195:**64–66.

118. Sharova, O. K., Bomologova, N. N., Koptyaeva, I. B., Gordienko, N. M., and Rozina, E. E., 1987, Pathomorphologic characterization of CNS damage in monkeys infected with a persistent variant of measles virus vaccine strain L-16, *Acta Virol.* **31;**336–351.

119. Giraudon, P., and Wild, T. F., 1985, Correlation between epitopes on hemagglutinin of measles virus and biological activities: Passive protection by monoclonal antibodies is related to their hemagglutination inhibiting activity, *Virology* **144:**46–58.

120. Örvell, C., Sheshberadaran, H., and Norrby, E., 1985, Preparation and characterization of monoclonal antibodies directed against four structural components of canine distemper virus, *J. Gen. Virol.* **66:**443–456.

121. Bollo, E., Zurbriggen, A., Vandevelde, M., and Frankhauser, R., 1986, Canine distemper virus clearance in chronic inflammatory demyelination, *Acta Neuropathol. (Berl.)* **72:**69–73.

122. Kimoto, T., 1986, *In vitro* and *in vivo* properties of the virus causing natural canine distemper encephalitis, *J. Gen. Virol.* **67:**487–503.

123. Vandevelde, M., and Kristensen, B., 1977, Observations on the distribution of canine distemper virus in the central nervous system of dogs with demyelinating encephalitis, *Acta Neuropathol. (Berl.)* **40:**233–236.

124. Vandevelde, M., Zurbriggen, A., Higgins, R. J., and Palmer, D., 1985, Spread and distribution of viral antigen in nervous canine distemper virus, *Acta Neuropathol. (Berl.)* **67:** 211–218.

125. McCullough, B., Krakowa, S., Koestner, A., and Shadduk, J., 1974, Demyelinating activity of canine distemper virus isolates in gnotobiotic dogs, *J. Infect. Dis.* **130:**343–350.

126. Lyons, M. J., Hall, W. W., Petito, C., Cam, V., and Zabriskie, J. B., 1980, Induction of chronic neurologic disease in mice with canine distemper virus, *Neurology* **30:**92–98.

127. Gilden, D. H., Wellish, M., Rorke, L. B., and Wroblewska, Z., 1981, Canine distemper virus infection in weanling mice. Pathogenesis of CNS disease, *J. Neurol. Sci.* **52:**327–339.

128. Bernard, A., Wild, T. F., and Tripier, M. F., 1983, Canine distemper infection in mice: Characterization of a neuro-adapted virus strain and its long-term evolution in the mouse, *J. Gen. Virol.* **64:**1571–1579.

129. Vandelvelde, M., Zurbriggen, A., Dumas, M., and Palmer, D., 1985, Canine distemper virus does not infect oligodendrocytes *in vitro, J. Neurol. Sci.* **69:** 133–137.

130. Zurbriggen, A., Vandevelde, M., and Dumas, M., 1986, Secondary degeneration of oligodendrocytes in canine distemper virus infection *in vitro, Lab. Invest.* **54:**424–431.

131. Zurbriggen, A., Vandevelde, M., Dumas, M., Griot, C., and Bollo, E., 1987, Oligodendroglial pathology in canine distemper virus infection *in vitro, Acta Neuropathol. (Berl.)* **74:**366–373.

132. Nishikawa, K., Isomura, S., Suzuki, S, Watanabe, E., Hamaguchi, M., Yoshida, T., and Nagai, Y., 1983, Monoclonal antibodies to the HN glycoprotein of Newcastle disease virus. Biological characterization and use for strain comparison, *Virology* **130:**318–330.

133. Stevens, J. G., Nakamura, R. M., Cook, M. L., and Wilczynski, S. P., 1976, Newcastle disease as a model for paramyxovirus-induced neurological syndromes: Pathogenesis of the respiratory disease and preliminary characterization of the ensuing encephalitis, *Infect. Immun.* **13:**590–599.

134. Burks, J. S., Narayan, O., McFarland, H. F., and Johnson, R. T., 1976, Acute encephalopathy caused by defective virus infection. I. Studies of Newcastle disease virus infections in newborn and adult mice, *Neurology* **26:**584–588.

135. Nagai, Y., Klenk, H.-D., and Rott, R., 1976, Proteolytic cleavage of the viral glycoproteins and its significance for the virulence of Newcastle disease virus, *Virology* **72:**494–508.

136. Scheid, A., and Choppin, P. W., 1974, Identification of biological activities of paramyxo-virus glycoproteins. Activation of cell fusion, hemolysis and infectivity by proteolytic cleavage of an inactive precursor protein of Sendai virus, *Virology* **57**:475–490.
137. Roos, R. P., and Wollman, R., 1979, Non-productive paramyxovirus infection: Nariva virus infection in hamsters, *Arch. Virol.* **62**:229–240.
138. Moreno-López, J., Correa-Giron, P., Martinez, A., and Ericsson, A., 1986, Characterization of a paramyxovirus isolated from the brain of a piglet in Mexico, *Arch. Virol.* **91**:221–231.
139. Bachmann, P. A., ter Meulen, V., Jentsch, G., Appel, M., Iwasaki, Y., Meyermann, R., Koprowski, H., and Mayr, A., 1975, Sporadic bovine meningoencephalitis—isolation of a paramyxovirus, *Arch. Virol.* **48**:107–120.

11

Human Papovaviruses

JC Virus, Progressive Multifocal Leukoencephalopathy, and Model Systems for Tumors of the Central Nervous System

EUGENE O. MAJOR, DOMINICK A. VACANTE, and SIDNEY A. HOUFF

1. INTRODUCTION

After more than a decade of clinical descriptions of a human demyelinating disease known as progressive multifocal leukoencephalopathy (PML),[1-3] a virus was finally isolated from affected brain tissue by Padgett et al.[4] in 1971. Named JC virus after the initials of the patient, the agent quickly became grouped in the family of Papovavirus on the basis of the viral architecture, size, and genome content. Concurrently, another human papovavirus was isolated by Gardner and her colleagues[5] from the urine of a renal transplant recipient and also named after the initials of the patient, BK. Although BK virus (BKV) is still not clearly identified with a specific disease, JC virus (JCV) is now recognized as the causative agent for PML and has been repeatedly isolated from demyelinated

EUGENE O. MAJOR, DOMINICK A. VACANTE, and SIDNEY A. HOUFF • Laboratory of Viral and Molecular Pathogenesis, National Institute of Neurological Disorders and Stroke, National Institutes of Health, Bethesda, Maryland 20892.   *Present address of D.A.V.:* Microbiological Associates, Rockville, Maryland 20852.   *Present address of S.A.H.:* Department of Neurology, Washington Veterans Administration Hospital, Washington, DC 20422.

Neuropathogenic Viruses and Immunity, edited by Steven Specter *et al.* Plenum Press, New York, 1992.

brain tissue of suspected PML patients. In order to identify the different isolates made from human brain tissue,[6] the designation of the location of the isolation (in Madison, Wisconsin) and a serial number of the isolate were used, e.g., Mad-1 the prototype; Mad-4. The long delay from suspicion of a viral etiology to the final viral isolation was caused largely by the host restriction for growth of JCV in cell culture. The successful isolation made by Padgett *et al.* required the use of human fetal brain cells and several months in culture. Such cultures are composed of macroglial cells and their precursors, which become the target cells for infection in the adult and undergo demyelination. Neurotropism is one of the chief characteristics of JCV and is the singular feature of this viral pathogen that has generated the most experimental attention.

However, the close association of JCV with other members of the Papovavirus family such as the tumor viruses SV40 of rhesus monkeys and the mouse polymavirus, created an interest in the oncogenic properties of these viruses. Because this group of viruses has long been known for the ability to produce tumors in experimental animals and induce malignant transformation in tissue culture,[7] both newly isolated human polyomaviruses were tested for similar properties.

This review highlights the biological, pathological, and oncogenic charac- teristics of human JCV and describes the mechanism of pathogenesis for the acute infection in the human brain resulting in the demyelinating disease PML. In the description of PML, the key role of cells of the immune system that introduce the infection into the brain is discussed. Also, attempts to determine whether human polyomaviruses are involved in neoplasms of the human brain and nervous system are described in the context of rodent and primate models used for these studies.

2. BIOLOGY OF JC VIRUS

2.1. Host Range Properties for Growth

The neurotropic nature of JCV became apparent as investigators tried to grow the virus in cells derived from tissues other than brain. Although BKV could be cultured in human and simian epithelial cells and with some difficulty in fibroblast cells, JCV was initially successfully grown only in human fetal brain cells. Padgett[8] and Frisque[9] and their co-workers described experiments on the host range properties of the virus using several different isolates. A limited infection could be established in human embryonic kidney cells using the Mad-4 variant as well as in other human cell types such as human amnion and uroepithelial cells.[10] Cells derived from nonhuman tissues would not support the production of JCV but could become infected, as evidenced by the synthesis of one of the viral nonstructural proteins, T antigen (see below). Infectivity studies done with the viral genome[11] also indicated that JCV showed a neurotropism that reflected the disease it was associated with, PML.

Early descriptions of the cell type in which JCV could multiply included the so-called spongioblast cell, a primitive cell type thought to be a precursor to the myelinating oligodendrocyte.[12] Suggestions were also made that JCV infection of another glial cell in the human brain, the astrocyte, would lead to defective virus[13] with an altered genome content.[14] This observation was supported by clinical evidence from PML brain tissue indicating that astrocytes in demyelinating lesions were abortively infected.[15,16] More recently, Major and Vacante[17] examined and characterized the cell types found in cultures of human fetal brain. Their results showed that astrocytes were as permissive as oligodendrocytes for JCV multiplication. Another strong argument suggesting the permissiveness of the astrocyte came from the two astrocyte cell lines from human fetal brain established by Major et al.[18] and Mandl et al.[19] Each cell line was produced using the transforming gene product of either JCV itself or the closely related SV 40 virus. In both cell lines, immortalized as continuously growing cultures, JCV could replicate and produce infectious progeny. Further studies of brain tissues of PML cases, particularly those derived from AIDS patients, also revealed the infectivity of JCV for the astrocyte.[20] Other neural cell types, particularly the neuron, have not been shown to become infected with JCV.[13,21a] However, current evidence suggests that the Schwann cell, the myelin-producing cell of the peripheral nervous system, can be infected with JCV, produce T protein, and replicate the viral genome. Infectious virions have recently been demonstrated from such cultures.[21b]

There is strong evidence to implicate nonneural cells as targets for JCV infections since virus has been isolated from the urine of renal and bone marrow transplant patients,[22] pregnant women,[23] and cancer patients.[24] Also, JCV can be cultured in human embryonic kidney and transitional epithelial cells.[25] These cell types, however, produce very few progeny virions over a long period of time. There also is evidence suggesting that JCV grown in human kidney leads to genomic alterations.[26]

Experiments designed to examine the host range of JCV have generally resulted in the conclusion that although JCV can multiply to some limited extent in nonglial cells in vivo or in vitro, the macroglia of the human brain remain the most efficient host for JCV production, thus resulting in JCV's reputation as a tissue-specific neurotropic agent in the human population. The nature of the mechanism that accounts for this tissue specificity posed questions for the study of the molecular structure and regulation of the viral genome.

2.2. Genetics of the Viral Genome

Shortly following the first isolation of the prototype strain of JCV, Mad-1, the viral genome was extracted and characterized. Because the virus size and shape indicated that it belonged to the Papovavirus family, isolation of the genome was easily made. With techniques already applied to other Papovaviruses such as Hirt extraction of nucleic acids from infected cells, agarose gel electrophoresis, and restriction endonuclease analysis, the JCV genome was identified as a closed, circular, supercoiled DNA of 5.1 kb with a physical map similar to SV40 and BKV.[27,28] This by itself could not account for the neurotropic nature of the virus.

It was not until Frisque and his colleagues[29] determined the entire nucleotide sequence of the genome that some insight into this property of JCV started to become more clear. Serological tests already showed that the nonstructural or early protein and the capsid proteins of JCV were antigenically related but maintained some differences from those of SV40 and BKV.[30,31] The DNA sequence data revealed that the regulatory region of the genome that contains the signals for transcription and replication was very different from those of other primate papovaviruses and gave investigators their best clue to approach experimentally the mechanisms of tissue specificity.

Unlike either SV40 or BKV, JCV possessed a tandem direct repeat of 98 base pairs that included a duplication of the TATA sequences, a unique genomic arrangement among mammalian DNA viruses. Also, there was no clear region that could be identified as a transcriptional promoter or a consensus sequence for a transcriptional enhancer. These 98-bp repeats therefore were markedly different from the SV40 regulatory sequences, which demonstrate a highly organized TATA sequence, a GC-rich area used as a transcriptional promoter, and a very efficient enhancer sequence that was active in many cell types and species.[32] Martin and his collaborators[33] also produced sequence data on the regulatory regions of several JCV variants. Their results indicated that these sequences could be highly variable compared to the Mad-1 prototype sequence. Of particular interest was the insertion of a GGG purine-rich series at a point just upstream from the TATA region in several isolates that resembled the promoter sequence of SV40 that binds the Sp1 transcription factor. This insertion was found in all the variants except the Mad-1 and Mad-4 strains. A deletion was also found in the Mad-4 strain that eliminated the second TATA box in the distal 98-bp repeat. The biological significance of this is yet to be clarified. However, the Mad-4 strain has been shown to be highly neurooncogenic in hamsters.[34] Also Amirhaeri and colleagues[35] demonstrated that the viral genome has an unusual non-B conformational structure in the regulatory region but is not influenced by the presence of the second TATA sequence. Whether the conformation of the JCV genome plays a role in its cell specific regulation has yet to be tested in a biological system.

In order to test the tissue specificity of the JCV 98-bp repeats, Kenney and colleagues[36] cloned those sequences upstream from the reporter gene for chloramphenicol acetyl transferase and introduced the new recombinant constructions into several human and simian cell types. The conclusion was that the 98-bp repeats could act as transcriptional enhancer-promoters most efficiently in human fetal glial cells. Feigenbaum *et al.*[37] also demonstrated that the JCV genome was transcriptionally regulated by factors present in human or rodent glial cells, allowing the synthesis of the early T protein. They also produced evidence that suggested that the JCV genome was regulated in a cell-specific manner for transcription but in a species-specific manner for DNA replication. These data confirmed the biological data from cell culture experiments and clinical studies of PML and contributed another insight into the mechanism of the cell-specific regulation demonstrated by JCV. It now appeared that the efficiency of JCV infection in glial cells was related to the cells' recognition of the regulatory sequences, perhaps in a similar manner to immunoglobulin gene recognition by B cells, implicating nuclear transcription factors as the important molecules.

Using the molecular techniques of gel mobility shift assays to identify DNA binding proteins, Khalili et al.[38] described the presence of several proteins from extracts of human fetal brain that specifically bound the regulatory sequences of the JCV genome. In these experiments, nonglial cells also contained nuclear proteins that could bind JCV DNA in the 98-bp region. However, these factors either played no role in JCV gene expression or perhaps down-regulated viral transcription. Combining both gel shift assays and DNase footprinting, Amemiya et al.[39] identified a nuclear factor(s) from human fetal glial cells that specifically bound DNA sequences in similar regions shown by Khalili et al.[38] Competitive DNA binding tests with specific oligonucleotides and DNase footprints identified these sequences as very closely related to the targets for the NF-1 transcription factor. Again, the other nonglial cells also contained proteins that could bind the JCV sequences. However, data from both Khalili et al. and Amemiya et al. supported the conclusion that the cell specificity of JCV can be controlled at the transcriptional level through the interaction of cellular protein factors with DNA sequences in the JCV 98 bp repeats.

Another experimental approach to understanding the role of the regulatory sequences in controlling host range was presented by the data of Vacante et al.[40] A chimeric viral genome was constructed that included all the JCV sequences with the addition of the SV40 promoter and transcriptional enhancer sequences. Biological activity of this unique viral genome indicated that the JCV proteins were functional in human embryonic kidney cells as well as in simian fetal or adult glial cells in culture. Thus, the cell type and species host range for JCV was extended as compared to the prototype virus. The ability of the JCV proteins to function in these cell types, particularly the T protein, indicated that restriction of host range may not involve the functioning of the viral proteins in the cell but rather how efficiently these proteins are produced. Examination of the function of JCV proteins in the background of different cell types could then clarify the complicated biology of this human viral pathogen.

2.3. Protein Products of the Viral Genome

Like the other primate polyomaviruses, JCV produces the nonstructural large-T and small-t proteins from the early region of the genome and the virion capsid proteins from the late region. Frisque et al.[29] examined the coding sequences for these proteins and found a range of 70–80% homology with SV40 and BKV for both the early and late proteins, with slightly closer relatedness with BKV, the other human polyomavirus. Serological data[41] had already predicted this close relationship both from epidemiologic studies of sera from populations across the world and from sera raised from animal inoculations.

The capsid proteins are used for virion attachment and are able to agglutinate human type O erthrocytes, properties similar to BKV. The hemagglutination assay is still the only reproducible quantitative test for JCV since a plaque assay has not yet been developed. Epidemiologic studies used a hemagglutination-inhibition assay to show that seroconversion occurs very early in life and indicated that JCV infection is widespread throughout the world. Antibodies have been made that react with a common polyomavirus capsid antigen[42] as well as with antigens unique to the JC virions. Antibodies to virion antigens are neutralizing and have been useful reagents in detecting virus in cell culture and clinical tissues.[43,44]

The large T protein shares many functions with that of other papovaviruses. It is necessary for the initiation of DNA replication by binding specific sequences in the regulatory region of the genome and is responsible for a malignant infection. Biochemical studies of the T protein have not been done because JCV T is not produced in high concentrations in lytically or malignantly infected cells. In human fetal glial cells *in vitro*, the T protein has been identified[45] as a 94-kDa protein with no evidence of the synthesis of the small-t protein, although the mRNA for the small-t protein has been identified in transformed cells.[46] Also, in the permissive glial cells, the T protein[45] does not seem to have affinity for the cellular p53 phosphoprotein which is thought to be involved in cell cycle regulation and may function as a nuclear oncogene.[47]

In hamster and rat transformed cells, however, JCV T protein does bind the cellular p53 protein.[48] The biological consequence of this binding is not known for JCV. By analogy to studies in SV-40-infected cells, the association of the papovavirus T protein with p53 may affect the efficiency of tumor formation. In tumors induced by JCV in nonhuman primates, the T protein has the ability to disrupt actin cable organization, leading to altered cell morphology, and to cause the secretion of plasminogen activator. These characteristics are malignant properties of JCV shared with other papovaviruses.[49] The JCV T protein has also been expressed in transgenic mice, where its presence coincides with a dysmyelination phenotype.[50] This suggests that the protein may interfere with the synthesis or maturation of myelin in the mature oligodendrocyte in the mouse model. T protein has also been detected in adrenal medulloblastomas and neuroblastomas in some of the transgenic mice.

To what extent the JCV T protein itself may contribute to the restricted host-range properties of JCV was analyzed by Chuke *et al.*[51] in Frisque's laboratory. These investigators constructed a series of hybrid genomes by exchanging the regulatory sequences among the primate polyomaviruses SV40, BKV, and JCV. These constructs were then transfected into either human fetal glial cells, which are permissive for all three viruses, human fibroblast WI-38, or monkey CV-1 cell lines. Their results indicated that the T proteins from the three viruses could interact with each of the three regulatory sequences. In human fetal glial cells, however, the JCV T protein under the control of the SV40 or BKV regulatory sequences was synthesized but did not result in lytic infection. Use of the JCV regulatory sequences controlling T protein synthesis of either SV40 or BKV, however, did result in lytic infection. These data suggested that regulation of the viral host range for multiplication is a property shared by both the T protein and the regulatory sequences used as signals for transcription and replication. A similar approach was used by these investigators in determining the role of the T protein in transformation of cells in culture. In addition to exchanging the regulatory sequences among the three viruses, Haggerty *et al.*[52] constructed chimeric coding sequences for the T protein that resulted in the synthesis of a hybrid protein among SV40, BKV, and JCV. These recombinant genomic constructions were tested for their ability to induce foci of transformed rat 2 cells in culture and the subsequent property of anchorage-independent growth. The efficiencies for transformation of these complex genomes are described as intermediate between the high efficiency of SV40 and the low efficiency of JCV. The transformation properties appeared more dependent on the nature of the T protein than on the influence of the regulatory sequences. Bollag *et al.*[53] from the same laboratory described the use of hybrid genomes that maintained the entire regulatory sequences from one of these viruses and directed the expression of a hybrid T protein. Transformation assays were done on the rat 2 cell line also. In these experiments it appeared that the JCV regulatory sequences and the amino terminus of the T protein coding sequences contributed to the low efficiency and restricted nature of JCV transforming properties.

3. PATHOLOGY OF JC VIRUS IN THE HUMAN NERVOUS SYSTEM: PROGRESSIVE MULTIFOCAL LEUKOENCEPHALOPATHY

3.1. Clinical Features

Progressive multifocal leukoencephalopathy is a subacute white matter disease resulting from JCV infection of astrocytes, and oligodendrocytes.[54] The disease occurs most often in patients who have an underlying disease that impairs cellular immunity. A small number of patients have developed PML without any known underlying disease. Although the disease was once thought always to be fatal, recent studies have confirmed that recovery can occur in a small number of patients, even those with the acquired immunodeficiency syndrome (AIDS).[55]

Patients with PML develop multifocal white matter lesions that lead to neurological symptoms and signs. However, any focal neurological deficit that results from demyelination can be seen in patients with PML. The presence of hemiparesis or hemiplegia, cortical blindness or homonymous hemianopia, dementia, and multifocal white matter lesions with neuroimaging techniques that do not contrast enhance, in a patient with a predisposing disease that compromises cellular immunity is generally considered pathognomonic for PML. Even with this clinical and radiographic picture, cerebral biopsy is required for confirmation of the diagnosis. Cerebellar signs occur in nearly one-third of patients by the time the diagnosis is confirmed. The incidence of cerebellar signs appears to be higher in patients who develop PML as a result of AIDS. Sensory symptoms and signs are not common in PML. Some patients have developed hemisensory deficits that reflect involvement of the subcortical white matter serving the postcentral gyrus. As the disease progresses, neurological symptoms and signs progress with new neurological deficits appearing throughout the course. Terminally, patients usually either succumb to their underlying disease or to intercurrent infections of the respiratory or genitourinary tracts.

3.2. Description of the Pathology

The pathology of PML reflects the multifocal nature of JCV infection of macroglia. Infection of oligodendrocytes, the cells that produce myelin, results in cell death and subsequent loss of myelin sheaths surrounding axons. The hallmarks of the pathology of PML include eosinophilic intranuclear inclusions present in infected oligodendrocytes, bizarre, often multinucleated astrocytes that have abundant cytoplasm, and multifocal areas of demyelination. Lesions are often present at the junction of the gray and white matter of the cerebral hemispheres. Coalescence of smaller lesions may lead to large areas of demyelination.

The development of *in situ* DNA : DNA hybridization using biotinylated JCV probes has increased the understanding of the pathology of PML.[56,57] With this technique, virus-infected cells can be detected even though they do not contain intranuclear inclusions or other signs of viral infection. Identification of the cell type infected can usually be made using the nonradioisotopic label, since immunocytochemical methods are used for generation of the hybridization signal. With *in situ* hybridization, several patterns of JCV infection of the brain have been found. Perivascular accentuation of JCV-infected cells has been noted in both brain biopsies and autopsy specimens, suggesting that the virus reaches the brain by the bloodstream[20] (see below). Areas of demyelination include numerous cells that contain JCV DNA. Infected cells are most frequently found at the borders of lesions, being less frequent toward the center of the lesion, possibly because of cell loss by viral cytolysis. The vast majority of the cells have the morphology of oligodendrocytes. Astrocytes containing JCV DNA are found throughout the lesions. The frequency of infected astrocytes appears to be inversely related to the severity of the underlying disease. Macrophages, which are frequently found in the center of the lesions, do not contain JCV DNA by *in situ* hybridization. Isolated cells containing JCV DNA are often found in the white matter in areas in which demyelination has not occurred. Although these cells have the morphology of oligodendrocytes, a number of them have been identified as B lymphocytes infected with JCV.[58a]

3.3. Mechanisms of Pathogenesis

Richardson and co-workers in their paper in 1958[1] first described PML in detail and postulated a viral etiology for the disease. ZuRhein was the first to demonstrate papova-like virions by electron microscopy in nuclei of oligodendrocytes of the brain of a patient with PML.[3] Padgett et al.[4] isolated JCV from the brain of a patient with PML using human fetal glial cell cultures, thus confirming a viral etiology for PML. Subsequently, an SV-40-like virus was isolated from the brains of two patients with PML and systemic lupus erythematosus.[58b] However, except for these isolates, JCV has always been identified in brain from PML patients when virus has been isolated.

Whether JC virus is latent in the brain of normal individuals is not known. All attempts to isolate JCV from normal brain have been unsuccessful.[59] Neither *T* antigen nor virion antigens are detectable in brains of normal patients. In several brains from immunosuppressed individuals without PML, JCV has not been isolated, nor have viral antigens been detected. These findings suggest that JCV is latent in systemic organs, reaching the brain just prior to the time of manifestation of acute infection, development of PML. As noted above, JCV can be isolated from the urine of patients with PML. This suggests that virus may be latent in the genitourinary tract. The exact cell type that allows replication has not been specifically identified but is thought to be the transitional epithelial cell. The significance of these findings in the pathogenesis of PML is not clear; JCV can be isolated from the urine of pregnant women and renal transplant patients receiving exogenous immunosuppressive drugs, but neither pregnant women nor renal transplant patients shedding JCV in the urine develop PML. This observation suggests that virus growth in the urinary tract and viral shedding in the urine are not always associated with PML and may not be necessary for the development of the disease.

Recently we have been able to detect JCV infection of mononuclear cells of the bone marrow in patients with PML.[20] Double-labeling experiments have shown these cells to be B lymphocytes. Similar experiments have demonstrated the presence of JCV-infected lymphocytes in the brains of patients with PML. Infected B cells have been shown in demyelinating lesions as well as in areas of the brain in which demyelination has yet to occur. At present these results suggest that JCV most likely infects B lymphocytes during the primary infection, remains latent in the bone marrow, and reaches the brain when replication of the virus is allowed to occur as host T-cell immunoregulatory cells are impaired by an underlying disease. The proposed pathogenesis may explain several pathological findings in PML that have remained difficult to understand. Activated lymphocytes can cross the blood–brain barrier nonspecifically. Therefore, viral antigens do not have to be present on the endothelial cell surface. The perivascular accentuation of JCV—infected cells detected by in situ hybridization most likely results from continued seeding of JCV into the brain by infected B lymphocytes. The multifocal nature of the disease, which is often at some distance from blood vessels, may be the result of activated B cells traversing brain parenchyma and releasing JCV at sites distant from blood vessels. Finally, the isolated JCV-infected cell in areas of normal brain may not be an infected oligodendrocyte but rather a B cell transporting virus to the brain.

The diagnosis of PML does require cerebral biopsy or autopsy confirmation of JCV infection of the brain. Serological testing is not helpful for several reasons. Most individuals who reach the age in which PML is most frequent have serum antibodies to JCV, confirming previous primary infection with the virus. Even though they have detectable antibodies to JCV, the vast majority of PML patients have an immunocompromising disease that prevents them from developing an immune response that could be detected as a rise in antibody titer. Routine laboratory and immunologic testing is not helpful except in confirming the inmunocompromised state resulting from the predisposing disease. JC virus has been isolated from the urine of patients with PML. Examination of infected cerebral tissue is necessary to confirm the diagnosis using neuroimaging techniques such as computerized tomography (CT) and magnetic resonance imaging (MRI) scans. *In situ* hybridization can be performed on frozen sections of the cerebral biopsy, allowing confirmation of the diagnosis of PML within 4 hr of surgery.[60] Further information can then be obtained from the hybridization tests in paraffin-embedded, formalin-fixed biopsy tissue when available. In our laboratory, the results of *in situ* hybridization for JCV on frozen sections are as reliable as those of the same procedure on formalin-fixed tissue. Virus isolation from brain biopsy or autopsy material is impractical, since fetal glial cell cultures are required for virus growth, the amount of tissue available for culture is extremely limited, and weeks are required to detect viral growth.

The course of PML was once thought to be uniformly fatal. The average duration of the disease has been 9 months to 2 years. The longest survivor of virologically proven PML lived 6 years from the time of diagnosis. The severity of the predisposing disease appears to have some effect on the clinical course of PML. Patients with severe immunosuppression secondary to HIV infection have a more rapid progression, with death often occurring within 2 months of the onset of symptoms. Patients who are receiving exogenous immunosuppressive drugs to avoid graft rejection of a transplanted kidney and who develop PML may have a course of 4 to 5 years. With the development of stereotactic cerebral biopsy and molecular methods of identifying JCV, survivors of PML have now been recognized. One patient who recovered from PML had no underlying disease, but two others had AIDS.[55,61] How this recovery occurred is unclear at the present time, but careful examination of these patients' immune cells may provide the clue to answer the lingering questions on JCV latency and the pathway by which the virus is transported to the central nervous system (CNS).

4. ONCOLOGY OF JC VIRUS

Because of the physical and genetic association of JCV and BKV with the simian and rodent polyomaviruses and the frequency of appearance of tumors in immunocompromised patients, both JCV and BKV were suspected as possible causative agents in the development of human neoplasms. Experiments were designed to study the significance of this association using both animal models for tumor formation and examination of human tumors for evidence of these viruses.

4.1. Animal Models for Tumor Induction with JC Virus

Because papovaviruses were found to be generally oncogenic agents in rodents,[7] JCV was originally inoculated into hamsters.[62] When the Mad-1 strain of JCV was inoculated intracerebrally into newborn hamsters, multiple gliomas were found in 83% of the animals. These tumors were characterized as glioblastomas, medulloblastomas, and unclassified primitive tumors. The JCV *T* protein was detected in the nuclei of the tumor cells, and virus was rescued by fusion with permissive cell types. As with BKV, different virus strains have been found to produce distinctive tumor types. Inoculation of the Mad-2 strain, for example, caused 19 of 20 hamsters to develop cerebellar medulloblastomas, whereas the Mad-4 strain resulted in 10 of 20 hamsters to develop pineal gland tumors.[34] The route of inoculation seemed to determine the tumor type produced, as described by Varkis et al.[63] Intraocular inoculation of the Mad-1 virus induced large abdominal neuroblastomas in 10 of 31 hamsters. In these animals metastases were seen in the liver, bone marrow, and lymph nodes. Neuroblastomas also occurred following combined subcutaneous and intraperitoneal injection of virus.

Characterization of the cerebellar medulloblastomas in JCV-injected hamsters indicated that the tumors originated from the cellular elements of the internal granular layer.[64] The tumor cells were similar to granular neurons. The Tokyo-1 strain of JCV, inoculated intracerebrally, induced cerebellar medulloblastomas in 20 of 21 hamsters.[65] These tumor cells in culture expressed the T protein but lost expression following transplantation. These tumor cells produced glial fibrillary acidic protein, marking them astrocytic in origin. Owl monkey astrocytomas induced by JCV may also lose T protein expression once placed in culture.

The cellular origin of the Tokyo-strain-induced cerebellar medulloblastomas was studied by in situ hybridization with an antisense mRNA probe to T protein.[66] Virus-inoculated animals were examined histologically. At 10 days postinoculation, migrating cells in the cerebellar molecular layer as well as cells in the internal granular layer hybridized to the JCV probe. The incipient medulloblastoma consisted of many JCV T-protein-positive cells and was found in the cerebellar internal granular layer. These observations suggested that cells infected in the external granular layer migrated normally and then proliferated in the internal granular layer to become a medulloblastoma. The origin of this JCV-induced tumor supports the view that human medulloblastomas arise from the cerebellar granular layer. However, there is no evidence for involvement of JCV in these human tumors. The same Tokyo strain was inoculated into newborn rats and resulted in the formation of undifferentiated neuroectodermal tumors in the cerebrum.[67] The tumor cells were JCV T-protein positive. In the more differentiated areas of the tumor, the astrocyte-specific glial fibrillary acidic protein could be detected. These rat brain tumors were thought to correspond to transformation of a glial cell precursor.

In other rodent model studies, transgenic mice were made with the early region of the JCV genome coding only for the T protein.[50] Adrenal neuroblastomas developed in four of the ten founder mice. Three mice produced offspring with a neurological abnormality characteristic of the shaking phenotype similar to mutant strains of mice termed quaking and jimpy, which demonstrate abnormalities in myelin formation. The mRNA for T protein was detected in other systemic tissues, and neuropathological analysis of the CNS indicated a dysmyelination.[68] Further characterization by immunocytochemistry and in situ hybridization demonstrated a pronounced reduction in myelin proteins, proteolipid protein, and myelin basic protein. These results were interpreted as an effect of the JCV T protein on the maturation of oligodendroglial cells at a posttranscriptional level.

One of the unique oncogenic properties of JCV is its ability to induce tumors in nonhuman primates. When inoculated intracerebrally into SV-40-seronegative owl and squirrel monkeys, JCV would induce grade 3–4 glioblastomas, which were further classified as astrocytomas.[69,70] Tumor development, however, usually took place 18–36 months following inoculation, making study of this model of neural oncology very difficult. In contrast to the rodent tumor growth, no metastasis was evident in the monkeys even if inoculated either subcutaneously or intravenously. Tumor cells expressed T protein and had the entire viral genome integrated in tandem copies, usually in different sites in the chromosome in different tumors.[71,72]

Tumor tissue from one explanted owl monkey astrocytoma was inoculated intracerebrally into four owl monkeys.[73] One owl monkey developed an astrocytoma 24 months later, which was then explanted into tissue culture. The tumor cells spontaneously shed infectious JCV in culture. The viral DNA was both integrated in the chromosome of the cells and free in an episomal state. The free DNA was cloned and sequenced, revealing a Mad-4 genome as the agent involved. The T protein of the virus being continually shed by the cells in culture differed biologically from the Mad-4 or Mad-1 strains. It maintained antigenic characteristics similar to the SV40 T protein and could bind the primate cellular protein p53. Mutations in the T protein could account for these differences and could have occurred during tumor development.

Because JCV transforms cells in culture with such a low frequency, there have been fewer studies in this area than with BKV. Primary hamster brain cells in culture can be transformed with the Mad-1, -2, -3, and -4 strains.[74] These tumor cells display the T protein, which could bind the hamster cellular protein p53, maintain viral DNA in an integrated state, and could be transplanted into recipient hamsters.[75] Human amnion cells in culture could also become transformed by JCV genomic DNA introduced by transfection.[76]

4.2. Animal Models for Tumor Induction with BK Virus

Depending on the strain and the route of infection, inoculation of hamsters with BKV results in various tumor types. Only one undifferentiated sarcoma appeared in a study with over 50 hamsters when inoculation was done subcutaneously or intravenously.[77] In contrast, 44 of 50 animals in a study developed ependymomas from intracerebral inoculation with a high-titer stock of virus.[78] Papillary ependymomas and pancreatic islet cell tumors resulted from intracerebral inoculation. When the same high-titer stock of virus was inoculated intravenously, ependymomas, insulinomas, and osteosarcomas developed in the majority of the animals.[79,80] Tumor cells are T-protein positive and have integrated copies of the genomic viral DNA. Tumors of the choroid plexus were also found with intracerebral inoculations. In contrast to JCV, BKV was not oncogenic in nonhuman primates, although the animals did produce antibodies to T protein.[69]

In a more detailed study, Chenciner *et al.*[81] found viral DNA only in tumor tissue, not in normal tissues. The viral DNA from four ependymomas was integrated in different sites in the cellular DNA. In two osteosarcomas and two insulinomas, both integrated and episomal, multiple copies of viral DNA were identified.

The first descriptions of the ability of BKV to cause transformation in culture came from Major and Di Mayorca[82] using the BHK 21 cell line, which had been the model cell for studies of anchorage-independent growth, and from the laboratories of Barbanti-Brodano[83] and Tanaka.[84] Other transformation studies quickly followed, but the distinction between the contribution of the regulatory sequences and the coding sequences of BKV for inducing a malignant infection has not been examined until recently.

To determine the efficiency and mechanism of transformation of cells in culture, a set of deletions was introduced into the three tandemly repeated regions of BKV regulator sequences.[85] Decreasing the number of repeated elements from three to two led to an increase in the transforming capacity for hamster cells. If only one element was present to direct the synthesis of the T protein, transformation still occurred with good efficiency.[86] The effect of these changes is to alter the level of T protein expression. Whether cis-acting effects may be a part of the mechanism to account for the changes in transformation efficiency remains an important question. The influence of the level of the T protein in the transformed cells is somewhat similar to JCV transformation efficiency: the higher the concentration of T protein, the more efficient the transformation. For example, hamster kidney cells that expressed low levels of T protein demonstrated phenotypes of a diffuse appearance of tumor surface antigen but were highly metastatic in newborn hamsters.[87] Transformed cells expressing high levels of T protein demonstrated high levels of tumor surface antigens but were poorly metastatic. Immune system recognition of virally expressed gene products clearly plays an important role in the malignant property of transformed cells.

Similar to studies with JCV, transgenic mice have been made using the BKV early region and regulatory sequences.[88] Surviving mice developed primary hepatocellular carcinomas and renal tumors. The mRNA for T protein was expressed in tumor tissue and also in normal brain, heart, and lung tissue. The BKV tissue tropism is more extensive than JCV, selecting epithelial cells as the most suitable host for viral infection.

4.3. Involvement of JC Virus in Human Brain Tumors

Although JCV is considered the causative agent of PML through its cytolytic infection of glial cells in the brain, its oncogenic potential in humans is unclear. JCV induces tumors of neuroectodermal tissues in rodents and nonhuman primates, but there is no evidence for this virus causing similar tumors in humans. Hybridization analysis of the DNA from 24 human brain tumors using a highly specific JCV DNA probe did not reveal the presence of the viral genome.[89] JCV also was not detected in a blot hybridization analysis of 11 human retinoblastomas.[90] Evidence for the transforming potential of JCV for human brain cells was demonstrated by Mandl et al.[19] in primary cultures of human fetal glial cells. Replication-deficient mutants of the JCV DNA were used to transform these cells without undergoing lytic infection. The T protein was expressed from viral DNA integrated into the cellular chromosome.

Any suggestion that JCV may be associated with human neoplasms, however, comes from circumstantial data. There are two reported cases where multiple gliomas were found in brains of PML patients.[91,92] The most notable feature of these cases that may suggest viral involvement was the observation that the malignant astrocytomas in the brain were adjacent to the PML lesions. Virus was detected by electron microscopy in nonneoplastic tissue but not in the tumor tissue itself. There also have been four cases of primary cerebral lymphoma associated with PML.[93–96] JCV was identified in the demyelinated lesions in three cases. The location of the lymphoma was central in one PML lesion and adjacent in two others. Two of the lymphomas were characterized as plasmacytoid with production of immunoglobulins. The association of these two specific pathologies may not be merely coincidental, since JCV has been described in the Virchow–Robbins space of the brain and in B lymphocytes in bone marrow and spleen as well as in infected B cells in PML brain tissue.[20]

The available evidence suggests that JCV is not an important factor for neural-associated neoplastic diseases. However, few of the tumors so far analyzed occur in the age group that seroconverts, indicating current infection with JCV. Childhood tumors may be a logical choice to examine, since tumors prevalent in this group resemble the type of tumors JCV induces in rodents and simian models.

With reagents that are now capable of detecting T protein in paraffin-embedded tissues,[97] an easy analysis of such tumors is possible. In severely immunocompromised patients such as those with AIDS, lymphomas in the brain are a common feature of neurological complications. Studies on the presence of JCV genome in the AIDS patient are becoming more common and may lead to important new evidence for the ability of JCV to participate in other pathologies in addition to demyelination in the human CNS.

4.4. Involvement of BK Virus in Human Tumors

Unlike JCV, BKV has been associated with human neoplasias. One of the first reports was the isolation of a virus similar to BKV from a reticulum cell sarcoma of the brain and urine of a patient with Wiskott–Aldrich syndrome.[98] However, viral antigens were not detected in the tumor cells. Fiori and Di Mayorca[99] described BKV genomic DNA sequences in five of a series of 12 human tumors and in several tumor cell lines using reassociation kinetic hybridization assays. Other studies of tumors did not show positive results for viral DNA or proteins.[100] Analysis of the serum of cancer patients was also negative for specific antibody to the BKV T protein.

Because of the cell tropism of BKV for epithelial cells and its isolation from the urinary tract, several other studies concentrated attention on systemic tumors. Pater *et al.*[101] examined 12 urogenital tract tumors and found nine with DNA fragments that hybridized to BKV DNA. But the data were difficult to interpret, since normal tissue from noncancer patients was also positive for BKV DNA. Shah and colleagues[102] did an extensive search for viral T protein in 123 primary urogenital tumors and normal tissues, and none showed evidence for the T antigen. Serological studies of these and other patients also were negative for anti-T-protein antibodies.

Caputo et al.[103] reported BKV DNA in a human adenoma of the pancreatic islet cells, an insulinoma. The DNA was episomal with no integrated copies. Virus could be rescued by transfection of the DNA in human embryonic kidney cells. BKV genomic sequences have also been found in an episomal state in Kaposi's sarcoma tissue in three of five Ugandan patients with AIDS.[104] The viral DNA from the Kaposi's cells was similar to the DNA found in the insulinomas. Cloned DNA from normal liver and a kidney carcinoma was indistinguishable from the prototype BKV in the early region, but the kidney carcinoma showed mutations in sequences of the later region.[105]

.Preservation of sequences in the T-protein-coding region is necessary for the transformation properties of the virus. There have been several studies of BKV associated with human brain tumors. Corallini et al.[106] reported that BKV DNA was detected by blot hybridization in 19 of 74 human brain tumors mostly in glioblastomas. Some tumors demonstrated multiple copies of the viral genome, up to 15. Again, the viral DNA was episomal and not integrated. The episomal DNA was shown to be infectious in permissive cell types. Dorries et al.[89] found viral DNA in 11 of 24 tumors but indicated that the DNA was only associated with high-molecular-weight or chromosomal DNA, characteristic of an integrated state. The tumors in these cases were neurinomas and meningiomas.

BKV is a frequent infection in the population, as demonstrated by the fact that antibody levels usually remain measurable throughout life. Finding virus or viral DNA associated with tumor tissue is an indication that this virus may be a factor in cancer but may not be the sole etiologic agent. BKV can productively infect many cells in humans and therefore may be present in the tumors as an active infection and not as an initiation of a malignant infection. Continued analysis of neoplastic diseases and the viruses associated with tumors of many types will eventually reveal the true importance of this group of human polyomaviruses as human oncogenic agents.

5. SUMMARY

As more experiments are done on the transcriptional regulation of the JCV genome, it seems clear that glial cells of the human brain contain factors that bind DNA and recognize sequences present in the JCV genome. Considering some of the experiments discussed here it is possible to suggest that a family of proteins like nuclear factor 1 are able to recognize genomes like JCV. However, it could be that only the cell-specific proteins in that family can positively regulate gene expression. Similar proteins from other cell types that are able to bind the regulatory sequences may actually negatively regulate the JCV genome and account for the host-restricted nature of infection. The observation that JCV can infect B lymphocytes, which could serve as the site of latency and provide the vehicle for virus entry into the brain, would require that these factors be present in a positive regulatory manner. The demonstration that JCV-infected B lymphocytes are present in PML brain tissue in AIDS and non-AIDS cases provides strong evidence that this actually is the true mechanism of pathogenesis of this disease. Experiments that reproduce a lytic infection of JCV in human B-cell lines in culture have been done. These same B-cell lines have DNA-binding nuclear protein factors that recognize the identical JCV DNA sequences as are recognized by glial cells (unpublished results). Therefore, it appears that the understanding of the neurotropic nature of JCV infection in the human population and the pathogenesis of the resulting demyelinating disease has finally reached a stage where sufficient data are available to draw reasonable conclusions. What triggers release of latently infected B cells in the immunocompromised host is the next major question to be answered for a more complete understanding of this virus–cell interaction. Analysis of cells of the immune system and their relationship with glial cells of the brain could be the next area for experimentation in order to find the final answers.

ACKNOWLEDGMENTS.   The authors express their appreciation to Mrs. Pauline Ballew and Mrs. Claire Gibson for preparation of the manuscript and to Mrs. Renee Traub for editorial assistance. We also thank Dr. John Sever for support during many of the studies described here as a part of the Infectious Diseases Branch, NINDS.

REFERENCES

1. Astrom, K.-E., Mancall, E. L., and Richardson, E. P., Jr., 1958, Progressive multifocal leukoencephalopathy, *Brain* **81**:93–127.
2. Richardson, E. P., Jr., 1961, Progressive multifocal leucoencephalopathy, *N. Engl. J. Med.* **265**:815–823.
3. ZuRhein, G. M., 1967, Polyma-like virions in a human demyelinating disease, *Acta Neuropathol. (Berl.)* **8**:57– 68.
4. Padgett, B. L., ZuRhein, G. M., Walker, D. L., and Eckroade, R. J., 1971, Cultivation of papova-like virus from human brain with progressive multifocal leucoencephalopathy, *Lancet* **1**:1257.
5. Gardner, S. D., Field, A. M., Coleman, D. V., and Hulme, B., 1971, New human papovirus (BK) isolated from urine after renal transplantation, *Lancet* **1**:1253–1257.
6. Padgett, B. L., Walker, D. L., ZuRhein, G. M., Hodach, A. E., and Chou, S. M., 1976, JC papovavirus in progressive multifocal leukoencephalopathy, *J. Infect. Dis.* **133**:686–690.
7. Tooze, J., 1980, *Molecular Biology of Tumor Viruses II. DNA Tumor Viruses*, Cold Spring Harbor Laboratory, Cold Spring Harbor, NY, pp. 205–296.
8. Padgett, B. L., Rogers, C. M., and Walker, D. L., 1977, JC virus, a human polyomavirus associated with progressive multifocal leukoencephalopathy: Additional biological characteristics and antigenic relationships, *Infect. Immun.* **15**:656–662.
9. Frisque, R. J., Rifkin, D. B., and Walker, D. L., 1980, Transformation of primary hamster brain cells with JC virus and its DNA, *J. Virol.* **35**:265–269.
10. Fareed, G. C., Takemoto, K. K., and Gimbrone, M. A., Jr., 1978, Interaction of simian virus 40 and human papovaviruses, BK and JC, with human vascular endothelial cells, in:

Microbiology (D. Schlessinger, ed.), American Society for Microbiology, Washington, DC, pp. 427–431.

11. Frisque, R. J., Martin, J. D., Padgett, B. L., and Walker, D. L., 1979, Infectivity of the DNA from four isolates of JC virus, *J. Virol.* **32**:476–482.

12. Shein, H. M., 1956, Propagation of human fetal spongioblasts and astrocytes in dispersed cell cultures, *Exp. Cell Res.* **40**:554–569.

13. Walker, D. L., 1985, Progressive multifocal leucoencephalopathy, in: *Demyelinating Diseases. Handbook of Clinical Neurology. Revised Series 3* (J. C. Koetsier, ed.), Elsevier, Amsterdam, pp. 503–524.

14. Martin, J. D., Padgett, B. L., and Walker, D. L., 1983, Characterization of tissue culture-induced heterogeneity in DNAs of independent isolates of JC virus, *J. Gen. Virol.* **64**:2271–2280.

15. Mazlo, M., and Tariska, I., 1982, Are astrocytes infected in progressive multifocal leuko-encephalopathy (PML)? *Acta Neuropathol. (Berl.)* **56**:45–51.

16. Sangalang, V. E., and Embil, J. A., 1984, Emergence of papovavirus in long-term cultures of astrocytes from progressive multifocal leukoencephalopathy patients, *J. Neuropathol. Exp. Neurol.* **43**:553–567.

17. Major, E. O., and Vacante, D. A., 1989, Human fetal astrocytes in culture support the growth of the neurotropic human polyomavirus, JCV, *J. Neuropathol. Exp. Neurol.* **48**:425–436.

18. Major, E.O., Miller, A. E., Mourrain, P., Traub, R., de Widt, E., and Sever, J. L., 1985, Establishment of a line of human glial cells that supports JC virus multiplication, *Proc. Natl. Acad. Sci. U.S.A.* **82**:1257–1261.

19. Mandl, C., Walker, D. L., and Frisque, R. J., 1987, Derivation and characterization of POJ cells, transformed human fetal glial cells that retain their permissivity for JC virus, *J. Virol.* **61**:755–763.

20. Houff, S. A., Major, E. O., Katz, D. A., Kufta, C. V., Sever, J. L., Pittaluga, S., Roberts, J. R., Gitt, J., Saini, N., and Lux, W., 1988, Involvement of JC virus-infected mononuclear cells from the bone marrow and spleen in the pathogenesis of progressive multifocal leuko-encephalopathy, *N. Engl. J. Med.* **318**:301–305.

21a. Aksamit, A. J., Mourrain, P., Sever, J. L., and Major, E. O., 1985, Progressive multifocal leukoencephalopathy: Investigation of three cases using *in situ* hybridization with JCV virus biotinylated DNA probe, *Ann. Neurol.* **18**:490–496.

21b. Assouline, J. and Major, E. O., 1991, Human fetal Schwann cells support JC virus multiplication. *J. Virol.* **65**:1174–1179.

22. Hogan, T. F., Padgett, B. L., Walker, D. L., Bordon, E. C., and McBain, J. A., 1980, Rapid detection and identification of JC virus and BK virus in human urine by using immuno-fluorescence microscopy, *J. Clin. Microbiol.* **11**:178–183.

23. Daniel, R., Shah, K., Madden, D., and Stagno, S., 1981, Serological investigation of the possible congenital transmission of papovavirus JC, *Infect. Immun.* **33**:319–321.

24. Arthur, R. A., Shah, K. V., Charache, P., and Saral, R., 1988, BK and JC virus infections in recipients of bone marrow transplants, *J. Infect. Dis.* **158**:563–569.

25. Beckman, A., and Shah, K. V., 1983, Propagation and primary isolation of JCV and BKV in urinary epithelial cell cultures, in: *Polyomaviruses and Human Neurological Diseases* (J. L. Sever, ed.), Alan R. Liss, New York, pp. 3–14.

26. Yoshiike, K., Miyamura, T., Chan, H. W., and Takemoto, K. K., 1982, Two defective DNAs of human polyomavirus JC adapted to growth in human embryonic kidney cells, *J. Virol.* **42**:395–401.

27. Osborn, J. E., Robertson, S. M., Padgett, B. L., ZuRhein, G. M., Walker, D. L., and Weisblum, B., 1974, Comparison of JC and BK viruses with simian virus 40: Restriction endonuclease digestion and gel electrophoresis of resultant fragments, *J. Virol.* **13**:614–622.

28. Martin, J. D., Brackmann, K. H., Grinnell, B., and Frisque, R. J., 1982, Recombinant JC virus DNA: Verification and physical map of prototype, *Biochem. Biophys.* **109**:70–77.

29. Frisque, R. J., Bream, G. L., and Cannella, M. T., 1984, Human polyomavirus JC virus genome, J. Virol. **51**:458–469.

30. Simmons, D., and Martin, M., 1978, Methionine-tryptic peptides near the amino-terminal end of primate papovavirus tumor antigens, Proc. Natl. Acad. Sci. U.S.A. **73**:1131–1135.

31. Padgett, B. L., Walker, D. L., ZuRhein, G. M., Hodach, A. E., and Chou, S. M., 1976, JC papovavirus in progressive multifocal leucoencephalopathy, J. Infect. Dis. **113**:686–699.

32. Gruss, P., and Khoury, G., 1981, The SV 40 tandem repeats as an element of the early promoter, Proc. Natl. Acad. Sci. U.S.A. **78**:943–947.

33. Martin, J. D., King, D. M., Slauch, J. M., and Frisque, R. J., 1985, Differences in regulatory sequences of naturally occurring JC virus variants, J. Virol. **53**:306–311.

34. Padgett, B. L., Walker, D. L., ZuRhein, G. M., and Varakis, J. N., 1977, Differential neurooncogenicity of strains of JC virus, a human polyoma virus, in newborn Syrian hamsters, Cancer Res. **37**:718–720.

35. Amirhaeri, S., Wohlrab, R., Major, E. O., and Wells, R. D., 1988, Unusual DNA structure in the regulatory region of the human papovavirus JC virus, J. Virol. **62**:922–931.

36. Kenney, S., Natarajan, V., Strike, D., Khoury, G., and Saltzman, N. P., 1984, JC virus enhancer-promotor active in human brain cells, Science **226**:1337–1339.

37. Feigenbaum, L., Khalili, K., Major, E. O., and Khoury, G., 1987, Regulation of the host range of human papovavirus JCV, Proc. Natl. Acad. Sci. U.S.A. **84**:3695–3698.

38. Khalili, K., Rappaport, J., and Khoury, G., 1988, Nuclear factors in human brain cells bind specifically to the JCV regulatory region, EMBO J. **7**:1205–1210.

39. Amemiya, K., Traub, R., Durham, L., and Major, E. O., 1989, Interaction of a nuclear factor-1-like protein with the regulatory region of the human polyomavirus JC virus, J. Biol. Chem. **264**:7025–7032.

40. Vacante, D. A., Traub, R., and Major, E. O., 1989, Extension of JC virus host range to monkey cells by insertion of a simian virus 40 enhancer into the JC virus regulatory region, Virology **170**:353–361.

41. Brown, P., Tsai, T., and Gajdusek, D. C., 1975, Seroepidemiology of human papoviruses: Discovery of virgin populations and some unusual patterns of antibody prevalence among remote peoples of the world, Am. J. Epidemiol. **102**:331–340.

42. Gerber, M. A., Shah, K. V., Thung, S. N., and ZuRhein, G. M., Immunohistochemical demonstration of common antigen of polyomaviruses in routine histologic tissue sections of animals and man, Am. J. Clin. Pathol. **73**:794–797.

43. Padgett, B. I., and Walker, D. L., 1973, Prevalence of antibodies in human sera against JC virus, an isolate from a case of progressive multifocal leukoencephalopathy, J. Infect. Dis. **127**:467–470.

44. Walker, D. L., 1978, Progressive multifocal leucoencephalopathy: An opportunistic viral infection of the central nervous system, in: Handbook of Clinical Neurology, Volume 34 (P. J. Vinken and G. W. Bruyn, eds.), Elsevier–North Holland, Amsterdam, pp. 307–329.

45. Major, E. O., and Traub, R. G., 1986, JC virus T protein during productive infection in human fetal brain and kidney cells, Virology **148**:221–225.

46. Frisque, R., 1983, Regulatory sequences and virus–cell interactions of JR virus, in: Polyomaviruses and Human Neurological Diseases (J. Sever and D. Madden, eds.), Alan R. Liss, New York, pp. 41–59.

47. Eliyahu, D., Raz, A., Gruss, P., Givol, D., and Oren, M., 1984, Participation of p53 cellular tumour antigen in transformation of normal embryonic cells, Nature **312**:646–649.

48. Mandl, C. W., and Frisque, R. J., 1986, Characterization of cells transformed by the human polyomavirus JC virus, J. Gen. Virol. **67**:1733–1739.

49. Major, E. O., Mourrain, P., and Cummins, C., 1984, JC virus- induced owl monkey glioblastoma cells in culture: Biological properties associated with the viral early gene product, Virology **136**:359–367.

50. Small, J., Scangos, G. A., Cork, L., Jay, G., and Khoury, G., 1986, The early region of human papovavirus JC induces dysmyelination in transgenic mice, Cell **46**:13–18.

51. Chuke, W.-F., Walker, D. L., Peitzman, L. B., and Frisque, R. J., 1986, Construction and characterization of hybrid polyomavirus genomes, *J. Virol.* **60**:960–971.
52. Haggerty, S., Walker, D. L., and Frisque, R. J., 1989, JC virus–simian virus 40 genomes containing heterologous regulatory signals and chimeric early regions: Identification of regions restricting transformation by JC virus, *J. Virol.* **63**:2180–2190.
53. Bollag, B., Chuke, W.-F., and Frisque, R. J., 1989, Hybrid genomes of the polyomaviruses JC virus, BK virus, and simian virus 40: Identification of sequences important for efficient transformation, *J. Virol.* **63**:863–872.
54. Brooks, B. R., and Walker, D. L., 1984, Progressive multifocal leukoencephalopathy, *Neurol. Clin.* **2**:299–313.
55. Berger, J. R., and Mucke, L., 1988, Prolonged survival and partial recovery in AIDS-associated progressive multifocal leukoencephalopathy, *Neurology* **38**:1060–1065.
56. Aksamit, A. J., Sever, J. L., and Major, E. O., 1986, Progressive multifocal leukoencephalopathy, *Neurology* **36**:499–504.
57. Aksamit, A. J., Major, E. O., Ghatak, N. R., Sidhu, G. S., Parisi, J. E., and Guccion, J. G., 1987, Diagnosis of progressive multifocal leukoencephalopathy by brain biopsy with biotin labeled DNA : DNA *in situ* hybridization, *J. Neuropathol. Exp. Neurol.* **46**:556–566.
58a. Major, E. O., Amemiya, K., Elder, G., and Houff, S. A., 1990, Glial cells of the human developing brain and B cells of the immune system share a common DNA binding factor for the recognition of the regulatory sequences of the human polyomavirus, JCV. *J. Neurosci. Res.* **27**:461–471.
58b. Weiner, L. P., Herndon, R. M., Navayan, O., Johnson, R. T., Shah, K., Rubenstein, T. J., Prefiosi, T. J., and Conley, F. K., 1972, Isolation of virus related to SV40 from patients with progressive multifocal leucoencephalopathy, *N. Engl. J. Med.* **288**:1103–1110.
59. Chesters, P. M., Heritage, J., and McCance, D. J., 1983, Persistence of DNA sequences of BK virus and JC virus in normal human tissues and in diseased tissues, *J. Infect. Dis.* **147**: 676–682.
60. Houff, S. A., Katz, D., Kufta, C., and Major, E. O., 1989, A rapid method for *in situ* hybridization for viral DNA in brain biopsies from patients with acquired immunodeficiency syndrome, *AIDS* **3**:843–845.
61. Schlitt, M., Morawetz, R. B., Bonnin, J., Chandra-Sekar, B., Curtiss, J. J., Diethelm, A. G., Whelchel, J. D., and Whitley, R. J., 1986. Progressive multifocal leukoencephalopathy: Three patients diagnosed by brain biopsy, with prolonged survival in two, *Neurosurgery* **18**: 407–414.
62. Walker, D. L., Padgett, B. L., ZuRhein, G. M., Albert, A. E., and Marsh, R. F., 1973, Human papovavirus (JC): Induction of brain tumors in hamsters, *Science* **181**:674–676.
63. Varakis, J., ZuRhein, M., Padgett, B. L., and Walker, D. L., 1978, Induction of peripheral neuroblastomas in Syrian hamsters after injection as neonates with JC virus, a human polyoma virus, *Cancer Res.* **38**(6):1718–1722.
64. ZuRhein, G. M., and Varakis, J. N., 1979, Perinatal indulction of medulloblastomas in Syrian golden hamsters by a human polyoma virus (JC), *Natl. Cancer Inst. Monogr.* **51**: 205–208.
65. Nagashima, K., Yasui, K., Kimura, J., Washizu, M., Yamaguchi, K., and Mori, W., 1984, Induction of brain tumors by a newly isolated JC virus (Tokyo-1 strain), *Am. J. Pathol.* **116**: 455–463.
66. Matsuda, M., Yasui, K., Nagashima, K., and Mori, W., 1987, Origin of the medulloblastoma experimentally induced by human polyomavirus JC, *J. Natl. Cancer Inst.* **79**:585–591.
67. Ohsumi, S., Motoi, M., and Ogawa, K., 1986, Induction of undifferentiated tumors by JC virus in the cerebrum of rats, *Acta Pathol.* **36**:815–825.
68. Trapp, B. D., Small, J. A., Pulley, B. A., Khoury, G., and Scangos, G. A., 1988, Dysmyelination in transgenic mice containing JC virus early region, *Ann. Neurol.* **23**:38–48.
69. London, W. T., Houff, S. A., Madden, D. L., Fuccillo, D. A., Gravell, M., Wallen, W. C., Palmer, A. E., Sever, J. L., Padgett, B. L., Walker, D. L., ZuRhein, G. M., and Ohashi, T.,

1978, Brain tumors in owl monkeys inoculated with a human polyomavirus (JC virus), *Science* **201:**1246–1249.

70. London, W. T., Houff, S. A., McKeever, P. E., Wallen, W. C., Sever, J. L., Padgett, B. L., and Walker, D. L., 1983, Viral-induced astrocytomas in squirrel monkeys, *Prog. Clin. Biol. Res.* **105:**227–237.

71. Miller, N. R., McKeever, P. E., London, W. T., Padgett, B. L., Walker, D. L., and Wallen, W. C., 1984, Brain tumors of owl monkeys inoculated with JC virus contain the JC virus genome, *J. Virol.* **49:**848–856.

72. Major, E. O., 1983, JC virus T protein expression in owl monkey tumor cell lines, in: *Polyomavirus and Human Neurological Diseases* (J. L. Sever and D. L. Madden, eds.), Alan R. Liss, New York, pp. 289–298.

73. Major, E. O., Vacante, D. A., Traub, R. G., London, W. T., and Sever, J. L., 1987, Owl monkey astrocytoma cells in culture spontaneously produce infectious JC virus which demonstrates altered biological properties, *J. Virol.* **61:**1435–1441.

74. ZuRhein, G., 1983, Studies of JC virus-induced nervous system tumors in the Syrian hamster: A review, in: *Polyomaviruses and Human Neurological Diseaess* (J. L. Sever and D. L. Madden, eds.), Alan R. Liss, New York, pp. 205–221.

75. Mandl, R., and Frisque, R., 1986, Characterization of cells transformed by the human polyomavirus JC virus, *J. Gen. Virol.* **67:**1733–1739.

76. Howley, P. M., Rentier-DelRue, F., Heilman, C. A., Law, M. F., Chowdhury, K., Israel, M. A., and Takemoto, K. K., 1980, Cloned human polymavirus JC DNA can transform human amnion cells, *J. Virol.* **36:**878–882.

77. Shah, K. V., Daniel, R. W., and Strandberg, J. D., 1975, Sarcoma in a hamster inoculated with BK virus: A human papovavirus, *J. Natl. Cancer Inst.* **54:**945–950.

78. Corallini, A., Barbanti-Brodano, G., Bortoloni, W., Nenci, I., Cassai, E., Tampieri, M., Portolani, M., and Borgatti, M., 1977, High incidence of ependymomas induced by BK virus: A human papovavirus, *J. Natl. Cancer Inst.* **59:**1561–1564.

79. Uchida, S., Watanabe, S., Aizawa, T., Furuno, A., and Muto, T., 1979, Polyoncogenicity and insulinoma-inducing ability of BK virus, a human papovavirus, in Syrian golden hamsters, *J. Natl. Cancer Inst.* **63:**119–126.

80. Corallini, A., Altavilla, G., Cecchetti, M. G., Fabris, G., Grossi, M. P., Balboni, P. G., Lanza, G., and Barbanti-Brodano, G., 1978, Ependymomas malignant tumors of pancreatic islets and osteosarcomas induced in hamsters by BK virus, a human papovavirus, *J. Natl. Cancer Inst.* **61:**875–883.

81. Chenciner, N., Meneguzzi, G., Corallini, A., Grossi, M. P., Grassi, P., Barbanti-Brodano, G., and Milanesi, G., 1980, Integrated and free viral DNA in hamster tumors induced by BK virus, *Proc. Natl. Acad. Sci. U.S.A.* **77:**975–979.

82. Major, E. O., and Di Mayorca, G., 1973, Malignant transformation of BHK-21 clone 13 cells by BK virus: A human papovavirus, *Proc. Natl. Acad. Sci. U.S.A.* **70:**3210–3212.

83. Corallini, A., Barbanti-Brodano, G., Portolani, M., Balboni, P. G., Grossi, M. P., Possati, L., Honorati, C., La Placa, M., Mazzoni, A., Caputo, A., Veronesi, U., Orefice, S., and Cardinali, G., 1976, Antibodies to BK virus structural and tumor antigens in human sera from normal persons and from patients with various diseases, including neoplasia, *Infect. Immun.* **13:**1684–1691.

84. Tanaka, R., Koprowski, H., and Iwsaki, Y., 1976, Malignant transformation of hamster brain cells *in vitro* by human papovavirus BK, *J. Natl. Cancer Inst.* **56:**671–673.

85. Hara, K., Oya Y., and Yogo, Y., 1985, Enhancement of the transforming capacity of BK virus by partial deletion of the 68-base-pair tandem repeats, *J. Virol.* **55:**867–869.

86. Watanabe, S., and Yoshike, K., 1985, Decreasing the number of 68-base-pair tendem repeats in the BK virus transcriptional control region reduces plaque size and enhances transforming capacity, *J. Virol.* **55:**823–825.

87. Rosciani, C., Rubini, C., and Possati, L., 1988, Correlation between tumor antigens and

malignancy in BKV-transformed hamster cells, *Clin. Exp. Metastasis* **6**:325–332.

88. Small, J. A., Khoury, G., Jay, G., Howley, P. M., and Scangos, G. A., 1986, Early regions of JC virus and BK virus induce distinct and tissue-specific tumors in transgenic mice, *Proc. Natl. Acad. Sci. U.S.A.* **83**:8288–8292.

89. Dorries, K., Loeber, G., and Meixensbarger, J., 1987, Association of polyomaviruses JC, SV40, and BK with human brain tumors, *Virology* **160**:268–270.

90. Ueno, T., Suzuki, N., Kaneko, A., and Fujinaga, K., 1987, Analysis of retinoblastoma for human adenovirus and human JC virus genome integration, *Jpn. J. Ophthalmol.* **31**: 274–283.

91. Castaigne, P., Roudot, P., Escourolle, T., Ribadeau-Dumas, J.-L., Carthala, F., and Hauw, J.-J., 1974, Leucoencephalopathie multifocale progressive et "gliomes" multiple, *Rev. Neurol. (Paris)* **130**:379–392.

92. Sima, A. A. F., Finkelstein, S. D., and McLachlan, D. R., 1983, Multiple malignant astrocytomas in a patient with spontaneous progressive multifocal leukoencephalopathy, *Ann. Neurol.* **14**:183–188.

93. Gia Russo, M. H., and Koeppen, A. H., 1978, Atypical progressive multifocal leukoencephalopathy and primary malignant lymphoma, *J. Neurol. Sci.* **35**:391–398.

94. Egan, J. D., Ring, B. L., Reding, M. J., and Wells, I. C., 1980, Reticulum cell sarcoma and progressive multifocal leukoencephalopathy following renal transplantation, *Transplantation* **29**:84–86.

95. Ho, K.-C., Garancis, J. C., Paegli, R. D., Gerber, M. S., and Borkowski, W. J., 1980, Progressive multifocal leukoencephalopathy and malignant lymphoma of the brain in a patient with immunosuppressive therapy, *Acta Neuropathol. (Berl.)* **52**:81–83.

96. Liberski, P. P., Alwasiak, J., and Wegrzyn, F., 1982, Atypical progressive multifocal leucoencephalopathy and primary cerebral lymphoma, *Neuropathol. Pol.* **20**:413–419.

97. Greenlee, J. E., and Keeney, P. M., 1986, Immunoenzymatic labelling of JC papovavirus T antigen in brains of patients with progressive multifocal leukoencephalopathy, *Acta Neuropathol.* **71**:150–153.

98. Takemoto, K. K., Rabson, A. S., Mullarkey, M. F., Blaese, R. M., Garon, C. F., and Nelson, D., 1974, Isolation of papovavirus from brain tumor and urine of a patient with Wiskott–Aldrich syndrome, *J. Natl. Cancer Inst.* **53**:1205–1207.

99. Fiori, M., and Di Mayorca, G., 1976, Occurrence of BK virus DNA in DNA obtained from certain human tumors, *Proc. Natl. Acad. Sci. U.S.A.* **73**:4662–4666.

100. Israel, M. A., Martin, M. A., Takemoto, K. K., Howley, P. M., Aaronson, A., Soloman, D., and Khoury, G., 1978, Evaluation of normal and neoplastic human tissue for BK virus, *Virology* **15**:187–196.

101. Pater, H. M., Pater, A., and diMayorca, G., 1983, BK virus and its varients: Association with tumors and transformed cells, *Prog. Clin. Biol. Res.* **132D**:485–494.

102. Shah, K., Daniel, R. W., Stone, K. R., and Elliot, A. Y., 1978, Investigation of human urogenital tract tumors of papovavirus etiology; brief communication, *J. Natl. Cancer Inst.* **60**:579–582.

103. Caputo, A, Corallini, A., Grossi, M. P., Carra, L., Balboni, P. G., Negrini, M., Milanesi, G., Federspil, G., and Barbanti-Brodano, G., 1983, Episomal DNA of a BK virus variant in a human insulinoma, *J. Med. Virol.* **12**:37–49.

104. Barbanti-Brodano, G., Pagnani, M., Viadana, P., Beth-Giraldo, E., Giraldo, G., and Corallini, A., 1987, BK virus DNA in Kaposi's sarcoma, *Antibiot. Chemother.* **38**:113–120.

105. Knepper, J. E., and diMayorca, G., 1987, Cloning and characterization of BK virus-related DNA sequences from normal and neoplastic human tissues, *J. Med. Virol.* **21**:289–299.

106. Corallini, A., Pagnani, M., Viadana, P., Silini, E., Mottes, M., Milanesi, G., Gerna, G., Vettor, R., Trapella, G., and Silvani, V. E. A., 1987, Association of BK virus with human brain tumors and tumors of pancreatic islets, *Int. J. Cancer* **39**:60–67.

12

Neurological Aspects of Human Immunodeficiency Virus Infection

HOWARD E. GENDELMAN and SEYMOUR GENDELMAN

1. INTRODUCTION

The human immunodeficiency virus (HIV), the etiologic agent of the acquired immune deficiency syndrome (AIDS),[1-3] belongs to a taxonomic group of non-oncogenic retroviruses termed lentivirinae that share biological, biochemical, and molecular features for viral persistence.[4-11] The hallmark of HIV infection is a relentless and profound immunosuppression mediated by a selective depletion of helper/inducer T lymphocytes.[12-15] However, the virus-infected macrophage is also intimately involved in HIV pathogenesis. Primary HIV-induced disease revolves around a near-exclusive replication of virus in multinucleated and mono-nucleated macrophages typified by central nervous system (CNS) infection.[16-24] Brain macrophages are likely infected soon after viral exposure[22-25] and contribute to CNS-related neuronal injury through the secretion of neurotoxins (of viral

HOWARD E. GENDELMAN • HIV-Immunopathogenesis Program, Department of Cellular Immunology, Walter Reed Army Institute of Research, Rockville, Maryland 20850, and Henry M. Jackson Foundation for the Advancement of Military Medicine, Uniformed Services University Health Science Center, Bethesda, Maryland 20814.   SEYMOUR GENDELMAN • Department of Neurology, Mt. Sinai Medical Center, New York, New York 10028.

Neuropathogenic Viruses and Immunity, edited by Steven Specter *et al.* Plenum Press, New York, 1992.

or cellular origin) and/or by coexistent opportunistic infections. Indeed, the highest levels of HIV gene expression found in brain macrophages are associated with clinical disease in infected individuals.[25] The mechanisms by which HIV infection results in immunosuppression and progressive neurological disease are the primary focus of this chapter.

2. LENTIVIRUS–HOST INTERACTIONS: AN OVERVIEW

The lentivirinae include the ruminant viruses visna and caprine arthritis encephalitis viruses,[9–11] equine infectious anemia virus,[26] the simian immunodeficiency virus (SIV),[27–29] and the feline immunodeficiency virus (FIV).[30] HIV, which has 2 serotypes (I and II), shares similar biological, morphological, biochemical, and molecular features with all the other lentiviruses.[4–11,31,32]

Biologically, these viruses survive poorly outside their defined hosts and are rarely disseminated as aerosols. Their spread is species specific, from host to host during exchange of body fluids. *In vivo* replication is restricted. Characteristically, a noncytolytic infection of monocyte/macrophages and CD4+ lymphocytes predominates. Productive viral gene expression occurs at only low levels during any observed time period.[33–35] In contrast to replication *in vivo*, lentivirus infection in culture results in a high proportion of productively infected cells with concomitant syncytium formation and cell lysis. The mechanisms surrounding persistent *in vivo* infection during long periods of subclinical infection and disease are poorly understood but are thought to involve subversion of both nonspecific and specific immunologic responses. In several distinct ways, HIV is a typical lentivirus. Failure of production of high-titer neutralizing antibodies, antigenic drift of the proteins in the viral envelope, and enhancement of viral infection with specific antibodies are some examples of the ability of HIV to subvert the forces rallied against it.[36–41] The end result of continual replication of HIV in monocyte/macrophages and CD4+ T lymphocytes is degenerative and inflammatory destruction of the immune system and the CNS.

Under the electron microscope, all lentiviruses appear similar.[42–45] The virion particle, 110 nm in diameter, has a dense elongated central cylindrical core containing *gag* (group-specific antigen) structural elements and two copies of the RNA genome of approximately 10 kilobases. Virus-encoded enzymes required for efficient replication (i.e., the reverse transcriptase and integrase) are packaged into the virus particle. The lentivirus genome is considerably larger than those of oncogenic retroviruses. The virion is encased by a lipid envelope acquired as the virion buds from the cell surface or from vacuoles within the infected cell.

Biochemical similarities among lentiviruses include the large size and variability of the external viral envelope glycoprotein, the use of tRNA lysine as a primer for reverse transcription, the presence of genes that up-regulate viral gene expression (*trans*-activating proteins) and other novel open reading frames in the genome.[46–50] Homology of the *gag* and *pol* genes at the nucleotide level are present between HIV and other lentiviruses and attest to their molecular similarities.

3. HIV INFECTION DURING DISEASE: MODES OF IMMUNOSUPPRESSION

The hallmark of immunodeficiency in AIDS is a selective depletion of CD4+ helper/inducer lymphocytes.[11–13] The CD4+ T lymphocyte is at the center of the immune response and closely involved with monocyte/macrophages, cytotoxic T cells, natural killer cells, and B-cell functions. In addition, these cells elaborate regulatory factors including interleukin-2 (IL-2) that are trophic for lymphocytes as well as for other cells of the myeloid series. The selective loss of this particular cell in AIDS results in opportunistic infections, neoplasms, and inevitably death. CD4+ T-lymphocyte depletion may be explained by productive infection and subsequent killing and/or by indirect mechanisms. In support of direct infection as the mechanism for T-cell depletion is the demonstration of high numbers of HIV-infected leukocytes in blood ($\leq 1\%$).[51,52] In contrast, the number of productively infected cells in blood is low. HIV-specific RNA is detected in one in 10,000 to one in a million circulating leukocytes.[33] The paucity of virus-expressing cells in blood make other mechanisms for CD4+ T-lymphocyte depletion possible. Indeed, in view of the normal turnover of T lymphocytes in the body, the T-cell pool should compensate for the numbers of infected and subsequently destroyed CD4+ T lymphocytes.

Indirect mechanisms for the immunologic defects and CD4+ T-cell destruction in AIDS are supported by experimental analyses. One mechanism for accelerated cell death is the accumulation of unintegrated HIV DNA in infected cells. This is based on the known association between unintegrated DNA and cytopathogenicity in cells infected with avian and spleen leukosis viruses.[53–55] Recent polymerase chain reaction (PCR) analyses of peripheral blood mononuclear cells (PBMCs) from HIV-infected individuals demonstrate on average one to two copies of proviral DNA per infected cell and makes this a less attractive hypothesis.[51] Alternatively, investigators speculated that HIV may induce terminal differentiation of the infected T4 cell, leading to a shortened lymphocyte life span.[56,57] However, there is no direct evidence for this *in vivo*, and the observation that HIV infection of CD4+ HeLa cells results in cell lysis does not support this theory.[58]

Additional mechanisms of CD4+ T-lymphocyte depletion, however, may be operative. First, the interactions between the CD4 molecule and the virus envelope[58–61] may be important in CD4+ T-lymphocyte cytopathogenicity. Infected or uninfected CD4+ cells may be coated with free gp120, which could be recognized as foreign and then cleared by the immune system.[57] This could explain the pancytopenia commonly seen in patients with AIDS. Interactions between the HIV envelope glycoprotein present on the surface of infected antigen-presenting cells (e.g., monocyte/macrophages or CD4+ T lymphocytes) and uninfected CD4+ cells could lead to the elimination of the latter.[62,63]

Monocyte/macrophages express low levels of CD4 on their surface, yet these cells can bind to and be infected with HIV. The fact that HIV does not induce a significant cytopathic effect in monocytes suggests that the density of CD4 receptor expression is important in determining cytopathic effects elicited by the viral envelope.[58–61] Indeed, superinfection of human T-cell leukemia virus type I (HTLV-1)-transformed T-cell clones of either the CD4 or CD8 phenotype with HIV results in a productive infection. However, cytopathogenicity occurs only in the CD4+ clones.[64] It is possible that cell loss may result from the formation of intracellular toxic complexes of CD4 and the HIV envelope. Another explanation of CD4+ lymphocyte depletion involves the expression or alteration of cellular or viral epitopes in virus-susceptible cells. HIV-infected CD4+ T cells can alter their HLA class II major histocompatibility complex (MHC) phenotype and thereby become more susceptible to immune clearance.[65] The HIV envelope binds to the CD4 molecule and may mimic the configuration of a portion of the class II MHC antigen. Alternatively, viral epitopes expressed on the surface of immune-stimulated and virus-infected cells may precipitate their own demise. Here, host antibody and cytotoxic lymphocyte responses against HIV-specific epitopes clear virus-infected CD4+ T lymphocytes.[66] HIV-infected lymphocytes may also become more susceptible to superinfection by other pathogens. Cytomegalovirus (CMV) can abortively infect T cells and through dual CMV/HIV infection lead to an accelerated depletion of CD4+ T lymphocytes.[67]

HIV may preferentially infect a small population of precursor cells that is responsible for growth of other CD4+ cells.[15] This possibility has recently fallen into disfavor with the inability to find infected precursor cells *in vivo*. A selective depletion of a critical subset of CD4+ T lymphocytes could result in the elimination of all cells carrying this phenotype. Subsets of CD4+ cells that recognize and respond to soluble antigen are selectively deficient in patients with AIDS. This deficiency occurs early in the course of disease and is quite common. B-cell abnormalities, consisting of polyclonal activation with high immunoglobulin levels and a poor antibody response to novel antigens, are common in AIDS.[68–71] Last, substances released as a consequence of viral infection, such as soluble cell factors, viral proteins other than gp120, or other toxic elements, might ultimately destroy CD4+ T lymphocytes.[15]

4. ROLE OF MONOCYTES/MACROPHAGES IN THE PERSISTENCE AND DISSEMINATION OF LENTIVIRUS INFECTIONS

Perhaps the most important shared properties among lentiviruses involve target cell tropism. In each instance, infected monocyte/macrophages serve as a viral reservoir, evade host immune surveillance, and initiate fulminant disease in virus-target tissues.[72] Monocyte/macrophages are also efficient host cells for the isolation and propagation of HIV during the course of infection and disease. These cells continually harbor virus in intracytoplasmic vacuoles (Fig. 12-1) and provide an important means for escape from immune surveillance.[73,74] The viral life cycle in the monocyte/macrophage is regulated by physiological factors involved in maturation and differentiation of the cells from their precursors in bone marrow. In bone marrow or blood of infected animals or people, the

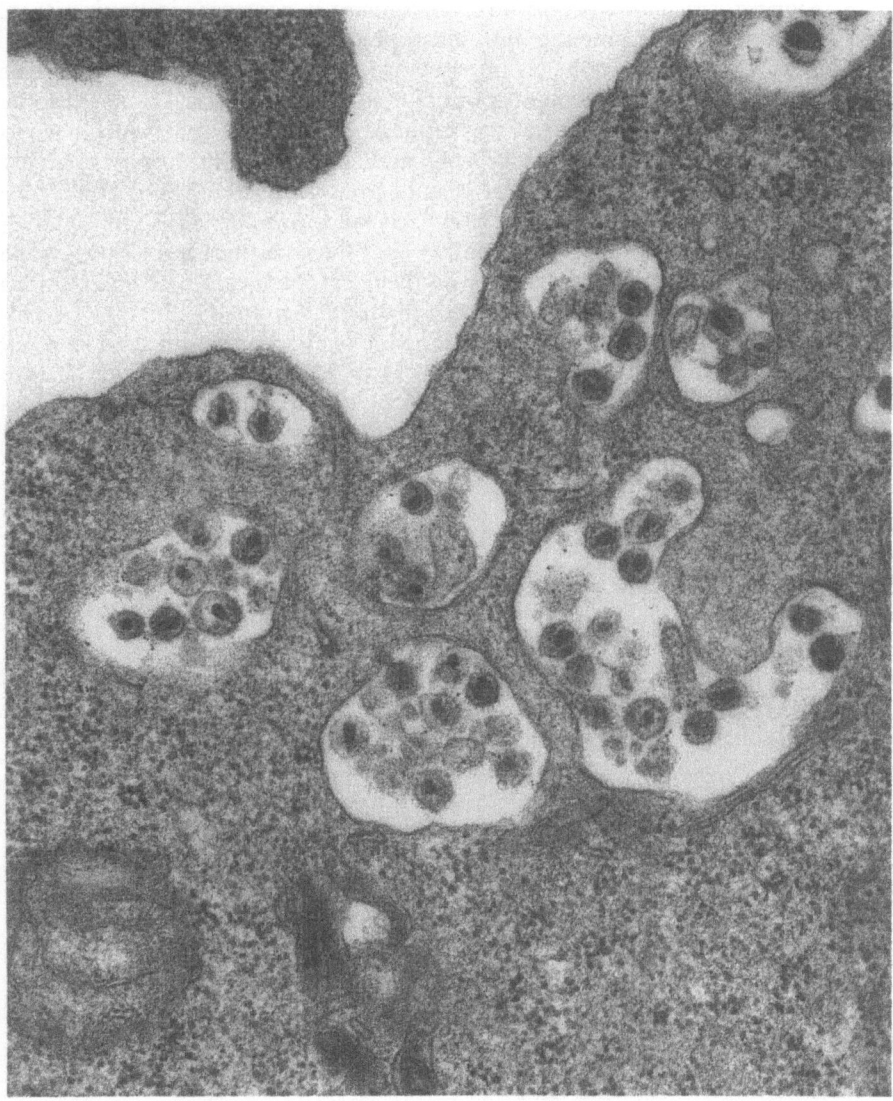

FIGURE 12-1. Intracytoplasmic HIV-1 particles are seen within a cultured monocyte. Infected cells were stained with antibodies to gp120 linked to gold particles. Virions are seen with surrounding black grains indicative of gp120 spikes.

number of cells expressing viral RNA is very low. After these infected monocytes migrate from blood and mature into tissue macrophages, viral gene expression increases several thousandfold, and the virus life cycle goes to completion (e.g., mature virus particles are produced).[75] A similar phenomenon occurs *in vitro* as infected monocytes differentiate into macrophage-like cells.[76]

Not all mature macrophages in tissue are equally permissive for lentivirus infections. The specific susceptibility of tissues to the pathological process can be traced to permissiveness of local macrophage populations that support virus replication. In visna virus infections, brain and alveolar macrophages are highly permissive, but the mature Kupffer cells in liver, the histiocytes of connective tissue, and the Langerhans' cells in skin each fail to support viral replication.[77] Moreover, the lung and brain are primary sites for virus-induced lesions, while skin and liver tissues are not targets for pathological changes. These observations are consistent with the notion that genetically predetermined cellular transcriptional factors (factors that vary with macrophage phenotype, maturation, and cell differentiation) may regulate viral gene expression and/or virus cell surface receptors. Such factors are found in the subpopulations of monocyte/macrophages that support viral replication and ultimately provide the molecular basis for the unique tissue tropism that underlies viral pathogenesis and the symptomatology of lentivirus infections.[77]

5. MONOCYTE/MACROPHAGES AND HIV INFECTION

In the infected human host, HIV has been demonstrated in or recovered from CD4+ T lymphocytes,[51,52] monocytes in blood,[23,24] brain[17,78–82] (Fig. 12-2) and spinal cord[21] macrophages, alveolar macrophages,[19,83] Langerhans'/dendritic cells,[18] and follicular dendritic cells of germinal centers.[84] Macrophages in tissue play a pivotal role in the persistence and pathogenesis of HIV.[22] Infected alveolar macrophages play prominent roles in the high incidence of *Pneumocystis carinii* pneumonia in AIDS patients.[85] The lymphocytic interstitial pneumonitis, spinal cord myelopathy, and AIDS-associated encephalopathy are all strongly associated with HIV-infected tissue macrophages. Here, as with the ruminant lentivirus infections, the virus-producing macrophage serves as a primary perpetrator and a vehicle for dissemination in disease.

6. NEUROLOGICAL MANIFESTATIONS OF HIV INFECTION: CLINICAL ASPECTS

6.1. Clinical and Neuropathological Observations: Introduction

Neurological abnormalities are quite common in AIDS and occur to varying degrees in up to 75% of patients.[86–88] Although the CNS may be affected by a number of opportunistic conditions, infectious and neoplastic, primary infection by HIV is the most frequent and important cause of CNS morbidity in HIV-

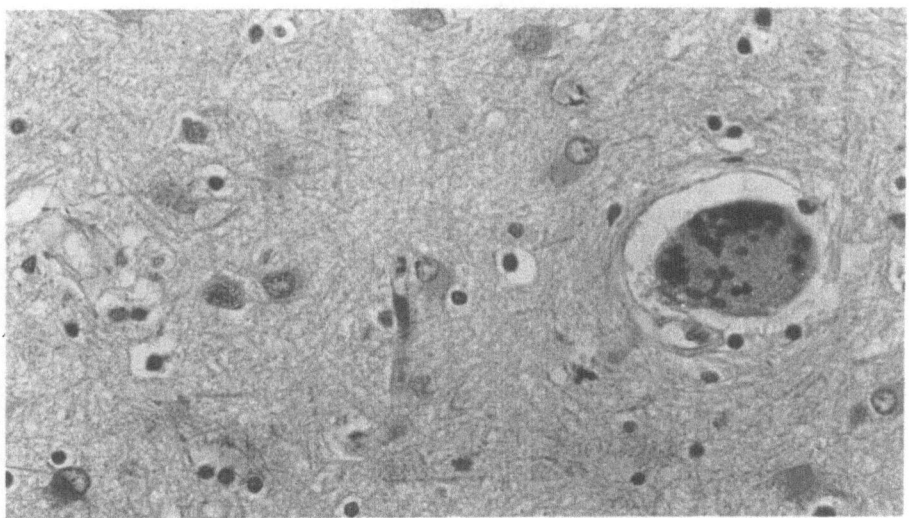

FIGURE 12-2. CNS histopathological changes in the AIDS–dementia complex. A multinucleated macrophage, reactive astrocytes, and myelin pallor are illustrated.

infected individuals.[86,87] Primary HIV infection of the CNS can manifest itself in different forms: an aseptic meningitis, an AIDS-associated dementia complex (ADC), and a vacuolar myelopathy.

In addition to primary HIV infection of the brain and spinal cord, opportunistic infections are a common cause of CNS morbidity (with frequencies in symptomatic patients of >30%).[89] The extent of overlap between primary HIV disease and opportunistic infection makes a specific CNS diagnosis on the basis of clinical and radiographic examinations difficult. Indeed, the list of secondary opportunistic infections in the CNS of AIDS patients is expansive and incudes fungal, mycobacterial, parasitic, and viral diseases. The AIDS-related CNS diseases are listed in Table 12-1.

6.2. Neurological Manifestations of Primary HIV Infection: Aseptic Meningitis

Although aseptic meningitis occurs most often during acute seroconversion, it can be seen at any point during the course of disease.[90–94] Acute meningitis commonly occurs following CNS exposure to HIV and before significant immunosuppression. Clinically, it is indistinguishable from self-limited viral meningitis. A diagnosis of HIV-1 infection must be considered in any patient with aseptic meningitis, particularly in high-risk individuals. Signs and symptoms include headache, photophobia, and low-grade fever.[93] Examination of the cerebrospinal fluid (CSF) reveals a mononuclear pleocytosis, a normal glucose, and a slightly

TABLE 12-1
Neurological Involvement in AIDS

Direct neurological infection by HIV-1
    Acute aseptic meningitis or meningoencephalitis
    AIDS–dementia complex (subacute encephalopathy)
    Vacuolar myelopathy
Neuromuscular syndromes associated with but not caused by HIV-1 infection
    Peripheral neuropathies
    Mononeuropathy multiplex
    Myopathy–myositis
Opportunistic infection
    Viral
        Adenovirus, type 2
        Cytomegalovirus
        Herpes simplex virus, types I and II
        Herpes zoster (varicella) virus
        Papovavirus (progressive multifocal leukoencephalopathy)
    Fungal
        Aspergillus fumigatus
        Candida albicans
        Coccidioides immitis
        Cryptococcus neoformans
        Histoplasma capsulatum
        Mucormycosis
        Nocardia asteroides
        Rhizopus species
    Bacterial
        Listeria monocytogenes
        Mycobacterium avium intracellulare
        Mycobacterium tuberculosis
        Treponema pallidum
    Parasitic
        Toxoplasma gondii
Cerebrovascular
    Hemorrhage
    Infarction
    Vasculitis
Neoplastic
    Lymphoma—primary CNS
    Lymphoma—secondary—metastatic systemic
    Sarcoma—Kaposi's—metastatic

elevated protein.[95–97] Oligoclonal bands and HIV-specific antibodies are usually present in CSF.[98] The isolation of HIV from CSF often signals a poor prognosis.[99,100] Signs and symptoms usually clear spontaneously and likely reflect a rising antibody titer. In select cases, an acute meningitis may persist to chronicity and be characterized by headache and a low- grade CSF pleocytosis. An encephalitis with cranial nerve involvement and/or long-tract signs may also occur as a primary complication of HIV infection.[101]

6.3. AIDS–Dementia Complex (Subacute Encephalopathy)

The AIDS–dementia complex (ADC) is the most common primary neurological manifestation of HIV infection and the most important in terms of morbidity. Up to 75% of all patients with AIDS show at autopsy some manifestation of this CNS abnormality.[87] Although ADC can occur at any time during HIV infection, including the presenting manifestation,[102–105] it is most commonly seen during advanced disease (either alone or in association with opportunistic infections). Initial clinical findings include mental and physical slowing, diminished concentration, forgetfulness, and behavioral changes.[87] Symptoms may be confirmed by family, friends, or colleagues. Signs are often subtle. The neurological examination may be totally normal. Early in the disease course, the patient is alert, oriented, and aware of current events, is able to abstract, and shows good insight.

After many months symptoms worsen and become objectively apparent, and cognitive, motor, and behavioral changes ensue. Symptoms of cognitive dysfunction include difficulty with concentration and performance of sequential mental activities, complaints of memory loss, and difficulty in reading or carrying out complex tasks at work. Early motor symptoms frequently involve the legs more than the arms, with clumsiness and, to lesser degrees, weakness of gait. Behavioral changes including apathy, loss of spontaneity, social withdrawal, and change in personality are frequent. However, distinction among psychiatric syndromes, depression, and anxiety, may be difficult. Prescription drugs including antidepressants and nonprescription medications are frequently used or abused by this patient population and may further cloud interpretation of neurological examinations and testing. Indeed, the signs and symptoms of ADC, including psychomotor slowing, blunted affect, apathy, anorexia, weight loss, poor sleep, or excessive lethargy, often mimic depression and social withdrawal. Anxiety and substance abuse may be manifested with agitation, excitement, or hyperactivity. Preexisting or coexisting medical conditions may further complicate a diagnosis. In toto, ADC is difficult to diagnose in its early stages, and care must be taken to separate its signs and symptoms from other ongoing clinical events.[87,102,103]

In later stages, making a diagnosis of ADC is less difficult. Cognitive impairment is seen in performance of daily activities. The patient is unable to perform simple calculations or to write legible script. Lists of daily activities are made and lost. Work performance suffers, and the patient becomes easily distracted. During advanced disease social withdrawal, a diminished vocabulary, delusions, and disorientations become prominent. Progression continues to dementia, incontinence, hallucinations, seizures, and coma.[87,102,103,106–108] Although ADC is distinctive, it must always be considered in the differential diagnosis of the demented elderly patient.

In addition to cognitive dysfunction, ADC is often characterized by motor abnormalities, all characteristic of a "subcortical dementia."[109–111] Symptoms begin with an almost imperceptible mechanical slowing progressing to akinetic mutism. During progressive disease gait and balance are impaired, and fine motor movements, for example handwriting, become disorganized. Clumsiness may be followed by objective signs on neurological examination. These include enhanced reflexes, gait dystaxia, spasticity, and frontal release signs such as snout and suck responses. In addition, tremor, myoclonus, extrapyramidal rigidity, hemiballism, limb dystonia, supranuclear ophthalmoplegia, irregular dysconjugate gaze, and nystagmus can occur.[112–114] Ultimately, the patient becomes progressively paraparetic, incontinent, and bedridden with a relentless deterioration to akinetic mutism, coma, and eventually death.

The clinical syndrome in children differs only slightly from that of the adult, and is based predominantly on age and the stage of brain development at the time of HIV infection.[115–117] Typically, onset of disease following *in utero* or prenatal exposure to HIV, from 2 to 5 months, is characterized by a loss or delay in anticipated milestones. A progressive encephalopathy develops as evidenced by impaired brain growth and microcephaly.

Making a laboratory diagnosis of ADC is often difficult. The electroencephalogram generally shows slowing of basic rhythms but may have paroxysmal activity.[106–108] Examination of the CSF shows a mild mononuclear cell pleocytosis, a mildly elevated protein (40–100 mg/dl), and a usually normal or mildly reduced glucose.[97] Although an elevated IgG or IgG index may be present, oligoclonal bands are infrequent (<20% of all cases) regardless of neuropsychiatric symptomatology. HIV and HIV-specific antibodies are detected in CSF from > 50% of patients. Viral antigen often correlates with progressive disease.[98] Computerized tomography usually reveals atrophy, and magnetic resonance imaging shows both atrophy and an occasional increased signal in white matter and periventricular regions of the CNS. Neuropsychological testing is usually abnormal and may be helpful in following patients through the evolution of behavioral changes.[119–121] In all instances, care must be taken to evaluate all patients for coexistent treatable psychiatric illness or coexistent opportunistic infections.

The major pathological abnormalities in ADC are found in the cerebral white matter and, as the disease progresses, in subcortical gray structures and cerebral cortex. Diffuse pallor of myelin associated with an astrocytosis and infiltration of macrophages and lymphocytes develops with increased disease severity. Multinucleated macrophage cells, the hallmark of this disease process, are present in white matter. Foci of demyelination, glial nodules, and collections of mononuclear cell infiltrations are also found in the parenchyma or adjacent to small blood vessels (Fig. 12-2). Neuronal loss with reactive astrocytosis may be localized to subcortical structures (e.g., the caudate and putamen).[88]

Treatment of CNS disorders associated with HIV infection is directed toward the virus, the pathogenic mechanisms of disease, or concomitant opportunistic infections or neoplasms. The commonly employed antiretroviral agents 3'-azido-2',3'-dideoxythymidine (AZT), 2'3'-dideoxyinosine (ddI), and 2',3'-dideoxycytidine (ddC) may cross the blood–brain barrier and produce at least temporary clinical improvement.[122,123]

6.4. Vacuolar Myelopathy

Spinal cord disease occurs during HIV infection and is usually seen in concert with late ADC or opportunistic infections.[124–128] Signs include paraparesis and ataxia. Pathologically there is predilection for the posterior columns of the dorsal spinal cord, mimicking subacute combined degeneration. Infiltration with mononucleated and multinucleated giant cells occurs, but less frequently than in ADC. Vacuolar myelopathy is not found in children.

The disease presents as a spastic paresis of variable severity. Sensory abnormalities can occur, and progression is variable. The clinical evaluation is often complicated by the coexistence of neuropathy, debilitating dementia, and opportunistic infections in late-stage HIV disease. Subclinical HIV infection of the spinal cord may be suspected by early and serial evoked-potential tests. Progressive disease is often associated with significant gait abnormalities. The most severe forms are manifest by progressive spastic–ataxic paraparesis with urinary incontinence.

Laboratory examinations are frequently of little use in making a diagnosis of vacuolar myelopathy. The CSF findings are similar to those described for ADC. Myelography, computerized tomography, and magnetic resonance tests are usually normal.[125]

Pathologically, vacuolar myelopathy[126–128] is characterized by vacuolation in the spinal cord white matter with infiltration by macrophages. The initial lesions consist of scattered vacuoles caused by swelling of the myelin sheaths and macrophage infiltration, progressing to extensive vacuolation and secondary axonal degeneration. Vacuolar myelopathy is frequently found, in 10–30% of patients with AIDS, during postmortem examination.[128]

6.5. Neuromuscular Syndromes Associated with HIV Infection

Neuromuscular complications of HIV-1 infection are common and are usually not a result of direct HIV infection.[129–144] Indeed, these disorders are likely associated with autoimmune-mediated damage. The various syndromes are classified on the basis of anatomic involvement and pathological features of disease (Table 12-2). These include acute and chronic inflammatory polyneuropathies (Guillain–Barré syndrome or AIDP and CIDP), polymyositis, cranial neuropathies, mononeuritis multiplex, distal symmetrical polyneuropathy, and progres-

TABLE 12-2
Peripheral Neuropathies with HIV-1 Infection

Acute inflammatory demyelinating polyradiculoneuropathy (Guillain–Barré syndrome, AIDP)
Chronic progressive inflammatory demyelinating polyradiculoneuropathy (CIDP)
Mononeuritis multiplex
Distal symmetrical polyneuropathy (DSP)
Ataxic ganglioneuropathy (AGN)
Progressive inflammatory polyradiculopathy (PIP)
Polymyositis
Other
    Rod (nemaline) myopathy
    Proximal myopathy with type II muscle atrophy

sive inflammatory polyradiculopathy. Complications occur throughout the course of HIV infection.

Acute demyelinating inflammatory polyradiculoneuropathy (Guillain–Barré syndrome or AIDP) may occur at any stage of HIV infection, though it is more common during the early stages of disease.[141] The most common clinical presentation is progressive distal limb weakness associated with a mild sensory disturbance. Neurological examination shows a flaccid motor neuropathy, diminished deep tendon reflexes, and a mild glove-and-stocking sensory deficit. A CSF examination demonstrates a mild pleocytosis and HIV antigen and antibodies, which differentiates this from other forms of Guillain–Barré syndrome. Electromyography shows a segmental demyelination. Concomitant CMV inclusions may be found in Schwann cells and nerve roots in affected patients. Disease course is often unpredictable, ranging from spontaneous remission to fulminate disease and death.

Successful treatment of this disease by plasmapheresis has been reported.[143] The mechanisms of disease pathogenesis are thought to involve a postviral autoimmune demyelination involving the myelin directly or the Schwann cell. HIV-infected patients with AIDP or CIDP may have circulating antibodies reacting with normal peripheral nerve or myelin. The signs and symptoms of CIDP are similar to those of AIDP, but they differ in duration of disease. Distal symmetrical polyneuropathy is the most frequent neuropathy in AIDS and predicts a poor clinical outcome. Sensory signs and symptoms predominate. In advanced disease electrophysiological and pathological evidence of axonal degeneration are evident.

Other less common manifestations of neuromuscular disease include a progressive inflammatory polyradiculoneuropathy and polymyositis.[138–140] The former manifests as a flaccid paraplegia with sensory impairment. A CSF examination demonstrates a pleocytosis with predominant polymorphonuclear leukocytes, elevated protein, and hypoglycorrhachia. Pathological examinations show marked inflammation of nerve roots and ganglia with CMV inclusions. Treatment including acyclovir, corticosteroids, and azidothymidine is ineffective. Polymyositis is an inflammatory myopathy found during subclinical HIV infection or AIDS. Presenting signs and symptoms include muscle cramps, myalgias, weakness, and high serum creatine kinase levels. Electromyography may reveal myopathy and denervation. Corticosteroid therapy is an effective treatment during early stages of disease.[145]

7. BIOLOGY AND PATHOGENESIS OF CNS DISEASE

HIV is detected in brain tissue by a variety of techniques, including Southern blot, in situ hybridization, immunohistochemical staining, electron microscopy, and viral isolation assays.[22] Virus is expressed almost exclusively in cells of macrophage lineage (brain macrophages, microglia, and multinucleated giant cells). Up to 15% of brain macrophages express HIV-specific RNAs (Fig. 12-3). Infection of cerebral endothelial cells may also occur and lead to blood–brain barrier permeability abnormalities. No clear evidence of *in vivo* HIV infection in neurons or neuroglia (astrocytes or oligodendrocytes) is seen. This is underscored by experimental observations of explant cultures of adult human brains inoculated with HIV. Only selective viral replication in microglial cells among neurons and neuroglia is demonstrated.[146] Furthermore, patients with the most severe clinical symptoms usually have the most intense pathology and the highest levels of HIV infection in macrophages.[25]

The CNS may be affected early in the course of virus infection. Invasion by HIV into the CNS may occur at the time of seroconversion-related mononucleosis illness or during the subclinical phase of infection. Symptoms and signs of CNS disease soon after exposure to HIV include headache, encephalitis, aseptic meningitis, ataxia, and myelopathy and were listed previously. Laboratory evidence for an early invasion of HIV into the brain includes (1) inflammatory cells, cellular proteins, and oligoclonal immunoglobulin bands in CSF, (2) detectable levels of intrathecal antibodies to HIV, and (3) progeny HIV and viral antigens in CSF.[98–100] Of all the laboratory tests, the recovery of virus from CSF of neurologically asymptomatic individuals at the time of or subsequent to seroconversion most strongly supports early virus invasion of the CNS.[99,100]

An important question in the pathogenesis of CNS infection is whether productive HIV infection in brain macrophages induces disease or whether CNS disease is part of a broader metabolic dysfunction. Many investigators support a theory that low-level infection of neurons and neuroglia produces the neurological impairment associated with AIDS. Proposed theories of CNS dysfunction include a restricted noncytopathic infection of neurons and neuroglia, coexistence of opportunistic herpesviral or fungal infections, growth factor blockade mediated by gp120, CNS toxicity related to cytokine secretion, and direct neuron killing by gp120. Low-level infection in neurons and glia may occur at levels capable of inducing cellular aberrations but below the levels detectable by immunocytochemical and *in situ* hybridization assays. Numerous published reports do indicate that cultured neuronal and astroglial cells support HIV replication.[147–150] This infection may perturb their function and produce the cognitive dysfunctions characteristic of ADC. Furthermore, previous works demonstrate that low levels of CD4 mRNA are present in astrocytic cell lines susceptible to HIV.[150] This does suggest that brain cells may be susceptible to HIV infection *in vivo*. Activation of viral infection in brain cells might occur during superinfection with herpesvirus or fungal pathogens. Indeed, several DNA-group viruses, present as opportunistic pathogens in the brain, transactivate HIV LTR-directed gene expression (for example, CMV, herpes simplex virus type 1, and JC virus).[151]

Other infectious processes seen in association with neurologic disease include fungal infection and toxoplasmosis, which could further up-regulate viral replication by immune stimulation. Coinfection with multiple interacting pathogens may act in a concerted fashion to augment HIV expression and precipitate disease. Once productive infection is established, the precipitation of neurological disease may be a consequence of toxicities elicited by viral proteins. For example, gp120 may antagonize normal vasoactive intestinal peptide (VIP) function in brain tissue.[152] Previously, HIV was found to mimic VIP-binding activities. Other

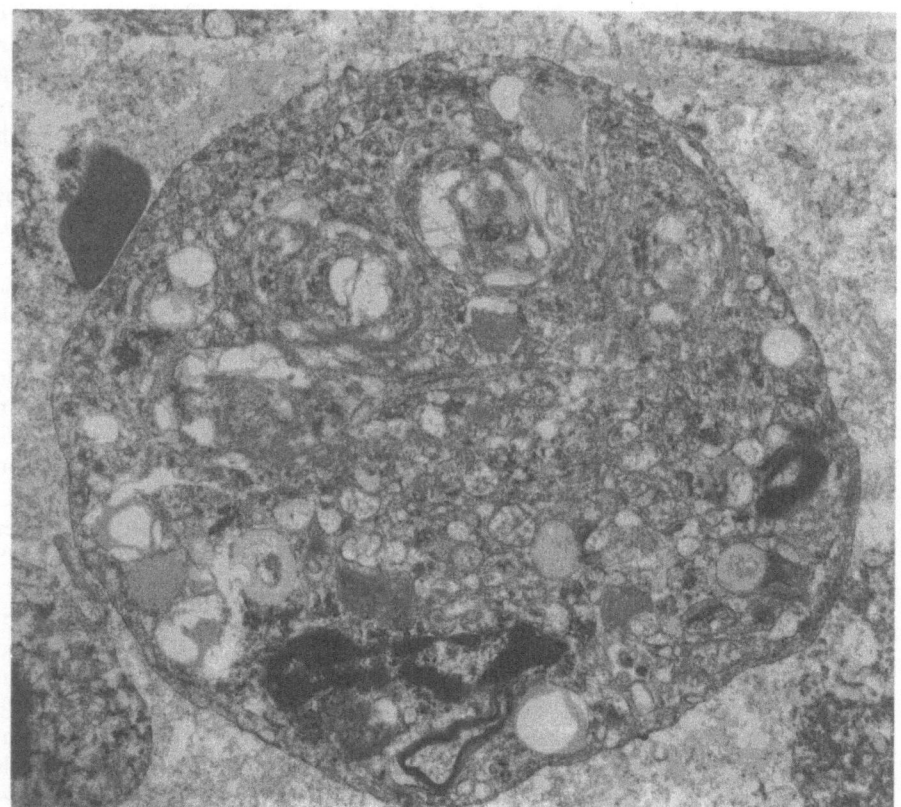

A

FIGURE 12-3. The brain macrophage is the predominant cell in the CNS supporting HIV replication. (A) Mature CNS macrophage is demonstrated with budding viral particles. (B) High-powered view of free virions in the CNS of a patient who died of the AIDS–dementia complex.

studies demonstrate a more direct neurotoxicity mediated by HIV gp120. In these experiments gp120 induced neurotoxicity by increasing free Ca^{2+} in cultured neurons.[153] The effect was prevented by Ca^{2+} channel antagonists. Recently, Sabatier et al.[154] demonstrated that intracerebroventricular injections of tat could induce toxic effects in rodents. Here, radiolabeled tat bound to rat brain synaptic nerve endings in a dose-dependent manner and induced a large depolarization modifying neural cell permeability. The neurotoxicity of tat was also demonstrated on glioma and neuroblastoma cells. Thus, HIV gene products are directly toxic to primary neuronal cells and may play pivotal roles in the pathogenesis of ADC.

Further works suggest that secretory products from HIV-infected monocytes

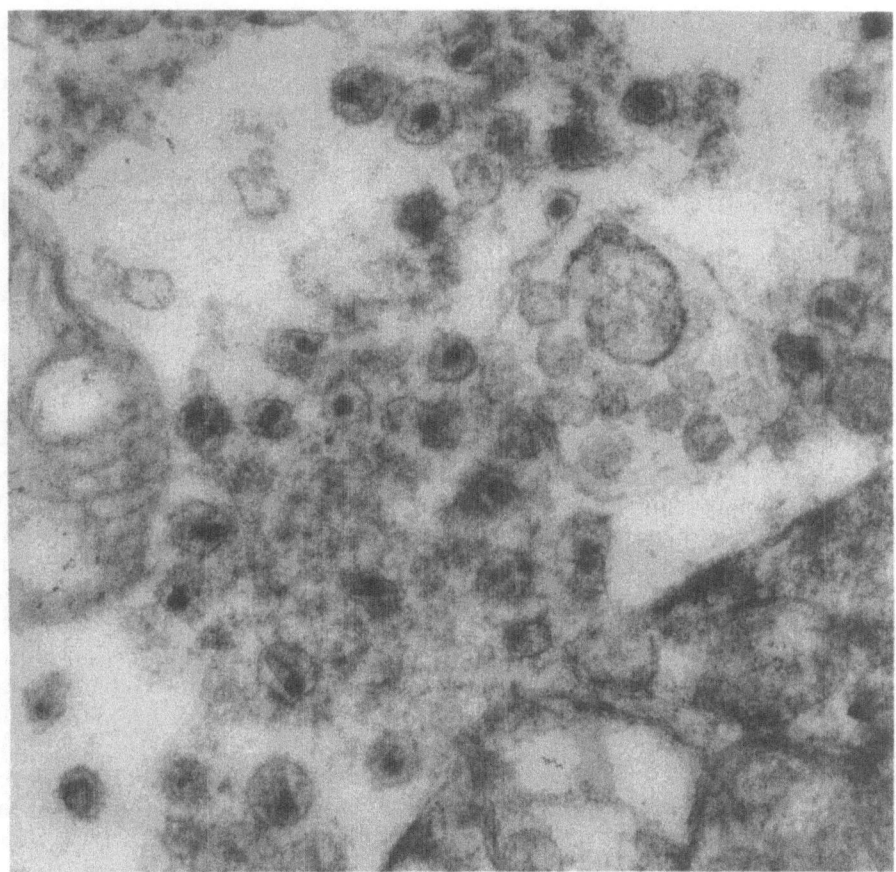

B

FIGURE 12-3. *(Cont.)*

affect neuronal cell viabilities and function. Because the macrophage is the predominant cell type productively infected by HIV in the brain, investigators theorized that brain dysfunction was related in part or whole to cell-coded toxins generated from HIV-infected macrophages.[155–157] Indeed, macrophages play a preeminent role during steady state and inflammation in the regulation of tissue function. The regulatory role of macrophages is promulgated through release of hundreds of secretory molecules made under different physiological conditions. Changes in the secretion or release of certain of these mediators could lead to disease and contribute to the symptomatology of AIDS. For example, disordered secretion of the monokine tumor necrosis factor (TNF) or cachectin has been postulated as the basis for "slim disease," a wasting syndrome unrelated to opportunistic infection commonly seen in African AIDS.[155] Enhanced release of interleukin-1 (IL-1) or TNF could explain chronic fever in AIDS patients. Both are endogenous pyrogens produced by monocytes. Chemotactic factors released from the infected monocytes could lead to infiltration of brain substance with inflammatory cells and work in concert with other cytokines that directly precipitate disease by altering brain cell function. There is contradicting evidence, however, that blood monocytes and macrophages are depleted and/or functionally impaired in AIDS.[158–161] In one study, the number of Langerhans' cells in the skin of patients with AIDS was reduced by 50%,[158] and in others no changes in Langerhans' cell numbers were observed.[159]

Published works describe monocyte bactericidal activity and chemotaxis as both normal and defective.[160,161] Monocyte secretion of IL-1 in the absence of exogenous stimuli was normal or elevated. One central problem in these analyses is that blood monocytes were the most frequently analyzed cell population. The short circulating half-life of monocytes, <30 hr, the unidirectional migration of these cells into tissues, and the low frequency of infection *in vivo* suggest that they comprise a poor study population. However, and in contrast to results obtained with circulating monocytes, tissue macrophages play a preeminent role in HIV disease. The fact that cytolytic infections of neurons or neuroglia by HIV are lacking further suggests an indirect macrophage-mediated mechanism for CNS dysfunction. Indeed, recent studies suggest that disordered secretion of one or more monokines from HIV-infected monocyte/macrophages may initiate CNS cell damage.[156] In one report HIV-infected human monocytoid cells, but not infected human lymphoid cells, released toxic factors that destroyed chick and rat neurons in culture. The monocyte-produced neurotoxins were heat stable and protease resistant and acted by way of N-methyl-D-asparate receptors. The authors hypothesized that the presence of chronically HIV-infected macrophages in brain continuously disrupt neurological function through the release of neuron-killing factors until the death of the patient. Other reports refute these observations.[157]

The origins of infection in the brain macrophage and the *in vivo* mechanisms underlying disease remain poorly understood. Infection may begin from activation or expansion of latent HIV infection of monocytes carried into the brain during cell maturation (the "Trojan horse" hypothesis).[9] Alternatively, but not mutually exclusive, is the notion that monocyte/macrophage infection occurs in brain by infection of microglia through contact with infected capillary endothelial cells or T lymphocytes. Whether cell-free virus crosses the blood–brain barrier or virus enters by way of infected monocytes or CD4+ T cells remains an open debate. Endothelial cells may provide a conduit of infection between blood and brian. That HIV enters into the CNS through capillary endothelial cells is supported by the work of Sharer *et al.*[81] Two weeks following SIV infection of macaques, animals demonstrated inflammation and multinucleated giant cells in the leptomeninges and brain parenchyma (around blood vessels). This suggests that entry is mediated through the choroid plexus, leptomeninges, and/or capillary endothelial cells.

Productive HIV replication in brain macrophages is influenced by cell maturational factors. The regulation of HIV gene expression by maturational and activation factors influences HIV gene synthesis acquired through cell activation/differentiation and may play pivotal roles in the permissive nature of the brain macrophage for HIV. For example, in the T-cell system, the mitogens phorbol myristate acetate and phytohemagglutinin positively stimulate HIV LTR-directed gene synthesis by increasing the synthesis of a cellular DNA-binding protein, NF-κB.[162] In HIV-infected persons, leukocytes can be induced to produce infectious virus only after exposure to similar T-cell mitogens. The high percentage of HIV-infected brain macrophages suggests that these cells are already stimulated and have acquired the necessary transcriptional factors for sustained efficient viral replication and possibly as mediators of disease. A schematic illustration of the neuropathogenesis of HIV infection is illustrated in Fig. 12-4.

8. CONCLUSION

Infection with HIV results in a persistent, productive infection in monocyte/macrophage and CD4+ T cells despite an often vigorous but ineffective host immune response. Ultimately, most infected individuals develop immunologic and neurological abnormalities leading to opportunistic infections and a variety of CNS disorders. The availability of treatment and preventive measures will

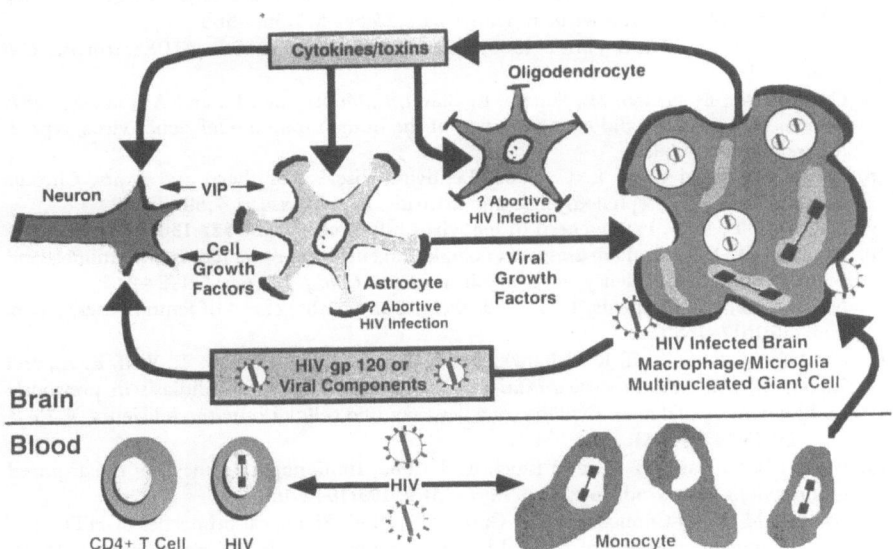

FIGURE 12-4. Illustration of the possible mechanisms underlying the neuropathogenesis of HIV infection.

ultimately rely on a better understanding of the mechanisms of viral persistence and disease pathogenesis.

ACKNOWLEDGMENTS. The authors thank members of the Military Medical Consortium for Applied Retroviral Research (MMCARR) for excellent patient management and continuing support and Ms. Victoria Hunter for excellent graphics. Dr. H. E. Gendelman is a Carter–Wallace Fellow of the Johns Hopkins University School of Public Health and Hygiene. The opinions expressed by the authors are not necessarily those of the Department of Defense or the U.S. Army.

REFERENCES

1. Barre-Sinoussi, F., Chermann, J. C., Rey, F., Nugeryre, M. T., Chamaret, S., Gruest, J., Dauget, C., Axler-Blin, C., Vezinet-Brun, F., Rouzioux, C., Rosenbaum, W., and Montagnier, L., 1983, Isolation of a T-lymphotropic retrovirus from a patient at risk for acquired immune deficiency syndrome (AIDS), *Science* **220:**868–871.
2. Popovic, M., Sarngadharan, M. G., Read, E., and Gallo, R. C., 1984, Detection, isolation, and continuous production of cytopathic retroviruses (HTLV-III) from patients with AIDS and pre-AIDS, *Science* **224:**497–500.
3. Levy, J. A., Hoffman, A. D., Kramer, S. M., Landis, J. A., Shimabukuro, J. M., and Oshiro, L. S., 1984, Isolation of lymphocytopathic retroviruses from San Francisco patients with AIDS, *Science* **225:**840–842.
4. Gonda, M. A., Wong-Staal, F., Gallo, R. C., Clements, J. E., Narayan, O., and Gilden, R. V., 1985, Sequence homology and morphologic similarity of HTLV III and visna virus, a pathogenic lentivirus, *Science* **227:**173–177.
5. Chiu, I. M., Yaniv, A., and Dahlberg, J. E., 1985, Nucleotide sequence evidence for relationship of AIDS retrovirus to lentiviruses, *Nature* **317:**366–368.
6. Rabson, A. B., and Martin, M. A., 1985, Molecular organization of the AIDS retrovirus, *Cell* **40:**477–480.
7. Guyader, M., Emerman, M., Sonigo, P., Clavel, F., Montagnier, L., and Alizon, M., 1987, Genome organization and transactivation of the human immunodeficiency virus, type 2, *Nature* **326:**662–669.
8. Narayan, O., and Cork, L. C., 1985, Lentiviral diseases of sheep and goats: Chronic pneumonia leukoencephalomyelitis and arthritis, *Rev. Infect. Dis.* **7:**89–98.
9. Haase, A. T., 1986, Pathogenesis of lentivirus infections, *Nature* **322:**130–136.
10. Narayan, O., 1990, Lentiviruses are etiological agents of chronic diseases in animals and acquired immunodeficiency syndrome in humans. *Can. J. Vet. Res.* **54:**42–48.
11. Narayan, O., and Clements, J. E., 1989, Biology and pathogenesis of lentirviruses, *J. Gen. Virol.* **70:**1617–1639.
12. Gottlieb, M. S., Schroff, R., Schanker, H.M., Weisman, J. D., Fan, P. T., Wolf, R. A., and Saxon, A., 1981, *Pneumocystis carinii* pneumonia and mucosal candidiasis in previously healthy homosexual men: Evidence of a new acquired cellular immunodeficiency, *N. Engl. J. Med.* **305:**1425–1431.
13. Bowen, D. L., Lane, H. C., and Fauci, A. S., 1985, Immunopathogenesis of the acquired immunodeficiency syndrome, *Ann. Intern. Med.* **103:**704–709.
14. Popovic, M., Read-Connole, E., and Gartner, S., 1986, Biological properties of HTLV-III/ LAV: A possible pathway of natural infection *in vivo, Ann. Inst. Pasteur. Immunol.* **137D:** 413–417.
15. Fauci, A. S., 1988, The human immunodeficiency virus: Infectivity and mechanisms of pathogenesis, *Science* **239:**617–622.

16. Gartner, S., Markovits, P., Markovitz, D. M., Kaplan, M. H., Gallo, R. C., and Popovic, M., 1986, The role of mononuclear phagocytes in HTLV-III/LAV infection, *Science* **233**:215–219.

17. Koenig, S., Gendelman, H. E., Orenstein, J. M., Dal Canto, M. C., Pezeshkpour, G. M., Yungbluth, M., Janotta, F., Aksamit, A., Martin, M. A., and Fauci, A. S., 1986, Detection of AIDS virus in macrophages in brain tissue from AIDS patients with encephalopathy, *Science* **233**:1089–1093.

18. Tschacler, E., Groh, V., Popovic, M., Mann, D. L., Konrad, K., Safai, B., Eron, L., diMarzo Veronese, F., Wolff, K., and Stingl, G., 1987, Epidermal Langerhans cells—a target for HTLV-III/LAV infection, *J. Invest. Dermatol.* **88**:233–237.

19. Salahuddin, S. Z., Rose, R. M., Groopman, J. E., Markham, P. D., and Gallo, R. C., 1986, Human T lymphotropic virus type III infection of human alveolar macrophages, *Blood* **68**: 281–284.

20. Gendelman, H. E., Orenstein, J. M., Martin, M. A., Ferrua, C., Mitra, R., Phippa, T., Wahl, L. M., Lane, H. C., Fauci, A. S., Burke, D. S., Skillman, D., and Meltzer, M. S., 1988, Efficient isolation and propagation of human immunodeficiency virus on CSF-1 stimulated human monocytes, *J. Exp. Med.* **167**:1428–1441.

21. Eilbott, D. J., Peress, N., Burger, H., LaNeve, D., Orenstein, J., Gendelman, H. E., Seidman, R,. and Weiser, B., 1989, Human immunodeficiency virus expression and replication in macrophages in the spinal cords of AIDS patients with myelopathy, *Proc. Natl. Acad. Sci. U.S.A.* **86**:3337–3341.

22. Gendelman, H. E., Orenstein, J. M., Baca, L. M., Weiser, B., Burger, H., Kalter, D. C., and Meltzer, M. S., 1989, Editorial review: The macrophage in the persistence and pathogenesis of HIV infection, *AIDS* **3**:475–495.

23. Schuitemaker, H., Kootstra, N. A., Goede, R. E. Y., De Wolf, F., Miedema, F., and Termette, M., 1991, Monocytotropic human immunodeficiency virus type 1 (HIV-1) variants detectable in all stages of HIV-1 infection lack T-cell line tropism and syncytium-inducing ability in primary T-cell culture, *J. Virol.* **65**:356–363.

24. McElrath, M. J., Steinman, R. M., and Cohn, A. Z., 1991, Latent HIV-1 infection in enriched populations of blood monocytes and T cells from seropositive patients, *J. Clin. Invest.* **87**: 27–30.

25. Price, R. W., Brew, B., Sidtis, J. Rosenblum, M., Scheck, A. C., and Cleary, P., 1988, The brain in AIDS: Central nervous system HIV-1 infection and AIDS dementia complex, *Science* **239**:586–592.

26. Cheevers, W. P., and McGuire, T. C., 1985, Equine infectious anemia virus: Immunopathogenesis and persistence, *Rev. Infect. Dis.* **7**:83–88.

27. Daniel, M. D., Letvin, N. L., King, N.W ., Kannagi, M., Sehgal, P. K., Hunt, R. D., Kanki, P. J., Essex, M., and Desrosiers, R. C., 1985, Isolation of T-cell tropic HTLV-III-like retrovirus from macaques, *Science* **228**:1201–1204.

28. Kanki, P. J., McLane, M. F., King, N. W., Jr., Letvin, N. L., Hunt, R. D., Sehgal, P., Daniel, M. D., Desrosiers, R. C., and Essex, M., 1985, Serologic identification and characterization of a macaque T-lymphotropic retrovirus closely related to HTLV-III, *Science* **228**:1199–201.

29. Letvin, N. L., Daniel, M. D., Sehgal, P. K., Desrosiers, R. C., Hunt, R. D., Waldorn, L. M., Mackey, J. J., Schmidt, D. K., Chalifaux, L. V., and King, N. W., 1985, Induction of AIDS-like disease in macaque monkeys with T-cell tropic retrovirus STLV-III, *Science* **230**:71–73.

30. Pederson, N. C., Ho, E. W., Brown, M. L., and Yamamoto, J. K., 1987, Isolation of a T-lymphotropic virus from domestic cats with an immunodeficiency-like syndrome, *Science* **235**:790–793.

31. Sonigo, P., Alizon, M., Staskus, K., Klatzmann, D., Cole, S., Danos, O., Retzel, E., Tiollois, P., Haase, A., and Wain-Hobson, S., 1985 Nucleotide sequence of the visna lentivirus: Relationship to the AIDS virus, *Cell* **42**:369–382.

32. Stephens, R. M., Casey, J. W., and Rice, N. R., 1986, Equine infectious anemia virus *gag* and *pol* genes: Relatedness to visna and AIDS virus, *Science* **231**:589–594.

33. Harper, M. E., Marselle, L. M., Gallo, R. C., and Wong-Staal, F., 1986, Detection of lymphocytes expressing human T-lymphotropic virus type III in lymph nodes and peripheral blood from infected individuals by *in situ* hybridization, *Proc. Natl. Acad. Sci. U.S.A.* **83:** 772–776.

34. Narayan, O., Kennedy-Stoskopf, S., and Zink, C. M., 1988, Lentivirus–host interactions: Lessons from visna and caprine arthritis-encephalitis viruses, *Ann. Neurol.* **23**(Suppl.):S95–S100.

35. Gendelman, H. E., Leonard, J. M., Dutko, F. J., Koenig, S., Khillan, J. S., and Meltzer, M. S., 1988, Immunopathogenesis of human immunodeficiency virus infection in the central nervous system, *Ann. Neurol.* **23**(Suppl.):S78–S81.

36. Clements, J. E., Pedersen, F. S., Narayan, O., and Haseltine, W. A., 1980, Genomic changes associated with antigenic variation of visna virus during persistent infection, *Proc. Natl. Acad. Sci. U.S.A.* **77:**4454–4458.

37. Narayan, O., Clements, J. E., Kennedy-Stoskopf, S., and Royal, R., 1987, Mechanisms of escape of visna lentiviruses from immunological control, *Contrib. Microbiol. Immunol.* **8:** 60–76.

38. Kennedy-Stoskopf, S., and Narayan, O., 1986, Neutralizing antibodies to visna lentivirus: Mechanism of action and possible role in virus persistence, *J. Virol.* **59:**37–44.

39. Francis, D. P., and Petricciani, J. C., 1985, The prospects for and pathways toward a vaccine for AIDS, *N. Engl. J. Med.* **313:**1586–1590.

40. Hahn, B. H., Shaw, G. M., Taylor, M. E., Redfield, R. R., Markham, P. D., Salahuddin, S. Z., Wong-Staal, F., Gallo, R. G., Parks, E. S., and Parks, W. P., 1986, Genetic variation in HTLV-III-LAV over time in patients with AIDS or at risk for AIDS, *Science* **232:**1548–1553.

41. Ho, D. D., Sarnagaharan, M. G., Hirsch, M. S., Schooley, R. T., Rota, T. R., Kennedy, R. C., Chanh, T. C., and Sato, V. L., 1987, Human immunodeficiency virus neutralizing antibodies recognize several conserved domains on the envelope glycoproteins, *J. Virol.* **61:** 2024–2028.

42. Bouillant, A. M. P., and Becker, S. A. W., 1984, Ultrastructural comparison of oncovirinae (type C), spumavirinae, and lentivirinae: Three subfamilies of retroviridae found in farm animals, *J. Natl. Cancer Inst.* **72:**1075–1084.

43. Dahlberg, J. E., Gaskin, J. M., and Perk, K., 1981, Morphological and immunological comparison of caprine arthritis encephalitis and ovine progressive pneumonia viruses, *J. Virol.* **39:**914–919.

44. Gonda, M. A., Charman, H. P., Walker, J. L., and Coggins, L., 1978, Scanning and transmission electron microscopic study of equine infectious anemia virus, *Am. J. Vet. Res.* **39:**731–740.

45. Munn, R. J., Marx, P. A., Yamamoto, J. K., and Gardner, M. B., 1985, Ultrastructural comparison of the retroviruses associated with human and simian acquired immunodeficiency syndromes, *Lab. Invest.* **53:**194–199.

46. Wain-Hobson, S., Sonigo, P., Danos, O., Cole, S., and Alizon, M., 1985, Nucleotide sequence of the AIDS virus, LAV, *Cell* **40:**9–17.

47. Ratner, L., Haseltine, W., Patarca, R., Livak, K. J., Starcich, B., Jacobs, S. F., Doran, E. R., Rafalski, J. A., Whitehorn, E. A., Baumeister, K., Ivanoff, L., Petteway, S. R., Pearson, M. L., Lautenberger, J. A., Papis, T. S., Ghrayeb, J., Chang, N. T., Gallo, R. C., and Wong-Staal, F., 1985, Complete nucleotide sequence of the AIDS virus, HTLV-III, *Nature* **313:**277–284.

48. Starcich, B., Ratner, L., Josephs, S. F., Okamoto, T., Gallo, R. C., and Wong-Staal, F., 1985, Characterization of long terminal repeat sequences of HTLV-III, *Science* **227:**538–540.

49. Muesing, M. A., Smith, D. H., Cabradilla, C. D., Benton, C. V., Lasky, L. A., and Capon, D. J., 1985, Nucleic acid structure and expression of the human AIDS/lymphadenopathy retrovirus, *Nature* **313:**450–458.

50. Sanchez-Pescador, R., Power, M. D., Barr, P. J., Steimer, K. S., Stemfeien, M. M., Brown-Shimer, S. L., Gee, W. W., Bernard, A., Randolph, A., Levy, J. A., Dina, D., and Luciw, P. A.,

1985, Nucleotide sequence and expression of an AIDS-associated retrovirus (ARV-2), *Science* **227**:484–492.

51. Schnittman, S. M., Psallidopoulos, M. C., Lane, H.C ., Thompson, L., Baseler, M., Massari, F., Fox, C. H., Salzman, N. P., and Fauci, A. S., 1989, The reservoir for HIV-1 in peripheral blood is a T cell that maintains expression of CD4, *Science* **245**:305–308.

52. Ho, D. D., Moudgil, T., and Alam, M., 1989, Quantitation of human immunodeficiency virus type 1 in the blood of infected persons, *N. Engl. J. Med.* **321**:1621–1625.

53. Varmus, H., 1988, Retroviruses, *Science* **240**:1427–1435.

54. Weller, S. K., Joy, A. E., and Temin, H. M., 1980, Correlation between cell killing and massive second-round superinfection by members of some subgroups of avian leukosis virus, *J. Virol.* **33**:494–506.

55. Keshet, E., and Temin, H. M., 1979, Cell killing by spleen necrosis virus is correlated with a transient accumulation of spleen necrosis virus DNA, *J. Virol.* **31**:376–388.

56. Zagury, D., Bernard, J., Leonard, R., Cheynier, R., Feldman, M., Sarin, P. S., and Gallo, R. C., 1986, Long-term cultures of HTLV-III-infected T cells: a model of cytopathology of T-cell depletion in AIDS, *Science* **231**:850–853.

57. Klatzmann, D., and Gluckman, J. C., 1986, HIV infection: Facts and hypotheses, *Immunol. Today* **7**:291–296.

58. Maddon, P. J., Dalgleish, A. G., McDougal, J. S., Clapham, P. R., Weiss, R. A., and Axel, R., 1986, The T4 gene encodes the AIDS virus receptor and is expressed in the immune system and the brain, *Cell* **47**:333–348.

59. McDougal, J. S., Kennedy, M. S., Sligh, J. M., Cort, S. P., Mawle, A., and Nicholson, J. K. A., 1986, Binding of HTLV-III/LAV to T4+ T cells by a complex of the 110K viral protein and the T4 molecule, *Science* **231**:382–385.

60. Dalgleish, A. G., Beverley, P. C. L., Clapham, P.R., Crawford, D. H., Greaves, M. F., and Weiss, R. A., 1984, The CD4(T4) antigen is an essential component of the receptor for the AIDS retrovirus, *Nature* **312**:763–767.

61. Klatzmann, D., Champagne, E., Chamaret, S., Gruest, J., Guefard, D., Hersend, T., Gluckman, J-C., and Montagnier, L., 1984, T-lymphocyte T4 molecule behaves as the receptor for human retrovirus LAV, *Nature* **312**:767–768.

62. McDougal, J. A., Mawle, A., Cort, S. P., Nicholson, J. K. A., Cross, G. D., Scheppler-Campbell, J. A., Hicks, D., and Sligh, J., 1985, Role of T cell activation and expression of the T4 antigen, *J. Immunol.* **135**:3151–3162.

63. Ho, D. D., Pomerantz, R. J., and Kaplan, J. C., 1987, pathogenesis of infection with human immunodeficiency virus, *N. Engl. J. Med.* **317**:278–286.

64. DeRossi, A., Franchini, G., Aldonvini, A., del Mistro, A., Chieco-Bianchi, L., Gallo, R. C., and Wong-Staal, F., 1986, Differential response to the cytopathic effects of human T-cell lymphotropic virus type III (HTLV-III)-superinfection in T4+ (helper) and T8+ (suppressor) T-cell clones transformed by HTLV-1, *Proc. Natl. Acad. Sci. U.S.A.* **83**:4297–4301.

65. Ziegler, J. L., and Stites, D. P., 1986, Hypothesis: AIDS is an autoimmune disease directed at the immune system and triggered by a lymphotropic retrovirus, *Clin. Immunol. Immunopathol.* **41**:305–314.

66. Stricker, R. B., McHugh, T. M., Moody, D. J., Morrow, W. J. W., Stites, D. P., Shuman, M. A., and Levy, J. A., 1987, An AIDS-related cytotoxic autoantibody reacts with a specific antigen on stimulated CD4+ T cells, *Nature* **327**:710–713.

67. Schrier, R. D., Nelson, J. A., and Oldstone, M. B. A., 1985, Detection of human cytomegalovirus in peripheral blood lymphocytes in a natural infection, *Science* **230**:1048.

68. Lane, H. C., Depper, J. M., Greene, W. C., Gail Whalen, B. S., Waldmann, T. A., and Fauci, A. S., 1985, Qualitative analysis of immune function in patients with the acquired immunodeficiency syndrome evidence for a selective defect in soluble lantigen recognition, *N. Engl. J. Med.* **313**:79–85.

69. Lane, H. C., Masur, H., Edgar, L. C., Whalen, G., Rook, A. H., and Fauci, A. S., 1983, Abnormalities of B-cell activation and immunoregulation in patients with the acquired immunodeficiency syndrome, *N. Engl. J. Med.* **309**:453–458.

70. Margolick, D. J., Volkman, D. J., Folks, T. M., and Fauci, A. S., 1987, Amplification of HTLV-III/LAV infection by antigen-induced activation of T cells and direct suppression by virus of lymphocyte blastogenic responses, *J. Immunol.* **138**:1719–1723.

71. Schnittman, S. M., Lane, H. C., Higgins, S. E., Folks, T., and Fauci, A. S., 1986, Direct polyclonal activation of human B lymphocytes by the acquired immune deficiency syndrome virus, *Science* **233**:1083–1086.

72. Narayan, O., and Zink, C., 1987, Role of macrophages in lentivirus infections, in: *Immunodeficiency Disorders and Retroviruses* (K. Perk, ed.), Academic Press, New York.

73. Orenstein, J. M., Meltzer, M. S., Phipps, T., and Gendelman, H. E., 1988, Cytoplasmic assemble and accumulation of human immunodeficiency virus types 1 and 2 in recombinant human colony-stimulating factor-1-treated human monocytes: An ultrastructural study, *J. Virol.* **62**:2578–2586.

74. Ringler, D. J., Hunt, R. D., Desrosiers, R. C., Daniel, M. D., Chalifoux, L. V., and King, N. W., 1988, Simian immunodeficiency virus-induced meningoencephalitis: Natural history and retrospective study, *Ann. Neurol.* **23**:S101–S107.

75. Gendelman, H. E., Narayan, O., Kennedy-Stoskopf, S., Kennedy, P. G. E., Ghotbi, Z., Clements, J. E., Stanley, J., and Pezeshkpour, G. H., 1986, Tropism of sheep lentiviruses for monocytes: Susceptibility to infection and virus gene expression increases during maturation of monocytes to macrophages, *J. Virol.* **58**:67–74.

76. Gendelman, H. E., Narayan, O., Molineaux, S., Clements, J. E., and Ghotbi, Z., 1985, Slow persistent replication of lentiviruses: Role of macrophages and macrophage precursors in bone marrow, *Proc. Natl. Acad. Sci. U.S.A.* **82**:7086–7090.

77. Gendelman, H. E., Narayan, O., Kennedy-Stoskopf, S., Clements, J. E., and Pezeshkpour, G. H., 1984, Slow virus macrophage interactions: Characterization of a transformed cell line of sheep alveolar macrophages that express a marker for susceptibility to ovine–caprine lentivirus infections, *Lab. Invest.* **51**:547–555.

78. Stoler, M. H., Eskin, T. A., Benn, S., Angerer, R. C., and Angerer, L. M., 1986, Human T-cell lymphotropic virus type III infection of the central nervous system—a preliminary *in situ* analysis, *J.A.M.A.* **256**:2360–2364.

79. Wiley, C. A., Schrier, R. D., Nelson, J. A., Lampert, P. W., and Oldstone, M. B.A., 1986, Cellular localization of human immunodeficiency virus infection within the brains of acquired immune deficiency syndrome patients, *Proc. Natl. Acad. Sci. U.S.A.* **83**:7089–7093.

80. Vazeux, R., Brousse, N., Jarry, A., Henin, D., Marche, C., Vedrenne, C., Mikol, J., Wolff, M., Michon, C., Rozenbaum, W., Bureau, J.-F., Montagnier, L., and Brahic, M., 1987, AIDS subacute encephalitis; identification of HIV-infected cells, *Am. J. Pathol.* **126**:403–410.

81. Michaels, J., Sharer, L. R., and Epstein, L. G., 1988, Human immunodeficiency virus type 1 (HIV-1) infection of the nervous system: a review, *Immunodef. Rev.* **1**:71–104.

82. Gabuzda, D. H., Ho, D. D., de al Monte, S. M., Hirsch, M. S., Rota, T. R., and Sobel, R. A., 1986, Immunohistochemical identification of HTLV-III antigen in brains of patients with AIDS, *Ann. Neurol.* **20**:289–295.

83. Chayt, K. J., Harper, M. E., Marselle, L. M., Lewin, E. B., Rose, R. M., Oleske, J. M., Epstein, L. G., Wong-Staal, F., and Gallo, R. C., 1986, Detection of HTLV-III RNA in lungs of patients with AIDS and pulmonary involvement, *J.A.M.A.* **256**:2356–2359.

84. Le Tourneau, A., Audouin, J., Diebold, J., Marche, C., Tricottet, V., and Reynes, M., 1986, LAV-like viral particles in lympho node germinal centers in patients with the persistent lymphadenopathy syndrome and the acquired immunodeficiency syndrome-related complex: An ultrastructural study of 30 cases, *Hum. Pathol.* **17**:1047–1053.

85. Bender, B. S., Davidson, B. L., Kline, R., Brown, C., and Quinn, T., 1988, Role of the mononuclear phagocyte system in the immunopathogenesis of human immunodeficiency

virus infection and the acquired immunodeficiency syndrome, *Rev. Infect. Dis.* **10:**1142–1154.

86. Navia, B. A., Jordan, B. D., and Price, R. W., 1986, The AIDS dementia complex: I. Clinical features, *Ann. Neurol.* **19:**517–524.

87. Navia, B. A., Cho, E.-S., Petito, C. K., and Price, R. W., 1986, The AIDS dementia complex: II. Neuropathology, *Ann. Neurol.* **19:**525–535.

88. de la Monte, S. M., Ho, D. D., Schooley, R. T., Hirsh, M. S., and Richardson Jr., E. P., 1987, Subacute encephalomyelitis of AIDS and its relation to HTLV-III infection, *Neurology* **37:** 562–569.

89. Levy, R. M., Bredesen, D. E., and Rosenblum, M. L., 1988, Opportunistic central nervous system pathology in patients with AIDS, *Ann. Neurol.* **23**(Suppl.):S7–S12.

90. Snider, W. D., Simpson, M. D., Nielsen, S., Gold, J. W. M., Metroka, C. E., and Posner, J. B., 1983, Neurological complications of acquired immune deficiency syndrome: Analysis of 50 patients, *Ann. Neurol.* **14:**403–418.

91. Bredesen, D. E., Lipkin, W. I., and Messing, R., 1983, Prolonged recurrent aseptic meningitis with prominent cranial nerve abnormalities: A new epidemic in gay men? *Neurology* **33**(Suppl.):85.

92. Cooper, D. A., Gold, J., Maclean, P., Donovan, B., Finlayson, R., Barnes, T. G., Michelmore, H. M., Brooke, P., and Penny, R., 1985, Acute AIDS retrovirus infection: Definition of a clinical illness associated with seroconversion, *Lancet* **1:**537–540.

93. Hollander, H., and Stringari, S., 1987, Human immunodeficiency virus-associated meningitis: Clinical course and correlations, *Am. J. Med.* **83:**813–816.

94. Ho, D. D., Sarngadharan, M. G., Resnick, L., DiMarzo-Veronese, F., Rota, T. R., and Hirsch, M. S., 1985, Primary human T-lymphotropic virus type III infection, *Ann. Intern. Med.* **103:** 880–883.

95. Griffin, D. E., McArthur, J. C., and Cornblath, D. R., 1991, Neopterin and interferon-gamma in serum and cerebrospinal fluid of patients with HIV associated neurologic disease, *Neurology* **4:**69–74.

96. Bredesen, D. E., and Messing, R., 1983, Neurological syndromes heralding the acquired immune deficiency syndrome, *Ann. Neurol.* **14:**141.

97. McArthur, J. C., Cohen, B. A., Farzedegan, H., Cornblath, D. R., Selnes, O. A., Ostrow, D., Johnson, R. T., Phair, J., and Polk, B. F., 1988, Cerebrospinal fluid abnormalities in homosexual men with and without neuropsychiatric findings, *Ann. Neurol.* **23**(Suppl.): S34–S37.

98. Goudsmit, J., Wolters, E. C., Bakker, M., Smit, L., van der Noordaa, J., Hische, E. A. H., Tutuarima, J. A., and van der Helm, H. J., 1986, Intrathecal synthesis of antibodies to HTLV-III in patients without AIDS or AIDS related complex, *Br. Med. J.* **292:**1231–1234.

99. Levy, J. A., Shimabukuro, J., Hollander, H., Mills, J., and Kaminsky, L., 1985, Isolation of AIDS-associated retroviruses from cerebrospinal fluid and brain of patients with neurological symptoms, *Lancet* **2:**586–588.

100. Ho, D. D., Rota, T. R., Schooley, R. T., 1985, Isolation of HTLV-III from cerebrospinal fluid and neural tissue of patients with neurologic syndromes related to the acquired immunodeficiency syndrome, *N. Engl. J. Med.* **313:**1493–1497.

101. Levy, R. M., Bredesen, D. E., and Rosenblum, M. C., 1985, Neurological manifestations of the acquired immunodeficiency syndrome (AIDS): Experience at UCSF and review of the literature, *J. Neurosurg.* **62:**475–495.

102. Kieburtz, K., and Schiffer, R. B., 1989, Neurologic manifestations of human immunodeficiency virus infections, *Neurol. Clin.* **7:**447–468.

103. Price, R. W., Sidtis, J., and Rosenblum, M., 1988, The AIDS dementia complex: Some current questions, *Ann. Neurol.* **23**(Suppl.):S27–S33.

104. Berger, J. R., Moskowitz, L., and Fishl, M., 1984, The neurologic complications of AIDS: Frequently the initial manifestation, *Neurology* **34**(Suppl. 1):134–135.

105. Navia, B. A., and Price, R. W., 1987, The acquired immunodeficiency syndrome dementia complex as the presenting or sole manifestation of human immunodeficiency virus infection, *Arch. Neurol.* **44:**65–69.

106. Parisi, A., Strosselli, M., DiPerri, G., Silvano, C., Minoli, L., Bono, G., Moglia, A., and Nappi, G., 1989, Electroencephalography in the early diagnosis of HIV-related subacute encephalitis: analysis of 185 patients. *Clin EEG* **20:**1–5.

107. Gabuzda, D. H., Levy, S. R., and Chiappa, K. H., 1988, Electroencephalography in AIDS and AIDS-related complex, *Clin Electroencephalogr.* **19:**1–6.

108. Parisi, A., Stossell, M., Pan, A., Maserati, R., and Minoli, L., 1991, HIV-related encephalitis presenting as convulsant disease, *Clin. Electroencephalogr.* **22:**1–4.

109. Cummings, J. L., and Benson, D. F., 1984, Subcortical dementia—review of an emerging concept, *Arch. Neurol.* **41:**874–879.

110. Cummings, J. L., and Benson, F., 1983, Dementia: Definition, prevalence, classification, and approach to diagnosis, in: *Dementia: A Clinical Approach*, Butterworths, Stoneham, MA, pp. 7–10.

111. Fraser, M., 1987, *Dementia: Its Nature and Management*, Bath Press, Avon, Great Britain.

112. Nath, A., Jancovic, J., and Pettigrew, L. C., 1987, Movement disorders and AIDS, *Neurology* **37:**37–41.

113. Metzer, W. S., 1987, Movement disorders with AIDS encephalopathy: A case report, *Neurology* **37:**1438.

114. Gabuzda, D. H., and Hirsch, M. S., 1987, Neurologic manifestations of infection with human immunodeficiency virus. Clinical features and pathogenesis, *Ann. Intern. Med.* **107:** 383–391.

115. Epstein, L. G., Sharer, L. R., and Goudsmit, J., 1988, Neurological and neuropathological features of human immunodeficiency virus infection in children, *Ann. Neurol.* **23**(Suppl.): S19–S23.

116. Epstein, L. G., Sharer, L. R., Joshi, V. V., Fojas, M. M., Koenigsberger, M. R., and Oleske, J. M., 1985, Progressive encephalopathy in children with acquired immune deficiency syndrome, *Ann. Neurol.* **17:**488–496.

117. Iannetti, P., Falconieri, P., and Imperato, C., 1989, Acquired immune deficiency syndrome in childhood. Neurological aspects, *Child. Nerv. Syst.* **5:**281–287.

118. Pumarola-Sune, T., Navia, B. A., Cordon-Cardo, C., Cho, E.-S., and Price, R. W., 1987, HIV antigen in the brains of patients with the AIDS dementia complex, *Ann. Neurol.* **21:** 490–496.

119. Grant, I., Atkinson, J. H., Hesselink, J. R., Kennedy, C. J., Richman, D. D., Spector, S. A., and McCutchan, J. A., 1987, Evidence for early CNS involvement in AIDS and other HIV infections, *Ann. Intern. Med.* **107:**828.

120. Saykin, A. J., Janssen, R. S., Sprehn, G. C., Kaplan, J. E., Spira, T. J., and Weller, P., 1984, Neuropsychological dysfunction in AIDS-related complex, *Neurology* **37**(Suppl. 1):374.

121. Rubinow, D. R., Berrettini, C. H., Brouwwers, P., and Lane, H. C., 1988, Neuropsychiatric consequences of AIDS, *Ann. Neurol.* **23**(Suppl.):S24–26.

122. Yarchoan, R., Thomas, R. V., Grafman, J., Wichman, A., Dalakas, M., McAtee, N., Berg, G., Fishl, M., Perno, C. F., Klecker, R. W., Buchbinder, A., Tay, S., Larson, S. M., Myers, C. E., and Broder, S., 1988, Long-term administration of 3'-azido-2',3'- dideoxythymidine to patients with AIDS-related neurological disease, *Ann. Neurol.* **23**(Suppl.):S82–S87.

123. Johnson, V. A., and Hirsch, M. S., 1990. New developments in antiretroviral drug therapy for human immunodeficiency virus infections, in: *AIDS Clinical Review 1990* (P. Volberding and M. A. Jacobson, eds.), Marcel Dekker, New York, pp. 235–272.

124. Johnson, R. T., and McArthur, J. C., 1987, Myelopathies and retroviral infections, *Ann. Neurol.* **21:**113–116.

125. Griffin, J. W., McArthur, J. C., and Cornblath, D. R., 1990, Peripheral nerve and spinal cord disease in human retrovirus infections, *Curr. Opin. Neurol. Neurosurg.* **3:**697–703.

126. Petito, C. K., Navia, B. A., Cho, E.-S., and Jordan, B. D., 1985, Vacuolar myelopathy pathologically resembling subacute combined degeneration in patients with acquired immunodeficiency syndrome (AIDS), *N. Engl. J. Med.* **312**:874–879.

127. Sharer, L. R., Epstein, L. G., Cho, E. S., and Petito, C. D., 1986, HTLV-III and vacuolar myelopathy, *N. Engl. J. Med.* **315**:62–63.

128. Petito, C., 1988, Review of central nervous system pathology in human immunodeficiency virus infection, *Ann. Neurol.* **23**(Suppl.):S54–S57.

129. Miller, R. G., Kiprov, D. D., Parry, G., and Bredesen, D. E., 1988, Peripheral nervous system dysfunction in acquired immunodeficiency syndrome, in: *AIDS and the Nervous System* (M. L. Rosenblum, R. M. Levy, and D. E. Bredesen, eds.), Raven Press, New York, pp. 65–78.

130. Roman, G. C., 1987, Retrovirus associated myelopathies, *Arch. Neurol.* **44**:659–663.

131. Dalakas, M. C., and Pezeshkpour, G. H., 1988, Neuromuscular diseases associated with human immunodeficiency virus infection, *Ann. Neurol.* **23**(Suppl.):S38–S48.

132. Lipkin, W. I., Parry, G., Kiprov, D., and Abrams, D., 1985, Inflammatory neuropathy in homosexual men with lymphadenopathy, *Neurology* **35**:1479–1483.

133. Eidelberg, D., Sotrel, A., Vogel, H., Walker, P., Kleefield, J., and Crumpacker III, C. S., 1986, Progressive polyradiculopathy in acquired immune deficiency syndrome, *Neurology* **36**:912–916.

134. So, Y. T., Holtzman, D. M., Abrams, M. I., and Olney, R. K., 1988, Peripheral neuropathy associated with acquired immunodeficiency syndrome, *Arch. Neurol.* **48**:945–948.

135. Cohen, J. A., and Laudenslager, M., 1989, Autonomic nervous system involvement in patients with human immunodeficiency virus infection, *Neurology* **39**:1111–1112.

136. Parry, G. J., 1988, Peripheral neuropathies associated with human immunodeficiency virus infection, *Ann. Neurol.* **23**(Suppl.):S49–S53.

137. Gabbai, A. A., Schmidt, B., Costelo, A., Oliveira, A. S. B., and Lima, J. G. C., 1990, Muscle biopsy in AIDS and ARC: Analysis of 50 patients, *Muscle Nerve* **13**:541–544.

138. Cornblath, D. R., McArthur, J. C., and Griffin, J. W., 1986, The spectrum of peripheral neuropathies in HTLV-III infection, *Muscle Nerve* **9**:76.

139. Cornblath, D. R., McArthur, J. C., Kennedy, G. E., Witte, A. S., and Griffin, J. W., 1987, Inflammatory demyelinating peripheral neuropathies associated with human T-lymphotropic virus type III infection, *Ann. Neurol.* **21**:32–40.

140. Chaunu, M. P., Ratinahirana, H., Raphael, M., Henin, D., Leport, C., Brun-Vezinet, F., Leger, J.-M., Brunet, P., and Hauw, J.-J., 1989, The spectrum of changes on 20 nerve biopsies in patients with HIV infection, *Muscle Nerve* **12**:451–459.

141. Vendrell, J., Heredia, C., Pujol, M., Asjo, B., and Fenyo, E. M., 1987, Guillain–Barré syndrome associated with seroconversion for anti-LAV/HTLV-III, *Neurology* **37**:544.

142. Rance, N. E., McArthur, J. C., Cornblath, D. R., Landstrom, D. L., Griffin, J. W., and Price, D. L., 1988, Gracile tract degeneration in patients with sensory neuropathy and AIDS, *Neurology* **38**:265–271.

143. Miller, R. G., Parry, G., Lang, W., Lippert, R., and Kiprov, D., 1985, AIDS-related inflammatory polyradiculoneuropathy: Prediction of response to plasma exchange with electrophysiologic testing, *Muscle Nerve* **8**:626.

144. Miller, R. G., Storey, J. R., and Greco, C. M., 1990, Ganciclovir in the treatment of progressive AIDS-related polyradiculopathy, *Neurology* **40**:569–574.

145. Dalakas, M. C., Pezeshkpour, G. H., Gravell, M., and Sever, J. L., 1986, Polymyositis associated with AIDS retrovirus, *J.A.M.A.* **256**:2381–2383.

146. Watkins, B. A., Dorn, H. H., Kelly, W. B., Armstrong, R. C, Potts, B., Michaels, F., Kufta, C. V., and Dubois-Dalcq, M., 1990, Specific tropism of HIV-1 for microglial cells in primary human brain cultures, *Science* **249**:549–553.

147. Chiodi, F., Fuerstenberg, S., Gidlund, M., Asjö, B., and Fenyö, E. M., 1987, Infection of

brain-derived cells with the human immunodeficiency virus, *J. Virol.* **61**:1244–1247.

148. Popovic, M., Mellert, W., Erfle, V., and Gartner, S., 1988, Role of mononuclear phagocytes and accessory cells in human immunodeficiency virus type I infection of the brain, *Ann. Neurol.* **23**(Suppl.):S74–S77.

149. Cheng-Mayer, C., Rutka, J. T., Rosenblum, M. L., McHugh, T., Stites, D. P., and Levy, J. A., 1987, Human immunodeficiency virus can productively infect cultured human glial cells, *Proc. Natl. Acad. Sci. U.S.A.* **84**:3526–3530.

150. Funke, I., Hahn, A., Rieber, E. P., Weiss, E., and Reithmuller, G., 1987, The cellular receptor (CD4) of the human immunodeficiency virus is expressed on neurons and glial cells in human brain, *J. Exp. Med.* **165**:1230–1235.

151. Gendelman, H. E., Phelps, W., Fiegenbaum, L., Adachi, A., Ostrove, J. M., Howley, P. M., Khoury, G., Ginsberg, H. S., and Martin, M. A., 1986, Transactivation of the human immunodeficiency virus long terminal repeat sequence by DNA viruses, *Proc. Natl. Acad. Sci. U.S.A.* **83**:9759–9763.

152. Brenneman, D. E., Westbrook, G. L., Fitzgerald, S. P., Ennist, D. L., Elkins, K. L., Ruff, M. R., and Pert, C. B., 1989, Neuronal cell killing by the envelope protein of HIV and its prevention by vasoactive intestinal peptide, *Nature* **335**:639–642.

153. Dreyer, E. B., Kaiser, P. K., Offermann, J. T., and Lipton, S. A., 1990, HIV-1 coat protein neurotoxicity prevented by calcium channel antagonists, *Science* **248**:364–367.

154. Sabatier, J.-M., Vives, E., Mabrouk, K., Benjouad, A., Rochat, H., Duval, A., Hue, B., and Bahroui, E., 1991, Evidence for neurotoxic activity of tat from human immunodeficiency virus type 1, *J. Virol.* **65**:961–967.

155. Serwadda, D., Mugerwa, R. D., Sewankambo, N., Lwegaba, A., Carswell, J. W., Kirya, G. B., Bayley, A. C., Downing, R. G., Tedder, R. S., Clayden, S. A., Weiss, R. A., and Dalgleish, A. G., 1985, Slim disease: A new disease in Uganda and its association with HTLV-III infection, *Lancet* **2**:849–852.

156. Giulian, D., Vaca, K., and Noonan, C. A., 1991, Secretion of neurotoxins by mononuclear phagocytes infected with HIV-1, *Science* **250**:1593–1596.

157. Bryant, H., Burgess, S., Gendelman, H. E., Meltzer, M. S., Holaday, J., and Berton, E., 1991, Neuronotropic activity associated with monocyte growth factors and products of stimulated monocytes, in: *Peripheral Signalling of the Brain in Neuroimmune and Cognitive Function*, (R. C. A. Frederickson, ed.), Hogrefe and Huber, Toronto, pp. 83–99.

158. Belsito, D. V., Sanchez, M. R., Baer, R. L., Valentine, F., and Thorbecke, G. J., 1984, Reduced Langerhans' cell Ia antigen and ATPase activity in patients with the acquired immunodeficiency syndrome, *N. Engl. J. Med.* **310**:1279–1282.

159. Kanitakis, J., Marchacd, C., Su, H., Thivolet, J., Zambruno, G., Schmitt, D., and Gazzolo, L., 1989, Immunohistochemical study of normal skin of HIV-1-infected patients shows no evidence of infection of epidermal Langerhans' cells by HIV, *AIDS Res. Hum. Retrovir.* **5**: 293–302.

160. Roy, S., and Wainberg, M. A., 1988, Role of the mononuclear phagocyte system in the development of acquired immunodeficiency syndrome (AIDS), *J. Leuk. Biol.* **43**:91–97.

161. Pauza, C. D., 1988, HIV Persistence in monocytes leads to pathogenesis and AIDS, *Cell. Immunol.* **112**:414–419.

162. Nabel, G., and Baltimore, D., 1987, An inducible transcription factor activates expression of human immunodeficiency virus in T cells, *Nature* **326**:711–713.

13

Alphaviruses, Flaviviruses, and Bunyaviruses

DIANE E. GRIFFIN

1. INTRODUCTION

The alphaviruses, flaviviruses, and bunyaviruses constitute three different families of enveloped RNA viruses that share the ability to be transmitted by insect vectors. They can be classified into a broader category of arthropod-borne viruses or arboviruses, which are able to replicate in both their vertebrate and invertebrate hosts. Some members of each of these families can also cause central nervous system (CNS) infection. Infection is usually initiated by subcutaneous or intravenous inoculation of virus into vertebrates by injection of infected saliva from the insect vector. For maintenance of infection in nature the vertebrate host must develop a viremia of sufficient magnitude to infect the insect vector during a blood meal. Because of a low-level viremia man is often a dead-end host, infected only when vector populations are high, and unimportant to maintenance of virus in nature.

Over 20 arboviruses cause CNS disease. Because of the need for vector transmission, all are geographically restricted, and disease has a distinct seasonal distribution. Therefore, in any one part of the world only a few of these viruses are found. Many are named for the geographical site of the original virus isolation, and individual virus groups have many distinctive characteristics.[1]

DIANE E. GRIFFIN • Departments of Medicine and Neurology, The Johns Hopkins University School of Medicine, Baltimore, Maryland 21205.

Neuropathogenic Viruses and Immunity, edited by Steven Specter *et al.* Plenum Press, New York, 1992.

2. ALPHAVIRUSES

2.1. Virus Biology

The neurotropic alphaviruses are members of the *Togaviridae*, are primarily agents of meningoencephalitis in the New World, and include such important causes of encephalitis as Eastern equine (EEE), Western equine (WEE), and Venezuelan equine encephalitis (VEE) viruses (Table 13-1). Other members of the alphavirus family—Sindbis, Chikungunya, Onyong-nyong, and Ross River viruses—primarily cause syndromes of rash and arthritis rather than CNS disease.[2] Animal models of alphavirus encephalitis have concentrated on studies of encephalitis caused by Sindbis, VEE, and Semliki Forest viruses.[3]

2.1.1. Virus Structure and Replication

Alphaviruses are small, enveloped, positive-stranded RNA viruses approximately 60–65 nm in diameter. The lipid envelope contains two virus-specified glycoproteins (E_1 and E_2), which form heterodimers and project as trimeric spikes from the cell surface. These spikes are able to attach to specific cell receptors, which have not yet been identified on neurons. Virus enters susceptible cells through receptor-mediated endocytosis. Lowered pH in the endosome causes a conformational change in the virion spike glycoproteins that results in fusion of viral and cellular membranes and release of virion RNA into the cytoplasm.[4]

The virion RNA is capped, polyadenylated, and associated in the virion with the capsid protein in an icosahedral structure.[5,6] The RNA is infectious and on release into the cytoplasm serves as the message for the nonstructural proteins needed for subsequent steps in replication. Synthesis of structural proteins occurs off a subgenomic RNA species representing the 3' one-third of the genome[6] (Fig. 13-1). Assembly occurs at the plasma membrane, and mature virus is released by budding. Infection is lytic in vertebrate cells but is often nonlytic in mosquito cells, resulting in persistent infection.[7]

2.1.2. Epidemiology

In North and South America the alphaviruses EEE, WEE, and VEE are important causes of anthropod-borne encephalitis (Fig. 13-2). All are transmitted

TABLE 13-1
Alphaviruses Causing Encephalitis

Virus	Vector	Geographic location
Eastern equine	*Culiseta, Aedes* mosquitoes	Eastern and Gulf coasts of U.S., Caribbean, and South America
Western equine	*Culiseta, Culex* mosquitoes	Western U.S. and Canada
Venezuelan equine	*Aedes, Culex*, and other mosquitoes	South and Central America, Florida, and southwestern U.S.

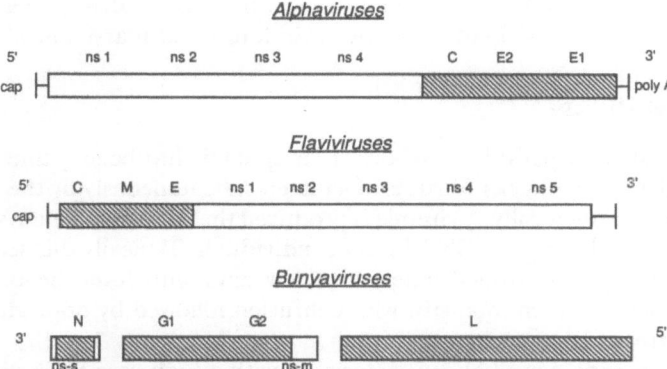

FIGURE 13-1. Genomic organization of the alphaviruses, flaviviruses, and bunyaviruses. All have genomes consisting of RNA of approximately 11,000 nucleotides. The RNA of alphaviruses and flaviviruses is message-sense, whereas the RNA of bunyaviruses is negative-sense and segmented.

by mosquitoes and have birds as their primary vertebrate hosts.[3,8,9] The EEE virus is endemic along the eastern and Gulf coasts of the United States, in the Caribbean, and in South America and causes localized outbreaks of equine, pheasant, and human encephalitis during the summer.[9] The primary enzootic vector to birds is *Culiseta melanura*, a swamp-dwelling mosquito.[10] The WEE virus produces epidemics of equine and human encephalitis primarily in the western and midwestern United States.[9] The virus is maintained in a natural cycle by *Culex tarsalis* mosquitoes, which breed in flood pools and irrigated areas.[8] The VEE virus occurs in the northern parts of South America, Central America, and the southern United States with considerable regional variation in the types of disease associated with infection.[8,11] Both endemic and epidemic disease occurs. Epidemic strains can be transmitted by various species of mosquitoes and are generally more virulent than endemic strains, which are transmitted primarily by

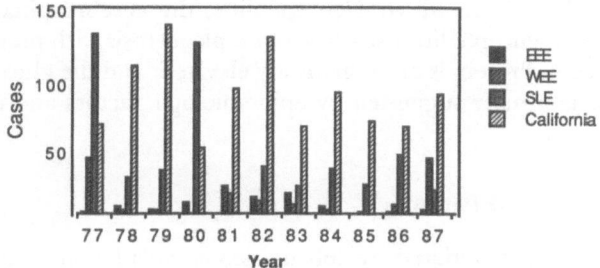

FIGURE 13-2. Cases of arbovirus encephalitis in the United States from 1977 to 1987 caused by eastern equine encephalitis (EEE), western equine encephalitis (WEE), St. Louis encephalitis (SLE), and California serogroup viruses. Data from the Centers for Disease Control.

Culex mosquitoes to rodents.[12-15] Equine epizootics arise in drier areas, whereas enzootic disease usually occurs in wetter rain forest and marshlands.[9]

2.2. Clinical Disease

The incubation period for arboviral encephalitis has been estimated to be between 4 days and 3 weeks. Virus is inoculated subcutaneously by the mosquito vector and replicates locally. A viremia is produced that lasts about 4 days, seeding target organs, including the CNS, in some individuals. Typically this sequence of events results in a prodromal illness of a few days with fever, headache, and malaise. Symptoms then intensify, with confusion followed by obtundation, seizures, and frequently coma.[1]

Eastern equine encephalitis is associated with a high case fatality rate (50–75%) in individuals of all ages. Serological surveys suggest that there are approximately 23 inapparent infections for every case of recognized encephalitis, but this declines to only eight to one for children under 4.[16] The encephalitis tends to be fulminant and is associated with fever, headache, altered consciousness, and seizures. Sequelae are common, with more than 80% of survivors having significant neurological residua including paralysis, seizures, and mental retardation.[17]

The WEE virus causes encephalitis in horses and in man with signs and symptoms similar to those of EEE but a lower case fatality rate of 10%. Western encephalitis is associated with fever, headache, irritability, tremors, and seizures along with signs and symptoms of meningitis such as nuchal rigidity and photophobia.[18] Severe disease, fatal encephalitis, and significant sequelae are more likely to occur in infants and young children than in older children and adults.[7,19] Transplacental transmission can occur.[20]

The VEE virus causes severe, frequently fatal, disease in horses but usually mild disease in man. Infection can occur by the respiratory route as well as by mosquito transmission, as evidenced by a number of laboratory infections.[21] Illness in adults is usually manifested by fever, headache, myalgias, and pharyngitis. Encephalitis is infrequent. More severe disease, including fulminant reticuloendothelial infection and encephalitis, may occur in young children.[8,9,11,21] Fetal abnormalities have been reported with infection during pregnancy in both man[22] and monkeys.[23] Children with encephalitis may be left with neurological deficits.[24]

As with other forms of viral encephalitis, the cerebrospinal fluid (CSF) during alphavirus encephalitis usually shows a pleocytosis with predominance of mononuclear cells. Protein is often modestly elevated, and the glucose is normal. The diagnosis is usually suggested by epidemiologic factors and confirmed by serology.[1,7]

2.3. Pathogenesis and Pathology

Virus infection is initiated by subcutaneous inoculation of infected saliva from the transmitting mosquito, and there is local replication in muscle or subcutaneous tissue that progresses to a viremia. Spread to the CNS is usually through the blood,[1,3] although in experimental infection entry by the olfactory route has been suggested.[25] Initial CNS infection in experimental animals is of the capillary endothelial or choroid epithelial cells, and spread within the CNS can be cell to cell or through the CSF.[26,27] The target cells within the CNS are the neurons, and the damage to these cells may be severe and irreversible. Histopathology demonstrates a diffuse meningoencephalitis with widespread neuronal destruction, neuronophagia, gliosis, and perivascular inflammation.[17,28,29]

The immune response has been studied most thoroughly in animal models. Antiviral antibody can usually be detected in serum within days after infection and its appearance correlates with the termination of the plasma viremia.[3,14] Monoclonal antibody studies have shown that epitopes capable of eliciting neutralizing antibody are present on both the E_1 and E_2 surface glycoproteins.[3] Passively transferred antibody can protect against fatal disease in experimental systems, and several lines of data suggest the importance of antibody for recovery.[3]

A cellular immune response is also elicited and is manifest by the mononuclear inflammatory response in the CNS to infection. The cells present in the CNS during experimental alphavirus infection include CD4+ and CD8+ T cells, B cells, macrophages and natural killer (NK) cells.[30,31] However, the disease appears to be caused by virus destruction of targeted cells, not by immunologically mediated damage, and immunosuppression speeds rather than retards the time to death.[3]

2.4. Persistence

The virus persists in the mosquito vector, and a late demyelinating disease associated with persistent infection can be induced in mice after infection with Semliki Forest virus.[3,32] Arthritis can be prolonged and recurrent after Ross River, Ockelbo, and Chikungunya virus infections,[33–36] but it is not clear whether this is because of persistent infection or immune-mediated events. There is, however, no evidence for persistence or delayed CNS disease in man.

2.5. Prevention and Treatment

Vaccines for EEE, WEE, and VEE viruses are available for horses and, on an experimental basis, for investigators working with these agents in the laboratory.[9] Therapy of alphavirus encephalitis is primarily limited to supportive care. Remarkable recoveries can occur even after prolonged coma, so vigorous supportive therapy and treatment of complications are essential.

3. FLAVIVIRUSES

3.1. Virus Biology

3.1.1. Virus Structure and Replication

The flaviviruses are small, enveloped, positive-stranded RNA viruses that are morphologically similar to the alphaviruses but have a distinctive genomic organization (Fig. 13-1) and replication cycle.[37,38] The RNA is infectious and consists of a single long open reading frame. The structural proteins C, M, and E are encoded in the 5' one-quarter of the genome.[37] M and E are glycosylated envelope proteins, and C is associated with the virion RNA. Virions are spherical and have a diameter of about 40–50 rm. The outer surface of the virion envelope contains projections 5–10 nm long made up of the E glycoprotein and through which the virion attaches to cells.

Flaviviruses replicate in a wide variety of vertebrate and arthropod cell cultures, often without producing cytopathic effect. The virus enters by adsorptive endocytosis, but the exact mechanism of RNA delivery into the cytoplasm is not clear, since pH-dependent fusion has not been clearly demonstrated.[38] Virus entry into cells with Fc receptors is enhanced, rather than retarded, by the presence of virus-specific antibody, and this phenomenon may play a role in the pathogenesis of some flavivirus-mediated diseases.[39–41] The 40 S plus-stranded RNA is capped but not polyadenylated, is infectious and serves as template for minus-strand synthesis, mRNA for protein synthesis, and genome for encapsidation into virions.[37] Unlike the alphaviruses, there is no subgenomic RNA produced during infection. Processing of structural and nonstructural proteins occurs post- or cotranslationally.[37,38] During virus assembly the RNA is encapsidated by C. Virus assembly occurs in the cytoplasm with budding into intracellular vesicles; final maturation and release are at the plasma membrane.[38]

3.1.2. Epidemiology

Many members of this group cause neurological disease, primarily encephalitis, whereas others cause hepatitis, rash, and fever. Flavivirus encephalitis has been reported from all continents except Antarctica. Viruses can be grouped antigenically and by vector. Encephalitic strains are transmitted either by mosquitoes or by ixodid ticks (Table 13-2).

The mosquito-borne flaviviruses that cause encephalitis are all members of the West Nile antigenic complex and are found worldwide. The largest numbers of cases of encephalitis occur in Asia, where Japanese encephalitis is endemic and epidemic. Sporadic outbreaks also occur in the Americas (St. Louis, Rocio), Australia (Murray Valley), and Africa (West Nile).[1,42]

Japanese encephalitis is widely distributed in Asia including Japan, China, the Soviet Union, the Philippines, and all of Southeast Asia and India. Japanese encephalitis is the most important of the arbovirus-induced encephalitides in terms of worldwide morbidity and mortality, with tens of thousands of cases occurring annually.[42] The primary vector is *Culex tritaeniorhynchus*, a mosquito that breeds in rice paddies and other standing water and tends to feed on large mammals and birds at dusk and dawn.[43] Other species of *Culex* mosquitoes can serve as vectors, but all breed in rural or sylvan habitats, feed on domestic animals or birds, and are night-biting. Several vertebrate species can become infected, and birds (particularly herons and egrets) and pigs develop a viremia of sufficient magnitude to infect feeding mosquitoes and thus serve to amplify infection in the environment (Fig. 13-3). Japanese encephalitis virus is neurotropic in mosquitoes

TABLE 13-2
Flaviviruses Causing Encephalitis

Virus	Vector	Geographic location
West Nile complex		
St. Louis	*Culex* mosquitoes	Widespread in U.S.
Japanese	*Culex* mosquitoes	Japan, China, Southeast Asia and India
Murray Valley	*Culex* mosquitoes	Australia and New Guinea
West Nile	*Culex* mosquitoes	Africa and Middle East
Ilheus	*Psorophora* mosquitoes	South and Central America
Rocio	? mosquitoes	Brazil
Tick-borne complex		
Far Eastern	*Ixodes* ticks	Eastern USSR
Central European	*Ixodes* ticks	Central Europe
Kyasanur Forest	*Haemophysalis* ticks	India
Louping-ill	*Ixodes* ticks	United Kingdom
Powassan	*Ixodes* ticks	Canada and northern U.S.
Negishi	? ticks	Japan

as well as man.[44] Field studies in Thailand suggest that infection results in altered behavior of the mosquito, which could affect the likelihood of virus transmission.[44] Humans are not preferred hosts for the vector so man tends to become infected only when the mosquito population is high[43,44] (Fig. 13-3). In endemic areas such as southern Thailand, children are most often affected because of preexisting immunity in older age groups. In epidemic areas such as China, all ages are susceptible and equally affected, and cases occur in outbreaks, usually beginning in late summer.[42,45]

St. Louis encephalitis is the most common flavivirus-induced encephalitis in the United States (Fig. 13-2) and is widely distributed. Numerous outbreaks have been documented, usually in August, September, and October, somewhat later than outbreaks for most other arboviruses.[42] The virus can have both urban (epidemic) and rural (endemic) cycles.[46] The vector for epidemic disease in the

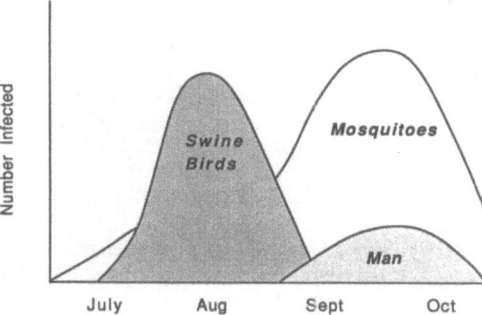

FIGURE 13-3. Appearance of Japanese encephalitis virus in vector (mosquito), amplifying hosts (birds and swine), and man as the cycle occurs during epidemics in temperate climates.

midwestern and eastern United States is usually *Culex pipiens*, in which the virus can overwinter,[47] whereas endemic disease in the western United States is usually spread by *Culex tarsalis*.[48]

Rocio virus is endemic in Brazil, where it has caused outbreaks of encephalitis. Murray Valley encephalitis virus causes infrequent outbreaks of encephalitis in Australia, Asia, and New Guinea during the summer after high rainfall for two consecutive years. West Nile virus is widely distributed throughout Africa, the Middle East, Europe, the Soviet Union, India, and Indonesia. Infection is common, but meningitis and encephalitis are rare complications occurring primarily in the elderly.[42]

The tick-borne flaviviruses form a separate antigenic group and are most prominent in Europe and Africa. Transmission by ingestion of infected goat milk in addition to tick transmission has been documented.[42] The tick-borne encephalitis complex includes six closely related viruses that cause encephalitis in man: Kyasanur Forest, louping-ill, Powassan, Negishi, and two strains of tick-borne encephalitis, Far Eastern (Russian spring–summer) and Central European encephalitis viruses.

The Far Eastern and Central European strains of tick-borne encephalitis virus are endemic over a wide area of Europe and the Soviet Union (Fig. 13-4) and cause thousands of cases of encephalitis each year, primarily in adults working or vacationing in wooded areas.[42] Other member of the tick-borne complex of flaviviruses are associated with only occasional cases of encephalitis. Louping-ill is endemic in the British Isles and causes encephalitis in sheep. Most human infections have followed laboratory exposure, but transmission by ticks and by direct contact with sick sheep also occurs. No human deaths have been reported. Powassan virus has a widespread distribution in the United States and Canada but has been associated with only 20 cases of encephalitis. Negishi virus is a rare cause

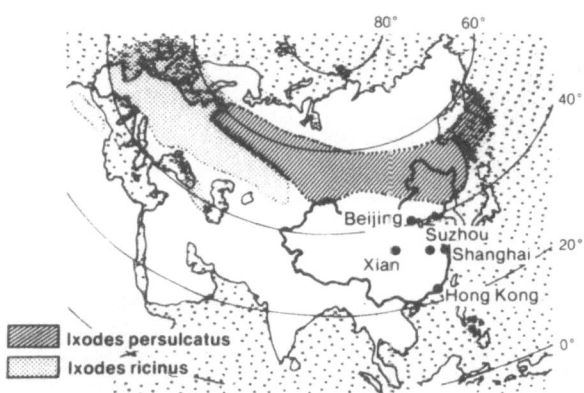

FIGURE 13-4. Geographic distribution of the tick vectors of Far Eastern and Central European encephalitis. (Reproduced with permission of The Johns Hopkins University Medical Grand Rounds, Vol. VIII, Program 7.)

of encephalitis in Japan. Kyasanur Forest virus causes a hemorrhagic fever syndrome occasionally complicated by encephalitis in the Mysore state of India.[42]

Flavivirus strains vary considerably in neurovirulence when tested in laboratory animals. For instance, Japanese encephalitis strains isolated from humans and pigs are almost always neurovirulent, whereas 10% of strains from mosquitoes have low virulence.[41] It has been speculated that maintenance of neurovirulent strains in nature may require cycling through hosts such as young birds that sustain CNS infection and that attenuated strains may arise by persistent infection of mosquitoes.[49] St. Louis encephalitis virus strains from various sources and geographic locations have also been reported to vary in virulence and ability to produce high-titer viremia in birds. Strains isolated from birds (the usual viremic host) are generally virulent, whereas those from more unusual hosts are attenuated. Virus strains recovered during major epidemics in the eastern United States are usually highly virulent, whereas those·recovered in the western United States are relatively attenuated and correlate with the lower human case fatality rates in the West.[41,48,50]

3.2. Clinical Disease

Infection with Japanese encephalitis virus may be asymptomatic or manifested by fever alone, aseptic meningitis or encephalitis. The onset of encephalitis is rapid, beginning with a 2- to 3-day prodrome of headache, fever, chills, malaise, and nausea. In children abdominal symptoms may be prominent. The acute stage lasts 2–4 days and is marked by sustained fever (usually >104°F), meningismus, photophobia, confusion, and delirium. Characteristic neurological signs include a parkinsonian-like picture of mask-like facies, rigidity, and involuntary movements, altered consciousness, and generalized or localized paralysis. Seizures are frequent in children but occur in <10% of adults. Tremor is present in 90% of patients and is moderately coarse and accentuated by fatigue and fine purposive movements, with fingers, tongue, and eyelids most frequently affected.[51,52]

St. Louis encephalitis is usually initiated by a prodrome of several days followed by the abrupt onset of severe generalized headache, nausea, and vomiting followed by disorientation, irritability, and stupor. Low serum sodium as a result of inappropriate secretion of antidiuretic hormone and pyuria with elevated blood urea nitrogen are relatively common laboratory findings.[53–55] The risk of illness after infection increases markedly with age, and mortality is low in the young but over 20% in the elderly.[53,55] Convalescence may be prolonged, and significant neurological residua are present in 20% of survivors.[42]

The onset of Far Eastern encephalitis is gradual, progressing from fever and headache to paralysis and seizures over several days. Evidence of lower motor neuron disease with loss of tone and reflexes and localized muscle weakness, usually limited to the upper extremities, reflects selective involvement of the motor neurons of the cervical spinal cord. The case-fatality rate is approximately 20%. Neurological sequelae occur in 30–60% of survivors. Especially characteristic is a residual flaccid paralysis of the shoulder girdle and arms (Fig. 13-5).[42,56,57]

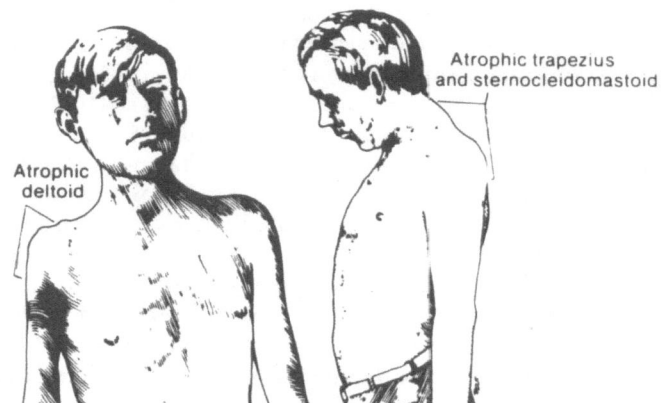

FIGURE 13-5. Drawing illustrating the residual paralysis characteristic of previous infection with Far Eastern encephalitis virus. (Drawing based on Smorodintsev[56] and reproduced with permission of The Johns Hopkins University Medical Grand Rounds, Vol. VIII, Program 7.)

Central European encephalitis is milder than Far Eastern and exhibits a typical biphasic course in half the cases. The first phase is a nonspecific flu-like illness lasting about a week, followed by a 1- to 3-day remission. The neurological phase may manifest as aseptic meningitis or encephalitis with tremor, diplopia, altered mental status, and paresis. The case fatality rate is 1–5%, and approximately 20% of survivors have mild neurological residua.[42,58,59]

3.3. Pathogenesis and Pathology

Most flavivirus infections are initiated by subcutaneous or intravenous inoculation of virus by the arthropod vector. Initial virus replication is in subcutaneous tissue. Tick-borne encephalitis can also be acquired by the oral route through consumption of unpasteurized milk or cheese from infected sheep or goats.[42,58] After initial infection virus replicates in muscle, connective tissue, and reticuloendothelial cells near the site of inoculation. Newly synthesized virus may be carried to the bloodstream by lymphatic channels.[42] The process by which flaviviruses enter the CNS across the blood–brain barrier is uncertain. Many strains can infect capillary endothelial cells, but viral antigen is rarely found in brain capillaries.[41,60,61] Some experimental studies suggest that initial CNS infection may occur in olfactory neurons, which are susceptible to infection but unprotected by the blood–brain barrier.[41,62]

In experimental models three patterns of pathogenesis have been defined[41,63,64]: (1) fatal encephalitis usually preceded by early viremia and extensive extraneural replication, (2) subclinical encephalitis, usually preceded by a low viremia, late establishment of brain infection, and clearance with minimal destructive pathology, and (3) inapparent infection with trace viremia, limited extraneural replication, and no neuroinvasion. Generally immune suppression converts subclinical experimental infection to fatal encephalitis but may delay death in otherwise fatal infection, suggesting that there is an immunopathological component to severe disease.[41]

During flavivirus encephalitis there is usually a mildly elevated peripheral white blood count (10,000–20,000 cells/mm[3]) with an initial neutrophilia and decrease in T lymphocytes.[65,66] The CSF shows a moderate lymphocytosis (10–500 cells/mm[3]), mildly elevated protein (50–100 mg%), and normal glucose.[52] Most of the CSF cells are T lymphocytes of the CD4 (helper-inducer) subset,[67] and there is often a persistence of the pleocytosis well beyond the acute phase of the disease.[52]

Virus can rarely be isolated from blood or CSF in individuals early in disease. Pathological studies of Japanese, Rocio, and St. Louis encephalitis have shown that the virus replicates primarily in neurons.[60,63,68,69] Infection is usually multifocal, consistent with entry into the CNS from the blood. In Japanese encephalitis the greatest involvement is in the thalamus and brainstem.[60] The brainstem distribution readily explains the profound coma and respiratory failure, and the thalamic involvement is consistent with the tremors, dystonia, and parkinsonian facies[52] frequently seen during the acute disease. Patients who die years after acute Japanese encephalitis often have residual thalamic and substantia nigra scars.[70]

Tick-borne encephalitis is also associated primarily with neuronal infection with a predilection for neurons of the motor strip and cervical spinal cord as well as thalamus, cerebellum, and brainstem. The tendency to infect motor neurons is consistent with the clinical picture of poliomyelitis that frequently occurs as a part of the clinical syndrome[63,71] (Fig. 13-5).

Pathological changes indicative of a cellular immune response to infection are also present in the brains of those dying of flavivirus encephalitis. These changes include a mononuclear inflammatory response in the leptomeninges and perivascular areas of the brain parenchyma. Infiltration of inflammatory cells into the parenchyma, neuronophagia and formation of glial nodules are seen in gray matter areas.[60,63,68,69,72] In Japanese encephalitis the inflammatory cells are primarily CD4+ T lymphocytes but include significant numbers of macrophages and B lymphocytes as well.[60]

The antiviral antibody response is usually present early and appears to be crucial to recovery. By a sensitive antibody-capture immunoassay to detect Japanese encephalitis virus-specific IgM or IgG[73] CSF specimens from 75% of patients are found positive for antibody at the time of admission, and the rest become positive within a few days.[74] Lack of CSF antibody, often accompanied by recovery of virus from CSF, is associated with a poor prognosis.[75] Similar tests have been developed for the diagnosis of St. Louis and tick-borne encephalitides.[76–78]

3.4. Persistence

In nature persistent infection of certain species, such as bats, hedgehogs, snakes, and lizards, with Japanese, tick-borne, and St. Louis encephalitis viruses may contribute to overwintering of these viruses.[41] Persistence in experimental animals is correlated with failure to develop neutralizing antibody and subsequent recovery of relatively avirulent strains of virus.[38,41] However, there is no evidence for persistence in man of most flaviviruses in either the nervous system or other tissues. Even examination of persons with prolonged IgM responses after Japanese encephalitis has revealed no evidence of persistent infection.[79]

The exception to this generalization is Far Eastern encephalitis, where there is suggestive evidence from study of both monkeys[80–82] and man[83] that virus can persist in the nervous system. There are numerous reports in the literature from the Soviet Union of chronic neurological diseases with delayed onset and progressive course following tick-borne encephalitis with estimates of frequency varying from 1–20%. Seizure disorders are most frequent, and chronic tick-borne encephalitis may account for many cases of epilepsy partialis continua in the Soviet Union. Virus has been isolated only rarely from such patients, but active inflammation and local production of antibody to the virus suggest chronic infection.[83,84]

3.5. Prevention and Treatment

Several flavivirus vaccines are in use in various parts of the world. Formalin-inactivated Japanese encephalitis virus vaccines are produced in China, Japan, India, and Korea, and several large-scale field trials have demonstrated vaccine efficacy.[45,85] Immunization is recommended for individuals traveling to endemic areas for 3 weeks or more during the Japanese encephalitis season (June to October in temperate climates). As with other mosquito-borne diseases, exposure to the vector should be minimized by avoiding outdoor exposure at dawn and dusk, sleeping in screened or netted quarters, and using insect repellents.[42] Pasteurization of milk prevents oral transmission of tick-borne encephalitis, and vaccines, consisting of inactivated or egg-grown virus, are available in the Soviet Union and eastern Europe[42,86] for persons at high risk of exposure to this virus.

Therapy of flavivirus encephalitis is not well established. There is some experience using antiserum in the Soviet Union,[87] but no trials with other agents have been reported, and animal studies have not been encouraging.[88] As with alphavirus infection, survival depends in part on supportive care. Routine use of steroids, other than to control increased intracranial pressure, has not proven beneficial.[1]

4. BUNYAVIRUSES

4.1. Virus Biology

The bunyavirus family has more than 200 members, but the most important causes of CNS disease belong to the California serogroup[89] (Fig. 13-2). This serogroup includes LaCrosse virus, the most frequent arbovirus cause of encephalitis in the United States, Jamestown Canyon, snowshoe hare, and California encephalitis viruses, which are found in North America, and Inkoo and Tahyna viruses, found in Europe (Table 13-3).

TABLE 13-3
Bunyaviruses Causing Encephalitis

Virus	Vector	Geographic location
California	*Aedes* mosquitoes	Western U.S.
La Crosse	*Aedes* mosquitoes	Middle and eastern U.S
Jamestown Canyon	*Culiseta* mosquitoes	Alaska
Snowshoe hare	*Culiseta* mosquitoes	Canada, Alaska, northern U.S
Tahyna	*Aedes*, *Culiseta* mosquitoes	Czechoslovakia, Yugoslavia, Italy, France
Inkoo	? mosquitoes	Finland

4.1.1. Virus Structure and Replication

Bunyaviruses are enveloped, spherical, and 90–100 nm in diameter. Virus particles have three internal nucleocapsids, each consisting of viral nucleoprotein, a unique single-stranded negative-sense RNA (S, M, and L), and a transcriptase enzyme (Fig. 13-1).[89] Because of the segmented genome bunyaviruses are capable of forming reassortants. Reassortment in nature may contribute to the emergence of new strains.[90] Virions have two surface glycoproteins, G_1 and G_2, present in equimolar amounts that induce and bind neutralizing antibodies.[89]

Infection is initiated by binding of virus glycoprotein to an as yet unidentified cell surface receptor. The G_1 protein also influences the efficiency of infection of the invertebrate vector.[91,92] Penetration of the host cell is believed to depend on phagocytosis followed by low-pH-dependent fusion of the viral envelope with endosomal membranes and release of nucleocapsids into the cell cytoplasm. This results in activation of the transcriptase and synthesis of mRNA from the three viral nucleocapsids.[89] Virus morphogenesis occurs in the Golgi, a unique feature of this virus group, but the mechanism of release has not been identified. The virus is cytopathic for most mammalian cells but not for most insect cells.[89]

4.1.2. Epidemiology

All of the California serogroup bunyaviruses associated with encephalitis are endemic in their particular locations, and most cause encephalitis primarily in children.[93,94] Both urban and rural cases of LaCrosse encephalitis occur, since the primary vector, *Aedes triseriatus*, can breed in urban containers such as old tires as well as rural locations such as tree holes.[89] Virus is maintained in the mosquito population by transovarial and venereal transmission.[95,96] The recent introduction and spread in the United States of another container-breeding mosquito, *Aedes albopictus*, an efficient vector for the California serogroup viruses, has led to concern about the potential for a significant increase in these infections.[97,98] Amplification of most viruses occurs in small mammals such as squirrels, chipmunks, and rabbits,[89] although deer and moose are important hosts for Jamestown Canyon and Inkoo viruses.[89] Numbers of cases usually peak during late summer and early fall.

Bunyaviruses may also infect the nervous system of the arthropod vector and modify behavior. There is evidence that *Aedes triseriatus* mosquitoes infected with LaCrosse virus exhibit enhanced probing responses during breeding,[99] behavior that would favor transmission.

4.2. Clinical Disease

Many infections are inapparent. When encephalitis occurs there is often a prodrome of malaise and fever for 1–4 days, but occasionally the presentation is acute. Encephalitis caused by LaCrosse virus is the most common and occurs nearly exclusively among children.[89,100] The case-fatality rate is <1%. Snowshoe hare encephalitis is similar clinically but less common than La Crosse. During the acute illness, seizures occur in 50%, and focal weakness, paralysis, or other neurological signs occur in 25%. The acute illness typically lasts 7 days with gradual recovery. Some neurological residua (usually seizure disorders) are noted in approximately 10%.[100] In contrast to LaCrosse encephalitis, Jamestown Canyon encephalitis is usually a disease of adults, and prodromal symptoms tend to be fever and respiratory illness.[89]

The diagnosis of bunyavirus encephalitis cannot be made on clinical grounds. La Crosse-virus-specific IgM is usually present at the time of presentation. Therefore, a rapid means of making the diagnosis uses a capture enzyme immunoassay to identify IgM antibody in serum or CSF.[101]

4.3. Pathogenesis and Pathology

In the rare fatal cases of LaCrosse encephalitis, lesions of perivascular infiltration are most prominent in the cerebral hemispheres. After subcutaneous inoculation of mice, virus replicates in muscle and fibroblasts[102] and has been found at neuromuscular junctions, suggesting the possibility that virus enters the CNS by retrograde axonal transport.[89] Virus also appears to be able to enter the CNS from the blood by infecting the endothelial cells of small cerebral vessels and then spreading to the brain parenchyma.[102] Ability to infect neuromuscular junctions and cause encephalitis appears to be a property of the virus surface glycoproteins.[102] In the CNS, neurons are the primary cell infected.[102,103]

4.4. Persistence

There is no evidence of persistent infection in experimental animals or in man. Persistence in the environment is probably accomplished by persistent infection of the vector and overwintering of the virus in infected mosquito eggs.[95]

4.5. Prevention and Treatment

No vaccine is available for prevention of infection with the California serogroup bunyaviruses. Vector control has been accomplished in some areas by cleaning up the environment to eliminate breeding areas, such as old tires and other containers, from urban and suburban areas.[89]

As with other causes of arbovirus encephalitis, there is no specific treatment. Supportive care including seizure control and treatment for cerebral edema are important to successful recovery.

ACKNOWLEDGMENTS. Work from the author's laboratory was supported by a grant from the G. Harold and Leila Y. Mathers Charitable Foundation and grant NS29234 from the National Institutes of Health.

REFERENCES

1. Johnson, R. T., 1989, Arboviral encephalitis, in: *Tropical and Geographical Medicine* (K. S. Warren and A. A. F. Mahmoud, ed.), McGraw–Hill, New York, pp. 691–700.
2. Tesh, R. B., 1982, Arthritides caused by mosquito-borne viruses, *Annu. Rev. Med.* **33:** 31–40.
3. Griffin, D. E., 1986, Alphavirus pathogenesis and immunity, in: *The Togaviridae and Flaviviridae* (S. Schlesinger and M. J. Schlesinger, eds.), Plenum Press, New York, pp. 209–249.
4. Kielian, M., and Helenius, A., 1986, Entry of alphaviruses, in: *The Togaviridae and Flaviviridae* (S. Schlesinger and M. J. Schlesinger, eds.), Plenum Press, New York, pp. 91–119.
5. Harrison, S. C., 1986, Alphavirus structure, in: *The Togaviridae and Flaviviridae* (S. Schlesinger and M. J. Schlesinger, eds.), Plenum Press, New York, pp. 21–34.
6. Strauss, E. G., and Strauss, J. H., 1986, Structure and replication of the alphavirus genome, in: *The Togaviridae and Flaviviridae* (S. Schlesinger and M. J. Schlesinger, eds.), Plenum Press, New York, pp. 35–90.
7. Brown, D. T., and Condreay, L. D., 1986, Replication of alphaviruses in mosquito cells, in: *The Togaviridae and Flaviviridae* (S. Schlesinger and M. J. Schlesinger, eds.), Plenum Press, New York, pp. 171–207.
8. Shope, R. E., 1985, Alphaviruses, in: *Virology* (B. N. Fields, D. M. Knipe, R. M. Chanock, J. L. Melnick, B. Roizman, and R. E. Shope, eds.), Raven Press, New York, pp. 931–953.
9. Russell, P. K., 1985, Alphavirus (Eastern, Western and Venezuelan equine encephalitis), in: *Principles and Practice of Infectious Diseases*, 2nd ed. (G. L. Mandell, R. G. Douglas, Jr., and J. E. Bennett, eds.), John Wiley & Sons, New York, pp. 917–920.
10. Hayes, R. O., 1961, Host preferences of *Culiseta melanura* and allied mosquitoes, *Mosquito News* **21:**179–187.
11. Ehrenkranz N. J., and Ventura, A. K., 1974, Venezuelan equine encephalitis virus infection in man, *Annu. Rev. Med.* **25:**9–14.
12. Grayson, M. A., and Galindo, P., 1968, Epidemiologic studies of Venezuelan equine encephalitis virus in Almirante, Panama, *Am. J. Epidemiol.* **88:**80–96.
13. Sudia, W. D., Newhouse, V. F., and Henderson, B. E., 1971, Experimental infection of horses with three strains of Venezuelan equine encephalomyelitis virus. II. Experimental vector studies, *Am. J. Epidemiol.* **93:**206–211.
14. Monath, T. P., Calisher, C. H., Davis, M., Bowen, G. S., and White, J., 1974, Experimental studies of rhesus monkeys infected with epizootic and enzootic subtypes of Venezuelan equine encephalitis virus, *J. Infect. Dis.* **129:**194–200.
15. Sudia, W. D., and Newhouse, V. F., 1975, Epidemic Venezuelan equine encephalitis in North America: A summary of virus–vector–host relationships, *Am. J. Epidemiol.* **101:**1–13.
16. Goldfield, M., Welsh, J. N., and Taylor, B. F., 1968, The 1959 outbreak of Eastern

encephalitis in New Jersey. 5. The inapparent infection: Disease ratio, *Am. J. Epidemiol.* **87**: 32–38.

17. Farber, S., Hill, A., Connerly, M. L., and Dingle, J. H., 1940, Encephalitis in infants and children caused by the virus of the Eastern variety of equine encephalitis, *J.A.M.A.* **114**: 1725–1731.

18. Kokernot, R. H., Shinefield, H. R., and Longshore, W. A., 1953, The 1952 outbreak of encephalitis in California, *Calif. Med.* **79**:73–77.

19. Centers for Disease Control, 1988, Arboviral infections of the central nervous system— United States, 1987, *Morbid. Mortal Weekly Rep.* **37**:506–515.

20. Shinefield, M. R., and Townsend, T. E., 1953, Transplacental transmission of western equine encephalitis, *J. Pediatr.* **43**:21–25.

21. Lennette, E. H., and Koprowski, H., 1943, Human infection with Venezuelan equine encephalomyelitis virus: A report on eight cases of infection acquired in the laboratory, *J.A.M.A.* **1231**:1088–1095.

22. Wenger, F., 1977, Venezuelan equine encephalitis, *Teratology* **16**:359–362.

23. London, W. T., Levitt, N. H., Kent, S. G., Wong, V. G., and Sever, J. L., 1977, Congenital cerebral and ocular malformations induced in rhesus monkeys by Venezuelan equine encephalitis virus, *Teratology* **16**:285–296.

24. Leon, C. A., Jaramillo, R., Martinez, S., Fernandez, F., Tellez, H., Lasso, B., and de Guzman, R., 1975, Sequelae of Venezuelan equine encephalitis in humans: A four year follow-up, *Int. J. Epidemiol.* **4**:131–140.

25. Danes, L., Kufner, J., Hruskova, J., and Rychterova, V., 1973, The role of the olfactory route of infection of the respiratory tract with Venezuelan equine encephalomyelitis virus in normal and operated *Macaca rhesus* monkeys. I. Results of virological examination, *Acta Virol.* **17**:50–56.

26. Liu, C., Voth, D. W., Rodina, P., Shauf, L. R., and Gonzalez, G., 1970, A comparative study of the pathogenesis of Western equine and Eastern equine encephalomyelitis viral infections in mice by intracerebral and subcutaneous inoculations, *J. Infect. Dis.* **122**:53–63.

27. Jackson, A. C., Moench, T. R., and Griffin, D. E., 1987, The pathogenesis of spinal cord involvement in the encephalomyelitis of mice caused by neuroadapted Sindbis virus infection, *Lab. Invest.* **56**:418–423.

28. Noran, H. H., and Baker, A. B., 1945, Western equine encephalitis: The pathogenesis of the pathological lesions, *J. Neuropathol. Exp. Neurol.* **4**:269–276.

29. Johnson, K. M., Shelokov, A., Peralta, P. H., Dammin, G. J., and Young, N. A., 1968, Recovery of Venezuelan equine encephalomyelitis virus in Panama, *Am. J. Trop. Med. Hyg.* **17**:432–440.

30. Moench, T. R., and Griffin, D. E., 1984, Immunocytochemical identification and quantitation of mononuclear cells in cerebrospinal fluid, meninges, and brain during acute viral meningoencephalitis, *J. Exp. Med.* **159**:77–88.

31. Griffin, D. E., and Hess, J. L., 1986, Cells with natural killer activity in the cerebrospinal fluid of normal mice and athymic nude mice with acute Sindbis virus encephalitis, *J. Immunol.* **136**:1841–1845.

32. Suckling, A. J., Pathak, S., Jagelman, S., and Webb, H. E., 1978, Virus-associated demyelination: A model using avirulent Semliki Forest virus infection of mice, *J. Neurol. Sci.* **39**:147–154.

33. Kennedy, A. C., Fleming, J., and Solomon, L., 1980, Chikungunya viral arthropathy: A clinical description, *J. Rheumatol.* **7**:231–236.

34. Rosen, L., Gubler, D. J., and Bennett, P. H., 1981, Epidemic polyarthritis (Ross River) virus infection in the Cook Islands, *Am. J. Trop. Med. Hyg.* **30**:1294–1302.

35. Aaskov, J. G., Mataika, J. U., Lawrence, G. W., Rabukawaga, V., Tucker, M. M., Miles, J. A. R., and Dalglish, D. A., 1981, An epidemic of Ross River virus infection in Fiji, 1979, *Am. J. Trop. Med. Hyg.* **30**:1053–1059.

36. Espmark, A., and Niklasson, B., 1984, Ockelbo disease in Sweden: Epidemiological, clinical, and virological data from the 1982 outbreak, *Am. J. Trop. Med. Hyg.* **33**:1203–1211.
37. Rice, C. M., Strauss, E. G., and Struass, J. H., 1986, Structure of the flavivirus genome, in: *The Togaviridae and Flaviviridae* (S. Schlesinger and M. J. Schlesinger, eds.), Plenum Press, New York, pp. 279–326.
38. Brinton, M. A., 1986, Replication of flaviviruses, in: *The Togaviridae and Flaviviridae* (S. Schlesinger and M. J. Schlesinger, eds.), Plenum Press, New York, pp. 327–374.
39. Peiris, J. S. M., and Porterfield, J. S., 1979, Antibody-mediated enhancement of flavivirus replication in macrophage-like cell lines, *Nature* **282**:509–511.
40. Halstead, S. B., Porterfield, J. S., and O'Rourke, E. J., 1980, Enhancement of dengue infection in monocytes by flavivirus antisera, *Am. J. Trop. Med. Hyg.* **29**:638–642.
41. Monath, T. P., 1986, Pathobiology of the flaviviruses, in: *The Togaviridae and Flaviviridae* (S. Schlesinger and M. J. Schlesinger, eds.), Plenum Press, New York, pp. 375–440.
42. Monath, T. P., 1985, Flaviviruses, in: *Virology* (B. N. Fields, D. M. Knipe, R. M. Chanock, J. L. Melnick, B. Roizman, and R. E. Shope, eds.), Raven Press, New York, pp. 955–1004.
43. Scherer, W. F., and Buescher, E. L., 1959, Ecologic studies of Japanese encephalitis virus in Japan I–IX, *Am. J. Trop. Med. Hyg.* **8**:644–722.
44. Leake, C. J., and Johnson, R. T., 1987, The pathogenesis of Japanese encephalitis virus in *Culex tritaeniorhynchus* mosquitoes, *Trans. R. Soc. Trop. Med. Hyg.* **81**:681–685.
45. Umenai, T., Krzysko, R., Bektimirov, T. A., and Assaad, F. A., 1985, Japanese encephalitis: Current worldwide status, *Bull. WHO* **63**:625–631.
46. Kokernot, R. H., Hayes, J., Will, R. L., Tempelis, C. H., Chan, D. H. M., and Radivojivic, B., 1969, Arbovirus studies in the Ohio–Mississippi basin, 1964–1967. II. St. Louis encephalitis virus, *Am. J. Trop. Med. Hyg.* **18**:750–761.
47. Bailey, C. L., Eldridge, B. F., Hayes, D. E., Watts, D. M., Tammariello, R. F., and Dalrymple, J. M., 1978, Isolation of St. Louis encephalitis virus from overwintering *Culex pipiens* mosquitoes, *Science* **199**:1346–1349.
48. Monath, T. P., and Tsai, T. F., 1987, St. Louis encephalitis: Lessons from the last decade, *Am. J. Trop. Med. Hyg.* **37**(Suppl.):40S–59S.
49. Huang, C. H., 1982, Studies of Japanese encephalitis in China, *Adv. Virus Res.* **27**:71–101.
50. Monath, T. P., Cropp, C. B., Bowen, G. S., Kemp, G. E., Mitchell, C. J., and Gardner, J. J., 1980, Variation in virulence for mice and rhesus monkeys among St. Louis encephalitis virus strains of different origin, *Am. J. Trop. Med. Hyg.* **29**:948–962.
51. Sabin, A. B., 1947, Epidemic encephalitis in military personnel, *J.A.M.A.* **133**:281–293.
52. Dickerson, R. B., Newton, J. R., and Hasen, J. E., 1952, Diagnosis and immediate prognosis of Japanese B encephalitis, *Am. J. Med.* **12**:277–288.
53. Southern, P. M., Smith, J. W., Luby, J. P., Barnett, J. A., and Sanford, J. P., 1969, Clinical and laboratory features of epidemic St. Louis encephalitis, *Ann. Intern. Med.* **71**:681–690.
54. White, M. G., Carter, N. W., Rector, F. C., and Seldin, D. W., 1969, Pathophysiology of epidemic St. Louis encephalitis. I. Inappropriate secretion of antidurietic hormone, *Ann. Intern. Med.* **71**:691–702.
55. Tsai, T. F., Canfield, M. A., Reed, C. M., Flahnery, V. L., Sullivan, K. H., Reeve, G. R., Bailey, R. E., and Poland, J. D., 1988, Epidemiological aspects of a St. Louis encephalitis outbreak in Harris County, Texas, 1986, *J. Infect. Dis.* **157**:351–356.
56. Smorodintsev, A. A., 1958, Tick-borne spring–summer encephalitis, *Prog. Med. Virol.* **1**: 210–248.
57. Graščenkov, N. I., 1964, Tick-borne encephalitis in the USSR, *Bull. WHO* **30**:187–196.
58. Blaškovič, D., 1967, The public health importance of tick-borne encephalitis in Europe, *Bull. WHO* **36**:5–13.
59. Cruse, R. P., Rothner, A. D., Erenberg, G., and Calisher, C. H., 1979, Central European tick borne encephalitis: An Ohio case with a history of foreign travel, *Am. J. Dis. Child.* **133**: 1070–1071.

60. Johnson, R. T., Burke, D. S., Elwell, M., Leake, C. J., Nisalak, A., Hoke, C. H., and Lorsmrudee, W., 1985, Japanese encephalitis: Immunocytochemical studies of viral antigen and inflammatory cells in fatal cases, *Ann. Neurol.* **18**:567–573.

61. Albrecht, P., 1968, Pathogenesis of neurotropic arbovirus infections, *Curr. Top. Microbiol. Immunol.* **43**:44–91.

62. Monath, T. P., Cropp, C. P., and Harrison, A. K., 1983, Mode of entry of a neurotropic arbovirus into the central nervous system. Reinvestigation of an old controversy, *Lab. Invest.* **48**:399–410.

63. Jervis, G. A., and Higgins, G. H., 1953, Russian spring–summer encephalitis (Clinicopathologic report of a case in the human), *J. Neuropathol. Exp. Neurol.* **12**:1–10.

64. Weiner, L. P., Cole, G. A., and Nathanson, N., 1970, Experimental encephalitis following peripheral inoculation of West Nile virus in mice of different ages, *J. Hyg.* **68**:435–446.

65. Chaturvedi, U. C., Mathur, A., Tandon, P., Natu, S. M., Rajvanshi, S., and Tandon, H. O., 1979, Variable effect on peripheral blood leukocytes during JE virus infection of man, *Clin. Exp. Immunol.* **38**:492–498.

66. D' Souza, M. B., Nagarkatli, P. S., and Rao, K. M., 1979, Subpopulation of peripheral blood lymphocytes in human encephalitis caused by group B arboviruses (Dengue, West Nile and Japanese B encephalitis), *J. Hyg. Epidemiol. Microbiol.* **23**:59–66.

67. Johnson, R. T., Intralawan, P., and Puapanwatton, S., 1986, Japanese encephalitis: Identification of inflammatory cells in cerebrospinal fluid, *Ann. Neurol.* **20**:691–695.

68. Reyes, M. G., Gardner, J. J., Poland, J. D., and Monath, T. P., 1981, St. Louis encephalitis: Quantitative histologic and immunofluorescent studies, *Arch. Neurol.* **38**:329–334.

69. Rosemberg, S., 1980, Neuropathology of S. Paulo south coast epidemic encephalitis (Rocio flavivirus), *J. Neurol. Sci.* **45**:1–12.

70. Ishii, T., Matsushita, M., and Hamada, S., 1977, Characteristic residual neuropathological features of Japanese B encephalitis, *Acta Neuropathol. (Berl.)* **38**:282–286.

71. Mazlo, M., and Szanto, J., 1978, Morphological demonstration of the virus of tick borne encephalitis in the human brain, *Acta Neuropathol. (Berl.)* **43**:251–253.

72. Kornyey, S., 1978, Contribution to the histology of tick-borne encephalitis, *Acta Neuropathol. (Berl.)* **43**:179–183.

73. Burke, D. S., Nisalak, A., and Ussery, M. A., 1982, Antibody capture immunoassay detection of Japanese encephalitis virus immunoglobulin M and G antibodies in cerebrospinal fluid, *J. Clin. Microbiol.* **16**:1034–1042.

74. Burke, D. S., Nisalak, A., Ussery, M. A., Laorakpongse, T., and Clantavibul, S., 1985, Kinetics of IgM and IgG antibodies to Japanese encephalitis virus in human serum and cerebrospinal fluid, *J. Infect. Dis.* **151**:1093–1099.

75. Burke, D. S., Losomrudee, W., Leake, C. J., Hoke, C. H., Nisalak, A., Chongswasdi, V., and Laorakpongse, T., 1985, Fatal outcome in Japanese encephalitis, *Am. J. Trop. Med. Hyg.* **34**:1203–1210.

76. Granstrom, M., Grandien, M., and Saikku, P., 1978, Early diagnosis of tick-borne encephalitis (TBE) by demonstration of specific IgM antibodies, *Scand. J. Infect. Dis.* **10**:97–100.

77. Hofmann, H., Frisch-Niggemeyer, W., and Heinz, F., 1979, Rapid diagnosis of tick-borne encephalitis by means of enzyme-linked immunosorbent assay, *J. Gen. Virol.* **42**:505–511.

78. Monath, T. P., Nystrom, R. R., Bailey, R. E., Calisher, C. H., and Muth, D. J., 1984, Immunoglobulin M antibody capture enzyme-linked immunosorbent assay for diagnosis of St. Louis encephalitis, *J. Clin. Microbiol.* **20**:784–790.

79. Edelman, R., Schneider, R. J., Vejjajiva, A., Pornpibal, R., and Voodhikul, P., 1976, Persistence of virus-specific IgM and clinical recovery after Japanese encephalitis, *Am. J. Trop. Med. Hyg.* **25**:733–738.

80. Ilienko, V. I., Komandenko, N. I. Platonov, V. G., Prozorova, I. N., and Panov, A. G., 1974, Pathogenetic study on chronic forms of tick-borne encephalitis, *Acta Virol.* **18**:341–346.

81. Pogodina, V. V., Frolova, M. P., Malenko, G. V., Fokina, G. I., Levnia, L. S., Mamonenko,

L. L., Koreshkova, G. V., and Ralf, N. M., 1981, Persistence of tick-borne encephalitis virus in monkeys. I. Features of experimental infection, *Acta Virol.* **25**:337–343.

82. Pogodina, V. V., Levina, L. S., Fokina, G. I., Koreshkova, G. V., Malenko, G. V., Bochkova, N. G., and Pzhakhova, O. E., 1981, Persistence of tick-borne encephalitis virus in monkeys. III. Phenotypes of the persisting virus, *Acta Virol.* **25**:352–360.

83. Asher, D. M., 1980, Chronic encephalitis, in: *Search for the Cause of Diseases of the Central Nervous System* (A. Boese, ed.), Verlag Chemie, Weinheim, pp. 272–279.

84. Asher, D. M., 1971, Focal neurological disease with chronic encephalitis in children and an experimental primate model, in: *Proceedings 13th International Congress of Pediatrics, III*, Volume 2, Verlag der Wiener Medizinischen Akademie, Vienna, pp. 379–384.

85. Hoke, C. H., Nisalak, A., Sangawhipa, N., Jatanasen, S., Laorakopongse, T., Innis, B. L., Kolchasenee, S. O., Gingrich, J. B., Latendresso, J., Fukai, K., and Burke, D. S., 1988, Protection against Japanese encephalitis by inactivated vaccines, *N. Engl. J. Med.* **319**: 608–624.

86. Kung, C., Hofmann, H., Heinz, F. X., and Dippe, H., 1980, Efficacy of vaccination against tick-borne encephalitis (TBE), *Wein. Klin. Wochschr.* **92**:809.

87. Glukhov, B. N., Jerusalimsky, A. P., Canter, V. M., and Salganik, R. I., 1976 Ribonuclease treatment of tick-borne encephalitis, *Arch. Neurol.* **33**:598–603.

88. Huggins, J. W., 1990, RNA viruses causing hemorrhagic encephalitic and febrile disease, in: *Antiviral Agents and Viral Diseases of Man*, 3rd ed. (G. Galasso, R. J. Whitley, and T. C. Merigan, eds.), Raven Press, New York, pp. 691–726.

89. Shope, R. E., 1985, Bunyaviruses, in: *Virology* (B. N. Fields, D. M. Knipe, R. M. Chanock, J. L. Melnick, B. Roizman, and R. E. Shope, eds.), Raven Press, New York, pp. 1055–1082.

90. Beaty, B. J., Sundin, D. R., Chandler, L. J., and Bishop, D. H. L., 1985, Evolution of bunyaviruses by genome reassortment in dually infected mosquitoes (*Aedes triseriatus*), *Science* **230**:548–550.

91. Beaty, B. J., Holterman, M., Tabachnick, W., Shope, R. E, Rozhon, E. J., and Bishop, D. H. L., 1981, Molecular basis of bunyavirus transmission by mosquitoes: Role of the middle-sized RNA segment, *Science* **211**:1433–1435.

92. Sundin, D. R., Beaty, B. J., Nathanson, N., and Gonzalez-Scarano, F., 1987, A G_1 glycoprotein epitope of LaCrosse virus: A determinant of infection of *Aedes triseriatus*, *Science* **235**:591–593.

93. Johnson, K. P., Lepow, M. L., and Johnson, R. T., 1968, California encephalitis. I. Clinical and epidemiological studies, *Neurology* **18**:250–254.

94. Parkin, W. E., Hammon, W. McD., and Sather, G. E., 1972, Review of current epidemiological literature on viruses of the California arbovirus group, *Am. J. Trop. Med. Hyg.* **21**: 964–978.

95. Balfour, H. H., Jr., Edelman, C. K., Cook, F. E., Barton, W. I., Buzicky, A. W., Siem, R. A., and Bauer, H., 1975, Isolates of California encephalitis (LaCrosse) virus from field-collected eggs and larvae of *Aedes triseriatus*: Identification of the overwintering site of California encephalitis, *J. Infect. Dis.* **131**:712–716.

96. Thompson, W. H., and Beaty, B. J., 1979, Venereal transmission of LaCrosse virus from male to female *Aedes triseriatus*, *Am. J. Trop. Med. Hyg.* **27**:187–196.

97. Centers for Disease Control, 1987, Update: *Aedes albopictus* infestation—United States, *Morbid. Mortal. Weekly Rep.* **36**:769–773.

98. Centers for Disease Control, 1987, Arboviral infections of the central nervous system— United States, *Morbid. Mortal. Weekly Rep.* **36**:450–455.

99. Grimstead, P. R., 1983, Mosquitoes and the incidence of encephalitis, *Adv. Virus. Res.* **28**: 157–233.

100. Centers for Disease Control, 1988, LaCrosse encephalitis in West Virginia, *Morbid. Mortal. Weekly Rep.* **37**:79–82.

101. Jamnback, T. L., Beaty, B. J., Hildreth, S. W., and Brown, K. L., 1982, Capture immuno-

globulin M system for rapid diagnosis of LaCrosse (California encephalitis) virus infections, *J. Clin. Microbiol.* **16:**577–580.

102. Johnson, K. P., and Johnson, R. T., 1968, California encephalitis. II. Studies of experimental infection in the mouse, *J. Neuropathol. Exp. Neurol.* **27:**390–400.
103. Janssen, R. S., Nathanson, N., Endres, M. J., and Gonzalez-Scarano, F., 1986, Virulence of LaCrosse virus is under polygenic control, *J. Virol.* **59:**1–7.
104. Jortner, B. S., Shope, R. E., and Manuelidis, E. E., 1971, Neuropathologic and virus assay studies of experimental California virus encephalitis in mouse, *J. Neuropathol. Exp. Neurol.* **30:**91–98.

IV

Perspectives

14

Antiglycolipid Immunity

Possible Viral Etiology of Multiple Sclerosis

H. E. WEBB

1. INTRODUCTION

The concept that infection plays a part in the pathogenesis of Multiple Sclerosis (MS) has been considered for over 100 years.[1] It is therefore appropriate to review the evidence to see why this concept is still very much in the forefront of present-day investigation into the etiology of this disease and to look at new ideas that might contribute to solving this elusive problem.

2. EPIDEMIOLOGIC EVIDENCE

Full reviews of the important factors in the epidemiology of MS are available.[2] In summary, over 200 prevalence studies of MS indicate that high-incidence areas of >30 cases/100,000 population are present throughout northern and central Europe, southern Canada, the northern United States, New Zealand, and southeast Australia, with some evidence of clustering that remains static. Rates of 5–30/100,000 are medium-frequency areas and include those areas around the Mediterranean, Israel, the Siberian and Ural areas of the U.S.S.R., southwest Norway, northern Scandinavia, most of Australia, and pos-

H. E. WEBB • Department of Neurovirology, UMDS, The Rayne Institute, St. Thomas' Hospital, London SE1 7EH, England.

Neuropathogenic Viruses and Immunity, edited by Steven Specter *et al.* Plenum Press, New York, 1992.

sibly Tunisia. Incidence less than 5/100,000 is seen in areas around the equator, the Caribbean, Mexico, Asia, Alaska, and Greenland.

The Caucasian populations of the high- and medium-risk areas are particularly prone to this disease. Migrants from high- to low-risk areas moving after the age of 15 years appear to carry with them the risk of their birthplace. Similarly, migrants under the age of 15 going from a high- to a low-risk area tend to acquire the incidence of the low-risk area. There is also evidence that those going from the low- to the high-risk areas increase their chance of acquiring MS. Epidemics of MS may have occurred in two areas, the Faroe Islands and Iceland.[2] In this report it was also established that there is a susceptibility related to genetic determinants, particularly as some families in which several cases of MS have occurred possessed a common gene. HLA typing suggests that people with the HLA class II antigen D/DR2 particularly may have a higher incidence of MS in the white population of North and Central Europe, North America, and Australia. Class I antigens A3 and B7 may not be as important as originally indicated. Thus, the geographical distribution and possibly the migrant data suggest that MS may be caused by acquired, exogenous environmental factors, the most likely being some sort of infection. The findings also suggest a long incubation period for the disease to manifest itself.

An etiologic role for virus(es) in the pathogenesis of this disease would seem to implicate a long-term, perhaps immunologic, damaging process rather than any direct cytolytic effect viruses may have. It also seems likely, if a long incubation period is relevant, that the immunologic stimulus may be subtle rather than obvious, so that a longer period is necessary for a damaging immunopathological reaction to occur.

3. EVIDENCE FROM ANTIVIRAL ANTIBODY STUDIES AND VIRUS ISOLATION

Many antiviral antibody studies have been done in both sera and cerebrospinal fluid (CSF). What has become quite clear is that no single virus is particularly associated with MS. Adams et al.[3,4] and Salmi et al.[5] report that measles virus might be important in relation to MS. Haire[6] discussed the significance of antiviral antibodies in MS, particularly those to measles. She discussed the significance of IgM activity to the membranes of measles-virus-infected cells. Elevated serum antibodies also have been found against herpes simplex virus (HSV),[7] canine distemper,[8] rubella,[9] and corona[10] viruses. Antimeasles antibodies, as well as antibodies against rubella and vaccinia, have been shown to be synthesized within the central nervous system (CNS) of MS patients.[11,12] Antibodies to simian virus 5 (SV 5) have been found in MS patients.[13]

Salmi et al.[14] showed evidence of intrathecal antibody synthesis to a far wider range of viruses. These include rubella, measles, parainfluenza 2, respiratory syncytial, influenza A and B, mumps, adeno-, HS and varicella–zoster, parainfluenza 3, corona OC43 and 229E, rota-, polio, and cytomegaloviruses. This has been confirmed for the first nine of these viruses (up to HSV).[15] Antibodies against up to 11 different viruses were synthesized simultaneously in the CNS in the same patients in the study of Salmi *et al*.[14] The intrathecal IgG index indicated that the antibody was being produced within the CNS. When they tried to relate intrathecal antibody synthesis to the clinical data in MS patients, they came to the conclusion that the bulk of the intrathecal synthesis of immunoglobulins and specific viral antibodies is not relevant to the pathogenesis of MS. However, they also concluded that random and continuous intrathecal antibody synthesis is a characteristic and unique feature in MS patients and that possibly a minor fraction of the antibody specificities may play a pathogenic role in the disease process. Vandvik *et al*.[16] showed that a small fraction of oligoclonal IgG bands in MS carry measles-specific activity. Nordal *et al*.[17] showed local synthesis of antibodies in the CNS of MS patients against measles, mumps, rubella viruses, and HSV, with antibodies to more than one virus present at the same time. However, they also found these antibodies in normal patients.

Viruses isolated from MS material include HSV,[18] parainfluenza virus type 1,[19] and coronavirus.[20] Paramyxovirus-like inclusions in MS brains have been seen by electron microscopy.[21,22]

The frequency of all these findings concerning the possible roles of different viruses as causes of MS tended to detract from the concept that they were involved in the etiology. However, if there was some common factor present that could produce CNS inflammation and demyelination among the many viruses incriminated in MS, then it might be possible to correlate disease with an etiology involving infection with more than one virus.

Many of the viruses that have been related to MS are enveloped budding viruses (Table 14-1). Enveloped budding viruses take host cell glycolipid into their coat (Fig. 14-1). This glycolipid presented in the virus coat to the immune system may be much more antigenic than glycolipid on its own. The fact that it can be derived from the cells of the CNS, including oligodendrocytes, and can be

TABLE 14-1
Enveloped Budding Viruses
that Have Been Reported in
Relation to MS

Herpes simplex	DNA
Varicella zoster	DNA
Vaccinia	DNA
Measles	RNA
Canine distemper	RNA
Mumps	RNA
Parainfluenza	RNA
Influenza	RNA
Corona	RNA
Rubella	RNA
Retroviruses	RNA

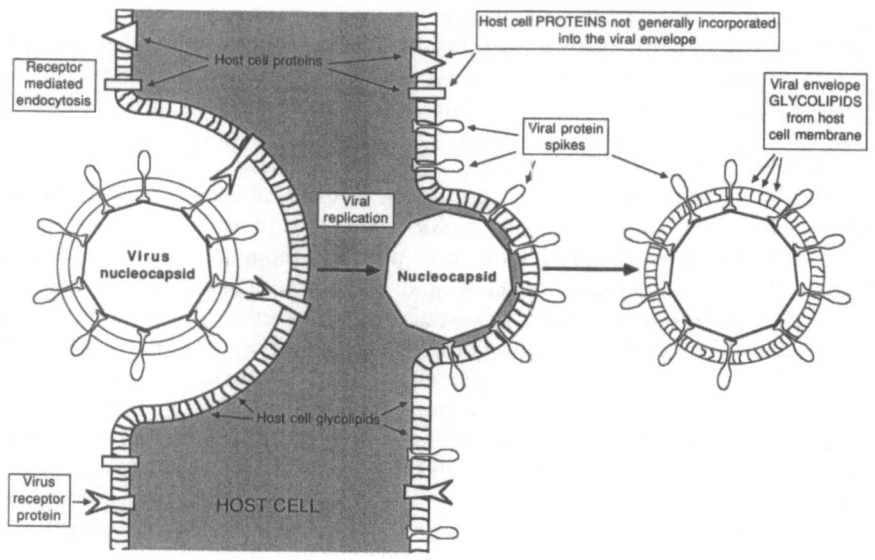

FIGURE 14-1. Incorporation of host cell glycolipid into the envelope of a budding virus.

presented from this partially immunologically privileged site to the peripheral immune system might provoke both cell-mediated and humoral antibody attacks on glycolipid in CNS cells and, in MS particularly, oligodendrocytes and myelin. The concept of this type of autoimmune damage has been put forward previously[23] and is discussed in detail later.

For viruses to cause the damage seen in MS, they have to be able to get to the brain. This may be accomplished with little difficulty, as most virus infections have a viremia, and the virus is likely to get into the brain transported in coated vesicles through endothelial cells.[24] Whether these viruses then do any harm is dependent on many factors, such as suitable cells for replication, the immune state of the host at the time of infection, and probably the HLA type of the individual infected. This latter point has been made clear in animal models in which the pathology may differ considerably depending on the host species and strains and their different major histocompatibility complex genes, some showing complete resistance to the disease process and others showing intense pathology. Theiler's murine encephalomyelitis virus is a good example in that there are susceptible and resistant strains of mice to this infection.[25] Viruses can also enter the CNS along nerves, e.g., HSV and rabies virus, but this is probably not relevant in the context of MS. It is of interest that Rogers *et al.*[26] isolated several viruses from kuru-infected chimpanzee brains. These animals were kept in very strict isolation and certainly were not expected to have latent virus infections in their brain cells in addition to the kuru agent, which had been the only pathogen inoculated. This indicates the vulnerability of the brain to virus infection but does not tell us how many of the viruses that may be present might be associated with a disease process.

4. GENERAL MECHANISMS BY WHICH VIRUS DAMAGE COULD OCCUR

Once in the brain, viruses may cause damage directly by cytolysis. This is unlikely in MS, as virus would be much easier to find in the brain, and the damage would be more acute and inflammatory than that usually seen.

Viruses may set up a chronic infection of cells, altering their metabolism, e.g., of oligodendrocytes, which, as a result, might cause demyelination. Once myelin is broken down, secondary inflammation is likely to occur in relation to the removal of breakdown products. If this mechanism was relevant, virus would certainly have been found with the availability of modern probing and isolation techniques.

Viruses might induce an immune response to virus antigen that is presented at the host cell membrane surface. This response also would be likely to be inflammatory in nature, destroying virus and perhaps the cell from which the virus is originating. Secondary inflammation also would occur. Again, this is unlikely in MS for the reasons stated above.

Virus could persist in cells in an unusual form, causing metabolic disturbances, for example in oligodendrocytes, which would upset the production and support for the myelin they produce. The form of these viruses may be such that we have not as yet developed the technology to recognize them. Once myelin breaks down, a secondary, inflammatory reaction would occur.

Finally viruses could set up an autoimmune reaction against cells of the CNS, e.g., against oligodendrocytes and/or myelin, thus producing demyelination and subsequent secondary inflammation. This may be an example of "molecular mimicry." Another way might be by the enveloped budding viruses,mentioned previously, presenting glycolipid host cell membrane from an immunologically privileged site to the peripheral immune system and setting up an antiglycolipid immunopathogenic response. It is possible that both mechanisms might act together.

5. MOLECULAR MIMICRY AND AUTOIMMUNITY RELATED TO VIRUSES AND THEIR POSSIBLE ROLE IN THE IMMUNOPATHOGENESIS OF DEMYELINATION

By direct comparisons of amino acid sequences, viruses have been shown to share common polypeptide sequences with certain host cell components. These studies have been done using computer analysis but do not indicate whether the sequences are at a site that might have biological significance. These polypeptide sequences could produce autoimmune reactions. Relevant examples of auto-immunity related to both the CNS and other organs of the body may provide insight into mechanisms of MS pathogenesis. It is presently uncertain, in some cases, whether the mechanism involved in each case is molecular mimicry or presentation of host cell membrane in the envelope of budding viruses. Fujinami and Oldstone[27] showed that hepatitis B virus polymerase (HBVP) shares six consecutive amino acids with the encephalitogenic site of rabbit myelin basic protein (MBP). Rabbits immunized with selected peptides from HBVP produced antibody that reacted with the predetermined sequences of HBVP and MBP. Peripheral blood mononuclear cells from these rabbits show a proliferative response when incubated with either MBP or HBVP. The rabbits develop a pathological change in their CNS somewhat similar to experimental allergic encephalomyelitis (EAE) following immunization with MBP.

Tardieu et al.,[28] using reovirus type I, report evidence of an autoimmune reaction. they confirm that many autoantibodies are produced that react with a large variety of normal tissues and that there are antigenic structures shared between viral determinants and normal tissue. Lane and Hoeffler[29] have shown that the SV 40 T antigen mimics a structure on a host cell protein. This cross-reactive protein is located within the nucleus of all of the mammalian cell types examined. Fujinami et al.,[30] using monoclonal antibodies (MAb), showed that the phosphoprotein of measles virus and a protein of HSV type 1 cross-react with an intermediate-filament protein, probably vimentin of human cells. It is difficult for autoantibody to react with intracellular antigens; although prior disruption of the cell by some other lytic action might expose antigens, subsequently an autoantibody reaction could result in further damage. Jahnke et al.[31] discuss sequence homology between viral proteins and the proteins MBP and P_2 related to encephalomyelitis and neuritis, particularly mentioning measles, Epstein–Barr, influenza A and B, and other viruses that cause upper respiratory infections. They point out that postinfectious or postvaccinial neuritis may be caused by immunologic cross-reactions evoked by specific viral antigenic determinants that are homologous to regions in the target myelins of the central and peripheral nervous systems (PNS).

Kagnoff et al.,[32] working on the pathogenesis of celiac disease, suggested that a human adenovirus type 12 (Ad 12) might be involved. They showed that α-gliadin, a component of wheat and an activator of celiac disease, shares a region of amino acid sequence homology with the 54-kDa E_1G protein of Ad 12, which is usually isolated from the intestinal tract. They proposed that this Ad 12 amino acid sequence could act similarly to the α-gliadin in wheat and activate the disease process.

Srinivasappa et al.,[33] in an analysis of over 600 MAbs raised against many DNA and RNA viruses, found that approximately 4% showed some cross-reaction with host determinants expressed on uninfected tissues. Several MAbs reacted with antigens in more than one organ. Although this may be an example of autoimmunity occurring through molecular mimicry, some MAbs are likely to have been produced by the presentation of host cell membranes by the viruses themselves because of their mode of replication by budding. Many of these would be directed against the glycolipid component of the host cell, because this is what a budding virus mainly picks up when it leaves the cell.

Miller et al.,[34] using the nonenveloped Theiler's murine encephalomyelitis virus, has shown by using functional T-cell analysis that there was a correlation with the extent of exact amino acid homology among the viral capsid proteins, the neuroantigens, purified rat and guinea pig MBP, human proteolipid protein, and related picornaviruses. Antigenic mimicry between measles virus and human T lymphocytes has also been shown.[35] The authors suggest that this might play a part in the immune suppression seen in measles.

Haspel et al.[36] have shown that mice inoculated with reovirus type 1 (non-enveloped virus) develop an autoimmune polyendocrine disease. They produced a large panel of hybridomas making monoclonal autoantibodies that reacted with cells in the islets of Langerhans, anterior pituitary, gastric mucosa, and with cell nuclei. Several of the autoantibodies recognized hormones, e.g., glucagon, growth hormone, and insulin. The exact antigenic determinants that were being recognized, i.e., protein, carbohydrate, or lipid, were not determined.

Huber and Lodge[37] have shown in mice that a coxsackievirus B type 3 (CVB-3) causes an extensive myocarditis. This work demonstrates that two distinct cytolytic T-lymphocyte (CTL) populations are present. One lyses uninfected myocytes (autoreactive), and the other lyses CVB-3-infected myocytes (virus specific). The lesions caused by the autoreactive CTL are more extensive and necrotizing than those caused by the virus-specific CTL. It is interesting that autoreactive CTL are not demonstrated in animals infected with a nonmyocarditic CVB-3 strain. Athymic nude mice do not develop myocarditis unless reconstituted with autoreactive CTL sensitized against the CVB-3 myocarditic strain, showing that the lesions are directly T-cell dependent. All these examples show how viruses can behave within and outside the nervous system. It is reasonable to consider the application of these findings to demyelinating nervous system pathology.

6. EXPERIMENTAL ALLERGIC ENCEPHALITIS, EXPERIMENTAL ALLERGIC NEURITIS, VIRUSES, AND DEMYELINATION

Reference in the previous section to viruses being able to mimic portions of MBP and P_2 raises the question of whether EAE and/or experimental allergic neuritis (EAN) induced by viruses could be an important mechanism of demyelination. The EAN is included because I believe the mechanism for the demyelination in the PNS is the same as that seen after virus infections. A similar viral etiologic mechanism could result in MS. The EAE model has been used in many laboratories for many years because it has been felt that the mechanisms involved in this disease might play a part in the etiology of MS. There are problems related to this concept, the first being that highly purified MBP produces inflammation with very little demyelination,[38–41] which is the major feature of MS. In fact, in the EAE model significant demyelination is more prominent and better seen when whole CNS white matter, particularly that of the spinal cord,[42] or MBP with cerebrosides (i.e., galactocerebroside) is inoculated.[43] Dubois-Dalcq et al.[44] and Fry et al.[45] show that rabbits immunized with galactocerebroside produce antiglial and demyelinating antibodies. Maggio and Cumar[46] find that only sulfatide antibodies are demonstrable after EAE has been induced by inoculation of MBP and adjuvant in animals.

Seil et al.[47] show that animals sensitized against MBP only usually lacked the in vitro demyelinating factor commonly found in animals given whole white matter. Similarly, Raine et al.[41] show that in experiments concerned with demyelination in vitro using sera against whole white matter, MBP, and galactocerebroside, the damage to myelin is associated with antigalactocerebroside activity and not anti-MBP antibody. Paterson[43] suggests that more attention needs to be paid to the role of cerebrosides not only in the production of EAE but also in demyelination. However, Zamvil et al.,[49] using T-cell clones specific for MBP, have induced chronic relapsing paralysis and demyelination in PL/SJ F_1 mice. Watanabe et al.[50] have shown that EAE-like lesions in rats can be induced by lymphocytes taken from Lewis rats infected with a coronavirus, but demyelination is not a feature. Prior to transfer these lymphocytes are restimulated with MBP. This model demonstrates that a virus infection of CNS tissue can initiate a pathological autoimmune response. This study would have been of greater interest had they used glycolipids as well as MBP to stimulate the lymphocytes, particularly with reference to causing demyelination.

As in demyelination in some animal virus models, the strain of animal has been of paramount importance in the production of EAE. Gasser et al.[51] and Williams and Moore,[52] by comparing the EAE-sensitive Lewis strain of rats to the resistant BN strain, show that it is likely that an autosomal dominant gene linked to the histocompatibility locus determines susceptibility to EAE by acting as an immune response gene. However, EAE can be produced in BN rats if rat or guinea pig spinal cord is used rather than MBP.[53] Perhaps this could be explained because some other constituent, possibly cerebrosides (glycolipids), in spinal cord tissue is contributing significantly to the EAE and the demyelination. Tsukada et al.[54] describe a chronic EAE-type lesion with demyelination induced by inoculation of guinea pigs with cerebral endothelial cell membrane known to be entirely free of MBP and proteolipid protein. They do not identify the factor that provokes this demyelinating EAE, nor do they appear to consider the possible role of membrane glycolipids as an etiologic agent.

In MS it seems that MBP is unlikely to be the significant factor for the demyelinating aspect of this disease. The MBP may contribute to the inflammation seen if molecular mimicry of MBP through virus infections can be established. Certainly for EAE-type pathogenesis to be relevant in MS, there has to be an initial "trigger factor" for the process to occur in humans. It appears that viruses could possibly do this, and if they do, it could be by molecular mimicry of MBP amino acid peptides. However, because molecular mimicry of MBP is unlikely to produce demyelination, there may be a simpler and more understandable mechanism, i.e., the presentation of nervous system host cell glycolipid by a variety of enveloped budding viruses (previously mentioned) and other organisms that are known to have cerebrosides in their envelope, e.g., mycoplasma,[55] causing an autoimmune demyelinating disease. This hypothesis is appealing and is deserving of further examination.

7. NERVOUS SYSTEM GLYCOLIPIDS AND MULTIPLE SCLEROSIS

Ideas on the etiology of MS are still fairly poorly developed. Whether glycolipids from the CNS presented to the peripheral immune system on virus envelopes can produce a response that leads to demyelinating damage that could account for the MS lesions seen in the CNS should be addressed. Most of the work to date concerning viruses has been centered on reactions against virus proteins and MBP, including molecular mimicry. Proteins are known to be very immunogenic, and the example of MBP producing EAE has been a major model for MS research.

A good short review of brain glycolipids as cell surface antigens is that by Leibowitz and Gregson.[56] Glycolipids, although a major constituent of myelin and indeed most host cell membranes, have not been properly examined because, compared with proteins, they are less immunogenic and more difficult to study. It is very difficult to get an optimum composition of a glycolipid antigenic mixture, as glycolipids are not soluble in water. The amount of antigen has to be determined empirically. Sensitization with glycolipids to produce a demyelinating disease has been considered only recently.

That glycolipids are immunogenic is unquestionable. Landsteiner[57] reports that the ceramide glycolipid of Forsmann antigen derived from type 1 pneumococci is antigenic without the aid of added protein. Heterologous anti-Forsmann antibodies introduced into a carotid artery produce a severe vascular lesion with hemorrhage, edema, and necrosis on the same side.[58] Landsteiner[57] also demonstrates that injecting brain into animals produces two sorts of immune sera: some react with proteins; others react with emulsions of an alcohol-soluble extract of brain and testicular tissue. One brain hapten described is soluble in hot alcohol. This suggests that the immune serum contains antibodies not only to proteins but against lipid material. This phenomenon of organ-specific antibodies reacting with alcoholic extracts is also obtained by injection of liver, lung, or leukocytes, often together with Wasserman antibodies. In each case it is likely that glycolipids are the antigens producing immunity.

Cerebrosides, sulfatides, and gangliosides constitute the major glycolipids of the brain.[59] The first two are myelin lipids, whereas ganglioside is mainly associated with the neuronal elements and only a little is present in glia and myelin. Niedieck and Palacios[60] state that these glycolipids are not complete antigens but are haptens and must be introduced with an immunogenic carrier to produce antibody. Czeonkowska and Leibowitz[61] show that a homologous carrier to which the animal is tolerant is ineffective in inducing antibodies to glycolipids, although generally they are immunogenic when presented in an intact membrane. Thus, it seems likely that the host cell glycolipid membrane taken by budding virus for its envelope and presented with highly antigenic virus protein will be highly immunogenic. Rapport et al.[62] state that antibody to glycolipid is directed to the carbohydrate part of the molecule, but it seems likely that the lipid plays some part in the reaction.[56]

8. EVIDENCE THAT ANTIGLYCOLIPID ACTIVITY OCCURS IN MULTIPLE SCLEROSIS

Arnon et al.[63] found antibodies to glycolipids present in 40% of MS patients. Since only a limited number of glycolipid antigens were used to test this, the percentage of positives might have been higher if more glycolipid antigens had been tested. Kasai et al.[64] found that anti-G_{M4} and antigalactocerebroside antibody titers were significantly raised in the CSF of MS patients as measured by a solid-phase radioimmunoassay but not in the sera. G_{M4} is a ganglioside with a long base chain that occurs mainly in human myelin. There was no rise of antibodies in the CSF to G_{M1} or MBP. A significant number of the anti-G_{M4} and antigalactocerebroside antibodies existed as immune complexes within the CNS. Duponey[65] has shown the presence of antigalactocerebroside antibodies in the serum of patients suffering from MS. Evidence of T-cell activity against gangliosides also has been shown in MS. Offner et al.[66] show that G_{T1} and G_{Q1b} are powerful stimulators of active E-rosetting lymphocytes from MS patients. Sela et al.[67] show raised ganglioside levels in serum and peripheral blood lymphocytes from MS patients in remission compared with controls. Ilyas and Davison[68] show hypersensitivity to gangliosides in MS with an E-rosetting technique. This reaction to glycolipids seems to be much more specific in the MS patients than the reaction to MBP, which also occurs in patients with other CNS disturbances. Davison and Ilyas[69] report inhibition of E- rosette formation by gangliosides by cyclosporine A, which blocks receptors for HLA-DR antigens on T lymphocytes. Bellamy et al.[70] show that the gangliosides G_{M1}, G_{D1A}, G_{D1B}, and G_{Q1b} stimulate T_4 and T_8 lymphocytes from the CSF of MS patients. All these studies suggest both humoral and cell-mediated immunity to CNS glycolipids. The reaction against glycolipids seems more specific to MS than that found against MBP.

9. EVIDENCE THAT HOST-DERIVED ENVELOPE MEMBRANE IN VIRUSES IS ANTIGENIC

Some 25 years ago Harboe et al.[71] showed that fowl plague and influenza viruses derived from the entodermal cells of chick chorioallantoic membrane can be neutralized by serum from rabbits that had been immunized against normal, uninfected chick chorioallantoic membrane. The biologically active nature of this rabbit-derived immune serum against the chick-derived virus is important. Feinsod et al.[72] show that Sindbis virus derived from Aedes aegypti mosquitoes is neutralized by immune serum made against whole-body extracts of uninfected A. aegypti, i.e., the mosquito cell line in which the virus replicates. This same serum does not neutralize Sindbis virus derived from Vero cells, indicating the potency and specificity of the serum made against the uninfected mosquito cell line on the virus that had grown in the same cells. Almeida and Waterson[73] show that only the viral protein spikes of corona virus derived from chicken fibroblasts can be labeled by immune serum made to the virus in chickens. However, when the coronavirus derived from chickens is used to make immune serum in rabbits, both the viral protein spikes and the intermediate membrane envelope of the virus are able to be labeled, showing that the chicken-derived virus is able to produce in the rabbit very considerable antichicken host-membrane antibody. This is an excellent example of the immunogenicity of host-cell membrane when presented in a viral envelope. Friend leukemia virus (a retrovirus) budding from erythrocyte membrane takes erythrocyte membrane antigen into its coat[74] and can lead to an immune reaction that causes hemolysis of normal, uninfected erythrocytes.[75]

Steck et al.[76] inoculated mice with the neurotropic strain of vaccinia virus and produced antibodies that bound to normal, uninfected myelin and oligodendrocytes, indicating that the virus presents to the immune system myelin and oligodendrocytic membrane components that are antigenic. This does not happen if the dermatropic strain of virus is used. Lindenmann and Klein[77] made use of viruses to present tumor tissue so that it became more immunogenic. They homogenized and lyophilized Ehrlich's ascites tumor cells (EAC) and gave them to Swiss A_2G mice and showed that they are not immunogenic when presents in this manner, and no protection to the mice occurs following challenge with live EAC. However, if the neurotropic strain (WSA) of influenza virus is inoculated into the EAC cells and the resulting progeny virus are used to immunize mice, strong immunity against EAC is produced, and the mice are protected from challenge with these neoplastic cells. However, if the same virus is cultured in eggs and given to mice, no protection occurs to the EAC. This is an example of how well viruses can present host cell membrane (in this case EAC tumor cell membrane) so as to be very specifically immunogenic as compared to the inoculation of the cells only.

Rook and Webb[78] showed that lymphocytes sensitized to tick-borne encephalitis virus (TBE) kill not only TBE-infected glial cells but also a significant percentage of normal, uninfected glial cells. This indicates that cytotoxic lymphocytes may destroy normal cells directly, including oligodendrocytes, the producers of myelin, which might be very important in the pathology of MS. There are many other examples in the literature of anti-host-cell-membrane effects induced by the glycolipid envelope membrane of budding viruses.

10. IS THERE EVIDENCE THAT ANTIGLYCOLIPID ACTIVITY PRODUCES DEMYELINATION OF THE CENTRAL OR PERIPHERAL NERVOUS SYSTEM?

It is worth quoting some of the evidence that this can occur. Anticerebroside antibodies have been shown to produce demyelination of myelinated axons in cell cultures.[41,44,45,48] Raine et al.[41] showed that antibody against galactocerebroside or against whole white matter produces demyelination in mouse spinal cord cultures, whereas anti-MBP antibody does not. The demyelinating factor could be absorbed out by galactocerebroside. These workers feel that galactocerebroside (a marker for oligodendrocytes) is a major target for antibody-mediated demyelination. As stated previously, it seems to be true that pure anti-MBP antibody does not produce demyelination, and similarly, pure MBP, when used to produce EAE, produces inflammation and minimal demyelination. However, if glycolipids are inoculated with MBP there is good demyelination.[43,79]

Nagai et al.[80] report lesions of the CNS and PNS of rabbits and guinea pigs after immunization with ganglioside G_{M1} and G_{D1a}. Saida et al.[81] produce an EAN using purified galactocerebroside. Konat et al.[82] produce an MS-like disease in rabbits using bovine brain gangliosides. Saida et al.[81] show in vivo demyelination by the intraneural injection of antigalactocerebroside serum. Hughes and Powell[83] show enhancement of P_2-induced demyelination in Lewis rats by galactocerebroside and glucocerebroside added to the immunizing emulsion. Carroll et al.[84] produce demyelination of the optic nerve by intraneural injection of antigalactocerebroside serum. Roth et al.[85] show that cultures of spinal cord, when exposed to galactocerebroside and whole white matter antiserum, show myelin damage. Also, but to a lesser extent, there is some damage from antibodies to gangliosides G_{M1} and G_{M4}. Sergott et al.[86] show that antigalactocerebroside serum demyelinates the optic nerve in vivo.

11. CAN VIRUS INFECTIONS OF THE NERVOUS SYSTEM PRESENT GLYCOLIPIDS IN SUCH A WAY AS TO BE IMMUNOGENIC?

The viruses that have at one time or another been thought to be involved in MS have been mentioned previously. All of them are budding viruses and incorporate lipids from the membrane of the host cell into their envelopes. Certainly all the viruses mentioned can enter the CNS, replicate there, and, because of their mode of replication, take CNS host-cell glycolipid into their envelope. In the Paramyxoviridae, which include measles, canine distemper, mumps, parainfluenza types 1–4, and Sendai virus, release of mature virus takes place by budding, and 20–40% of the dry weight of the virion is lipid.[87] Particular interest must center on measles, since in acute measles encephalitis there is perivascular demyelination. This virus has the capacity for latency, and it buds from cells taking host-cell glycolipid into its coat.

Experimentally, Klenk and Choppin[88] cultured the paramyxovirus SV 5 in four different host cells with different lipid compositions and determined that the lipid composition of the viral envelope very closely resembled that of each of the host-cell membranes from which it was derived. Lipids in the viral envelope have been shown to resemble closely those of the host cells from which they were derived in mumps,[89] influenza,[90,91] Sindbis,[92,93] Venezuelan equine encephalomyelitis,[94] and Semliki Forest virus (SFV).[95–97] The SFV [the avirulent A7(74) strain] has been used by us as a model for virus-induced demyelination in mice.[98] Mice infected with this virus show a pronounced development of antiglycolipid activity, which aroused our interest.[99,100] The mechanism involved in the production of this antiglycolipid activity could be relevant to what is seen in MS. It is therefore worthwhile to point out some of the relevance of this model to the immunogenicity of glycolipids.

In the SFV model demyelination occurs throughout the CNS including the optic nerve and spinal cord, with maximal effects noted between days 14 and 21 after infection.[98,101,102] Remyelination is well in progress by day 35. The demyelination is dependent on T lymphocytes, as nude (T-cell-deficient) mice, in spite of high virus titers, do not demyelinate until they are given normal T cells from their *nude*/+ littermates.[103,104] Natural-killer-cell-deficient mice demyelinate well when SFV is used, and demyelination continues to occur normally in complement depleted mice. Repeated inoculation of immune sera, either intraperitoneally or intracerebrally, does not produce demyelination.[105] The first cells to enter the white matter from the perivascular cuffs are activated lymphoblastic-type cells, the majority of which do not stain with anti-IgG, -IgM, or -IgA. They can be seen in direct contact with the myelin just prior to the onset of demyelination and may be CTL.[106]

Amor and Webb[100] show a rise in antiglycolipid activity against both total neutral glycolipids and galactocerebrosides with SFV. This appears to give some protection to the mouse from further challenge with an antigenically unrelated encephalitogenic enveloped budding virus, TBE virus (Langat strain), which replicates in the same glial cells as SFV. The results suggest that cross-protection arises from immunity to common host glycolipid contained in the envelope of both viruses. However, further challenge with Langat virus not only increases the demyelination if given within 2 weeks of the SFV but delays the remyelination significantly if given later. Antiglucocerebroside, -ganglioside, and -galactocerebroside sera react with brain-derived SFV in an immunoenzymatic assay ELISA[107] and label brain-derived SFV budding virus[108] as observed by electron microscopy. Khalili-Shirazi et al.[99] have raised many MAbs against SFV, and these have been found to be against glycolipids and to label brain-derived "budding" SFV both by immunoelectron microscopy and by immunocytochemistry.[109] One of these anti-SFV glycolipid MAbs (308) also labels TBE virus in mouse brain, which is antigenically unrelated to SFV but buds from CNS cell membranes, again indicating a common host antigen.[110]

Khalili-Shirazi et al.[99] also raised MAbs against whole myelin. Some of these were against myelin proteins, and some against myelin glycolipids. One anti-myelin-glycolipid MAb (212) reacted with brain-derived SFV in an ELISA. Both the anti-SFV glycolipid MAb and myelin glycolipid MAb had some biological activity against SFV, producing either neutralization or steric hindrance. A recent MAb (555) raised against SFV, an antigalactocerebroside antibody, has very significant neutralizing activity against SFV and labels SFV as observed using electron microscopy. A further MAb (373), raised against brain-derived "inactivated" SFV, cross-reacts with sulfatides and galactocerebroside and also neutralizes SFV significantly.[111] This MAb 373 is of special interest as it labels SFV, influenza, and measles virus, which replicate in the same brain cell cultures from which the SFV is derived.[112]

This indicates that budding viruses take similar glycolipid into their coat if the original host cell of replication is the same. Influenza virus derived from mouse brain cell cultures is also labeled by antiglucocerebroside serum (S. Pathak, personal communication).

The immune response to glycolipids is thought to be T-cell independent, and most antiglycolipid antibodies are IgM, as are the antiglycolipid MAbs described above; IgG antiglycolipid antibodies are occasionally reported. Very little work has been done on the significance of IgM antibodies in MS.

The demyelination in this SFV model can be made significantly worse if a second inoculation of the avirulent strain of SFV, A7(74), is given within 14 days of the first injection,[113] even though the A7(74)-SFV-infected mice are completely protected against challenge with a lethal dose of the virulent L_{10} strain of SFV within 24 hr following the original A7(74) infection. This indicates that immunity to the lethal effect of the virulent virus develops very early.[114]

Even in the severely demyelinated mice following two injections of the avirulent strain, considerable repair of myelin takes place by day 45. For this to be able to occur, it seems likely that the oligodendrocytes are not destroyed, and this is confirmed by pathological examination. However, clearly, this cell or the myelin it supports must be under attack and physiologically not functioning properly. As yet, it is not clear whether this may be a result of persistent replication of virus within the cell or viral antigen present on the surface of oligodendrocytes reacting with anti-SFV antibody causing low-grade inflammation or of a direct attack on the oligodendrocyte/myelin membrane combination by CTL as mentioned previously.

Other mechanisms might be, in view of the rapid rise of antiglycolipid activity, a T-cell-mediated and/or antibody attack on glycolipids in the oligodendrocytic/ myelin membrane. Antiglycolipid activity is easily detectable by day 14.[100] Some research workers have suggested that the antiglycolipid activity seen is secondary to CNS damage and demyelination both in MS and in this model. We do not believe that to be the case in the SFV model, as mice given one dose of an inactivated brain-derived SFV vaccine develop very significant antiglycolipid activity in the blood by postvaccination day 18 without CNS damage. There is now some evidence that subsequent doses of SFV brain-derived vaccine will produce inflammation and demyelination.[110] Gliosis eventually occurs in the mouse brain following SFV infection.[115]

12. CONCLUSION

There now appears to be very considerable evidence that viruses can act as carriers for host-cell glycolipids and present them in an immunogenic manner. It seems a reasonable possibility that enveloped budding viruses can evoke an autoimmune reaction against nervous system glycolipid simply by the method of their replication, i.e., replicating in the brain and presenting host CNS glycolipid to the peripheral immune system, which in turn produces a cell-mediated and antibody response against nervous system glycolipids. This could apply not only to CNS glycolipids but also to PNS glycolipids. The Guillain–Barré syndrome

(GBS) following virus infection has still to be explained. Antiglycolipid antibodies occur in both GBS[116] and MS.[63] Many workers feel that an antiglycolipid immune response is probably occurring in these situations as a secondary reaction to damage of the nervous system and to myelin in particular. This is not necessarily the correct explanation. In fact, it is unlikely that glycolipids released in this way would produce an immune response with autoantibodies.

Both Allison[117] and Roitt[118] stress that in most cases release of tissue components directly into the circulation does not stimulate the production of autoantibodies. To produce autoimmunity the antigen must be presented in a manner acceptable to the immune system. Thyroiditis can be produced in rabbits if thyroid antigens are injected with Freund's adjuvant.[119] Konat et al.[82] produce EAE in rabbits by presenting the glycolipids in an inoculum with Freund's adjuvant. Although glycolipids are immunogenic, they can behave as haptens, and a carrier will increase their immunogenicity.[120,121] Although the carbohydrate moiety of the glycolipid is the most immunogenic, it must remain inconclusive at the moment whether the carbohydrate and/or the lipid is the most important immunogenic factor in the pathogenesis of demyelination. However, viruses can facilitate immunologic reactions in other ways, which may assist immunopathogenesis. In the ordinary state of events, budding viruses coated with host-cell glycolipids are known to be taken up by antigen-presenting cells. In this process they become uncoated, and their envelope, containing the glycolipid, is likely to be processed in the same way as viral proteins, thus allowing sensitization to the glycolipid of the cell of origin.

Viruses are known to induce the production of interferons α and β,[122] which in turn may stimulate class I antigen expression on brain cells.[123] For CTL attack on virally infected cells to occur, class I antigen expression is essential.[124] Suzumura et al.[125] show induction of class I antigens on oligodendrocytes and astrocytes following coronavirus infection; thus, these nervous system cells could be susceptible to CTL attack. Viruses, indirectly, can help to induce interferon γ by stimulating a T-cell immune response. It is activated T cells that induce interferon γ. Wong et al.[126] show that interferon γ induces a dramatic increase both in class I antigens on astrocytes, oligodendrocytes, microglia, and some neurons and class II antigens on some astrocytes. Coronavirus has been shown to induce class II antigens in astrocytes as well as class I.[127] Furthermore, class II antigens have been shown to be present on the surface of human oligodendrocytes and astrocytes.[128] Such activated cells could act as antigen-presenting cells. This function of viruses may indirectly render cells more vulnerable to immunologic recognition and damage. The retrovirus feline leukemia virus has been shown to incorporate class II antigens into its envelope, and therefore virions could present antigen.[129] If other viruses were to be shown to incorporate class II MHC antigens during the budding process, this would be a very significant finding.

The CNS, particularly, is considered a partially immunologically privileged site, as it lies within the blood–brain barrier. Foreign cells inoculated into the brain there are less easily rejected. Certainly tolerance to brain antigens by the vertebrate host is less highly developed than in many other tissues. Brain tissue is often damagingly immunogenic even if it is "self," as shown by Kabat et al.,[130] who took six individual monkeys and removed from each a portion of brain. The brain tissue was emulsified and reinoculated back peripherally with adjuvants into the monkey from which the respective brain had come. All five monkeys that survived the operation developed acute CNS lesions with demyelination. This sensitivity to brain tissue has also been a factor in the development of vaccines. The history of the brain-derived virus vaccines has been notoriously associated with paralytic incidences, e.g., the Pasteur and Semple-type rabies vaccines. This emphasizes that any mechanism that presents host-derived nervous tissue to the peripheral immune system can be dangerous in respect to neuroparalytic accidents with demyelination.

Enveloped viruses could theoretically incorporate MBP from some cell membranes, but it has never been shown to occur. However, the amino acid sequences of MBP associated with mimicry caused by viruses may be important if the site at which this occurs is appropriate to provoking a pathological immune response.

Myelin basic protein remains unlikely to be the cause of demyelination in MS for reasons previously stated. perhaps a combination of mimicry and glycolipid presentation by viruses may be the crucial factor in the inflammation and demyelination seen. However, to me the evidence tends toward the concept that nervous system glycolipids presented in budding virus envelope form a more likely initiator of this disease process. For a simple diagrammatic representation of the concept, see Fig. 14-2.

13. SUMMARY OF THE EVIDENCE THAT VIRUS-INDUCED ANTIGLYCOLIPID ACTIVITY MAY BE IMPORTANT IN MULTIPLE SCLEROSIS

1. Enveloped budding viruses take the host cell membrane glycolipid into their coat.[88-91,93-97]
2. Different viruses using the same cell for replication will have similar glycolipids in their coats.
3. The principal lipid haptens of the mammalian cell are glucoceramides,[56] and the major glycolipids of the CNS are cerebrosides, sulfatides, and gangliosides.[59]
4. Galactocerebroside is the major glycolipid of the CNS accounting for about 17% of the dry weight of adult human white matter and is an oligodendrocyte marker.[131]
5. Glycolipids are immunogenic, particularly as a component of an intact surface membrane.[56,57,132]
6. Galactocerebroside is exposed at the surface of myelin. Ganglioside is also present in myelin to a lesser extent.[56]
7. Immune reactions against both galactocerebroside and ganglioside can damage myelin and cells.[44,48,133,134]
8. Viruses can stimulate an antiglycolipid immune response.[99,100]
9. Enveloped budding viruses can be labeled by immunoelectron micros-

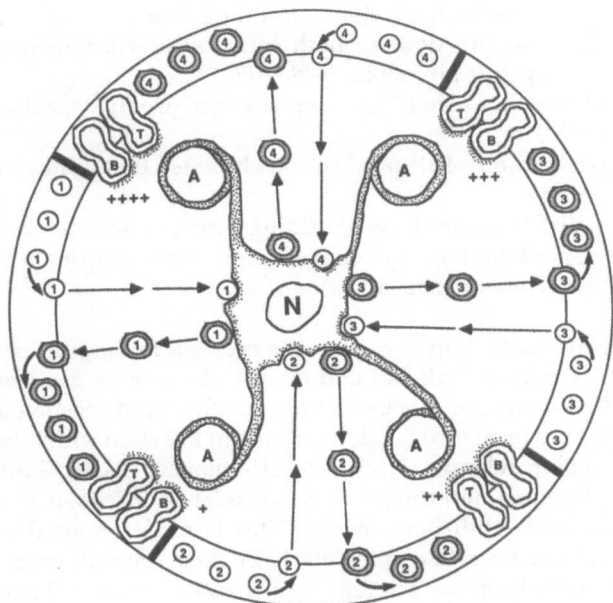

FIGURE 14-2. The possible role of recurrent infections of the CNS in MS in genetically susceptible individuals (HLA type). A neurotropic enveloped virus, for example measles (1), enters the brain and replicates in the cells of the CNS including, e.g., oligodendrocytes. The envelope of the budding virus is derived from the lipids of the host cell membranes. Glycolipids in the envelope of virions returning to the blood may be antigenic in association with the viral proteins, which may act as carrier determinants. Glycolipid-sensitized lymphocytes then enter the brian by diapedesis and attack either the myelin directly or the myelin-supporting cells. This results in demyelination and clinical relapse. After some time suppressor T cells are generated and control the reaction, resulting in remission. At a later date a second virus, e.g., a coronavirus (2) or an influenza virus (3), or a virus that has been latent and now become reactivated, enters, replicates in the brain, and returns to the circulation, presenting the same brain-specific glycolipid(s) in its envelope. The immune response is restimulated, resulting in a second, third, fourth, or fifth relapse. Remission intervenes as the T-suppressor cells control the response after each restimulation by virus. In this way any number of enveloped neurotropic viruses could be involved in initiating and restimulating an autoimmune response to the same brain cell membrane-specific glycolipid(s). Semliki forest virus (4) is included in the figure because it produces immune-mediated demyelination in experimental infection of mice. The figure represents a simplified concept of the foregoing hypothesis. The argument could be applied to other organisms, e.g., *Mycoplasma pneumoniae*, whose membrane constituents react with antibodies made against cerebrosides and indeed have been shown to react with antibodies produced in the CSF of multiple sclerosis patients.[55] Stippling, oligodendrocytic lipid membrane; N, nucleus of oligodendrocyte; T, T lymphocyte; B, B lymphocyte; A, axon.

copy and by immunocytochemistry and shown to react in an ELISA with various antiglycolipid sera and anti-RNA virus glycolipid MAb.[99,107–109,112]

10. Antiglycolipid MAb made against myelin will react in an ELISA with a demyelinating RNA enveloped budding virus.[99]

11. Patients with MS develop both humoral and cell-mediated immune reactions against glycolipids.[63,64,66–70]

12. Glycolipids inoculated into animals can produce demyelinating disease.[80,81,84–86]

13. Glycolipids and MBP produce much better demyelination than MBP alone.[41–43,47]

14. Many different enveloped budding viruses could be involved in this process, and this might account for some of the relapses in MS associated with intercurrent virus infection.

This concept would help to explain the previous findings of many different viruses being associated with MS and classify the disease as "a virus-induced antiglycolipid autoimmune disease." Once the glycolipid reaction against brain tissue glycolipid has been instigated, intercurrent infection with other organisms might also promote relapses. Husby et al.[135] showed that IgG antibody from children with rheumatic fever reacts with neuronal cytoplasm, particularly of the human caudate and subthalamic nuclei. This factor is removed by absorption with group A streptococci and particularly by their cell wall preparations. The antineuronal antibody appears to represent cross-reactions with antigens shared by group A streptococcal membranes and the neuronal cells.

Complement-fixing antibodies against *Mycoplasma pneumoniae* (MPN), more frequently IgM than IgG antibodies, are cross-reactive with nervous tissue, namely with cerebrosides.[136] Maida[55] has shown immunologic reactions against MPN in 18 cases of MS. The CSF titers are as high as or higher than the corresponding serum titers, indicating intrathecal antibody synthesis. I feel that his study does not rule out that these anticerebroside antibodies might be cross-reactive anti-MPN antibodies.

This is a concept that might be applied to the pathogenesis of other demyelinating diseases such as tropical spastic paraparesis associated with HTLV-I[137] and to the myelopathy found in the areas of Japan where HTLV-I leukemias are prevalent. Roman[138] draws attention to the fact that anti-HTLV-I-type antibodies have been found in patients given the definite clinical diagnosis of MS in Florida and Japan. He goes on to suggest that tropical spastic paraparesis, the Japanese myelopathy, and perhaps an MS-like neurological syndrome may represent clinical variants of the same disease, a retroviral myelopathy. The retroviruses may be of particular interest in this respect because HTLV-II can trigger transcription of mRNA for the interleukin 2 gene and the gene for its receptor.[139] This may assist stimulation of CTL.

I personally believe at this moment that the demyelinating disease multiple sclerosis occurs in people of suitable HLA type as a result of an enveloped virus infection acquired in early childhood. I believe that the initial damage to myelin may be produced by a CTL sensitized to brain myelin glycolipids migrating from the perivascular cuffs, the perivascular cuffs representing the earliest lesion seen in MS. Few CTL would be necessary to initiate the process. Once the damage has been initiated, secondary reactions will occur as a result of the breakdown of myelin, producing further damage and finally gliosis. As to whether molecular mimicry of MBP by viruses or MBP directly plays a part remains to be seen. The evidence for MBP alone playing a part in the pathogenesis of MS and the demyelination seen is very poor. Myelin basic protein is not expressed on the surface of myelin or oligodendrocytes.[140]

The general concept described here might also be applied to postviral PNS neuritis and perhaps postviral thyroiditis, pancreatitis, oophoritis, and testiculitis, the latter seen particularly after mumps virus infection. A recent paper by Fujinami *et al.*[141] describes a MAb produced by the CNS demyelinating Theiler's murine encephalomyelitis virus (TMEV) that reacts with both galactocerebroside and TMEV. This is of particular interest, as TMEV is a nonenveloped picornavirus. They suggest possible mechanisms by which a virus of this nature might produce this antigalactocerebroside response, one being that areas on picornavirus surfaces have hydrophobic pockets that could accommodate glycolipid-like structures (J. Hogle, Research Institute of Scripps Clinic, personal communication). This indicates the possibility that even in this group of viruses glycolipids can be presented on the surface in a significantly immunogenic fashion. Although glycolipids are very difficult to work with, this must not be made an excuse for not investigating this concept further: the work should be done to prove that these ideas are relevant or misfounded.

ACKNOWLEDGMENTS. I am very grateful to the MS Society of Great Britain and Northern Ireland, the Wellcome, Phillip Fleming, and Sir Jules Thorn Trusts, and the Charitable Funds of St. Thomas' Hospital for supporting work related to the virally induced antiglycolipid activity in relation to demyelinating disease; also to members of my own unit who have contributed so much, to distinguished scientific friends who have criticized this manuscript, and to my secretary Miss Joan Olejnik for all the typing involved.

I also wish to thank Professor J. B. Cavanagh for permission to publish Fig. 14-2 as it appeared in *Neuropathology and Applied Neurobiology*, 1984.

REFERENCES

1. Marie, P., 1884, Sclerose en plaque et maladies infectieuses, *Prog. Med.* **12**:287–289.
2. Kurtzke, J. F., 1983, Epidemiology of multiple sclerosis, in: *Multiple Sclerosis* (J. F. Hallpike, C. W. M. Adams, and W. W. Tourtellotte, eds.), Chapman and Hall, London, pp. 47–95.
3. Adams, J. M., and Imagawa, D. T., 1962, Measles antibodies in multiple sclerosis, *Proc. Soc. Exp. Biol. Med.* **111**:562–566.
4. Adams, J. M., Brooks, M. B., Fisher, E. D., and Tyler, C. S., 1970, Measles antibodies in patients with multiple sclerosis and with other neurological and non-neurological diseases, *Neurology* **20**:1039–1042.
5. Salmi, A. A., Norrby, E., and Panelius, M., 1972, Identification of different measles virus-

specific antibodies in the serum and cerebrospinal fluid from patient with subacute sclerosing panencephalitis and multiple sclerosis, *Infect. Immun.* **6**:248–254.

6. Haire, M., 1977, Significance of virus antibodies in multiple sclerosis, *Br. Med. Bull.* **33**:40–44.

7. Catalano, L. W., Jr., 1972, Herpesvirus hominis antibody in multiple sclerosis and amyotrophic lateral sclerosis, *Neurology* **22**:473–478.

8. Cook, S. D., Dowling, P. C., and Russell, W. C., 1979, Neutralizing antibodies to canine distemper and measles virus in multiple sclerosis, *J. Neurol. Sci.* **41**:61–70.

9. Horikawa, Y., Tsubaki, T., and Nakajima, M., 1973, Rubella antibody in multiple sclerosis, *Lancet* **1**:996–997.

10. Burks, J. S., Devald-Macmillan, B., Jankovsky, L., and Gerdes, J., 1979, Characterization of coronaviruses isolated using multiple sclerosis autopsy brain material, *Neurology* **29**:547.

11. Norrby, E., Link, H., Olsson, J. E., Panelius, M., Salmi, A. G., and Vandvik, B., 1974, Comparison of antibodies against different viruses in cerebrospinal fluid and serum samples from patients with multiple sclerosis, *Infect. Immun.* **10**:688–694.

12. Cremer, N. E., Johnson, K. P., Fein, G., and Litosky, W. H., 1980, Comprehensive viral immunology of multiple sclerosis II. Analysis of serum and CSF antibodies by standard serologic methods, *Arch. Neurol.* **37**:610–615.

13. Goswami, K. K. A., Randall, R. E., Lange, L. S., and Russell, W. C., 1987, Antibodies against the paramyxovirus SV5 in cerebrospinal fluids of some multiple sclerosis patients, *Nature* **327**:244–246.

14. Salmi, A., Reunanen, M., Ilonens, J., and Panelius, M., 1983, Intrathecal antibody synthesis to virus antigens in multiple sclerosis, *Clin. Exp. Immunol.* **52**:241–249.

15. Arnadottir, T., Reunanen, M., and Salmi, A., 1982, Intrathecal synthesis of virus antibodies in multiple sclerosis patients, *Infect. Immun.* **38**:399–407.

16. Vandvik, B., Norrby, E., Nordal, H., and Degre, M., 1976, Oligoclonal measles virus specific IgG antibodies isolated by virus immunoabsorption of cerebrospinal fluids, brain extracts, and sera from patients with subacute sclerosing panencephalitis and multiple sclerosis, *Scand. J. Immunol.* **5**:979–992.

17. Nordal, H. J., Vandvik, B., and Norrby, E., 1978, Multiple sclerosis: Local synthesis of electrophoretically restricted measles, rubella, mumps and herpes simplex virus antibodies in the central nervous system, *Scand. J. Immunol.* **7**(6):473–479.

18. Gudnadottir, M., Helgadottir, H., Bjarnason, O., and Jonsdottir, K., 1964, Virus isolated from the brain of a patient with multiple sclerosis, *Exp. Neurol.* **9**:85–95.

19. Meulen ter, V., Koprowski, H., Iwasaki, Y., Kackell, Y. M., and Muller, D., 1972, Fusion of cultured multiple-sclerosis brain cells with indicator cells: Presence of nucleocapsids and virion and isolation of parainfluenza-type virus, *Lancet* **1**:1–5.

20. Burks, J. S., DeVald, B. L., Jankovsky, L. D., and Gerdes, J. C., 1980, Two coronaviruses isolated from central nervous system tissue of two multiple sclerosis patients, *Science* **209**:933–934.

21. Prineas, J., 1972, Paramyxovirus-like particles associated with acute demyelination in chronic relapsing multiple sclerosis, *Science* **178**:760–763.

22. Pathak, S., and Webb, H. E., 1976, Paramyxovirus-like inclusions in brain of patient with severe multiple sclerosis, *Lancet* **2**:311.

23. Webb, H. E., and Fazakerley, J. K., 1984, Can viral envelope glycolipids produce autoimmunity, with reference to the CNS and multiple sclerosis? *Neuropathol. Appl. Neurobiol.* **10**:1–10.

24. Pathak, S., and Webb, H. E., 1974, Possible mechanisms for the transport of Semliki Forest virus into and within mouse brain, *J. Neurol. Sci.* **23**:175–184.

25. Rodriguez, M., Pease, L. R., and David, C. S., 1986, Immune-mediated injury of virus-infected oligodendrocytes, *Immunol. Today* **7**(12):359–362.

26. Rogers, N. G., Basnight, M., Gibbs, C. J., Jr., and Gajdusek, D. C., 1967, Latent viruses in chimpanzees with experimental kuru, *Nature* **216**:446–449.

27. Fujinami, R. S., and Oldstone, M. B. A., 1985, Amino acid homology between the encephalitogenic site of myelin basic protein and virus: Mechanism for autoimmunity, *Science* **230**:1043–1045.
28. Tardieu, M., Powers, M., Hafler, D. A., Hauser, S. L., and Weiner, H. L., 1984, Autoimmunity following viral infection: Demonstration of monoclonal antibodies against normal tissue following infection of mice with reovirus and demonstration of shared antigenicity between virus and lymphocytes, *Eur. J. Immunol.* **14**:561–565.
29. Lane, D. P., and Hoeffler, W. K., 1980, SV40 large T shares an antigenic determinant with a cellular protein of molecular weight 68,000, *Nature* **288**:167–170.
30. Fujinami, R. S., Oldstone, M. B. A., Wroblewska, Z., Frankel, M. E., and Koprowski, H., 1983, Molecular mimicry in virus infection: Cross reaction of measle virus phosphoprotein or of herpes simplex virus protein with human intermediate filaments, *Proc. Natl. Acad. Sci. U.S.A.* **80**:2346–2350.
31. Jahnke, U., Fischer, E. H., and Alvord, E. C., Jr., 1985, Sequence homology between certain viral proteins and proteins related to encephalomyelitis and neuritis, *Science* **229**:282–284.
32. Kagnoff, M. F., Austin, R. K., Hubert, J. J., Bernardin, J. E., and Kasarda, D. D., 1984, Possible role for a human adenovirus in the pathogenesis of celiac disease, *J. Exp. Med.* **160**:1544–1557.
33. Srinivasappa, J., Saegusa, J., Prabhakar, B. S., Gentry, M. K., Buchmeier, M. J., Wiktor, T. J., Koprowski, H., Oldstone, M. B. A., and Notkins, A. L., 1986, Molecular mimicry: Frequency of reactivity of monoclonal antiviral antibodies with normal tissues, *J. Virol.* **57**:397–401.
34. Miller, S. D., Clatch, R. J., Pevear, D. C., Trotter, J. L., and Lipton, H. L., 1987, Class II-restricted T cell responses in Theiler's murine encephalomyelitis virus (TMEV)-induced demyelinating disease, *J. Immunol.* **138**:3776–3784.
35. Bahmanyar, S., Srinivasappa, J., Casali, P., Fujinami, R. S., and Oldstone, M. B. A., 1987, Antigenic mimicry between measles and human T lymphocytes, *J. Infect. Dis.* **156**:526–527.
36. Haspel, M. V, Onodera, T., Prabhakar, B. S., Horita, M., Suzuki, H., and Notkins, A. L., 1983, Virus-induced autoimmunity: Monoclonal antibodies that react with endocrine tissues, *Science* **230**:304–306.
37. Huber, S. A., and Lodge, P. A., 1984, Coxsackie virus B-3 myocarditis in Balb/c mice. Evidence for autoimmunity to myocyte antigens, *Am. J. Pathol.* **116**:21–29.
38. Paterson, P. Y., 1976, Experimental autoimmune (allergic) encephalomyelitis. Induction, pathogenesis and suppression, in: *Textbook of Immunopathology* (P. A. Miescher and J. J. Mueller-Eberhard, eds.), Grune & Stratton, New York, pp. 225–251.
39. Raine, C. S., Traugott, V., Iqbal, K., Snyder, D. S., Cohen, S. R., Farooq, M., and Norton, W. T., 1978, Encephalitogenic properties of purified preparations of bovine oligodendrocytes tested in guinea pigs, *Brain Res.* **142**:85–96.
40. Tabara, T., and Endoh, M., 1985, Humoral immune response to myelin basic protein, cerebroside and ganglioside in chronic relapsing experimental allergic encephalomyelitis of the guinea pig, *J. Neurol. Sci.* **67**:201–212.
41. Raine, C. S., Traugott, U., Farooq, M., Bornstein, M. B., and Norton, W. T., 1981, Augmentation of immune-mediated demyelination by lipid haptens, *Lab. Invest.* **45**:174–182.
42. Hoffman, P.M., Gaston, D. D., and Spitler, L. E., 1973, Comparison of experimental allergic encephalomyelitis induced with spinal cord, basic protein, and synthetic encephalitogenic peptide, *Clin. Immunol. Immunopathol.* **1**:364–371.
43. Paterson, P. Y., 1977, Autoimmune neurological disease: Experimental animal system and implication for multiple sclerosis, in: *Autoimmunity* (N. Talal, ed.), Academic Press, New York, p. 658.
44. Dubois-Dalcq, M., Niedieck, B., and Buyse, M., 1970, Action of anti-cerebroside sera on myelinated nervous tissue cultures, *Pathol. Eur.* **5**:331–347.

45. Fry, J. M., Weissbarth, S., Lehrer, G. M., and Bornstein, M. B., 1974, Cerebroside antibody inhibits sulfatide synthesis and myelination and demyelinates in cord tissue cultures, *Science* **183**:540–542.

46. Maggio, B., and Cumar, F. A., 1975, Experimental allergic encephalomyelitis: Dissociation of neurological symptoms from lipid alterations in brain, *Nature* **253**:364–365.

47. Seil, F. J., Falk, G. A., Kies, M. W., and Alvord, E. C., 1968, The *in vitro* activity of sera from guinea pigs sensitized with whole CNS and with purified encephalitogen, *Exp. Neurol.* **22**: 545–555.

48. Saida, T., Saida, K., and Silberberg, D. H., 1979, Demyelination produced by experimental allergic neuritis serum and antigalactocerebroside antiserum in CNS cultures: An ultrastructural study, *Acta. Neuropathol. (Berl.)* **48**:19–25.

49. Zamvil, S. S., Nelson, P. A., Mitchell, D. J., Knobler, R. L., Fritz, R. B., and Steinman, L., 1985, Encephalitogenic T cell clones specific for myelin basic protein. An unusual bias in antigen recognition, *J. Exp. Med.* **162**:2107–2124.

50. Watanabe, R., Wege, H., and ter Meulen, V., 1983, Adoptive transfer of EAE-like lesions from rats with coronavirus-induced demyelinating encephalomyelitis, *Nature* **305**:150–153.

51. Gasser, D. L., Newlin, C. M., Palm, J., and Gomatas, N. K., 1973, Genetic control of susceptibility to experimental allergic encephalomyelitis in rats, *Science* **181**:872–873.

52. Williams, R. H., and Moore, M. J., 1973, Linkage of susceptibility to experimental allergic encephalomyelitis to the major histocompatibility locus in the rat, *J. Exp. Med.* **138**: 775–783.

53. Levine, S., and Sowinski, R., 1975, Allergic encephalomyelitis in the reputedly resistant brown Norway strain of rats, *J. Immunol.* **114**:597–601.

54. Tsukada, N., Koh, C.-S., Yanagisawa, N., Okano, A., Behan, W. M. H., and Behan, P. O., 1987, A new model for multiple sclerosis: Chronic experimental allergic encephalomyelitis induced by immunization with cerebral endothelial cell membrane, *Acta. Neuropathol. (Berl.)* **73**:259–266.

55. Maida, E., 1983, Immunological reactions against *Mycoplasma pneumoniae* in multiple sclerosis: Preliminary findings, *J. Neurol.* **229**:103–111.

56. Leibowitz, S., and Gregson, N. A., 1979, Brain glycolipids as cell-surface antigens, in: *Clinical Neuroimmunology* (F. C. Rose, ed.), Blackwell Scientific Publications, London, pp. 29–41.

57. Landsteiner, K., 1947, *The specificity of Serological Reactions*, Harvard University Press, Cambridge, pp. 210–240.

58. Leibowitz, S., Morgan, R. S., Berkinshaw-Smith, E. M. I., and Payling-Wright, G., 1961, Cerebral vascular damage in guinea-pigs induced by various heterophile antisera injected by the Forssman intracarotid technique, *Br. J. Exp. Pathol.* **42**:455–463.

59. Klenk, E., 1969, On cerebrosides and gangliosides, in: *Progress in the Chemistry of Fats and Other Lipids*, Volume 10, Part 4 (R. T. Holman, ed.), Pergamon, New York, pp. 411–431.

60. Niedieck, B., and Palacios, O., 1965, Uber die Produktion von Cerbrosidantikorpern nach intradermaler Injection v.n Cerebrosid-Protein-Adjuvans-Emulsionen, *Z. Immun. Forsch.* **129**:234–243.

61. Czeonkowska, A., and Leibowitz, S., 1974, The effect of homologous and heterologous carriers on the immunogenicity of the galactocerebroside hapten, *Immunology* **27**:1117–1126.

62. Rapport, M. M., Graft, L., and Yariv, J., 1961, Immunochemical studies of organs and tumor lipids. IX. Configuration of the carbohydrate residues in cytolipin H, *Arch. Biochem. Biophys.* **92**:438–440.

63. Arnon, R., Crisp, E., Kelly, R., Ellison, S. W., Myes, L. W., and Tourtellotte, W. W., 1980, Anti-ganglioside antibodies in multiple sclerosis, *J. Neurol. Sci.* **26**:179–186.

64. Kasai, N., Pachner, A. R., and Yu, R. K., 1986, Anti-glycolipid antibodies and their immune complexes in multiple sclerosis, *J. Neurol. Sci.* **75**:33–42.

65. Duponey, R., 1972, Role of the cerebrosides and a galactodiglyceride in the antigenic cross reaction between nerve tissue and treponema, *J. Immunol.* **109**:146–153.

66. Offner, H., Konat, G., and Sela, B.-A., 1981, Multi-sialo brain gangliosides are powerful stimulators of active E-rosetting lymphocytes from multiple sclerosis patients, *J. Neurol. Sci.* **52**:279–287.

67. Sela, B.-A., Konat, A. B., and Offner, H., 1982, Elevated ganglioside concentration in serum and peripheral blood lymphocytes from multiple sclerosis patients in remission, *J. Neurol. Sci.* **54**:143–148.

68. Ilyas, A. A., and Davison, A. N., 1983, Cellular hypersensitivity to gangliosides and myelin basic protein in multiple sclerosis, *J. Neurol. Sci.* **59**:85–95.

69. Davison, A. M., and Ilyas, A. A., 1982, Cyclosporin A inhibits ganglioside-stimulated lymphocyte rosette formation in multiple sclerosis, *Int. Arch. Allergy Appl. Immunol.* **69**:393–396.

70. Bellamy, A., Davison, A. N., and Feldmann, M., 1986, Derivation of ganglioside-specific T-cell lines of suppressor or helper phenotype from cerebrospinal fluid of multiple sclerosis patients, *J. Neuroimmunol.* **12**:107–120.

71. Harboe, A., Schoyen, R., and Bye-Hansen, A., 1966, Haemagglutination inhibition by antibody to host material of fowl plague virus grown in different tissues of chick chorioallantoic membranes, *Acta Pathol. Microbiol. Scand.* **67**:573–578.

72. Feinsod, F. M., Spielman, A., and Swaner, J. L., 1975, Neutralization of Sindbis virus by antisera to antigens of vector mosquitoes, *Am. J. Trop. Med. Hyg.* **24**:533–536.

73. Almeida, J. D., and Waterson, A. P., 1969, The morphology of virus antibody interaction, *Adv. Virus Res.* **15**:307–338.

74. Reilly, C. A., Jr., and Schloss, G. T., 1971, The erythrocyte as virus carrier in Friend and Rauscher virus leukemias, *Cancer Res.* **31**:841–846.

75. Cox, K. O., and Keast, D., 1973, Rauscher virus infection, erythrocyte clearance studies, and autoimmune phenomena, *J. Natl. Cancer Inst.* **50**:941–946.

76. Steck, A. J., Tschannen, R., and Schaefer, R., 1981, Induction of antimyelin and antioligodendrocyte antibodies by vaccinia virus—an experimental study in the mouse, *J. Neuroimmunol.* **1**:117–124.

77. Lindenmann, J., and Klein, P. A., 1967, Viral oncolysis: Increased immunogenicity of host cell antigen associated with influenza virus, *J. Exp. Med.* **126**:93–108.

78. Rook, G. A. W., and Webb, H. E., 1970, Anti-lymphocyte serum and tissue culture used to investigate the role of cell-mediated response in viral encephalitis in mice, *Br. Med. J.* **4**:210–212.

79. Madrid, R. E., Wisniewski, H. M., Iqbal, K., Pullarkat, R. K., and Lassmann, H., 1981, Relapsing experimental allergic encephalomyelitis induced with isolated myelin and with myelin basic protein plus myelin lipids, *J. Neurol. Sci.* **50**:399–411.

80. Nagai, Y., Momoi, T., Saito, M., Mitzuzawa, E., and Ohtani, S., 1976, Ganglioside syndrome, a new autoimmune neurologic disorder, experimentally induced with brain ganglioside, *Neurosci. Lett.* **2**:107–111.

81. Saida, T., Saida, K., Dorfman, S., Siblerger, D. H., Summer, A. J., Manning, M. C., Lisak, R. P., and Brown, M. N., 1979, Experimental allergic neuritis induced by sensitization with galactocerebrosides, *Science* **204**:1103–1106.

82. Konat, G., Offner, H., Lev-Ram, V., Cohen, O., Schwartz, M., Cohen, I. R., and Sela, B.-A., 1982, Abnormalities in brain myelin of rabbits with experimental auto-immune multiple sclerosis-like disease induced by immunization to gangliosides, *Acta Neurol. Scand.* **66**:568–574.

83. Hughes, R. A. C., and Powell, H. C, 1984, Experimental allergic neuritis: Demyelination induced by P_2 alone and non-specific enhancement by cerebroside, *J. Neuropathol. Exp. Neurol.* **43**:154–161.

84. Carroll, W. M., Jennings, A. R., and Mastaglia, F. L., 1984, Experimental demyelinating optic neuropathy induced by intra neural injection of galactocerebroside antiserum, *J. Neurol. Sci.* **65**:125–135.

85. Roth, G. A., Royitta, M., Yu, R. K., Raine, C. S., and Bornstein, M. B., 1984, Antisera to different glycolipids induce myelin alterations in mouse spinal cord tissue cultures, *Brain Res.* **339**:9–18.

86. Sergott, R. C., Brown, M. J., Silberberg, D., and Lisak, R. P., 1984, Antigalactocerebroside serum demyelinates optic nerve *in vivo*, *J. Neurol. Sci.* **64**:297–303.

87. Nakajima, H., and Obara, J., 1967, Physico-chemical studies of Newcastle disease virus 3. The content of virus nucleic acid and its sedimentation pattern, *Arch. Gem. Virusforsch.* **20**: 287–295.

88. Klenk, H. D., and Choppin, P. W., 1970, Glycosphingolipids of plasma membrane of cultured cells and an enveloped virus (SV5) grown in these cells, *Proc. Natl. Acad. Sci. U.S.A.* **66**:57–64.

89. Soule, D. W., Marinetti, G. V., and Morgan, H. R., 1959, Studies of the hemolysis of red blood cells by mumps virus. IV. Quantitative study of changes in red blood lipids and of virus lipids, *J. Exp. Med.* **110**:93–102.

90. Kates, M., Allison, A. C., Tyrrell, D. A. J., and James, A. T., 1961, Lipids of influenza virus and their relation to those of the host cell, *Biochim. Biophys. Acta* **52**:455–466.

91. Blough, H. A., and Merlie, J. P., 1970, The lipids of incomplete influenza virus, *Virology* **40**: 685–692.

92. Hirschberg, C. B., and Robbins, P. W., 1974, The glycolipids and phospholipids of Sindbis virus and their relation to the lipids of the host cell plasma membrane, *Virology* **61**:602–608.

93. Quigley, J. P., Rifkin, D. B., and Reich, E., 1971, Phospholipid composition of Rous sarcoma virus, host cell membranes and other enveloped RNA viruses, *Virology* **46**:106–116.

94. Heydrick, F. P., Comer, J. F., and Wachter, R. F., 1971, Phospholipid composition of Venezuelan equine encephalitis virus, *J. Virol.* **7**:642–645.

95. Renkonen, O., Kaarainen, L., Simons, K., and Gahmberg, C. G., 1971, The lipid class composition of Semliki Forest virus and of plasma membranes of the host cells, *Virology* **46**: 318–326.

96. Laine, R., Kettunen, M. L., Gahmberg, C. G., Kääriäinen, L., and Renkonen, O., 1972, Fatty chains of different lipid classes of Semliki Forest virus and host cell membranes, *J. Virol.* **10**:433–438.

97. Luukkonen, A., Kaariainen, L., and Renkonen, O., 1976, Phospholipids of Semliki Forest virus grown in cultured mosquito cells, *Biochim. Biophys. Acta* **450**:109–120.

98. Webb, H. E., Chew Lim, M., Jagelman, S., Oaten, S. W., Pathak, S., Suckling, A. J., and MacKenzie, A., 1979, Semliki Forest virus infections in mice as a model for studying acute and chronic central nervous system virus infections in mice, in: *Clinical Neuroimmunology* (F. Rose, ed.), Blackwell Scientific Publications, London, pp. 369–390.

99. Khalili-Shirazi, A., Gregson, N., and Webb, H. E., 1986, Immunological relationship between a demyelinating RNA enveloped budding virus (Semliki Forest) and brain glycolipids, *J. Neurol. Sci.* **76**:91–103.

100. Amor, S., and Webb, H. E., 1988, CNS pathogenesis following a dual virus infection with Semliki Forest (Alphavirus) and Langat (Flavivirus), *Br. J. Exp. Pathol.* **69**:197–208.

101. Kelly, W. R., Blakemore, W. F., Jagelman, S., and Webb. H. E., 1982, Demyelination induced in mice by avirulent Semliki Forest virus. II. An ultrastructural study of focal demyelination in the brain, *Neuropathol. Appl. Neurobiol.* **8**:43–53.

102. Illavia, S. J., Webb, H. E., and Pathak, S., 1982, Demyelination induced in mice by avirulent Semliki Forest virus. I. Virology and effects on optic nerve, *Neuropathol. Appl. Neurobiol.* **8**: 35–42.

103. Jagelman, S., Suckling, A. J., Webb, H. E., and Bowen, E. T. W., 1978, The pathogenesis of avirulent Semliki Forest virus infections in athymic nude mice, *J. Gen. Virol.* **41**:599–607.

104. Fazakerley, J. K., Amor, S., and Webb, H. E., 1983, Reconstitution of Semliki Forest virus infected mice, induces immune mediated pathological changes in the CNS, *Clin. Exp. Immunol.* **52**:115–120.

105. Fazakerley, J. K., 1985, The role of the immune system in protection and pathogenesis during Semliki Forest virus encephalitis, Ph.D. Thesis, University of London.

106. Pathak, S., Illavia, S. J., and Webb, H. E., 1983, The identification and role of cells involved in CNS demyelination in mice after Semliki Forest virus infection: An ultrastructural study, *Prog. Brain Res.* **59**:237–254.

107. Webb, H. E., Mehta, S., Leibowitz, S., and Gregson, N. A., 1984, Immunological reaction of the demyelinating Semliki Forest virus with immune serum to glycolipids and its possible importance to central nervous system viral auto-immune disease, *Neuropathol. Appl. Neurobiol.* **10**:77–84.

108. Evans, N. R. S., and Webb, H. E., 1986, Immunoelectron microscopic labeling of glycolipids in the envelope of a demyelinating brain derived RNA virus (Semliki Forest) by antiglycolipid sera, *J. Neurol. Sci.* **74**:279–287.

109. Khalili-Shirazi, A., Gregson, N., and Webb, H. E., 1988, Immunocytochemical evidence for Semliki Forest virus antigen persistence in mouse brain, *J. Neurol. Sci.* **85**:17–26.

110. Amor, S., 1988, An investigation into mixed Toga and Flavi virus infections and their effect on the pathogenesis of CNS disease, Ph.D. Thesis, University of London.

111. Khalili-Shirazi, A., 1988, A study of the antigenicity of the host central nervous system derived glycolipids in the envelope of Semliki Forest virus, Ph.D. Thesis, University of London.

112. Pathak, S., Illavia, S. J., and Khalili-Shirazi, A., 1990, Immunoelectron microscopical labelling of a glycolipid in the envelopes of brain cell derived budding viruses, Semliki Forest, influenza and measles, using a monoclonal antibody directed chiefly against galactocerebroside made by SFV immunization, *J. Neurol. Sci.* **96**:293–302.

113. Illavia, S. J., and Webb, H. E., 1988, The pathological effect on the central nervous system of mice following single and repeated infections of the demyelinating A7(74) strain of Semliki Forest virus, *Neuropathol. Appl. Neurobiol.* **14**:207–220.

114. Bradish, C. J., and Allner, K., 1972, The early responses of mice to respiratory or intraperitoneal infection by defined virulent and avirulent strains of Semliki Forest virus, *J. Gen. Virol.* **15**:205–218.

115. Zlotnik, I., Grant, D. P., and Batter-Hatton, D., 1972, Encephalopathy in mice following inapparent Semliki Forest virus infection, *Br. J. Exp. Pathol.* **53**:125–129.

116. Hughes, R. A. C., Gray, I. A., Gregson, N. A., Kadlubowski, M., Kennedy, M., Leibowitz, S., and Thompson, H., 1984, Immune responses to myelin antigens in Guillain–Barré syndrome, *J. Neuroimmunol.* **6**:303–312.

117. Allison, A. C., 1971, Unresponsiveness to self antigens, *Lancet* **2**:401.

118. Roitt, I., 1980, *Essential Immunology*, Blackwell Scientific Publications, Oxford, pp. 306–309.

119. Rose, N. R., and Witebsky, E., 1968, Thyroid autoantibodies in thyroid disease, *Adv. Metab. Dis.* **3**:231–277.

120. Marcus, D. M., and Schwarting, G. A., 1976, Immunochemical properties of glycolipids and phospholipids, *Adv. Immunol.* **23**:203–240.

121. Rapport, M. M., and Graf, L., 1969, Immunochemical reactions of lipids, *Prog. Allergy* **13**:273–331.

122. Toy, J. L., 1983,The interferons, *Clin. Exp. Immunol.* **54**:1–13.

123. Morris, A., Tomkins, P. T., Maudsley, D. J., and Blackman, M., 1987, Infection of cultured murine brain cells by Semliki Forest virus: Effects of interferon-β on viral replication, viral antigen display, major histocompatibility complex antigen display and lysis by cytotoxic T lymphocytes, *J. Gen. Virol.* **68**:99–106.

124. Zinkernagel, R. M., and Doherty, P. C., 1979, MHC-restricted cytotoxic T cells: Studies on the biological role of polymorphic major transplantation antigens determining T-cell restriction specificity, function and responsiveness, *Adv. Immunol.* **27**:52–180.

125. Suzumura, A., Lavi, E., Weiss, S. R., and Silberberg, D. H., 1986, Coronavirus infection induces H-2 antigen expression on oligodendrocytes and astrocytes, *Science* **232**:991–993.

126. Wong, G. H. W., Bartlett, P. F., Clark-Lewis, I., Battye, F., and Schrader, J. W., 1984, Inducible expression of H-2 and Ia antigens on brain cells, *Nature* **310**:688–691.
127. Massa, P. T., Dorries, R., and ter Meulen, V., 1986, Viral particles induce Ia antigen expression on astrocytes, *Nature* **320**:543–546.
128. Kim, S. U., Moretto, G., and Shin, D. H., 1985, Expression of Ia antigens on the surface of human oligodendrocytes and astrocytes in culture, *J. Neuroimmunol.* **10**(2):141–149.
129. Azocar, J., Essex, M., and Yunis, E. J., 1983, Cell culture density modulates the incorporation of HLA antigens by enveloped viruses, *Hum. Immunol.* **7**:59–65.
130. Kabat, E. A., Wolf, A., and Bezer, A. E., 1949, Studies on acute disseminated encephalomyelitis produced experimentally in rhesus monkeys, *J. Exp. Med.* **89**:395–398.
131. Martenson, E., 1969, Glycosphingolipids of animal tissues, in: *Progress in the Chemistry of Fats and Other Lipids*, Volume 10 (R. T. Holman, ed.), Pergaman Press, Oxford, pp. 365–407.
132. Gregson, N. A., Kennedy, M., and Leibowitz, S., 1977, Gangliosides as surface antigens of cells isolated from rat cerebellar cortex, *Nature* **266**:461–463.
133. Bornstein, M. B., and Raine, C. S., 1970, Experimental allergic encephalomyelitis in antiserum inhibition of myelination *in vitro*, *Lab. Invest.* **23**:536–542.
134. Gregson, N. A., Pytharas, M., and Leibowitz, S., 1977, the reactivity of anti-ganglioside antiserum with isolated cerebellar cells, *Biochem. Soc. Trans.* **5**:174–175.
135. Husby, G., Van De Rijn, I., Zabriskie, J. B., Abdin, Z. H., and Williams, R. C., Jr., 1976, Antibodies reacting with cytoplasm of subthalamic and caudate nuclei neurons in chorea and acute rheumatic fever, *J. Exp. Med.* **144**:1094–1110.
136. Biberfield, G., 1971, Antibodies to brain and other tissues in cases of *Mycoplasma pneumoniae* infection, *Clin. Exp. Immunol.* **8**:319–333.
137. Gessain, A., Barin, F., Varnant, J. C., Gout, O., Maurs, L., Calender, A., and De The, G., 1985, Antibodies to human T-lymphotrophic virus type I in patients with tropical spastic paraparesis, *Lancet* **2**:407–409.
138. Roman, G. C., 1987, Retrovirus associated myelopathies, *Arch. Neurol.* **44**(6):659–663.
139. Greene, W. C., Leonard, W. J., Wano, Y., Sveteik, P. B., Peffer, N. J., Sodroski, J. G., Rosen, C. A., Goh, W. C., and Haseltine, W. A., 1986, *Trans*-activator gene of HTLV-II induces IL-2 receptor and IL-2 cellular gene expression, *Science* **232**:877–880.
140. Sternberger, N. H., del Cerro, C., and Kies, M. W., 1985, Immunocytochemistry of myelin basic proteins in adult rats oligodendroglia, *J. Neuroimmunol.* **7**:355–363.
141. Fujinami, R. S., Zurbriggen, A., and Powell, H. C., 1988, Monoclonal antibody defines determinant between Theiler's virus and lipid-like structures, *J. Neuroimmunol.* **20**:25–32.

15

Viruses and Schizophrenia

JANICE R. STEVENS and LESLEY M. HALLICK

1. INTRODUCTION

Schizophrenia, a mental illness that affects approximately 1% of the U.S. population, is more common than diabetes, Alzheimer's disease, or AIDS. The cause remains unknown. None of the prevailing theories of etiology—genetic, biochemical, and psychosocial—can fully explain the epidemiology, clinical course, or pathology of the illness. The hypothesis that viruses or virus-like agents may have a role in the etiology of schizophrenia is not a new one. Following attacks of encephalitis lethargica (von Economo's encephalitis) and influenza in the pandemics following World War I, both European and American psychiatrists reported schizophrenia-like disorders as sequelae in a significant number of cases.[1-3] Although no agent associated with these pandemics was isolated, the clinical course and pathological changes in body and brain at autopsy implied that virus infections of unusual virulence had occurred. Cases typical of viral encephalitis accompanied by or followed by schizophrenia-like psychoses have been reported sporadically, but a common agent has not been identified, nor has the considerable effort of many investigators to detect virus or viral "footprints" from brains or cerebrospinal fluid (CSF) of schizophrenic patients been very fruitful. The majority of patients with schizophrenia do not have a history suggestive of closely antecedent infection.

Neuropathological studies of the brains of patients who died during acute or chronic schizophrenic illness have not, except in rare cases of malignant catatonia,

JANICE R. STEVENS • Neuropsychiatry Branch, National Institute of Mental Health, Neuroscience Center at St. Elizabeths Hospital, Washington, DC 20032, and Departments of Psychiatry and Neurology, Oregon Health Sciences University, Portland, Oregon 97201. LESLEY M. HALLICK • Department of Microbiology and Immunology, Oregon Health Sciences University, Portland, Oregon, 97201.

Neuropathogenic Viruses and Immunity, edited by Steven Specter *et al.* Plenum Press, New York, 1992.

demonstrated any inflammatory changes, although a variety of other abnormalities have been reported. Most of these changes have been ignored or dismissed as artifacts in the past.[4] However, recent studies with computerized tomography (CT) demonstrating enlarged ventricles and cortical sulci in a significant number of patients with schizophrenia[5-7] have stimulated new examinations of postmortem material. These investigations have also shown evidence of pathology, including cell loss, gliosis, and shrinkage of specific cerebral structures.[4-10]

Although genetic factors clearly play a significant part in schizophrenia, the fact that only 50–60% of monozygotic twins are concordant for schizophrenia, a figure similar to that for multiple sclerosis and tuberculosis, implies that additional factors are required and may indicate only an inherited predisposition to the illness.[11] This hypothesis is supported by recent studies utilizing CT[12a] or magnetic resonance imaging[12b] to examine brain structures in monozygotic twins discordant for schizophrenia showing that the schizophrenic twin of discordant pairs has larger ventricles. Biochemical abnormalities are strongly suspected in schizophrenia, in part because all effective pharmacological treatment blocks cerebral dopamine receptors and because such receptors may be increased in basal ganglia of schizophrenic brain. This abnormality may be intrinsic to the disease, drug related, or, as Knight has suggested, could reflect stimulation of dopamine receptors by a viral or autoimmune factor.[13]

2. EVIDENCE FOR A VIRAL ORIGIN

Current interest in the viral hypothesis of schizophrenia is based on a large number of circumstantial observations that suggest, but by no means prove, that schizophrenia or some schizophrenias have a viral or immunologic origin. These observations include:

1. Course and pathology of the illness
2. Epidemiology
3. Seasonal occurrence and season of birth
4. Immunologic studies
5. Viral studies

2.1. Course and Pathology of the Illness

Schizophrenia is a disorder of young adults with a peak age of onset between 18 and 35 years (range 15–45 years). The average age of onset for males is 4–5 years younger than that for females. Johnson has noted that similar unimodal ages of onset are typical of, although not unique to, infection acquired or activated in early adulthood.[14] The onset of schizophrenia is usually insidious or subacute, followed by a progressive deteriorating course in 20–30% of those affected. In another third, a pattern of remissions and exacerbations is observed that resembles the course of multiple sclerosis and results in increasing disability, and 30–

40% of those affected recover completely after some months or years. Prognosis is significantly worse for males than females, with nearly half of the latter demonstrating complete or nearly complete recovery, whereas fewer than one-third of males are similarly favored.[15] Reasons for the significant gender difference in age of onset (males, 23.5 years; females, 28.5 years) and the less favorable course in males have long been sought but remain unknown. Also intriguing is the observation that patients diagnosed with schizophrenia have a significantly better long-term prognosis in less developed countries than in Europe and North America. This finding is further discussed below.

2.2. Epidemiology

Differences in diagnostic requirements for schizophrenia in different parts of the world have gravely handicapped epidemiologic studies. Moreover, schizophrenia incidence and prevalence data reported from many developing countries, particularly in Africa, are generally based on hospital and clinic incidence rates rather than community- or countrywide surveys and are inaccurately low because of shortages of medical facilities and greater tolerance for some forms of schizophrenia in the community. However, even those relatively low numbers may be erroneously high in many areas of the developing world because of widespread diagnosis as schizophrenia of the brief reactive psychoses that are very common in Africa and southern Asia.[16,17] Diagnosed as schizophrenia, these brief psychoses, which resemble acute schizophrenia on presentation to hospital or clinic but remit within a few weeks or months, naturally inflate the incidence and prevalence figures and improve the prognosis.

In the absence of any firm laboratory or radiological criteria for the illness, psychiatrists have developed a group of clinical criteria in an attempt to achieve more uniform diagnoses. In a recent study undertaken in Harare, the capital and largest city in Zimbabwe, 40–50% of all admissions to the mental wards of two university-affiliated hospitals were diagnosed as schizophrenia in 1938–1984. However, when the current research diagnostic criteria for schizophrenia, which require at least 2 weeks of illness for diagnosis of schizophrenia,[18] were applied, more than half of the patients failed to meet the duration criterion and were accordingly diagnosed as "brief reactive psychosis." If the 6 months of illness specified by the American Psychiatric Association's *Diagnostic and Statistical Manual for Mental Disorders*[19] were required, only 10–15% of admissions met the criteria for schizophrenia.[17]

Although it is often stated that prevalence of schizophrenia is the same worldwide,[20] available data do not really permit this conclusion (Table 15-1). Indeed, the few population-based surveys that do exist indicate an approximately 10- to 20-fold variation with much lower rates in Africa and southern Asia than in western Europe or the United States.[21–23] As is also the case with multiple sclerosis, reported cases of schizophrenia appear to have their highest distribution in northern and eastern Europe, but the disease is comparatively less common in southern Europe and Africa, even in areas where trained European personnel have systematically looked for schizophrenia.[22]

TABLE 15-1
Published Prevalence Data for Schizophrenia (per 1000)

Northern U.S. (Rochester)	4.7	Japan	2.1–2.3
Southern U.S. (Baltimore)	2.9	Australia (Aborigines)	4.4
England (London)	3.4	Israel	1.4
Scotland	4.2	Ghana	0.5
Ireland (East)	3.5–4.0	India (W. Bengal)	4.3
Ireland (West)	7.1	India (Uttar Pradesh)	2.3
Sweden	8.2	Papua New Guinea (Highlands)	0.1
Western Europe	4.5	Papua New Guinea (Coastal)	0.5
Eastern Europe (NW Croatia)	7.4		

This trend remains true for schizophrenia as well as for multiple sclerosis when prevalence data are presented only for the population at risk (i.e., over age 15). Just as the British Isles and northern Europe have the highest prevalence of multiple sclerosis in the world, western Ireland, Sweden, and western Croatia reportedly have the highest prevalence of schizophrenia, with Ireland reportedly showing a rate 44 times that of Guyana.[24]

An unequal geographic distribution with pockets of very high incidence is compatible with the hypothesis that, as in multiple sclerosis, an environmental event (in addition to a genetic predisposition) best explains the known epidemiology of the illness. Although epidemiologic data for schizophrenia are less than adequate, the available information indicates that the dictum that schizophrenia is more or less equally distributed throughout the world is a premature extension of the observation that schizophrenia-like illnesses exist in all parts of the world.

The prognosis of schizophrenia is said to be significantly better in the developing world than in the developed (largely temperate) lands.[20,25] The better prognosis in developing countries, like the prevalence, may be erroneously exaggerated by inclusion of the brief schizophreniform psychoses that contribute significantly to admissions in psychiatric hospitals in less-developed countries. However, the better prognosis in these areas may also reflect different social conditions or greater immunity to schizophrenia, perhaps acquired by early and repeated exposure to the inciting agent or agents in populations with poor sanitary practices. Treatment also differs—there is much wider use of electroconvulsive therapies and perhaps better social support offered by families in developing countries.

For schizophrenia, as in multiple sclerosis, there are reports of clusters of increased prevalence in space and time. A recent survey in Micronesia presented evidence for clustering of cases on certain islands.[26] Fortes and Mayer[27a] studied a group of villages in Ghana in the 1950s and found only one case of schizophrenia in a population of 5000. Twenty-six years later, a resurvey of the area disclosed 13 cases per 5000. Between the first and second visits by the authors, the population had for the first time experienced considerable exposure to Western civilization. Clustering of cases was also noted recently in Ireland and from a study of West Indians in England.[27b,28a]

2.3. Season of Onset and Season of Birth

The principal importance of peak seasons of onset for diseases of unknown etiology is the possibility of relating such disorders to known fluctuations in incidence of recognized infections or toxic exposure. Analysis of admissions to mental hospitals in England and Wales during the 1970s demonstrated a 4.7% rise in the admission rate of schizophrenic patients in July and August and a 9.5% excess of admission for mania during the same period,[28b] both relative to patients with general neuroses. From Japan, Abe[29] reported a 9% excess of schizophrenia admissions in July among 90,000 patients admitted to a hospital between 1955 and 1961.

There has also been great interest in the unequal yearly distribution of birthdates in a number of disorders because this may indicate exposure of the fetus or infant to special risks at the time of conception, birth, or during a critical intrauterine period. An approximately 8% excess of births of future schizophrenic patients was shown during winter months, and a 4% excess in early spring months in the large English–Welsh population studies by Hare.[28b] A relative excess of winter–early spring births has also been reported for Scandinavian countries, the United States, Japan, and Australia.[30,31] In a recent study from Finland, Mednick and colleagues[32] found that the incidence of schizophrenia was nearly doubled in a cohort of schizophrenic patients who were in their second trimester of gestation during the widespread influenza A epidemic of 1957.

2.4. Immunologic Factors

Controversial arguments that schizophrenia may be an infectious or immunologic disorder stem from reports suggesting intrathecal IgG or IgM synthesis, elevated serum or CSF antibodies to specific viruses, or alterations in lymphocyte morphology, subtype distribution, and response to mitogens. Although Ahokas et al.[33] reported increased CSF oligoclonal IgG bands in 36% of diagnosed schizophrenics, these findings have not been reproduced by others.[34-36] In a finding reminiscent of the T-cell decline at the onset of exacerbations of multiple sclerosis, Coffey et al.[37] reported a reduction in T lymphocytes in patients with acute, including never-treated, schizophrenia and a rise to normal ratio following treatment with neuroleptics. DeLisi et al.[38] did not find a reduction in total T cells in chronically hospitalized schizophrenic patients (most of whom were taking neuroleptics) but observed a decrease in natural killer cells, a finding also reported for multiple sclerosis.[39] Decreased lymphocyte migration to measles and other antigens has been reported in schizophrenia by Vartanian et al.[40]

There are a number of reports of antibrain antibodies in schizophrenic serum.[41-48] This finding has been attributed to an autoimmune process, possibly initiated by tissue destruction caused by early viral encephalopathy followed by sensitization to brain proteins released by the destructive process. However, most recent studies have shown that similar numbers of schizophrenic patients and controls demonstrate antibrain antibodies in serum.[49-52] The search for serum antibodies against specific brain proteins in schizophrenic patients has thus far been negative,[53] and the antithymic antibodies in schizophrenic serum first reported by laboratories in the USSR[54] were not reproduced in a Canadian study.[55]

Increased serum and/or CSF titers against herpes simplex virus (HSV), influenza A, measles, or cytomegalovirus (CMV) have been reported in schizophrenia by some investigators[56–60] but have not been reproduced by others.[36,61–65] Tests for HIV antibody in schizophrenic serum and reverse transcriptase in CSF have thus far yielded negative results.[66,67] Serum interferon elevation detected in schizophrenic CSF by Preble and Torrey[68] and Libíková et al.[55] could not be replicated by Rimón et al.[53] Thus, it is not currently possible to correlate schizophrenia with a specific viral antibody response.

Cerebrospinal fluid proteins and cell counts are generally well within normal limits in schizophrenia.[35] Using a very sensitive radial immunodiffusion method, Kirch et al. found that eight of 24 (33%) patients with chronic schizophrenia demonstrated increased endogenous CNS IgG production and that one of these eight also demonstrated oligoclonal bands on high-resolution protein electrophoresis.[35] Using the same methods, Stevens et al. failed to find oligoclonal bands in any of eight patients with acute first break or exacerbated schizophrenia.[69a] Harrington et al.[69b] reported an unusual protein observed on two-dimensional gel electrophoresis of CSF from patients with schizophrenia, Creutzfeldt–Jakob disease, herpes encephalitis, and multiple sclerosis, but this protein has not yet been characterized.

Several laboratories have reported abnormal lymphocytes resembling those of infectious mononucleosis in the peripheral blood of schizophrenic patients.[70–74] This finding was not confirmed in a group of chronic drug-treated patients studied by DeLisi et al.[75] Ultrastructural abnormalities in lymphocyte nuclei were reported by Bonartsev et al.[76] Increased helper/suppressor cell ratio and increased CD5 expression in peripheral lymphocytes are reported.[36,77a] Decreased response of schizophrenic white blood cells to mitogenesis reported by Russian workers[77b] may have resulted from the effect of neuroleptic drugs used in the treatment of schizophrenia. The effects of varying medications and the great differences in the stage of illness in which patients are studied are likely to be very important barriers to achieving coherent results from these investigations.

2.5. Search for the Virus

Examination of brain specimens from patients who died during acute or chronic schizophrenia demonstrates no inflammatory reaction and no spongiform changes characteristic of the unconventional (slow) virus encephalopathies. However, several studies have reported focal chronic fibrillary gliosis, quantitative evidence of cell loss, and parenchymal atrophy or dystrophy indicative of an old, chronic, or indolent destructive process.[4,8,10]

Presence of virus-like particles in CSF and nasal secretions of schizophrenic patients was reported in 1954 by Russian workers, and extensive immunologic and virological investigations were carried out at that time, but no specific virus could be isolated.[78] In 1979, Tyrrell et al.[79] described a cytopathic effect (CPE) of CSF from schizophrenic patients on human embryonic fibroblasts. Similar effects were obtained with CSF from individuals with a number of neurodegenerative disorders. However, since the investigators were unable to passage the agent or inhibit the cytopathic effect with proteinases or nucleases, they concluded that their agent was a nonreplicating toxin.[80] Intracerebral inoculation of marmosets with the cytotoxic CSF reportedly caused certain behavioral changes.[81] Cytopathic effects of schizophrenic CSF and brain explants in tissue culture were also reported by Líbíkova.[57] However, no CPE was observed either by Mered et al.,[82] who attempted to repeat the experiments of Tyrrell et al. with CSF from a new group of patients, or by Cazzullo et al.[36]

Rajcáni, working with Líbíková and their colleagues, inoculated CSF and brain explants from schizophrenic patients onto several cell lines and into suckling mice.[83] They observed herpesvirus-like structures 100–120 nm in diameter in the brains of inoculated animals. In addition, Mesa-Castillo and Cabrero have reported that electron microscopic analysis of very fresh postmortem material from schizophrenic brains demonstrates round intracytoplasmic encapsulated structures resembling herpesviruses.[84]

In 1979, a group of researchers at the National Institutes of Health injected material from 20 brain specimens of patients diagnosed with schizophrenia (mostly chronic) and of 19 controls with other neurological disorders into guinea pigs and subhuman primates. Most of the animals were sacrificed or died 2 to 4 years later without developing clinical evidence of pathology. One guinea pig developed a paralysis of the hindlimbs 18 months after inoculation with material from a patient who committed suicide during the first year of his schizophrenia. Examination of CNS tissue from these animals with light and electron microscopy generally failed to reveal significant pathology.[85]

Following reports of increased CSF antibody to HSV[55] and of both IgG and IgM to CMV[56,57] in patients with schizophrenia, we undertook a search for HSV and CMV antigen in the brains of schizophrenic patients and controls, using immunocytochemical methods on both fixed and frozen material. Although our early studies with crude CMV antisera yielded positive results in a small number of schizophrenic patients compared with the controls,[86] in subsequent studies with affinity-purified CMV antisera and monoclonal antibodies, we could not reproduce our own initial positive results. Only one patient of the 22 fixed schizophrenic brains we have studied demonstrated immunoreactivity against CMV (in several large neurons of the vestibular nucleus), and none gave positive results for HSV 1 or HSV 2.[87] In a second study, the frozen brain specimens of ten schizophrenics and 13 control patients were tested with high-titer antisera to CMV, HSV 1 and 2, mumps, measles, rubella, and rubeola (Table 15-2). Although three schizophrenic patients reacted positively with two or more antisera, there was no convincing correlation of schizophrenia with any one virus antigen. In the case of a 21-year-old woman ill for less than 5 years before her death by suicide, there were scattered large neurons in the amygdala and hippocampus that showed a distinct diffuse granular staining of the cytoplasm but not of the nucleus following incubation with HSV 1 antibody.

During the past 2 years we have inoculated fresh and frozen brain and CSF from 24 schizophrenic and 20 control patients onto cultures of human dorsal root ganglia, neuroblastoma, and medulloblastoma cells. No cytotoxic changes have

TABLE 15-2
Immunoreactivity of Schizophrenic Brain Tissues with Specific Viral Antisera

	Age	Sex	Duration[a]	PMI[b]	CMV[c]	HSV1	HSV2	VZV	Mumps	Rubella	Rubeola
Schizophrenia											
33 B	25	M		21.5	=	–	–	±[d]	+[e]	–	±
32 D	23	M	6	16.5	–	–	–	–	–	–	–
32K	28	M		3.5	–	–	–	–	–	–	–
26F	21	F	4–5	14.0	±	±	+	±	–	–	–
A 16	28		?	6.7	–	–	–	–	–	–	–
26 IJ	31	M	12	20.0	–	–	–	–	–	–	–
A 264	45	F	20	24.0	–	–	–	–	–	–	–
1 M	23	M	2.5	18.0	–	–	–	–	–	–	–
28 2	30	F	1		–	–	–	–	–	–	–
21 D	31	M		2.5	–	±	±	–	–	–	–
Controls											
30 J	24	M		19.5	–	–	–	–	–	–	–
A 108	22	M		18.0	–	–	–	–	–	–	–
A 127	23	M		16.0	–	–	–	–	–	–	–
27 A	30	M		23.0	–	–	–	–	–	–	–
24 J	32	M		3.0	–	–	–	–	–	–	–
3 E	17	M		6.5	–	–	±	–	–	–	–
A 247	26	M		6.5	–	–	–	–	–	–	–
A 252	22	M		18.0	–	–	–	–	–	–	–
28 F	25	M		22.0	–	–	–	–	–	–	–
A 52	19	M		21.0	–	–	–	–	–	–	–
A 177	19	F		5.5	–	–	–	–	–	–	–
A 176	31	M		16.5	–	–	–	–	–	–	±
A 179	18	M		12.0	–	–	–	–	±	–	–

[a] Duration of schizophrenia (years).
[b] PMI, postmortem interval (hours).
[c] CMV, cytomegalovirus; HSV, herpes simplex virus; VZV, varicella–zoster virus.
[d] ±, equivocal immunoreactivity, not replicable on repeat trials.
[e] +, positive immunoreactivity in scattered neurons of amygdala and hypothalamus on repeated trials.

been observed. However, cells spun down from 10 ml of CSF from each of 11 drug-free patients with schizophrenia caused the neuroblastoma cell line (SH-EP) to grow to a 25–100% higher density than any of 14 control CSF-treated cultures ($P < 0.01$).[88] Cells from both schizophrenic and control CSF were incubated for 5 days with the SH-EP cells, after which the cells were passaged at 2-week intervals for 6 to 12 months. The ability to grow to increased density has continued to be expressed for up to 30 passages and can be transmitted by cell-free media passed through a 0.45- and 0.22-μm filter. Transformed cells from five of seven schizo-phrenic CSF-treated cultures, but only one of 13 control CSF-treated cultures, were able to grow to increased size and number of colonies in soft agar. Similar changes have been described in cell cultures inoculated with material from experimental Creutzfeldt–Jakob disease.[89]

In situ hybridization for the CMV genome has been negative in schizophrenic brain specimens in our laboratory and that of others,[90,91] although Moises *et al.* recently reported detection of CMV DNA in the temporal lobe of a schizophrenic patient.[92] A report by Sequiera *et al.*[93] of an HSV genome in schizophrenic brains may reflect specious cross-hybridization with human DNA.[94a] Studies in progress using the polymerase chain reaction (PCR) to probe for viral DNA sequences in post mortem brain specimens (herpes viruses, influenza A, measles) have thus far been negative (personal communication, David Asher and Alla Taller).[94b]

3. CONCLUSIONS

The unimodal age of onset, long duration, progressive deterioration, relaps-ing or remitting course, and unequal geographic distribution of schizophrenia resemble features of multiple sclerosis, a disorder for which a viral or immuno-logic etiology is also suspected. The data showing a summer peak of onset and winter peak of birth are consistent with an infectious disorder in at least some cases. The insidious onset and lack of inflammatory response in atrophic or gliosed brain regions are reminiscent of lentivirus infections. The agent or agents responsible for schizophrenia could be acquired at any time before or after birth and remain latent for a variable period. However, there is little evidence for horizontal transmission,[11] and the epidemiologic and pathological data appear to be more consistent with an early infection that causes little or mild damage at the time of initial exposure, remains dormant over months or years, and is then reactivated between late adolescence and middle adult life. Any infectious or immunologic theory must explain why males are vulnerable earlier than females and why males have a significantly more malignant course.

It is well known that viruses of the herpes family, especially HSV and CMV, are common latent infections in man, with a majority of adults having evidence of prior infection. HSV 1 remains dormant in spinal or cranial nerve ganglia and has a special predilection for limbic structures, particularly medial and inferior temporal lobe, regions for which there is mounting evidence of pathology in schizophrenia.[4,8] Cytomegalovirus causes an encephalitis in childhood and in immunosuppressed adults and frequently causes pathological changes in peri ventricular structures.[95] However, neither of these well-known viral encephalitides resembles schizophrenia, either clinically or pathologically.

If schizophrenia is caused by a herpesvirus, it is possible that in parallel to measles and SSPE, certain forms of HSV or CMV persist in brain and account for the few cases in which persistent elevation of IgG or IgM against these agents has been reported.[55–57] The possibility of an unconventional slow virus as etiologic agent for schizophrenia or some schizophrenias is equally plausible. These viruses are noted for long latency, persistence, an indolent course, and failure of a host inflammatory response or rapid clearance by the immune system.

In conclusion, it is not possible to implicate a specific viral agent or agents as responsible for schizophrenia on the basis of the currently available evidence in spite of a substantial effort to test this hypothesis. However, a viral etiology for schizophrenia remains an extremely attractive theory. To date, only a very few investigators well trained in modern virology and immunology have turned their attention to schizophrenia. Evidence from recent epidemiologic studies and development of the powerful polymerase chain reaction (PCR) technique for detection of viral sequences in CSF and brain provide impetus for further research.

REFERENCES

1. Menninger, K. A., 1928, The schizophrenic syndrome as a product of acute infectious disease, *Arch. Neurol. Psychiatry* **20**:464–481.
2. Goodall, E., 1932, The exciting cause of certain states, at present classified under "schizophrenia" by psychiatrists, may be infection, *J. Ment. Sci.* **78**:746–755.
3. Cooper, H. A., 1936, The mental sequelae of chronic epidemic encephalitis and their prognosis, *Lancet* **2**:677–679.
4. Stevens, J. R., 1982, Neuropathology of schizophrenia, *Arch. Gen. Psychiatry* **39**:1131–1139.
5. Johnstone, E. C., Crow, T. J., Frith, C. D., Husband, J., and Kreel, L., 1976, Cerebral ventricular size and cognitive impairment in chronic schizophrenia, *Lancet* **2**:924–926.
6. Weinberger, D. R., Torrey, E. F., Neophytides, A. N., and Wyatt, R. J., 1979, Lateral cerebral ventricular enlargement in chronic schizophrenia, *Arch. Gen. Psychiatry* **36**:735–739.
7. Tanaka, Y., Hazama, H., Kawahara, R., and Kobayashi, K., 1981, Computerized tomography of the brain in schizophrenic patients. A controlled study, *Acta Psychiatr. Scand.* **63**:191–197.
8. Bogerts, B., Meertz, E., and Schönfeld-Bausch, R., 1985, Basal ganglia and limbic system pathology in schizophrenia, a morphometric study of brain volume and shrinkage, *Arch. Gen. Psychiatry* **42**:784–791.
9. Falkai, P., and Bogerts, B., 1986, Cell loss in the hippocampus of schizophrenics, *Eur. Arch. Psychiatr. Neurol. Sci.* **236**:154–161.
10. Nieto, D., and Escobar, A., 1972, Major psychoses, in: *Pathology of the Nervous System*, Volume 3 (J. Minckler, ed.), McGraw-Hill, New York, p. 2654.
11. Crow, T. J., 1987, Genes and viruses in schizophrenia: The retrovirus/transposon hypothesis, in: *Viruses, Immunity and Mental Disorders* (E. Kurstak, Z. J. Lipowski, and P. V. Morozov, eds.), Plenum Press, New York, pp. 271–283.
12a. Reveley, A. M., Reveley, M. A., and Murray, R. M., 1984, Cerebral ventricular enlargement in non-genetic schizophrenia: A controlled twin study, *Br. J. Psychiatry* **144**:89–93.
12b. Suddath, R. L., Christison, G. W., Torrey, E. F., and Weinberger, D. R., 1990, Cerebral anatomical abnormalities in monozygotic twins discordant for schizophrenia. *N. Engl. J. Med.* **322**:789–794.

13. Knight, J. G., 1982, Dopamine-receptor-stimulating autoantibodies: A possible cause of schizophrenia, *Lancet* **2**:1073–1076.

14. Johnson, R. T., 1975, The possible viral etiology of multiple sclerosis, *Adv. Neurol.* **13**:1–48.

15. Watt, D. C., Katz, K., and Shepherd, M., 1983, The natural history of schizophrenia: A 5-year prospective follow-up of a representative sample of schizophrenics by means of a standardized clinical and social assessment, *Psychol. Med.* **13**:663–670.

16. German, G. A., 1972, Aspects of clinical psychiatry in sub-Saharan Africa, *Br. J. Psychiatry* **121**:461–479.

17. Stevens, J., 1987, Brief psychoses: Do they contribute to the good prognosis and equal prevalence of schizophrenia in developing countries? *Br. J. Psychiatry* **151**:393–396.

18. Spitzer, R. L., and Endicott, J., 1978, *Research Diagnostic Criteria (RDC) for a Selected Group of Functional Disorders*, New York State Department of Mental Hygiene, New York.

19. American Psychiatric Association, 1980, *Diagnostic and Statistical Manual of Mental Disorders*, 3rd ed., APA Press, Washington, DC.

20. Sartorius, N., Jablensky, A., Korten, A., Ernberg, G., Anker, M., Copper, J. E., and Day, R., 1986, Early manifestations and first-contact incidences of schizophrenia in different cultures, *Psychol. Med.* **16**:909–928.

21. Giel, R., and Van Luijk, J. N., 1969, Psychiatric morbidity in a small Ethiopian town, *Br. J. Psychiatry* **115**:149–162.

22. Torrey, E. F., 1980, *Schizophrenia and Civilization*, Jason Aronson, New York.

23. Eaton, W. W., 1985, Epidemiology of schizophrenia, *Epidemiol. Rev.* **7**:105–126.

24. Walsh, D., and Walsh, B., 1970, Mental illness in the Republic of Ireland—First admissions, *J. Ir. Med. Assoc.* **63**:365–370.

25. Sartorius, N., Jablensky, A., and Shapiro, R., 1978, Cross-cultural difference in the short-term prognosis of schizophrenic psychoses, *Schizophrenia Bull.* **4**:102–113.

26. Dale, P. W., 1981, Prevalence of schizophrenia in the Pacific Island populations of Micronesia, *J. Psychiatr. Res.* **16**:103–111.

27a. Fortes, M., and Mayer, D. Y., 1969, Psychosis and social change among the Tallensi of Northern Ghana, in: *Psychiatry in a Changing Society* (S. H. Foulkes and G. S. Prince, eds.), Tavistock, London.

27b. Youssef, H. A., Kinsella, A., and Waddington, J. L., 1990, Evidence for geographical variations in the prevalence of schizophrenia in rural Ireland. *Arch Gen. Psychiat.* **48**:254–258.

28a. Harrison, G., Holton, D. N., Owens, D., Boot, D., and Cooper, J., 1989, Severe mental disorder in Afro-Caribbean patients: some social demographic and service factors. *Psychol. Med.* **19**:683–686.

28b. Hare, E. H., 1983, Epidemiological evidence for a viral factor in the aetiology of the functional psychoses, *Adv. Biol. Psychiatry* **12**:52–75.

29. Abe, K.,, 1963, Seasonal fluctuation of psychiatric admissions, based on data for 7 prefectures of Japan for a 7-year period 1955–1961, with a review of the literature, *Folio Psychiatr. Neurol. Jpn.* **17**:102–111.

30. Dalen, P., 1975, *Season of Birth in Schizophrenia and Other Mental Disorders*, North-Holland, Amsterdam.

31. Parker, G., and Balza, B., 1977, Season of birth and schizophrenia—An equatorial study, *Acta Psychiatr. Scand.* **56**:143–146.

32. Mednick, S. A., Machon, R. A., Huttunen, M. O., and Bonett, D., 1988, Adult schizophrenia following prenatal exposure to an influenza epidemic, *Arch. Gen. Psychiatry* **45**:189–192.

33. Ahokas, A., Rimón, R., Koskiniemi, M., Vaheri, A., Julkunen, I., and Sarna, S., 1987, Viral antibodies and interferon in acute psychiatric disorders, *J. Clin. Psychiatry* **48**:194–196.

34. Roos, R. P., Davis, K., and Meltzer, H. Y., 1985, Immunoglobin studies in patients with psychiatric diseases, *Arch. Gen. Psychiatry* **42**:124–128.

35. Kirch, D. G., Kaufman, C. A., Papadopoulos, N. M., Martin, B., and Weinberger, D. R., 1985, Abnormal cerebrospinal fluid protein indices in schizophrenia, *Biol. Psychiatry* **20**:1039–1046.

36. Cazzullo, C. L., Caputo, D., Bellodi, L., Maffei, C., Ferrante, P., Bergamini, F., Landini, M. P.,

and La Placa, M., 1987, Schizophrenia: An epidemiologic, immunologic, and virological approach, in: *Viruses, Immunity and Mental Disorders* (E. Kurstak, Z. J. Lipowski, and P. V.Morozov, eds.), Plenum Press, New York, pp. 149–155.

37. Coffey, C. E., Sullivan, J. L., and Rice, J. R., 1983, T lymphocytes in schizophrenia, *Biol. Psychiatry* **18**:113–119.

38. DeLisi, L. E., Goodman, S., Neckers, L. M., and Wyatt, R. J., 1982, An analysis of lymphocyte subpopulations in schizophrenic patients, *Biol. Psychiatry* **17**:1003–1009.

39. Merrill, J., Jondal, M., Seeley, J., Ullberg, M., and Siden, A., 1982, Decreased NK killing in patients with multiple sclerosis: An analysis on the level of the single effector cell in peripheral blood and cerebrospinal fluid in relation to the activity of the disease, *Clin. Exp. Immunol.* **47**:419–430.

40. Vartanian, M. E., Kolyaskina, G. I., Lozovsky, D. V., Burbaeva, G. S., and Ignatov, S. A., 1978, Aspects of humoral and cellular immunity in schizophrenia, in: *Neurochemical and Immunologic Components in Schizophrenia* (D. Bergsma and A. L. Goldstein, eds.), Alan R. Liss, New York, pp. 339–364.

41. Lehmann, Facius, H., 1937, Uber die Liquor Diagnose der Schizophrenien, *Klin. Wochenschr.* **16**:1646–1648.

42. Heath, R. G., and Krupp, I. M., 1967, Schizophrenia as an immunologic disorder. I. Demonstration of antibrain globulins by fluorescent antibody techniques, *Arch. Gen. Psychiatry* **16**:1–9.

43. Kuznetoza, N. I., and Semenov, S. F., 1961, Detection of antibrain antibodies in the sera of patients with neuropsychiatric disorders, *Zh. Neuropatol. Psikhiatr.* **61**:869–873.

44. Kuritzky, A., Livni, E., Munitz, h., Englander, T., Tyano, S., Wijsenbeek, H., Joshua, H., and Kott, E., 1976, Cell-mediated immunity to human myelin basic protein in schizophrenic patients, *J. Neurol. Sci.* **30**:369–373.

45. Kolyaskina, G. I., 1983, Blood lymphocytes in schizophrenia—Immunological and virological aspects, *Adv. Biol. Psychiatry* **12**:142–149.

46. Fessel, W. J., and Hirata-Hibi, M., 1963, Abnormal leukocytes in schizophrenia, *Arch. Gen. Psychiatry* **9**:601–613.

47. Pandey, R. S., Gupta, A. K., and Chaturvedi, V. C., 1981, Autoimmune model of schizophrenia with special reference to antibrain antibodies, *Biol. Psychiatry* **16**:1123–1136.

48. Baron, M., Stern, M., Anavi, R., and Witz, I. P., 1977, Tissue-serum affinity in schizophrenia: A clinical and genetic study, *Biol. Psychiatry* **12**:199–219.

49. Whittingham, S., Mackay, I. R., Jones, I. H., and Davies, B., 1968, Absence of brain antibodies in patients with schizophrenia, *Br. Med. J.* **1**:347–348.

50. Logan, D. G., and Deodhar, S. D., 1970, Schizophrenia, an immunological disorder? *J.A.M.A.* **212**:1703–1704.

51. Boehme, D. H., Cottrell, J. C., Dohan, F. C., and Hillegass, L. M., 1974, Demonstration of nuclear and cytoplasmic fluorescence in brain tissues of schizophrenic and non-schizophrenic patients, *Biol. Psychiatry* **8**:89–94.

52. DeLisi, L. E., 1987, Immunologic studies of schizophrenic patients, in: *Viruses, Immunity, and Mental Disorders* (E. Kurstak, Z. J. Lipowski, and P. V. Morozov, eds.), Plenum Press, New York, pp. 271–283.

53. Rimón, R., Ahokas, A., and Palo, J., 1986, Serum and cerebrospinal fluid antibodies to cytomegalovirus in schizophrenia, *Acta Psychiatr. Scand.* **73**:642–644.

54. Luria, E. A., and Domashneva, I. V., 1974, Antibodies to thymocytes in sera of patients with schizophrenia, *Proc. Natl. Acad. Sci. U.S.A.* **71**:235–236.

55. Singal, D. P., Grof, P., and MacCrimmon, D., 1975, Antibodies to thymocytes in sera of normal controls and of patients with schizophrenia, *J. Immunol.* **114**:1425–1427.

56. Halonen, P. E., Rimón, R., Arohonka, K., and Jäntti, V., 1974, Antibody levels to herpes simplex type I, measles and rubella viruses in psychiatric patients, *Br. J. Psychiatry* **125**:461–465.

57. Libíková, H., 1983, Schizophrenia and viruses: Principles of etiologic studies, in: *Research on the Viral Hypothesis of Mental Disorders* (P. V. Morozov, ed.), S. Karger, Basel, pp. 20–51.

58. Albrecht, P., Torrey, E. F., Boone, E., Hicks, J. T., and Daniel, N., 1980, Raised cytomegalo-virus-antibody level in cerebrospinal fluid of schizophrenic patients, *Lancet* **2**:769–772.

59. Torrey, E. F., Yolken, R. H., and Winfrey, C. J., 1982, Cytomegalovirus antibody in cere-brospinal fluid of schizophrenic patients detected by enzyme immunoassay, *Science* **216**: 892–893.

60. Akohas, A., Koskimiemi, M.-L., Vaheri, A., and Rimón, R., 1985, Altered white cell count, protein concentration and oligoclonal IgG bands in the cerebrospinal fluid of many patients with acute psychiatric disorders, *Neuropsychobiology* **14**:1–4.

61. Lycke, E., Norrby, R., and Roos, B.-E., 1974, A serological study on mentally ill patients with particular reference to the prevalence of herpes virus infections, *Br. J. Psychiatry* **124**:273–279.

62. Gotlieb-Stematsky, T., Zonis, J., Arlazoroff, A., Mozes, T., Sigal, M., and Szekely, A. G., 1981, Antibodies to Epstein–Barr virus, herpes simplex type 1, cytomegalovirus and measles virus in psychiatric patients, *Arch. Virol.* **67**:333–339.

63. King, D. J., Cooper, S. J., Earle, J. A. P., Martin, S. J., McFerran, N. V., Rima, B. K., and Wisdom, G. B., 1985, A survey of serum antibodies to eight common viruses in psychiatric patients, *Br. J. Psychiatry* **147**:137–144.

64. Rimón, R. H., Halonen, P., Lebon, P., Heikkila, L., Frey, H., Karhula, P., and Hintikka Salmela, L., 1983, Antibrain antibodies and interferon in the serum and the cerebrospinal fluid of patients with schizophrenia, *Adv. Biol. Psychiatry* **12**:161–167.

65. DeLisi, L. E., Smith, S. B., Hamovit, J. R., Maxwell, M. E., Goldin, L. R., Dingman, C. W., and Gershon, E. S., 1986, Herpes simplex virus, cytomegalovirus and Epstein–Barr virus antibody titres in sera from schizophrenic patients, *Psychol. Med.* **16**:757–763.

66. DeLisi, L. E., and Sarin, P. S., 1985, Lack of evidence for retrovirus infection in schizophrenic patients, *Br. J. Psychiatry* **146**:674.

67. Robert-Guroff, M., Torrey, E. F., and Brown, M., 1985, Retroviruses and schizophrenia, *Br. J. Psychiatry* **146**:326.

68. Preble, O. T., and Torrey, E. F., 1985, Serum interferon in patients with psychosis, *Am. J. Psychiatry* **142**:1184–1186.

69a. Stevens, J. R., Papadopoulos, N. M., and Resnick, M., 1990, Oligoclonal bands in acute schizophrenia: a negative search. *Acta Psychiatr. Scand.* **81**:262–264.

69b. Harrington, M. G., Merrill, C. R., and Torrey, E. F., 1985, Differences in cerebrospinal fluid proteins between patients with schizophrenia and normal persons, *Clin. Chem.* **31**:722–726.

70. Hirata-Hibi, M., Higashi, S., Tachibana, T., and Watanabe, N., 1982, Stimulated lympho-cytes in schizophrenia, *Arch. Gen. Psychiatry* **39**:82–87.

71. Kamp, H. V., 1962, Nuclear changes in the white blood cells of patients with schizophrenic reactions, *J. Neuropsychiatry* **4**:1–3.

72. Fieve, R. R., Blumenthal, B., and Little, B., 1966, The relationship of atypical lymphocytes, phenothiazine, and schizophrenia, *Arch. Gen. Psychiatry* **15**:529–534.

73. Kolyaskina, G. I., Sekirina, T. P., Zozulya, A. A., Kushner, S. G., Zuzulkovskaya, M. Y., and Abramova, L. I., 1987, Some aspects of immunologic studies in schizophrenia, in: *Viruses, Immunity, and Mental Disorders* (E. Kurstak, Z. J. Lipowski, and P. V. Morozov, eds.), Plenum Press, New York, pp. 285–294.

74. Hollister, L. L., and Kosek, J. C., 1965, Abnormal lymphocytes in schizophrenia, *Int. J. Neuropsychiatry* **1**:559–560.

75. DeLisi, L. E., Ortaldo, J. R., Maluish, A. E., and Wyatt, R. J., 1983, Natural killer (NK) cell activity of lymphocytes from schizophrenic patients, *J. Neural Transm.* **58**:96–106.

76. Bonartsev, P. D., 1971, An ultrastructure study of lymphocytes of schizophrenia in conditions of short living cultures, *J. Neuropathol. Psychiatry* **9**:1363–1367.

77a. McAllister, C. G., Rapaport, M. H., Pickar, D., Podruchny, T. A., Christison, G., Alphs, L. D., and Paul, S. M., 1989, Increased numbers of CD5+ B lymphocytes in schizophrenic patients. *Arch. Gen. Psychiat.* **46**:892–894.

77b. Liedemann, R. R., and Prilipko, L. L., 1978, The behavior of T lymphocytes in schizo-phrenia, in: *Neurochemical and Immunologic Components in Schizophrenia* (D. Bergsma and

A. L. Goldstein, eds.), Alan R. Liss, New York, pp. 365–374.

78. Morozov, M. A., 1954, The virus-like agent in schizophrenia, *J. Neuropathol. Psychiatrie* **54**: 735–740.

79. Tyrrell, D. A. J., Parry, R. P., Crow, T. J., Johnstone, E., and Ferrier, I. N., 1979, Possible virus in schizophrenia and some neurological disorders, *Lancet* **1**:839–841.

80. Baker, H. F., Ridley, R. M., Crow, T. J., Bloxham, C. A., Parry, R. P., and Tyrrell, D. A. J., 1983, An investigation of the effects of intracerebral injection in the marmoset of cytopathic cerebrospinal fluid from patients with schizophrenia or neurological disease, *Psychol. Med.* **13**:499–511.

81. Mered, B., Albrecht, P., Torrey, E. F., Weinberger, D. R., Potkin, S. G., and Winfrey, C. J., 1983, Failure to isolate virus from CSF of schizophrenics, *Lancet* **2**:919.

82. Tyrrell, D. A. J., Parry, R., Davies, H., Bloxham, C., and Crow, T. J., 1983, Further studies of the cytopathic effect in tissue cultures inoculated with CSF from patients with schizophrenia and other nervous system diseases, *Br. J. Exp. Pathol.* **64**:445–450.

83. Rajcáni, J., Libíková, H., Smereková, J., Mucha, V., Kúdelová, M., Pogády, J., Breier, S., and Skodácek, I., 1987, Investigations on the possible role of viruses affecting the CNS in the etiology of schizophrenia and related mental disorders, in: *Viruses, Immunity, and Mental Disorders* (E. Kurstak, Z. J. Lipowski, and P. V. Morozov, eds.), Plenum Press, New York, pp. 135–148.

84. Mesa, C. S., and Cabrera, J. S., 1979, Estudio de las partículas semejantes a virus observadas en la esquizofrenia, *Rev. Hosp. Psiquiátr. Habana* **10**:725–736.

85. Kaufmann, C. A., Weinberger, D. R., Stevens, J. R., Asher, D. M., Kleinman, J. E., Sulima, M. P., Gibbs, C. J., Jr., and Gajdusek, D. C., 1988, Intracerebral inoculation of experimental animals with brain tissue from patients with schizophrenia. Failure to observe consistent or specific behavioral or neuropathological effects, *Arch. Gen. Psychiatry* **45**:648–652.

86. Stevens, J. R., Albrecht, P., Godfrey, L., and Krauthammer, E., 1983, Viral antigen in the brain of schizophrenic patients? *Adv. Biol. Psychiatry* **12**:76–96.

87. Stevens, J. R., Langloss, J. M., Albrecht, P., Yolken, R., and Wang, Y.-N., 1984, A search for cytomegalovirus and herpes viral antigen in brains of schizophrenic patients, *Arch. Gen. Psychiatry* **41**:795–801.

88. Schwartz, J. P., and Stevens, J. R., 1988, Transmissible agent in schizophrenic CSF, *Neurology* **38**(Suppl.):119.

89. Oleszak, E. L., Manuelidis, L., and Manuelidis, E. E., 1986, *In vitro* transformation elicited by Creutzfeldt–Jakob-infected brain material, *J. Neuropathol. Exp. Neurol.* **45**:489–502.

90. Aulakh, G. S., Kleinman, J. E., Aulakh, H. S., Albrecht, P., Torrey, E. F., and Wyatt, R. J., 1981, Search for cytomegalovirus in schizophrenic brain tissue, *Proc. Soc. Exp. Biol. Med.* **167**: 172–174.

91. Taylor, G. R., Crow, T. J., Higgins, T., and Reynolds, G., 1985, Search for cytomegalovirus in postmortem brain tissue from patients with Huntington's chorea and other psychiatric disease by molecular hybridization using cloned DNA, *J. Neuropathol. Exp. Neurol.* **44**:176–184.

92. Moises, H. W., Rüger, R., Reynolds, G. P., and Fleckenstein, B., 1988, Human cytomegalovirus DNA in the temporal cortex of a schizophrenic patient, *Eur. Arch. Psychiatry Neurol. Sci.* **238**:110–113.

93. Sequiera, L. W., Jennings, L. C., Carrasco, L. H., Lord, M. A., Curry, A., and Sutton, R. N. P., 1979, Detection of herpes-simplex viral genome in brain tissue, *Lancet* **2**:609–612.

94a. Jones, T. R., and Hyman, R. W., 1983, Specious hybridization between herpes simplex virus DNA and human cellular DNA, *Virology* **131**:555–560.

94b. Alexander, R., Spector, S., Casanova, M., Kleinman, J., and Wyatt, R. J., 1992, Search for cytomegalovirus in the post-mortem brains of schizophrenic patients using the polymerase chain reaction. *Arch. Gen. Psychiat.* (in press).

95. Hanshaw, J. B., Scheiner, A. P., Moxley, A. W., Gaev, L., Abel, V., and Scheiner, B., 1976, School failure and deafness after "silent" congenital cytomegalovirus infection, *N. Engl. J. Med.* **295**:468–470.

16

The Role of Viruses
in Dementia

ANNE M. DEATLY, ASHLEY T. HAASE,
and MELVYN J. BALL

1. INTRODUCTION

Dementia is a clinical state characterized by impaired cognition and behavior. In this chapter we briefly summarize the historical, clinical, and neurological aspects of dementias that are known or suspected to be caused by unconventional virus-like agents as well as large and small conventional viruses with RNA or DNA genomes. These different agents gain access to the central nervous system (CNS), an organ normally protected from systemic infections, to cause (either directly or indirectly) the pathological damage that results in a state of dementia by mechanisms involving complex host–viral interactions. Wherever possible, we discuss how a particular virus induces the neuropathological changes and offer conjectures on other aspects of pathogenesis. We also emphasize the role of host factors such as genetic predisposition and the immune regulatory system in the dementing diseases associated with viral infection.

ANNE M. DEATLY and ASHLEY T. HAASE • Department of Microbiology, University of Minnesota, Minneapolis, Minnesota 55455.   MELVYN J. BALL • Division of Neuropathology, Oregon Health Sciences University, Portland, Oregon 97201.

Neuropathogenic Viruses and Immunity, edited by Steven Specter *et al.* Plenum Press, New York, 1992.

2. VIRUSES OR VIRUS-LIKE AGENTS, DEMENTIA, AND CHRONIC NEUROLOGICAL DISEASE

2.1. Viruses Known to Cause Dementia

2.1.1. HIV-1 and Dementia of AIDS

Human immunodeficiency virus type 1 (HIV-1) is associated not only with acquired immunodeficiency syndrome (AIDS) but also with AIDS dementia complex (ADC).[1] ADC is a "subcortical dementia"[2,3] characterized by neurological abnormalities including progressive cognitive loss, impaired memory, impaired motor performance, and abnormal behavior.[4] It has also been referred to as AIDS-related subacute encephalopathy.[1] The major histopathological abnormalities involve subcortical structures, in particular the white matter of the cerebral centrum semiovale, the deep gray structures (including the basal ganglia, thalamus, and the brainstem),[5] and demyelination of the peripheral nerves. It seems likely that the CNS dysfunction is caused by HIV-1 and not by another potential opportunistic infectious agent of the CNS associated with HIV-1-induced immunosuppression, since HIV-1 retrovirus has been demonstrated in the CNS of ADC patients in several ways.[5,6] Integrated and nonintegrated proviral DNA copies have been identified in brains of patients with the ADC pathology. *In situ* hybridization and immunohistochemical studies have revealed the presence of viral nucleic acid and antigens in ADC patients. HIV-1 has been cultured directly from the brain and cerebrospinal fluid (CSF) of demented patients and HIV-1 virions have been detected in AIDS brain tissue by electron microscopy.[5]

There is much debate over the specific cell types involved in HIV-1 infection in the CNS. Multinucleated cells, the histological hallmarks of productive HIV-1 CNS infection, result from direct virus-induced cell fusion. The productive HIV-1 infection in the CNS appears primarily limited to macrophages, cells not usually resident in CNS, and to a lesser degree microglia. *In situ* hybridization studies (M. Zupancic, R. W. Price, and A. T. Haase, unpublished results) have detected high copy numbers of HIV-1 RNA in multinucleated cells, and studies are under way to determine if other cell types are involved that harbor reduced copy numbers of the viral nucleic acid.

Even though ADC does not usually become evident until the later stages of the HIV-1 systemic infection, it seems most likely that involvement of the CNS occurs in the early stages of viral infection. At the time of AIDS diagnosis, roughly one-third of the patients already exhibit some obvious form of the dementia complex, and one-fourth have subclinical forms.[5] Because HIV-1 specific antibodies are detected in the CSF of neurologically asymptomatic patients, it is possible that the HIV-1 infects the CNS at a very early stage of AIDS and remains latent[7] until a decline in host antiviral response results in rapid spread of virus, and ensuing neuropathological changes manifest as ADC. However, it is possible that different strains of HIV-1 may be responsible for the immune deficiency syndrome and the subcortical dementia syndrome[8] in view of the genetic diversity of HIV-1.[9–12]

By analogy to another lentivirus, visna virus, which causes a paralytic condition in sheep, HIV may enter the CNS by a "Trojan Horse" mechanism in which monocyte-derived macrophages infected peripherally migrate to the CNS to eventually cause the neurological disease.[13–15] Because the number of HIV-1-infected cells in the CNS is large, it is possible that the infected macrophages selectively migrate to the CNS, or the latent infection of the monocyte/macrophage is reactivated by the immunosuppression associated with the disease, or that HIV-1 spreads secondarily to permissive cells within the CNS.[5] In order to understand how this virus causes the unique form of dementia, it is important to determine how the virus enters the CNS, the types of cells it infects, and the extent of viral gene expression at different stages of the disease.

The HIV does not infect large numbers of neurons or oligodendrocytes. The pallor of white matter may be indirectly mediated by a generalized toxic process.[5] This toxicity could result from substances released from the infected cells, such as cytokines or enzymes, which might affect the neighboring cells.[16] Another possibility is that the HIV-1 infection causes an alteration in localized cellular metabolism, resulting in a generalized toxic effect. It is also possible that viral gene products are toxic to the nearby uninfected tissue. This latter possibility is supported by experiments demonstrating that the HIV-1 envelope glycoprotein (gp120) interferes with the activity of a neurotropic factor, neuroleukin.[17–19] This interference may or may not be related to the fact that these two proteins have partial sequence homology.

2.1.2. Transmissible Agents in the Spongiform Encephalopathies of Creutzfeldt–Jakob Disease and Kuru

Two human presenile dementias, Creutzfeldt–Jakob disease (CJD) and kuru, and two dementias of animals, scrapie in sheep and goats and transmissible mink encephalopathy, are classified as subacute spongiform encephalopathies because of similar lesions in the CNS consisting of vacuoles in pre- and postsynaptic processes, neuronal degeneration, spongiosis of gray matter, associated astrocyte alterations, and amyloid plaques.[20–21] These diseases are caused by unconventional viruses,[22,23] so named because they produce slow infections with long incubation times.

Kuru, now an extinct condition, first described by Gajdusek and Zigas in 1957,[24] was limited geographically to a mountainous region of the eastern highlands in Papua, New Guinea. Kuru presents clinically as cerebellar condition with ataxia and shivering-like tremors (kuru = dance) that progresses to complete immobility and death usually 1 year after onset.[24] The mode of transmission is believed to have been by ritual cannibalism.[25] The disease has been transmitted successfully from humans to primates and from primates to other primates by both intracerebral and peripheral routes.[20,23,26]

Creutzfeldt–Jakob disease, described in 1920 and 1921 by Creutzfeldt[27,28] and Jakob[29,30] is of worldwide distribution with an incidence of one to two per million per year.[31] It was first compared to kuru in 1959 by Klatzo et al..[32] This chronic dementing disease has a slow clinical course starting with symptoms limited to intellectual and behavioral abnormalities, followed by myoclonus, characteristic EEG waves, and complete mental and physical deterioration over months, a few years, or rarely even decades.[33] In general, CJD in humans affects the cortex and basal ganglia, and neuritic plaques are rare. It has also been transmitted successfully to primates from human[34,35] and from primates to primates,[20] guinea pigs, hamsters, and mice.[36]

What is known about the nature of these unconventional viruses? First, replication of these agents must occur, since new infectious agents are produced on inoculation into new animals.[37] Second, there must be different strains of the agent, since there are differences in incubation periods, clinical presentations of the disease, and differences in the types of lesions produced.[31] For example, in the Gerstman–Straussler–Scheinker syndrome, a variant of CJD, lesions are routinely located in the cerebellum, and plaque formation is the rule rather than the exception.[38,39] Third, different strains of the host affect the pathology of the disease. Fourth, there is a species barrier effect, with increases in incubation times when the virus is transmitted from one species to another; these times diminish with increasing serial passage within the same animal species. Fifth, presence of infectious materials in the blood has been demonstrated,[31] but virus or virus-like particles are never detectable by electron microscopy.

These five properties illustrate the similarities between unconventional viral agents and conventional viruses. Unconventional agents also differ in several respects: they rarely elicit an inflammatory response,[40] possibly because they are sequestered from the immune system or because their hosts are tolerant to them because they contain normal host proteins. Unconventional agents are also resistant to formalin, heat, nuclease treatment, and ionizing and ultraviolet radiation, treatments that would destroy conventional viruses. Phenol extraction, guanidinium HCl treatment, membrane-disrupting agents, protease treatment, sodium dodecyl sulfate, and diethylpyrocarbonate treatment all cause a reduction in titer, indicating that protein stability is essential for infectivity. Since nuclease treatment does not reduce infectivity, there is much debate about whether these unconventional agents are actually devoid of nucleic acid. If there is an unconventional agent genome, it is either protected by protein or exceptionally small, to account for the resistance to ionizing radiation. These enigmatic properties continue to stimulate efforts to understand the nature of these agents and how they cause disease.

2.2. Viruses and Chronic Neurological Disease

2.2.1. Polyomavirus and Progressive Multifocal Leukoencephalopathy

Progressive multifocal leukoencephalopathy (PML) is a rare chronic human demyelinating disorder of the brain characterized by multiple neurological deficits, most frequently involving both motor and mental disturbances or dementia.[41,42] The course of the disease is usually 3–6 months from onset to death.[42] Virus particles have been consistently associated with diseased brains since 1965,[41,43–45] and in 1971, the etiologic agent, a human papovavirus of the polyoma subgroups, designated JC virus, was cultured from diseased brain tissue.[46] Even though PML is rare, infection with JC virus is common; about 70% of tested adults are seropositive.[47] It is believed that a subclinical infection is acquired at an early age,[47,48] and the infection persists.[49] PML is associated primarily with other diseases that depress the patient's immune system, such as Hodgkin's disease, leukemia, and more recently and with increasing frequency, AIDS;[50–52] however, there are several reports of cases in which PML has occurred in the absence of immunosuppression.[53,54] It is not clear whether PML is caused by a reactivation of JC virus infection or by a primary infection in an immuno-compromised host.

The earliest pathological changes in PML disease are pinhead-sized foci of demyelination found along the cortical ribbon. These enlarge and coalesce into lesions that may eventually occupy a major part of the white matter of one hemisphere.[42] The widely disseminated demyelination[55] is a result of infection and lysis of oligodendrocytes, the cells that produce myelin basic protein and maintain the myelin sheath in the CNS. Astrocytes are also abortively infected and partially transformed, resulting in increased mitotic activity and alteration in the number and morphology of their nuclei.

2.2.2. Measles and Subacute Sclerosing Panencephalitis

Subacute sclerosing panencephalitis (SSPE) is a rare degenerative disease of the CNS that affects children and young adults. Subtle intellectual and psycho-logical dysfunction is followed by additional symptoms such as myoclonic epi-lepsy, which last from months to years. Progressive deterioration of sensory and motor function accompanying cerebral degeneration lead to death[56] from inter-current infection or vasomotor collapse.

The neuropathological changes include diffuse encephalitis of both white and gray matter (panencephalitis) with infiltrating lymphocytes, perivascular cuffing by inflammatory cells, and a dramatic neuronal loss in the cerebral cortex. Also characteristic are intranuclear inclusion bodies in nerve cells and oligo-dendrocytes.[57] Widespread destruction of cells produces demyelination, and the astrocytic response to injury causes the alteration in the brain substance sub-sumed by the term "sclerosing."

The viral etiology of SSPE was first suspected in 1933 when Dawson noted inclusion bodies in cells,[58] and subsequent studies have established measles virus or a measles virus variant as the cause of SSPE. The first evidence of this relationship was provided in 1965 by Bouteille and colleagues when they demon-strated structures within the inclusion bodies that resembled paramyxovirus nucleocapsids.[59] SSPE patients also have high titers of measles virus antibodies in their serum and CSF.[60,61] Finally, in 1969, measles virus was isolated from the brain and lymph nodes of SSPE patients using cocultivation techniques,[62–65] and later measles virus RNA was found in cells that make up foci or lesions.[66]

It is likely that the measles virus infection eventually responsible for SSPE occurs at an early age (about half of the SSPE patients had an acute measles infection before 2 years of age) with an average incubation period of 6–8 years before the first symptoms of SSPE develop.[67] Only one in 1,000,000 children who have had acute measles infection (and apparently fully recover) develop this slowly progressive neurological disease.[68] The frequency of SSPE has been reduced five- to 50-fold since the introduction of live measles virus vaccines,[69] presumably related to the fact that the vaccine virus has a decreased ability to infect the CNS.

Even though it appears that a persistent measles virus infection is ultimately responsible for SSPE, the mechanisms of pathogenesis are still unknown. The development of the disease may depend on whether the brain is infected during the acute episode of measles infection as well as on the type and number of cells infected at that time.[68] Because of the young age of most SSPE patients, other factors contributing to the disease may include an immature immune system, increased susceptibility to variant measles viruses, and other immunologic abnormalities.[57] The fact that 70–90% of peripheral mononuclear cells (as well as 15% of nerve cells) from three SSPE patients contained measles virus RNA (detected by in situ hybridization) provides an alternative reservoir for the infectious agent and a mechanism for dissemination to the brain.[66]

There are many reports that the measles virus isolated from cocultivation of tissue from SSPE patients is different from the wild-type strains of measles virus. In some cases the measles virus isolated from these patients contains a larger matrix protein (M), an internal membrane protein required for virus assembly and budding from the cell membrane, and larger associated mRNA than the corresponding molecules in the wild-type virus.[70,71] The sera of SSPE patients also lack antibody to M protein,[72,73] consistent with low levels of M protein antigen in the infected cell. Transcription of the M gene has been reported to be reduced,[74,75] translation of its mRNA is defective,[76] and the M protein may be rapidly degraded.[77–79] In other reports, M protein has been detected in the brain of SSPE patients but not in cultures infected by the virus isolated from those brains.[80] All of these studies point to abnormalities perhaps created by mutations. Cattaneo and colleagues[81] have estimated that one out of 100 bases of the measles virus genome is altered in a persistently infected SSPE cell line, reflecting the inaccuracy of the RNA-dependent RNA polymerases that lack a proofreading function.[82,83] The nature or position of a particular mutation in M protein sequence may determine whether the change involves instability, inactivity, altered translatability, or premature termination of the M protein. Because of the variability of results obtained from different SSPE patients and the rarity of the disease, it is very possible that mutants or variants of the measles virus that cause SSPE could arise independently within each individual. It is possible that genetic changes that may affect the measles M protein could lead to an altered biological property, resulting in persistence of the virus in the CNS undetected by the immune system. Certainly the absence of a functional M protein would result in an abortive infection, since no virus particles could be produced but other viral proteins could accumulate, and the glycoproteins present on the cell surface could stimulate high levels of antibodies.[68] Modulation of internal viral antigen expression by the presence of glycoprotein antibodies has been proposed as a mechanism for viral persistence in the presence of an immune response.[84]

Of course, a genetic change in the viral genome is not entirely responsible for the restricted measles gene expression; the host must play a restrictive role as well, since the restriction can be relieved on cocultivation or explantation of diseased tissues with cells permissive to measles virus infection. This restriction has been documented *in vivo* by quantitative *in situ* hybridization of SSPE, where Haase and colleagues demonstrated a decrease in the levels of both plus- and minus-strand measles virus RNA compared with the acutely infected controls.[85,86] By immunofluorescent studies, the nucleocapsid protein was detected, but M protein was not detectable in these tissues.[85,86] From these studies with SSPE and other slow virus infections, the more general view emerges that the reduced synthesis and expression of viral components enable the virus to go unnoticed by the immune system and reduce the cellular injury normally caused by unrestricted virus synthesis, thus allowing the neurological damage to accumulate slowly.

2.2.3. *Viliuisk Encephalomyelitis Virus and Viliuisk Encephalomyelitis*

Viliuisk encephalomyelitis (VE), a motor neuron disease with varying degrees of dementia, affects the Iakut people of the Viliui River Valley in Siberia. It was first reported in 1887 by a German ethnologist and explorer, R. K. Maak. About 1% of the 60,000 Iakuts of all ages are affected with VE, which accounts for 5% of all deaths among them.[87]

The acute stage of this disease is characterized by high fever, chills, severe headache, influenza-like muscle pains, and lethargy. Accompanying these symptoms are signs of cranial nerve dysfunction, visual problems, rigidity, and abnormal reflexes. Mental deficits include mild dementia and depression to hypochondria, aggressive behavior, and delirium. The acute stage lasts for a matter of days up to a month. From the acute through the chronic stages of VE, the CSF of patients exhibits an increased, albeit low, cell count and moderately elevated protein level. Following the acute stage of the disease, the patient may partially recover for weeks to a year or more, only to succumb to a progressive panencephalitis syndrome 3–5 years to as long as 10–20 years later. The manifestations of VE include progressive dementia, disturbances of speech, increasing spasticity, altered gait, rigidity, and signs of cranial nerve involvement.[87]

The neuropathology of VE disease involves all areas of the brain and spinal cord. Hydrocephalus, mild demyelination, severe cortical atrophy, diffuse gliosis, small areas of inflammation with neuronal loss, invasion of the brain with mononuclear cells in both white and gray matter, and amyloid plaques and neurofibrillary tangles have been described.[87]

Since the 1950s, the disease has spread to neighboring areas of the Viliui Valley. From studies of the limited migration of the Iakut people and the spread of the disease, VE appeared to be caused by an infectious agent of low communicability, with a pattern of dissemination and long incubation period and latency.[87]

The etiologic agent for VE disease is a virus named Viliuisk encephalomyelitis virus, which has been isolated from VE patients and inoculated into laboratory mice. Viliuisk encephalomyelitis virus is similar to mouse encephalomyelitis virus and less similar to encephalomyocarditis virus. One strain (KPN) isolated from one patient has been shown to be sensitive to lipid solvents (ether and chloroform), sodium deoxycholate, and formaldehyde. However, VE patients do not have any antibody titer to this virus. There is one case of a non-Iakut Russian laboratory worker who inoculated herself with CSF from a VE patient and subsequently developed VE clinical and neuropathological symptoms and later died.[87]

3. THE ROLE OF VIRUSES IN THE ETIOLOGY OF ALZHEIMER'S DISEASE

The etiology of Alzheimer's disease, the most common organic dementia and the fourth to fifth leading cause of death in the United States,[88,89] first described by Alois Alzheimer in 1907, is still unknown. The typical features of the disease include memory and other cognitive impairment and abnormal behavior that affect the patients' ability for self-care, interpersonal relationships, and adjustment in the community. Death usually occurs 5–10 years after the onset of symptoms.[90] Alzheimer's disease is diagnosed by exclusion and can only be confirmed accurately at autopsy, where the characteristic neurofibrillary tangles, neuritic plaques (containing β-amyloid cores), and congophilic (amyloid) angiopathy can be confirmed in the cortical and subcortical areas of the brain,[91] especially in neocortex, basal forebrain, and hippocampus. Although neuritic plaques and tangles occur during normal aging, these changes are quantitatively increased in Alzheimer's disease.[92,93] Additional pathological changes in Alzheimer's disease include dramatic neuronal loss, granulovacuolar degeneration and Hirano bodies in hippocampal cortex, brain atrophy, a decrease in the level of markers for acetylcholine, and lack of an inflammatory response.[88]

Several factors are thought to play a role either directly or indirectly in the etiology of this common dementia: genetics, neurotransmitters, aluminum, immune system dysfunction, and viruses. The only known risk factor is age: the incidence and prevalence increase with each decade after 60. It is possible that there is only one etiologic agent responsible for Alzheimer's disease, but on the other hand there may be multiple causative agents that are not similar but can cause similar clinical and pathological changes. It is also very possible that Alzheimer's disease is caused by an interaction of several such factors: pathological agents (either infectious or toxic), the environment (which determines the extent of exposure to potential etiologic agents), and the host (which may in some families have a certain genetic predisposition to this disease).[89]

Alzheimer's disease can be divided into two categories: familial and sporadic. In familial cases (Familial Alzheimer's disease, FAD), the age of onset[94] is generally earlier, the course of the disease is shorter, and the disease is more severe. Inheritance appears to be autosomally dominant.[95] There is also an association between Down's syndrome (a genetic disease of mental retardation usually caused by three copies of chromosome 21) and Alzheimer's disease, because if an individual with Down's syndrome lives to be 35 to 40 years or more, pathological lesions of the Alzheimer type invariably develop in identical severity and location.[96] It is possible that this association between the two diseases is not only a genetic one but an association of shared risk factors, since some studies show the maternal and paternal ages to be higher at birth both of Alzheimer's and of Down's syndrome patients than that of normal control individuals.[89] In a search for an Alzheimer (familial) gene on chromosome 21, two loci were found: the β-amyloid locus and the FAD locus, which may interact at the gene or protein level.[97,98] Also associated with FAD is a greater incidence of lymphoproliferative malignancy, indicating that the immune system may play a role in the etiology of this disease.[99]

The far more prevalent sporadic form of Alzheimer's disease does not appear to be strictly a genetic disorder, since parents or siblings are not at an increased risk. There is evidence, nevertheless, that suggests that genetic predisposition does play a role in sporadic Alzheimer's disease; for example, there is a correlation of a specific complement phenotype with Alzheimer's disease.[94]

The immune system, as already indicated, may also play a role in the etiology of this disease. There is some evidence that the function of lymphocytes is impaired in Alzheimer's disease, since antibodies to brain proteins have been found circulating in the bloodstream of Alzheimer patients. It is proposed that these antibodies bind to normal brain proteins and interfere with their normal function, perhaps causing cell death.

The hypothesis that aluminum causes Alzheimer's disease is now less attractive for four reasons. (1) Aluminum is present in normal aged brains in association with lesions unlike that of Alzheimer's type. (2) Experimental aluminum intoxication affects spinal cord neurons, whereas Alzheimer's disease does not. (3) Aluminum-induced tangles are composed of straight intraneuronal filaments, whereas the Alzheimer-type tangles are composed of paired helical filaments; the two types of filaments are probably composed of different proteins. (4) Dialysis dementia, which is associated with excess amounts of aluminum does not exhibit either the same histopathology or clinical symptoms as Alzheimer's disease patients.[91,100] It seems more likely that the small increase in aluminum in Alzheimer's disease can be attributed to concentration at sites of neuronal death.

The neurotransmitter hypothesis is based on the notion that if neurotransmitters function improperly, they could "disconnect" portions of the brain involved in memory, for example, the hippocampus. Two neurotransmitters, acetylcholine and somatostatin, have been studied extensively, since both are reduced in Alzheimer's disease, indicating loss of both pre- and especially postsynaptic elements. Moreover, acetylcholine- and somatostatin-containing nerve terminals participate in neuritic plaque formation. Cholinergic deficits are most closely correlated with memory disturbances. Choline acetyltransferase is known to be immunogenic, and if this enzyme leaked out because of injury, an autoimmune response could be induced. Antibodies produced against choline acetyltransferase could result in interference with normal neural transmission.[101]

If neurotransmitters play a primary role in Alzheimer's disease, then multiple transmitter systems must be involved because of the location of the neurofibrillary changes. In addition, it is just as likely that degeneration of the corticocortical system causes reduction or damage to the chemical neurotransmitters rather than their dysfunction as the primary pathological event.

Numerous attempts to transmit Alzheimer's disease to nonhuman primates and laboratory animals using brain tissue from the terminal stages of the disease have been unsuccessful.[102,103] However, recently the buffy coat of peripheral blood from one Alzheimer patient, but not four preclinical (unaffected) relatives, has been reported to induce[104] spongiform encephalopathy in these hamsters after a long latent interval. The argument that a transmissible subacute spongiform encephalopathy-like virus could cause Alzheimer's disease is also strengthened by analogous pathological changes such as neuritic plaques and tangles in CJD, kuru, and scrapie and the occurrence of CJD and Alzheimer's within the same families.[104–106] Although Alzheimer's disease does not typically include the spongiform lesion, a subtype of Alzheimer's disease has been reported in which vacuolar change histologically indistinguishable from CJD is restricted to mesial temporal cortex and amygdala.[107]

Conventional viruses, particularly herpes simplex virus (HSV) or another herpesvirus, have also been invoked as causative agents of Alzheimer's disease.[108–114] Herpesviruses, which infect the majority of the human population, remain latent in the neurons of the trigeminal ganglia and CNS between recurrent reactivation episodes. During the latent state, the HSV type 1 (HSV-1) DNA genome is not undergoing replication, but it is also not inactive. A restricted region of the HSV-1 genome is transcribed in relatively abundant amounts throughout the latent period.[115–120] It is presumed that the latent RNA(s) produced play a role in maintaining the latent state by preventing the transcriptional cascade of a normal lytic herpesvirus infection. One abundant latent RNA is transcribed from the strand complementary to that which encodes an important immediate early regulatory gene.[117] Since this latent RNA partially overlaps the gene,[121] it has been proposed that the latent RNA functions as an antisense message by preventing the immediate early stage of transcription and, thus, a productive herpesvirus infection.[117] It is possible that the latent RNA(s) is translated into a polypeptide that could also function in blocking the normal transcriptional cascade events.

During latency, herpesviruses may be reactivated by stress or other conditions. Virus produced from reactivation from peripheral nervous system neurons (trigeminal ganglia) could spread to the CNS, either establishing a latent infection there or destroying neurons. The lymphocytic infiltrates adjunct to latently infected cells are one indication of recurrent reactivation. Reactivation of herpesvirus infections in regions of the brain that are responsible for memory, for example, could destroy neurons or cause specific neurons to degenerate that are needed for these functions. These neurons could develop the neuritic plaques and tangles observed in Alzheimer's disease cases. It is also possible that a latency polypeptide could stimulate an immune reaction, resulting in destruction of the neuron, or a latency polypeptide could alter the gene expression of cellular proteins, resulting in, for example, β-amyloid accumulation in the neuron. With recurrent reactivation events and increasing numbers of neurons destroyed by acute or latent herpesvirus infections, there is reason to believe that this virus could cause similar neuropathology to that observed in brain tissues with Alzheimer's disease. The high proportion of human trigeminal ganglia with latent herpesvirus would also be consistent with the high incidence of Alzheimer's disease[120] and with the fact that a majority of normal-aged brains also show Alzheimer-type neuritic plaques and tangles.[108] There are also well-defined anatomic routes by which herpesviruses travel from the trigeminal ganglia to the temporal lobe and limbic structures such as the hippocampus and entorhinal cortex, which have been documented in a mouse model system.[116,122] These same regions also show the most extensive histopathology in Alzheimer's disease.

In support of the virus hypothesis are reports of HSV-1 DNA in human brain tissue with senile plaques and neurofibrillary tangles[123] and immunocytochemical studies demonstrating HSV-1-infected cells in Alzheimer's brain.[124,125] In an attempt to confirm the association of latent herpesviruses with Alzheimer's disease pathology, we have analyzed the hippocampus of two Alzheimer's disease patients who had HSV-1 latently infected neurons in their trigeminal ganglia. We hybridized every 15th to 20th tissue section from these hippocampi with a probe to detect the abundant HSV-1 latent RNA but were unable to find any hippocampal pyramidal neurons expressing HSV-1 RNA. Although these recent studies do not support the virus hypothesis, studies are confounded by the fact that in end-stage disease tissue may no longer contain the associated etiologic agent. For example, the neuronal loss in the hippocampus of one of the patients tested was 90%. For this reason we are surveying tissues harvested from patients in a much earlier phase of illness. Moreover, we are investigating the possibility of detecting HSV-1 latently infected cells in the brainstems of these patients with the rationale that the virus could initiate its damaging effects in an area along the pathway from the peripheral to CNS, at some distance from the sites of neuropathological damage. (A. M. Deatly, E. Lewis, A. T. Haase, and M. J. Ball, unpublished results).

4. NEUROLOGICAL DISORDERS WITH A POSSIBLE VIRAL ETIOLOGY

4.1. Tropical Spastic Paraparesis

Tropical spastic paraparesis (TSP) is a neurological disorder common in tropical areas with a prevalence as high as 1/1000.[126,127] The disease usually affects adults between 30 and 40 years of age. The symptoms of TSP are those of a slowly progressive (over a decade or more) predominantly spastic myelopathy characterized by spastic gait or paraplegia, lower back pain, hyperreflexia of legs and arms, extensor plantar responses, spastic bladder, severe constipation, and impotence (in males). Loss of sensory and mental function are minimal.[126–128] In the relatively few postmortem studies to date, the typical neuropathological feature of TSP is widespread chronic meningoencephalomyelitis with inflammatory changes predominantly in the spinal cord and to some extent in the midbrain, cerebellum, and cerebrum. The inflammatory changes involve perivascular cuffing with lymphocytes, plasma cells, and histiocytes. Another common neuropathological feature of TSP is demyelination and axonal loss in the posterior columns and pyramidal tracts. The spinocerebellar and spinothalamic tracts as well as the optic and auditory nerves may also be affected.[126]

Tropical spastic paraparesis has only recently been described, with the first cases reported in Martinique in 1952[129] and most of the studies performed since 1980. The disease is endemic to the tropics including Jamaica and Martinique in the Caribbean, Tumaco off the coast of Colombia, and the Seychelles Islands in the Indian Ocean. Cases have also been reported in Central and South America, India, and Africa.[126] TSP appears to be primarily a disease of people with black African ancestry, but cases have been described in Caucasians, Hindus, American Indians, and Orientals.[126]

Tropical spastic paraparesis has been attributed to toxins, environmental factors, vitamin B deficiencies, protein malnutritions, and infectious agents such as parasites, treponomes, and viruses. There is now a strong case implicating retroviruses. In 1985, Gessian and colleageus provided serological data linking TSP and infection by human T-cell lymphotropic virus type I (HTLV-I)[130] subsequently there have been confirmatory reports of this virus or a related retrovirus such as the HTLV-I-associated myelopathy (HAM) in Japan.[126,128,130] The serum and CSF of these patients contain viral antigens and high-titer HTLV-I antibodies, and the intrathecal synthesis of HTLV-I antibodies and the oligoclonal immunoglobulin bands in CSF that react with HTLV-I antigens suggest synthesis within the CNS. Virus has been isolated from peripheral blood and CSF mononuclear cells, which resemble those found in blood and CSF of patients with adult T-cell leukemia (a lymphotropic disease caused by HTLV-I); and McFarlin and colleagues have isolated an HTLV-I-like retrovirus from T-cell lines obtained from TSP patients.[131] HTLV-I nucleic acid sequences have been detected in the blood and CSF of patients,[132] and HTLV-I virions were detected by electron microscopy in spinal cord tissue of a Jamaican TSP patient known to have HTLV-I antibodies.[133]

Although these findings point to HTLV-I as the cause of TSP, only 80% of the patients have HTLV-I antibodies,[130,132] and no antibodies to the envelope protein have been detected.[134] We consider it quite conceivable that a neurotropic variant of HTLV-I is the etiologic agent for TSP and that the altered tropism is a consequence of recently documented deletions in the viral *env* gene. This could also rationalize the failure to find *env* antibodies.

Since HTLV-1 is markedly lymphotropic, it is very likely that the immune system may play a role in the etiology of this disease as well. An increase in the number of circulating activated T lymphocytes has been reported,[135] as have defects in the cellular immune response and a decrease in natural killer cell activity.[136] How the effects of HTLV-I on the immune function are related to neurological disease remains to be determined.

4.2. Postencephalitic Parkinsonism

Postencephalitic parkinsonism, first described in 1920–1921 after the influenza pandemic of 1917 and 1918,[137] is a disease frequently associated with dementia in which neurons in the substantia nigra show neurofibrillary changes (of Alzheimer type) rather than the Lewy bodies of Parkinson's disease.[138] There is a profound loss of neurons in the substantia nigra[139] and widespread lesions in basal ganglia and many other parts of the CNS.[140] The suspected etiologic agent of this dementia disease is influenza because of the history of acute encephalitis (encephalitis lethargica or von Economo's disease).[140]

5. DISCUSSION

One of the major problems in relating viral infection to the etiology of chronic human neurological and psychiatric disorders is that years or sometimes decades separate the onset of disease and examination of the central nervous tissues at death. Thus, at the end state of the disease or death, the virus may not even be present, and this may be the reason for the difficulty of successful experimental transmission of some of these diseases, for example, Alzheimer's disease. In addition, most of these disease states are not consistently associated with previous acute illnesses (as in SSPE and postencephalitic parkinsonism), so it is difficult to know how to implicate a particular group of viruses. Finally, even if viral sequences, antigens, and/or particles can be detected in damaged tissue (as in polyomavirus and PML and HIV-1 and ADC), this is insufficient evidence to establish the virus as the unequivocal cause of the disease.

Viral infections associated with chronic neurological disorders may be latent or slow infections characterized by complex virus–host interaction in which outcome is determined by the number and type of cells infected, the permissiveness of the cell with respect to viral replication, and the effectiveness of the nonspecific and specific immune defense mechanisms of the host. These in turn reflect the age of the host, genetic factors in both virus and host, and immune status. Generally speaking, the persistence of virus is a consequence of restricted replication and gene expression, which allow the infected cell to escape immune surveillance. The transition from a latent infection to disease may be the result of altered immunologic competence (e.g., PML). On the other hand, infection may provoke an immune and inflammatory response that is actually responsible for disease. Thus, viruses may damage the brain directly, by destroying or altering the function of the infected cell, or indirectly, by sensitizing the immune system to brain antigens, either by release or through shared epitopes of virus and host (molecular mimicry). Part of the fascination of this important area of virus research lies in dissecting these complex events and forging the linkages between the covert infections of a small number of cells and the generalized neuropathological alterations manifest as disease. This constitutes the exciting agenda for future studies.

ACKNOWLEDGMENTS. We would like to thank Dr. Richard Peluso for reading and Dana Clark for typing the manuscript. A.M.D. was supported by NIH grant 5T32CA09138. This work was supported in part by grants from the U.S. National Institutes of Health (AG03047) and the Atkinson Charitable Foundation of Toronto to M.J.B., and by NIH grant support to A.T.H.

REFERENCES

1. Snider, W. D., Simpson, D. M., Nielsen, S., Gold, J. W. M., Metroka, C. E., and Posner, J. B., 1983, Neurological complications of acquired immunodeficiency syndrome: Analysis of 50 patients, *Ann. Neurol.* **14**:403–418.
2. Albert, M. L., Feldman, R. G., and Willis, A. L., 1974, The "subcortical dementia" of progressive supranuclear palsy, *J. Neurol. Neurosurg. Psychiatry* **37**:121–130.
3. Cummings, J. L., and Benson, D. F., 1984, Subcortical dementia-A review of an emerging concept, *Arch. Neurol.* **41**: 874–879.
4. Navia, B. A., Jordan, B. D., and Price, R. W., 1986, The AIDS dementia complex. I. Clinical features, *Ann. Neurol.* **19**:517–524.
5. Price, R. W., Brew, B., Sidtis, J., Rosenblum, M., Scheck, A. C., and Cleary, P., 1988, The Brain in AIDS: Central nervous system HIV-1 infection and AIDS dementia complex, *Science* **239**:586–592.
6. Price, R. W., Sidtis, J., and Rosenblum, M., 1988, The AIDS dementia complex: Some current questions, *Ann. Neurol.* **23**:S27–S33.
7. Resnick, L., DiMarzo-Veronese, Schupbach, F., Tourtellotte, W. W., Ho, D. D., Müller, F., Shapshak, P., Vogt, M., Groopman, J. E., Markham, P. D., and Gallo, R. C., 1985, Intra-blood–brain-barrier synthesis of HTLV-III specific IgG in patients with neurological symptoms associated with AIDS or AIDS-related complex, *N. Engl. J. Med.* **313**:1498–1504.
8. Budka, H., Costanzi, G., Cristina, S., Lechi, A., Parravicini, C., Trabattoni, R., and Vago, L., 1987, Brain pathology induced by infection with the human immunodeficiency virus (HIV). A histological, immunocytochemical, and electron microscopical study of 100 autopsy cases, *Acta Neuropathol. (Berl.)* **75**:185–198.
9. Wong-Staal, F., Shaw, G. M., Hahn, B. H., Salahuddin, S. Z., Popovic, M., Markham, P., Redfield, R., and Gallo, R., 1985, Genomic diversity of human T-lymphotrophic virus type III (HTLV-III), *Science* **229**:759–762.
10. Benn, S., Rutledge, R., Folks, T., Gold, J., Baker, L., McCormick, J., Feorino, P., Piot, P., Quinn, T., and Martin, M., 1985, Genomic heterogeneity of AIDS retroviral isolates from North America and Zaire, *Science* **230**:949–951.
11. Hahn, B. H., Gonda, M. A., Shaw, G. M., Popovic, M., Hoxie, J. A., Gallo, R., and Wong-Staal, F., 1985, Genomic diversity of the acquired immune deficiency syndrome virus HTLV-III: Different viruses exhibit greatest divergence in their envelope genes, *Proc. Natl. Acad. Sci. U.S.A.* **82**:4813–4817.
12. Hahn, B. H., Shaw, G. M., Taylor, M. E., Redfield, R. R., Markham, P. P., Salahuddin, S. Z., Wong-Staal, F., Gallo, R., Parks, E. S., and Parks, W. P., 1986, Genetic variations in HTLV-III/LAV over time in patients with AIDS or at risk for AIDS, *Science* **232**:1548–1553.
13. Peluso, R. W., Haase, A. T., Stowring, L., Edwards, M., and Ventura, P., 1985, A Trojan horse mechanism for the spread of Visna virus in monocytes, *Virology* **147**:231–236.
14. Stowring, L., Haase, A. T., Petursson, G., Georgsson, G., Palsson, P., Lutley, R., Roos, R., and Szuchet, S., 1985, Detection of visna virus antigens and RNA in glial cells in foci of demyelination, *Virology* **141**:311–318.
15. Haase, A. T., 1986, Pathogenesis of lentivirus infections, *Nature* **322**:130–136.
16. Schelper, R. L., and Adrian, E. K., 1986, Monocytes become macrophages; they do not become microglia: A light and electron microscopic autoradiographic study using ^{125}-iododeoxyuridine, *J. Neuropathol. Exp. Neurol.* **45**:1–19.
17. Gurney, M. E., Heinrich, S. P., Lee, M. R., and Yin, H.-S., 1986, Molecular cloning and expression of neuroleukin, a neurotrophic factor for spinal and sensory neurons, *Science* **234**:566–574.
18. Lee, M. R., Ho, D. D., and Gurney, M. E., 1987, Functional interaction and partial homology between immunodeficiency virus and neuroleukin, *Science* **237**:1047–1051.

19. Apatoff, B. A., Lee, M. R., and Gurney, M. E., 1987, Trophic effects of neuroleukin on central neurons and functional interactions with HIV envelope protein, *Ann. Neurol.* **22**:156.

20. Gibbs, C. J., Jr., and Gajdusek, D. C., 1972, Neurological diseases of man with slow virus etiology, in: *Membranes and Viruses in Immunopathology, Proceedings Conference on Membranes, Viruses and Immune Mechanisms in Experimental and Unusual Disease, University of Minnesota,* Academic Press, New York, pp. 397–410.

21. Masters, C. L., Kakulas, B. A., Alpers, M. P., Gajdusek, D. C., and Gibbs, C. J., Jr., 1976, Preclinical lesions and their progression in the experimental spongiform encephalopathies (kuru and Creutzfeldt–Jakob disease) in primates, *J. Neuropathol. Exp. Neurol.* **35**:593–605.

22. Gibbs, C. J., Jr., and Gajdusek, D. C., 1970, Characterization and nature of viruses causing spongiform encephalopathies, in: *Proceedings VIth International Congress on Neuropathology,* Masson, Paris, pp. 779–801.

23. Gajdusek, D. C., 1977, Unconventional viruses and the origin and disappearance of kuru, *Science* **197**:943–960.

24. Gajdusek, D. C., and Zigas, V., 1957, Degenerative disease of the central nervous system in New Guinea. The endemic occurrence of "kuru" in the native population, *N. Engl. J. Med.* **257**:974–978.

25. Gajdusek, D. C., 1963, Kuru, *Trans. R. Soc. Trop. Med. Hyg.* **57**:151–169.

26. Gajdusek, D. C., Gibbs, C. J., Jr., and Alpers, M., 1966, Experimental transmission of kuru-like syndrome to chimpanzees, *Nature* **209**:794–796.

27. Creutzfeldt, H. G., 1920, Uber eine eigenartige herdformige Erkrankung der Zentral-nervensystems, *Z. Neurol. Psychiat.* **57**:1–18.

28. Creutzfeldt, H. G., 1921, Uber eine eigenartige herdformige Erkrankung der Zentral-nervensystems, in: *Histologische und Histopathologische Arbeiten uber die Grosshirnrinde* (F. Nissl and A. Alzheimer, eds.), Gustav Fischer, Jena, pp. 1–48.

29. Jakob, A., 1921, Uber eigenaritige Erkrankung des Zentralnervensystems mit bemerkens-wertem anatomischen Befunde, *Deut. Z. Nervenheilkd.* **70**:132–146.

30. Jakob, A., 1923. Die extrapyramidalen Erkrankungen mit besonderer Berucksichtigung der pathologischen Anatomie und Histologie und der Pathophysiologie der Bewegungs-storungen, in: *Monographien aus dem Gesamtgebiete der Neurologie und Psychiatrie,* Volume 37 (O. W. Foerster and K. Wilmanns, eds.), Springer, Berlin, pp. 215–345.

31. Manuelidis, E. E., 1985, Presidential address: Creutzfeldt–Jakob disease, *J. Neuropathol. Exp. Neurol.* **44**:1–17.

32. Klatzo, I., Gajdusek, D. C., and Zigas, V., 1959, Pathology of kuru, *Lab. Invest.* **8**:799–847.

33. Brown, P., Rodgers-Johnson, P., Cathala, F., Gibbs, C. J., Jr., and Gajdusek, D. C., 1984, Creutzfeldt–Jakob disease of long duration: Clinicopathological characteristics, trans-missibility and differential diagnosis, *Ann. Neurol.* **16**:295–304.

34. Gibbs, C. J., Jr., Gajdusek, D. C., Asher, D. M., Alpers, M. P., Beck, E., Daniel, P. M., and Matthews, W. B., 1968, Creutzfeldt–Jakob disease (spongiform encephalopathy): Trans-mission to the chimpanzee, *Science* **161**:388–389.

35. Gibbs, C. J., Jr., and Gajdusek, D. C., 1969, Infection as the etiology of spongiform encephalopathy (Creutzfeldt–Jakob disease), *Science* **165**:1023–1025.

36. Manuelidis, E. E., and Manuelidis, L., 1979, Observations on Creutzfeldt–Jakob disease propagated in small rodents, *Slow Trans. Dis. Nerv. Syst.* **2**:147–173.

37. Manuelidis, L., and Manuelidis, E. E., 1986, Recent developments in scrapie and Creutzfeldt–Jakob disease, *Prog. Med. Virol.* **33**:78–98.

38. Gerstmann, J., Straussler, E., and Scheinker, I., 1936, Uber eine eigenartige hereditar-familiare Erkrankung des Zentralnervensystems, zugleich ein Beitrag zur Frage des Vorzeitigen, *Alterns. Neurol.* **154**:736.

39. Masters, C. L., Gajdusek, D. C., and Gibbs, C. J., Jr., 1981, Creutzfeldt–Jakob disease virus

isolations from Gerstmann–Straussler syndrome with an analysis of the various forms of amyloid plaque deposition in the virus-induced spongiform encephalopathies, *Brain* **104:** 559–588.

40. Gajdusek, D. C., 1986, Chronic dementia caused by small unconventional viruses apparently containing no nucleic acid, in: *The Biological Substrates of Alzheimer's Disease, UCLA Forum in Medical Sciences, Number 27* (A. B. Scheibel and A. F. Wechsler, eds.), Academic Press, London, pp. 33–54.

41. Richardson, E. P., Jr., 1965, Progressive multifocal leukoencephalopathy, in: *Contemporary Neurology Symposia, I. The Remote Effects of Cancer on the Nervous System* (L. Brain and F. H. Norris, J., eds.), Grune & Stratton, New York, pp. 6–16.

42. ZuRhein, G. M., 1969, Association of papova-virions with a human demyelinating disease (progressive multifocal leukoencephalopathy), *Progr. Med. Virol.* **11:**185–247.

43. ZuRhein, G. M., and Chou, S. M., 1965, Particles resembling papovaviruses in human cerebral demyelinating disease, *Science* **148:**1477–1479.

44. Penny, J. B., Weiner, L. P. Herndon, R. M., Narayan, O., and Johnson, R. T., 1972, Virions from progressive multifocal leukoencephalopathy. Rapid serologic identification by electron microscopy, *Science* **178:**60–62.

45. Silverman, L., and Rubenstein, L. J., 1965, Electron microscopic observations on a case of progressive multifocal leukoencephalopathy, *Acta Neuropathol. Berl.* **5:**215–224.

46. Padgett, B. L., Walker, D. L., ZuRhein, G. M., Eckroade, R. J., and Dessel, B. H., 1971, Cultivation of papova-like virus from human brain with progressive multifocal leukoencephalopathy, *Lancet* **1:**1257–1260.

47. Padgett, B. L., and Walker, D. L., 1973, Prevalence of antibodies in human sera against JC virus, an isolate from a case of progressive multifocal leukoencephalopathy, *J. Infect. Dis.* **127:**467–470.

48. Padgett, B. L., and Walker, D. L., 1976, New human papovaviruses, *Prog. Med. Virol.* **22::**1–35.

49. Padgett, B. L., and Walker, D. L., 1980, Human papovavirus JCV: Natural history, tumorigenicity, and interaction with human cells in culture, in: *Viruses in Naturally Occurring Cancers* (M. Essex, G. Todaro, and H. zur Hausen, ed.), Cold Spring Harbor Laboratory, Cold Spring Harbor, NY, pp. 319–327.

50. Krupp, L. B., Lipton, R. B., Swedlow, M. L., Leeds, N. E., and Llena, J., 1985, Progressive multifocal leukoencephalopathy: Clinical and radiographic features, *Ann. Neurol.* **17:**344–349.

51. Stoner, G. L., Ryschkewitsch, C. F., Walker, D. L., and Webster, H. deF., 1986, JC papovavirus large tumor (T)-antigen expression in brain tissue of acquired immune deficiency syndrome (AIDS) and non-AIDS patients with progressive multifocal leukoencephalopathy, *Proc. Natl. Acad. Sci. U.S.A.* **83:**2271–2275.

52. Lane, H. C., and Fauci, A. S., 1985, Immunologic abnormalities in the acquired immunodeficiency syndrome, *Annu. Rev. Immunol.* **3:**477–500.

53. Rockwell, D., Ruben, F. L., Winkelstein, A., and Mendelow, H., 1976, Absence of immune deficiencies in a case of progressive multifocal leukoencephalopathy, *Am. J. Med.* **61:**433–436.

54. Van Horn, G., Bastian, F. O., and Moake, J. L., 1978, Progressive multifocal leukoencephalopathy: Failure of response to transfer factor and cytarabine, *Neurology* **28:**794–797.

55. Dorries, K., Johnson, R. T., and ter Meulen, V., 1979, Detection of polyoma virus DNA in PML-brain tissue by (*in situ*) hybridization, *J. Gen. Virol.* **42:**49–57.

56. ter Meulen, V., Stephenson, J. R., and Kreth, H. W., 1983, Subacute sclerosing panencephalitis, *Compr. Virol.* **18:**105–159.

57. Morgan, E. M. and Rapp, F., 1977, Measles virus and its associated diseases, *Bacteriol. Rev.* **41:**636–666.

58. Dawson, J. R., 1933, Cellular inclusions in cerebral lesions of lethargic encephalitis, *Am. J. Pathol.* **9:**7–16.

59. Bouteille, M., Fontaine, C., Verdrenne, C., and Delarue, J., 1965, Sur un cas d'encephalite subclique a inclusions. Etude Anatomoclinque et ultrastructurale, *Rev. Neurol.* 113:454–458.

60. Connolly, J. H., Allen, I. V., Hurwitz, L. J., and Miller, J. H. D., 1967, Measles-virus antibody and antigen in subacute sclerosing panencephalitis, *Lancet* 1:542–544.

61. ter Meulen, V., Katz, M., and Muller, D., 1972, Subacute sclerosing panencephalitis: A review, *Curr. Top. Microbiol. Immunol.* 57:1–38.

62. Payne, F. E., Baubis, J. V., and Itabashi, H. H., 1969, Isolation of measles virus from cell cultures of brain from a patient with subacute sclerosing panencephalitis, *N. Engl. J. Med.* 281:585–589.

63. Horta-Barbosa, L., Fuccillo, D. A., Sever, J. L., and Zeman, W., 1969, Subacute sclerosing panencephalitis: Isolation of measles virus from a brain biopsy, *Nature* 221:974.

64. Horta-Barbosa, L., Hamilton, R., Witting, B., Fuccillo, D. A., Sever, J. L., and Vernon, M. L., 1971, Subacute sclerosing panencephalitis: Isolation of suppressed measles virus from lymph node biopsies, *Science* 173:840–841.

65. Hall, W. W., and Choppin, P. W., 1981, Measles-virus proteins in the brain tissue of patients with subacute sclerosing panencephalitis: Absence of M protein, *N. Engl. J. Med.* 304:1152–1155.

66. Fournier, J.-G., Tardieu, M., Lebon, P., Robain, O., Ponsot, G., Rozenblatt, S., and Bouteille, M., 1985, Detection of measles virus RNA in lymphocytes from peripheral-blood and brain perivascular infiltrates of patients with subacute sclerosing panencephalitis, *N. Engl. J. Med.* 313:910–915.

67. ter Meulen, V., and Hall, W. W., 1978, Slow virus infections of the nervous system: Virological, immunological and pathogenetic considerations, *J. Gen. Virol.* 41:1–25.

68. Choppin, P. W., 1981, Measles virus and chronic neurological diseases, *Ann. Neurol.* 9:17–20.

69. Graves, M. C., 1984, Subacute sclerosing panencephalitis, *Neurol. Clin.* 2:267–280.

70. Wechsler, S. L., and Fields, B. N., 1978, Differences between the intracellular polypeptides of measles and subacute sclerosing panencephalities virus, *Nature* 272:458–460.

71. Hall, W. W., Kiessling, W., and ter Meulen, V., 1978, Membrane proteins of subacute sclerosing panencephalitis and measles virus, *Nature* 272:460–462.

72. Wechsler, S. L., Weiner, H. L., and Fields, B. N., 1979, Immune response in subacute sclerosing panencephalitis: Reduced antibody response to the matrix protein of measles virus, *J. Immunol.* 123:884–889.

73. Hall, W. W., Lamb, R. A., and Choppin, P. W., 1979, Measles and SSPE virus proteins: Lack of antibodies to the M protein in patients with SSPE, *Proc. Natl. Acad. Sci. U.S.A.* 76:2047–2051.

74. Baczko, K., Liebert, U. G., Billeter, M., Cattaneo, R., Budka, H., and ter Meulen, V., 1986, Expression of defective measles virus genes in brain tissue of patients of subacute sclerosing panencephalitis, *J. Virol.* 59:472–478.

75. Cattaneo, R., Rebmann, G., Schmid, A., Baczko, K., ter Meulen, V., Bellini, W. J., and Billeter, M. A., 1987, Altered transcription of a defective measles virus genome derived from a diseased human brain, *EMBO J.* 6:681–688.

76. Carter, M. J., Willcocks, M. M., and ter Meulen, V., 1983, Defective translation of measles virus matrix protein in a subacute sclerosing panencephalitis cell line, *Nature* 305:153–155.

77. Sheppard, R. D., Raine, C. S., Bornstein, M. B., and Udem, S. A., 1985, Measles virus matrix protein synthesized in a subacute sclerosing panencephalitis cell line, *Science* 228:1219–1221.

78. Sheppard, R. D., Raine, C. S., Bornstein, M. B., and Udem, S. A., 1986, Rapid degradation restricts measles virus matrix protein expression in a subacute sclerosing panencephalitis cell line, *Proc. Natl. Acad. Sci. U.S.A.* 83:7913–7917.

79. Young, K. K. Y., Heineke, B. E., and Wechsler, S. L., 1985, M protein instability and lack of H protein processing associated with nonproductive persistent infection of HeLa cells by measles virus, *Virology* 143:536–545.

80. Norrby, E., Kristensson, K., Brzosko, W. J., and Kapsenberg, J. G., 1985, Measles virus matrix protein detected by immune fluorescence with monoclonal antibodies in the brain of patients with subacute sclerosing panencephalitis. *J. Virol.* **56:**337–340.

81. Cattaneo, R., Schmid, A., Billeter, M. A., Sheppard, R. D., and Udem, S. A., 1988, Multiple viral mutations rather than host factors cause defective measles virus gene expression in a subacute sclerosing panencephalitis cell line, *J. Virol.* **62:**1388–1397.

82. Domingo, E., Martinez-Salas, E., Sobrino, F., de la Torre, J. C., Portela, A., Ortin, J., Lopez-Galindez, C., Perez-Brena, P., Villanueva, N., Najera, R., VandePol, S., Steinhauer, D., DePolo, N., and Holland, J., 1985, The quasispecies (extremely heterogeneous) nature of viral RNA genome populations: Biological relevance—A review, *Gene* **40:**1–8.

83. Holland, J., Spindler, K., Horodyski, F., Grabau, E., Nichol, S., and VandePol, S., 1982, Rapid evolution of RNA genomes, *Science* **215:**1577–1585.

84. Fujinami, R. S., and Oldstone, M. A., 1979, Antiviral antibody reacting on the plasma membrane alters measles virus expression inside the cell, *Nature* **279:**529–530.

85. Haase, A. T., Swoveland, P., Stowring, L., Ventura, P., Johnson, K. P., Norrby, E., and Gibbs, C. J., Jr., 1981, Measles virus genome in infections of the central nervous system, *J. Infect. Dis.* **144:**154–160.

86. Haase, A. T., Gantz, D., Eble, B., Walker, D., Stowring, L., Ventura, P., Blum, H., Wietgrefe, S., Zupancic, M., Tourtellotte, W., Gibbs, C. J., Jr., Norrby, E., and Rozenblatt, S., 1985, Natural history of restricted synthesis and expression of measles virus genes in subacute sclerosing panencephalitis, *Proc. Natl. Acad. Sci. U.S.A.* **82:**3020–3024.

87. Gajdusek, D. C., 1982, Foci of motor neuron disease in high incidence in isolated populations of East Asia and the Western Pacific, in: *Human Motor Neuron Diseases* (L. P. Rowland, ed.), Raven Press, New York, pp. 363–393.

88. Gibbs, C. J., Jr., and Gajdusek, D. C., 1978, Subacute spongiform virus encephalopathies: The transmissible virus dementias, in: *Alzheimer's Disease: Senile Dementia and Related Disorders, Aging*, Volume 7 (R. Katzman, R. D. Terry, and K. L. Bick, eds.), Raven Press, New York, pp. 559–575.

89. Mortimer, J. A., and Hutton, J. T., 1985, Epidemiology and etiology of Alzheimer's disease, in: *Senile Dementia of the Alzheimer Type, Neurology and Neurobiology*, Volume 18 (J. T. Hutton and A. D. Kenny, eds.), Alan R. Liss, New York, pp. 177–196.

90. Katzman, R., 1986, Alzheimer's disease, *N. Engl. J. Med.* **314:**964–973.

91. Ball, M. J., 1987, Histotopography of cellular changes in Alzheimer's disease, in: *Senile Dementia: A Biochemical Approach* (K. Nandy, ed.), Elsevier–North-Holland, Amsterdam, pp. 89–104.

92. Ball, M. J., and Nuttal, K., 1977, Neuronal loss, neurofibrillary tangles, and granovacuolar degeneration in the hippocampus with ageing and dementia. A quantitative study, *Acta Neuropathol. (Berl.)* **37:**111–118.

93. Roth, M., Tomlinson, B. E., and Blessed, G., 1966, Correlation between scores for dementia and counts of senile plaques in cerebral grey matter of elderly subjects, *Nature* **206:**109–110.

94. Hulette, C. M., and Walford, R. L., 1986, HLA associations in Alzheimer's disease, in: *The Biological Substrates of Alzheimer's Disease, UCLA Forum in Medical Sciences, Number 27* (A. B. Scheibel and A. F. Wechsler, eds.), Academic Press, London, pp. 161–165.

95. Fitch, N., Becker, R., and Heller, A., 1988, The inheritance of Alzheimer's disease: A new interpretation, *Ann. Neurol.* **23:**14–19.

96. Ball, M. J., Shapiro, M. B., and Rapoport, S. I., 1986, Neuropathological relationships between Down syndrome and senile dementia Alzheimer type, in: *The Neurobiology of Down Syndrome* (C. J. Epstein, ed.), Raven Press, New York, pp. 45–58.

97. Goldgaber, D., Lerman, M. I., McBride, O. W., Saffiotti, U., and Gajdusek, D. C., 1987, Characterization and chromosomal localization of a cDNA encoding brain amyloid of Alzheimer's disease, *Science* **235:**877–880.

98. St. George-Hyslop, P. H., Tanzi, R. E., Polinsky, R. J., Haines, J. L., Nee, L., Watkins, P. C., Myers, R. H., Feldman, R. G., Pollen, D., Brachman, D., Growdoin, J., Bruni, A., Foncin, J.-F., Salmon, D., Frommelt, P., Amaducci, L., Sorbi, S., Piacentini, S., Stewart, G. D., Hobbs, W. J., Conneally, P. M., and Gusella, J. F., 1987, The genetic defect causing familial Alzheimer's disease maps on chromosome 21, *Science* **235**:885–890.

99. Heston, L., 1985, Clinical genetics of Alzheimer's disease, in: *Senile Dementia of the Alzheimer Type, Neurology and Neurobiology*, Volume 18 (J. T. Hutton and A. D. Kenny, eds.), Alan R. Liss, New York, pp. 197–203.

100. Iqbal, K., Grundke-Iqbal, I., and Wisniewski, H., 1987, Alterations in the neuronal cytoskeleton in Alzheimer's disease, in: *Alterations of the Neuronal Cytoskeleton in Alzheimer's Disease and Related Conditions, Advances in Behavioral Biology*, Volume 34 (G. Perry, ed.), Plenum Press, New York, pp. 109–136.

101. Fillit, H., Luine, V. N., Reisberg, B., Amador, R., McEwen, B., and Zabriskie, J. B., 1985, Studies on the specificity of brain antibodies in Alzheimer's disease, in: *Senile Dementia of the Alzheimer Type, Neurology and Neurobiology*, Volume 18 (J. T. Hutton and A. D. Kenney, eds.), Alan R. Liss, New York, pp. 307–318.

102. Goudsmit, J., Morrow, C. H., Asher, D. M., Yanagikara, R. T., Marter, C. L. Gibbs, C. J., Jr., and Gajdusek, D. C., 1980, Evidence for and against the transmissibility of Alzheimer's disease, *Neurology* **30**:945–950.

103. Manuelidis, E. E., 1985, Creutzfeldt–Jakob disease, *J. Neuropathol. Exp. Neurol.* **44**:1–17.

104. Manuelidis, E. E., de Figueiredo, J. M. Kim, J. H., Fritch, W. W., and Manuelidis, L., 1988, Transmission studies from blood of Alzheimer disease patients and healthy relatives, *Proc. Natl. Acad. Sci. U.S.A.* **85**:4898–4901.

105. Masters, C. L., Gajdusek, D. C., Gibbs, C. J., Jr., Bernoulli, C., and Asher, D. M., 1979, Familial Creutzfeldt–Jakob disease and other familial dementias: An inquiry into possible modes of transmission of virus-induced familial diseases, in: *Slow Transmissable Diseases of the Nervous System*, Volume 1 (S. B. Pruisner and W. J. Hadlow, eds.), Academic Press, New York, pp. 143–194.

106. Ball, M. J., 1988, Features of Creutzfeldt–Jakob disease in brains of patients with familial dementia of Alzheimer type, *Can. J. Neurol. Sci.* **7**:51–57.

107. Smith, T. W., Anwer, U., DeGirolami, U., and Drachman, D. A., 1987, Vacuolar changes in Alzheimer's disease, *Arch. Neurol.* **44**:1225–1228.

108. Ball, M. J., 1982, Limbic predilection in Alzheimer dementia—Is reactivated herpes virus involves? *Can. J. Neurol. Sci.* **9**:303–306.

109. Ball, M. J., Nuttall, K., and Warren, K. G., 1982, Neuronal and lymphocytic populations in human trigeminal ganglia—implications for aging and for latent virus, *Neuropathol. Appl. Neurobiol.* **8**:177–187.

110. Ball, M. J., 1986, Herpesvirus in the hippocampus as a cause of Alzheimer's disease, *Arch. Neurol.* **43**:313.

111. Ball, M. J., 1987, Organic dementia of Alzheimer type—Possible role of reactivated herpes simplex virus? in: *Viruses Immunity and Mental Disorders* (E. Kurstak, Z. J. Lipowski, and P. V. Morozov, eds.), Plenum Press, New York, pp. 61–64.

112. M. J. Ball, Lewis, E., and Haase, A. T., 1987, Detection of herpes virus genome in Alzheimer's disease by *in situ* hybridization: A preliminary study, *J. Neural. Transm.* **24**: 219–225.

113. Libilova, H., Pogady, J., Wiedermann, V., and Breier, S., 1975, Search for herpetic antibodies in the cerebrospinal fluid in senile dementia and mental retardation, *Acta Virol.* **19**: 493–495.

114. Roberts, G. W., Taylor, G. R., Carter, G. I., Johnson, J. A., Bloxham, C., Brown, R., and Crow, T. J., 1986, Herpes simplex virus—A role in the etiology of Alzheimer's disease? *J.Neurol. Neurosurg. Psychiatry* **49**:216–219.

115. Deatly, A. M., Spivack, J. G., Lavi, E., and Fraser, N. W., 1987, RNA from an immediate early region of the HSV-1 genome is present in the trigeminal ganglia of latently infected mice, *Proc. Natl. Acad. Sci. U.S.A.* **84**:3204–3208.

116. Deatly, A. M., Spivack, J. G., Lavi, E., O'Boyle, D. R. II, and Fraser, N. W., 1987, Latent herpes simplex virus type 1 transcripts in peripheral and central nervous system tissues of mice map to similar regions of the viral genome, *J. Virol.* **62**:749–756.

117. Stevens, J. G., Wagner, E. K., Devi-Rao, G. B., Cook, M. L., and Feldman, L. T., 1987, RNA complementary to a herpesvirus α gene mRNA is prominent in latently infected neurons, *Science* **235**:1056–1059.

118. Puga, A., and Notkins, A. L., 1987, Continued expression of a poly(A+) transcript of herpes simplex type 1 in trigeminal ganglia of latently infected mice, *J. Virol.* **61**:1700–1703.

119. Rock, D. L., Nesburn, A. B., Ghiasi, H., Ong, J., Lewis, T. L., Lokensgard, J. R., and Wechsler, S. L., 1987, Detection of latency-related viral RNAs in trigeminal ganglia of rabbits latently infected with herpes simplex virus type 1, *J. Virol.* **61**:3820–3826.

120. Croen, K. D., Ostrove, J. M., Dragovic, L. J., Smialek, J. E., and Straus, S. E., 1987, Latent herpes simplex virus in human trigeminal ganglia, *N. Engl. J. Med.* **317**:1427–1432.

121. Wagner, E. K., Devi-Rao, G., Feldman, L. T., Dobson, A. T., Zhang, Y.-F., Flanagan, W. M., and Stevens, J. G., 1988, Physical characterization of the herpes simplex virus latency-associated transcript in neurons, *J. Virol.* **62**:1194–1202.

122. Stroop, W. G., Rock, D. L., and Fraser, N. W., 1984, Localization of herpes simplex virus in the trigeminal and olfactory systems of the mouse central nervous system during acute and latent infections by in situ hybridization, *Lab. Invest.* **51**:27–38.

123. Sequiera, L. W., Jennings, L. C., Carrasco, L. H., Lord, M. A., Curry, A., and Sutton, R. N. P., 1979, Detection of herpes-simplex viral genome in brain tissue, *Lancet* **2**:609–612.

124. Mann, D. M. A., Yates, P. O., Davies, J. S., and Hawkes, J., 1981, Viruses, parkinsonism and Alzheimer's disease, *J. Neurol. Neurosurg. Psychiatry* **44**:651.

125. Esiri, M., 1982, Herpes simplex encephalitis: An immunohistological study of the distribution of viral antigen within the brain, *J. Neurol. Sci.* **54**:209–226.

126. Roman, G. C., 1988, The neuroepidemiology of tropical spastic paraparesis, *Ann. Neurol.* **23**:S113–S120.

127. Zaninovic, V., Arango, C., Biojo, R., Mora, C., Rodgers-Johnson, P., Concha, M., Corral, R., Barreto, P., Borrero, I., Garruto, R. M., Gibbs, C. J., Jr., and Gajdusek, D. C., 1988, Tropical spastic paraparesis in Colombia, *Ann. Neurol.* **23**:S127–S132.

128. Roman, G. C., Schoenberg, B. S., Madden, D. L., Sever, J. L., Hugon, J., Ludolph, A., and Spencer, P. S., 1987, Human T-lymphotrophic virus type 1 antibodies in the serum of patients with tropical spastic paraparesis in the Seychelles, *Arch. Neurol.* **44**:605–607.

129. Vernant, J.-C., Maurs, L., Gout, O., Buisson, G., Plumelle, Y., Neisson-Vernant, C., Monplaisir, N., and Roman, G. C., 1988, HTLV-1-associated tropical spastic paraparesis in Martinique: A reappraisal, *Ann. Neurol.* **23**:S133–S135.

130. Gessain, A., Barin, F., Vernant, J.-C., Gout, O., Maurs, L., Calendar, A., and de The, G., 1985, Antibodies to human T-lymphotrophic virus type 1 in patients with tropical spastic paraparesis, *Lancet* **2**:407–409.

131. Jacobson, S., Raine, C. S., Mingioli, E. S., and McFarlin, D. E., 1988, Isolation of an HTLV-1-like retrovirus from patients with tropical spastic paraparesis, *Nature* **331**:540–543.

132. Brew, B. J., and Price, R. W., 1988, Another retroviral disease of the nervous system. Chronic progressive myelopathy due to HTLV-1, *N. Engl. J. Med.* **318**:1195–1197.

133. Liberski, P. P., Rodgers-Johnson, P., Char, G., Piccardo, P., Gibbs, C. J., Jr., and Gajdusek, D. C., 1988, HTLV-1-like viral particles in spinal cord cells in Jamaican tropical spastic paraparesis, *Ann. Neurol.* **23**:S185–S187.

134. Koprowski, H., and DeFreitas, E., 1988, HTLV-1 and chronic nervous system diseases: Present status and a look into the future, *Ann. Neurol.* **23**:S133–S135.

135. Dalakas, M. C., Stone, G., Elder, G., Ceroni, M., Madden, D., Roman, G., and Sever, J. L., 1988, Tropical spastic paraparesis: Clinical, immunological, and virological studies in two patients from Martinique, *Ann. Neurol.* **23**:S136–S142.
136. Johnson, R. T., Griffin, D. E., Arregui, A., Mora, C., Gibbs, C. J., Jr., Cuba, J. M., Trelles, L., and Vaisberg, A., 1988, Spastic paraparesis and HTLV-1 infection in Peru, *Ann. Neurol.* **23**: S151–S155.
137. Elizan, T. S., and Casals, J., 1987, The viral hypothesis in Parkinson's disease and Alzheimer's disease, in: *Viruses, Immunity and Mental Disorders* (E. Kurstak and Z. J. Lipowski, eds.), Plenum Medical, New York, pp. 47–59.
138. Iqbal, K., and Wisniewski, H. M., 1983, Neurofibrillary tangles, in: *Alzheimer's Disease, The Standard Reference* (B. Reisberg, ed.), Free Press, New York, pp. 48–56.
139. Appel, S. H., Tomozawa, Y., and Bostwick, R., 1986, Trophic factors and neurologic disease in Alzheimer's and Parkinson's disease, in: *Advances in Behavioral Biology*, Volume 29 (A. Fisher, I. Hanin, and C. Lachman, eds.), Plenum Press, New York, pp. 75–84.
140. Pollack, M., and Hornabook, R. W., 1966, The prevalence, natural history and dementia of Parkinson's disease, *Brain* **89**:429–448.

17

Continuing the Investigation
Viruses and Neurological Disorders

STEVEN SPECTER, MAURO BENDINELLI,
and HERMAN FRIEDMAN

1. UNANSWERED QUESTIONS

Numerous questions are left unanswered in this volume and provide most intriguing areas for further investigations. In spite of considerable progress, knowledge is far from satisfactory even in the field of acute neurological diseases for which a definite direct etiology is well established. For some of these diseases, Japanese encephalitis for example, we are in great need of efficient vaccines.[1] In addition, for such diseases it is especially urgent to develop better diagnostic procedures, since in even the most modern diagnostic virology facilities the etiologic agent remains unrecognized in over 30% of presumed cases of acute viral meningitis and in over 80% of presumed cases of acute viral encephalitis. Indeed, the precise etiologic diagnosis of viral infection of the nervous system (NS) in general is still fraught with difficulties. For example, the association of cerebral malformations and other neurological birth defects with antecedent viral infection of the mother is often beyond present methodologies.

We have much less understanding of the subacute and chronic diseases that recent breakthroughs have clearly shown to be caused by viral infections, as seen from several chapters in this volume. It will be many years before the new findings trickle down and are translated into better understanding of viral neuropathogenicity. Mention should be made here of the disorders collectively termed

STEVEN SPECTER and HERMAN FRIEDMAN • Department of Microbiology and Immunology, University of South Florida, College of Medicine, Tampa, Florida 33612-4799. MAURO BENDINELLI • Department of Biomedicine, University of Pisa, I-56127 Pisa, Italy.

Neuropathogenic Viruses and Immunity, edited by Steven Specter *et al.* Plenum Press, New York, 1992.

transmissible encephalopathies for which the causative agent remains a mystery, justifying the most "heretical" hypothesis that transmission may not involve nucleic acid (Chapter 7).

By and large it must be admitted that the entire process of viral invasion of the NS remains largely an enigma. (1) Under which conditions and how do viruses that generally cause inapparent infections cause disease of the NS? (2) How do viruses cause acute damage in the NS? (3) How do viruses persist in the NS? (4) How do viruses cause chronic damage to the NS? (5) What is the contribution of immune mechanisms in such damage? Nevertheless, because clarification of pathologies is critical to the design of rational therapies, each of these questions requires extensive study in the hope of providing a path that might be followed to future solutions.

Among the questions listed above, one of utmost importance is how and in which form viruses can persist in the NS. The unique anatomic features of the NS and the static, noncycling nature of most of its cell types are often invoked to explain why the NS represents a preferred milieu for persistence of many viruses. Although these and other assumptions might be correct, it seems important to recall that a preexisting defect of the immune defenses is a prerequisite to persistence and clinical disease of the NS by many viruses (enteroviruses, JC virus, rubella virus). Viruses that persist in the NS of hosts with an intact immune system are either immunologically inert (unconventional viruses) or endowed with powerful immunosuppressive properties (retroviruses, measles).

Perhaps with the recent recognition of the neuropathogenicity of human retroviruses we have entered a new era of interest in viral infections of the NS and rapid advancement in solving some of the mysteries. As described in Chapter 14 by Webb, the suggested association between the human T-lymphotropic virus (HTLV-1) and multiple sclerosis has led to a close scrutiny of this relationship. Furthermore, the causal relationship between HTLV-1, or a variant of this virus, and tropical spastic paraparesis can be considered very likely if not certain. The HTLV-1-associated myelopathy described in Japan may be a nontropical version of the same disease.

Although the acquired immunodeficiency syndrome (AIDS) has the ominous reputation of being the plague of our age, it has been a boon to scientific understanding of virus infection. The information obtained is likely to enhance our understanding of many neurological diseases directly or indirectly related to AIDS as a result of the aggressive approach adopted in the study of this disease. The AIDS epidemic has kindled great interest in neurological viral infections and psychiatric disorders. As discussed in Chapter 12, a large proportion of human immunodeficiency virus (HIV)-infected individuals develop dementia and a host of other neurological disorders, including multiple-sclerosis-like illnesses. Evidence indicates that HIV invades the brain early, probably at the time of initial infection in many if not all individuals.[2] Cognitive and behavioral changes can occur earlier, but the interval from initial invasion of the NS to the onset of overt neurological deterioration is usually long. What happens during this interval is presently the subject of extensive speculation and investigation. Much of the significant change seems to occur in the neurons, and yet cells of neuroectodermal origin are essentially free of replicating virus. This observation and the analogy to other lentiviral infections, particularly visna in sheep (discussed in Chapter 4), has led to a hypothesis that cortical atrophy and dementia may be mediated by disregulated production of cytokines or other indirect mechanisms.

The state of the art is uncertain regarding the large number of neuropsychiatric conditions for which a viral etiology has been implicated but the determination of an etiologic agent is still lacking. It is possible that most such conditions are multifactorial in origin and/or pathogenesis. For example, the causative virus might be instrumental only in triggering pathological mechanisms, such as trophic disturbances and autoimmunity, that, once initiated, might be self-maintained (Fig. 17-1). Should this "hit-and-run" interpretation be correct, then the development of disease might be subject to multifactorial regulation, thus explaining why the etiology of such diseases is proving so difficult to pinpoint by traditional biological and epidemiologic means. It is only recently that we have begun to appreciate that virus-induced organic changes can be responsible for psychiatric disorders, although it is now well established that certain viral infections can produce dementia. Although it is expected that a better understanding of the neurological consequences of AIDS will help to shed light on these afflictions as well, we must reemphasize the importance of continuing investigation of existing animal models and of developing new ones to evolve fresh concepts and ideas.

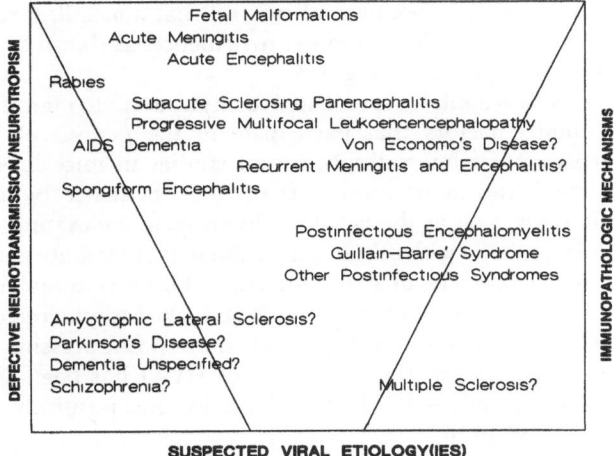

FIGURE 17-1. The viral etiologies and pathological mechanisms of several neurological diseases. Nearness to the top of the figure implies that viral etiology is determined. Neurological disease to the right-hand side has a more immunopathological etiology, whereas diseases to the left are caused by the neurological nature of the virus. Those diseases spanning the center of the figure have a greater balance of immunopathology and defective tropism. A question mark indicates that the viral etiology is still in question.

2. THE NEED FOR MODELS

The naturally occurring analogue of human diseases and the experimental animal models described herein (Chapters 4–7) epitomize comparative pathology as an essential approach to the study of human disease. In addition, the animal models of demyelinating diseases, such as that provided by Theiler's murine encephalomyelitis virus in selected strains of mice, have given considerable insight into the mechanisms involved in the disease process, although it is still controversial whether the process of demyelination is strictly virus induced or immune mediated. Transgenic animals that have selected viral genes, singly or in combination, are beginning to provide useful information in probing the complex pathogenesis of certain viral disease and will therefore also help to elucidate the function of specific viral molecules in NS diseases.[3] It has recently been reported that transgenic mice carrying the *tat* gene of HTLV-1 develop peripheral nerve lesions resembling von Recklinghausen's neurofibromatosis of man.[4]

There is also a need for more and versatile *in vitro* techniques. These may be especially useful for studying facets of neurovirulence and neuropathogenicity in conditions not influenced by host-induced reactions, such as virus uptake, transport, and replication in neural cells. Animal and *in vitro* models coupled with engineering of the viral genome may also further our knowledge of the molecular basis of virus-induced neurological diseases. Investigation of neurotropic and neuropathogenic variants of viruses can provide information on genomic and phenotypic features that permit invasion and damage of the NS. Recognizing the specific molecular domains in the outer proteins of a given virus responsible for invasion of the NS[5] may pave the way to the development of new treatments for neurological infections. It is, however, of interest that noncoding regions of the viral genome can also be implicated in neurovirulence, as shown by studies with polioviruses and other picornaviruses.[6-9]

Animal and *in vitro* models are essential to any attempt to identify the genetic and environmental cofactors that participate in the genesis of neurological damage by viruses. Most interestingly, recent studies in mice have implicated endogenous retroviruses as determinants of neurovirulence by an otherwise nonneurotropic virus such as the lactate-dehydrogenase-elevating virus,[10] thus suggesting that interaction with other viruses also may contribute in determining the neuropathogenic outcome of a viral infection. This is reminiscent of the fact that a higher incidence of progressive multifocal leukoencephalopathy has been noted in AIDS patients than in any other immunosuppressive disorder of comparable severity and that the *tat* protein of HIV-1 has been seen to *trans*-activate a late promoter of the JC virus in glial cells, thus positively affecting the lytic cycle of the JC virus.[11]

3. UNCONVENTIONAL THINKING

We are likely to see steady progress made toward a solution to many, if not all, of the questions posed in this volume. It seems an easy prediction that the present booming of neurobiology and neuroscience in general will also reflect itself beneficially on our understanding of infectious processes of the NS. Along with the application of modern and sophisticated technological advances, some unconventional thinking, perhaps applied to unconventional infectious agents, will probably help to solve some of the problems. In Chapter 16, Deatly, Haase, and Ball review current data on the possible involvement of viruses in Alzheimer's disease and other dementias of unknown origin. In Chapter 15, Stevens and Hallick summarize studies that attempt to tie schizophrenia to virus infection, although more evidence is needed to solidify this contention. The chapter on multiple sclerosis by Webb (Chapter 14) has taken an approach that has not been widely accepted as yet. Avenues such as immune responses to less conventional antigens[12] must be explored when the conventional approaches have resulted in the frustration of making little to no progress.

As virological and immunologic techniques become increasingly more powerful, the speed and certainty with which a suspected viral etiology will be proved or disproved will become greater. Even if viruses are eventually cleared of responsibility in most neuropsychiatric disorders for which a viral etiology has been suggested, the positive thinking and impetus to research given by the viral hypothesis will undoubtedly contribute to our understanding and mastery of many such diseases. In addition, it is to be expected that the flow of techniques and expertise will not be simply unidirectional. Advances in the area of viral infections of the NS may contribute to further our knowledge of the NS and its functioning. This is already exemplified by the use of neurotropic viruses to induce selected genes into mammalian neurons in order to assist in the analysis of physiology and ontogeny of NS cells.[13,14]

REFERENCES

1. Brandt, W. E., 1990, Development of dengue and Japanese encephalitis vaccines, *J. Infect. Dis.* **162**:577–583.
2. Berger, J. R., Sheremata, A., Resnick, L., Atherton, S., Fletcher, M. A., and Norenberg, M., 1989, Multiple sclerosis-like illness occurring with human immunodeficiency virus infection, *Neurology* **39**:324–329.
3. Trapp, B. D., Small, J. A., and Scangos, G. A., 1988, Dysmyelination in transgenic mice containing the early region of JC virus, in: *Virus Infection and the Developing Nervous System* (R. T. Johnson and G. Lyon, eds.), Kluwer, Dordrecht, pp. 21–35.
4. Hinrichs, S. H., Nerenberg, M., Reynolds, R. K., Khoury, G., and Jay, G., 1987, A transgenic mouse model for human neurofibromatosis, *Science* **237**:1340–1343.
5. Lustig, S., Jackson, A. C., Hahn, C. S., Griffin, D. E., Strauss, E. G., and Strauss, J. H., 1988, Molecular basis of Sindbis virus neurovirulence in mice, *J. Virol.* **62**:2329–2336.
6. Evans, D. M. A., Dunn, G., Minor, P. D., Schild, G. C., Cann, A. J., Stanway, G., Almond, J. W., Currey, K., and Maizel, W., Jr., 1985, Increased neurovirulence associated with a single nucleotide change in a noncoding region of the Sabin type 3 poliovirus genome, *Nature* **314**:548–550.
7. La Monica, N., Almond, J. W., and Racaniello, V. R., 1987, A mouse model for poliovirus neurovirulence identifies mutations that attenuate the virus for humans, *J. Virol.* **61**:2917–2920.

8. Duke, G. M., Osorio, J. E., and Palmenberg, A. C., 1990, Attenuation of Mengo virus through genetic engineering of the 5' noncoding poly(C) tract, *Nature* **343**:474–476.
9. Calendoff, M. A., Faaberg, K. S., and Lipton, H. L., 1990, Genomic regions of neurovirulence and attenuation in Theiler's murine encephalomyelitis virus, *Proc. Natl. Acad. Sci. U.S.A.* **87**:978–982.
10. Contag, C. H., and Plagemann, P. G. W., 1989, Age-dependent poliomyelitis in mice: Expression of endogenous retrovirus correlates with cytocidal replication of lactate dehydrogenase-elevating virus in motor neurons, *J. Virol.* **63**:4362–4369.
11. Tada, H., Rappaport, J., Lashgari, M., Amini, S., Wong-Staal, F., and Khalili, K., 1990, Trans-activation of the JC virus late promoter by the *tat* protein of type 1 human immunodeficiency virus in glial cells, *Proc. Natl. Acad. Sci. U.S.A.* **87**:3479–3483.
12. Anomasiri, W. T., Tovell, D. R., and Tyrell, D. L. J., 1990, Paramyxovirus membrane protein enhances antibody production to new antigenic determinants in the actin molecule: A model for virus-induced autoimmunity, *J. Virol.* **64**:3174–3184.
13. Galileo, D. S., Gray, G. E., Owens, G. C., Majors, J., and Sanes, J. R., 1990, Neurons and glia arise from a common progenitor in chicken optic tectum: Demonstration with two retroviruses and cell type-specific antibodies, *Proc. Natl. Acad. Sci. U.S.A.* **87**:458–462.
14. Geller, A. I., and Freese, A., 1990, Infection of cultured central nervous system neurons with a defective herpes simplex virus 1 vector results in stable expression of *Escherichia coli* beta-galactosidase, *Proc. Natl. Acad. Sci. U.S.A.* **87**:1149–1153.

Index

Acquired immunodeficiency syndrome
  (AIDS), 42, 104
  CD4 cell depletion in, 229–232, 234
  etiology, 229
  HIV-expressing immunocytes, 11
  neurological manifestations, 237, 239,
    240, 241, 340, 341
  neuropathies, 46–48, 50, 54, 209, 220
  prevalence, 303
  recovery from JCV infection in, 213, 216,
    222
  SRV-induced (SAIDS), 97, 98, 100, 103
  symptomatology, 243, 244
Acute inflammatory demyelinating
    polyneuropathy (AIDP), 41–43, 45,
    47–50, 239, 240
Acute measles encephalitis, 192, 288
Acute viral encephalitis, 339
Acute viral encephalomyelitis, 33
Adenoviruses, 2, 10, 236, 278
AIDS: see Acquired immunodeficiency
    syndrome
AIDS-dementia complex (ADC), 9, 235,
    236, 237–239, 242, 318, 329, 341
Allergic measles encephalitis, 192
Alphaviruses, 255–259, 260
Alzheimer's disease, 11, 303, 343
  viral etiology, 324–327
Amygdala, 309
β-amyloid accumulation, 326
Amyotrophic lateral sclerosis, 11, 54, 341
Anterior horn cells, 54
Anterior horn neurons, 95, 97
Anterior median fissure, 67

Antibodies, 27, 284
  activation by complement, 25
  antiAdenovirus, 278
  antibrain, 307
  antiCanine distemper virus, 187, 278
  anticerebroside, 287
  antiCoronavirus, 278
  antiCytomegalovirus, 278, 308, 309, 310
  antigalactocerebroside, 295
  antiglucocerebroside, 290
  antiHIV, 236
  antiHSV, 166, 278, 308, 309
  antiHSV-1, 309, 310, 311
  antiHSV-2, 309, 310
  antiHTLV-1, 100, 328
  antiInfluenza, 278, 308
  antiJCV, 215
  antiLaCrosse virus, 268
  antiMBP, 288
  antiMeasles specific, 187, 192, 193, 194,
    278, 279, 308, 309
  antiMumps, 190, 278, 295, 309, 310
  antiParainfluenzavirus, 278
  antiParamyxovirus, 184
  antiPoliovirus, 150, 278
  antiRespiratory syncytial virus, 278
  antiRotavirus, 278
  antiRubella virus, 278, 309, 310
  antiRubeola, 309, 310
  antithymic, 308
  antiVaccinia virus, 278
  antiVaricella zoster virus, 278, 310
  passive, 146
Aphthovirus, 139

345

Arboviruses, 255; *see also* specific arboviral infections
Arthropod-borne viruses, 255; *see also* specific arboviral infections
Astrocytes, 158, 193, 195, 197, 209, 291, 319
  hypertrophy, 184
  role in immune response, 30, 31
  viral gene expression in, 66, 70, 71, 73, 158, 217
  viral replication in, 158, 213
Astrocytosis, 82, 122, 238
Astroglial cells, 241
Ataxia, 195, 196, 241
Ataxic ganglioneuropathy (AGN), 239
Autoimmune factors
  in multiple sclerosis, 280, 281–283
  in schizophrenia, 304, 307
Autoimmunity, 6, 7, 41, 42, 49, 87, 239, 325, 341
Autonomic neuropathy, 54
Axons, 70, 83, 86, 144, 214, 287, 293

B cells, 16–18, 21, 32, 34–37, 117
  activation, 26–27
  deficiency, 165
  involvement in scrapie, 121
  latent infection of, 47, 64
  levels during arboviral infection, 259, 265
  surface antigens, 21–22, 24
  virus detection in, 220, 222
  virus infection of, 94, 95, 97, 98, 215
B virus, 2
Basal forebrain, 324
Behcet's disease, 11
BK virus (BKV), 2, 207–210, 212, 216, 218
  tumor induction with, 218–221
Blood–brain barrier, 28, 31, 32, 33, 186, 215, 240–241, 264, 291
Bone marrow, 20, 21, 215
  virus detection in, 220
Bovine spongiform encephalopathy, 4, 112
Brain, 67, 69, 98, 197, 209
  glycolipids, 285, 286, 288, 294
  macrophages, 240, 244
    HIV gene expression in, 229, 230
Brain stem, 33, 191, 265, 318
Bunyavirus, 2, 5, 255, 266–269
  California serogroup, 257, 266, 268

C-reactive protein, 25
Cachectin, 243
California encephalitis virus, 3, 266, 267

California serogroup virus, 5
Canine distemper virus (CDV), 178, 180, 181, 194–196, 279, 288
  antibodies to, 187, 278
  animal models, 183–184
  encephalitis, 195
  T cell response in, 188
  transmission, 180
Canine parainfluenza virus, 192
Caprine arthritis encephalitis virus (CAEV), 93, 101, 102, 103, 230
Cardiovirus, 139
Cardioviruses, 80
Cat-scratch disease, 11
Cauda equina, 50, 51
Caudate nucleus, 120, 121
Central European virus, 3, 261, 262
Central nervous system (CNS), 259
  antiviral antibodies, 3, 278, 279
  HIV infection, 235, 241
Cerebellar cortex, 33
Cerebellar lingula, 67
Cerebellar nuclei, 33
Cerebellar vermis, 144
Cerebellum, 67, 115, 120, 121, 195, 196, 265, 327
Cerebral centrum semiovale, 318
Cerebral cortex, 33, 119, 193, 194, 195
Cerebral edema, 269
Cerebrospinal fluid (CSF), 68, 184, 240
  B cell levels, 34, 35
  HIV-1 isolation from, 43, 235, 241, 318
  immunoglobulins in, 15, 31, 32, 98, 186–188, 195, 241, 286, 321
  mononuclear cell types in, 34–36, 235, 238, 258, 259
  virus recovery from, 43, 68, 140, 147, 181, 182, 190, 192, 265, 318, 323, 328
  in schizophrenics, 303, 309, 311
Cerebrum, 67, 327
Chickenpox virus, 3
Chikungunya virus, 256, 259
Chorioretinitis, 54
Choroid plexus, 97, 103, 185, 191
  epithelial cells, 98
    viral protein expression in, 71, 73
  sheep (SCP), 65, 66, 67, 68
Choroiditis, 191
Chronic acute hepatitis, 187
Chronic distemper meningoencephalitis, 188
Chronic encephalomyelitis, 188
Chronic fatigue syndrome, 157

Chronic inflammatory demyelinating polyneuropathy (CIDP), 48, 49, 239, 240
Chronic neuromyelopathy, 10
Chronic progressive panencephalitis, 9
Chronic wasting disease of mule deer and elk, 4, 112
Clinical polymyositis, 57
Colony stimulating factors, 23, 24, 25
Colorado tick fever virus, 3
Complement activation, 24, 25, 27
receptors, 21, 29
Concentric sclerosis, 11
Congenital cytomegalic inclusion disease, 9
Congophilic angiopathy, 324
Coronavirus, 279, 287, 291
antibodies to, 278
Coxsackie A virus, 3, 139, 140, 143
Coxsackie B virus, 3, 139, 140, 143, 150
Coxsackie B type 3 (CVB-3), 283
Coxsackieviruses, 139, 146
Cranial nerve ganglia, 312
Cranial neuropathies, 239
Creutzfeldt-Jakob disease (CJD), 4, 95, 112, 117, 121, 308, 311, 319, 320, 326
transmission, 113, 118
Cytolytic T-lymphocytes (CTL): see T cells, cytolytic
Cytomegalovirus (CMV), 2, 42, 47, 51, 52, 99, 157, 236, 311
antibodies to, 278, 308, 309, 310
encephalitis, 33, 312
human: see Human cytomegalovirus
inclusions, 50, 54, 240, 241
latent infections, 312
mouse: see Mouse cytomegalovirus
neuropathies, 8, 9
Cytotoxic cells: see T cells, cytotoxic

Delayed type hypersensitivity (DTH), 84, 86
Dementia, 101, 102, 239, 341, 343
subcortical, 237, 318
viral association with, 11, 317–329, 341
Demyelinating antibodies, 284
Demyelinating disease, 11, 68, 80, 81, 86, 87–88, 207, 294; see also Progressive multifocal leukoencephalopathy (PML)
animal models, 342
autoimmune, 285
Demyelinating lesions, 84, 85, 220
Demyelinating neuropathies, 45, 47–49, 54, 70, 73, 83, 95, 100, 101, 103

Demyelination, 158, 166, 192, 193, 195, 196, 208, 214, 321, 328
immunopathogenesis, 281–283
models of, 287–292
of peripheral nerves, 318
Dendritic cells, 16, 19, 22, 31, 162, 234
Distal symmetrical polyneuropathy (DSPN), 51, 239
Dorsal root ganglia, 46, 144
Dysesthesias, 51

Eastern equine encephalitis (EEE) virus, 3, 5, 256–259
Echovirus 11, 150
Echoviruses, 3, 9, 139, 140, 143, 147
Encephalitis, 2, 3, 36, 79, 166, 182, 188, 192, 193, 196, 236, 241, 262
acute, 194, 195, 341
animal models, 256
arboviral, 256, 258, 259, 260, 263–266, 268, 269
fatal, 157, 185
fetal, 191
necrotizing, 197
neonatal, 43
primary, 6
subacute, 9, 101, 194, 195, 237
viral, 33, 34, 35, 42, 71, 158, 161, 162, 166, 303, 312
Encephalitis lethargica, 303; see also Von Economo's disease
Encephalomyelitis, 2, 33, 63, 79, 189, 195, 282
Encephalomyocarditis virus (EMCV), 80, 323
Endothelial cells, 30, 94, 95
viral protein expression in, 71
Enterovirus 70, 3, 140, 143, 150
Enterovirus 71, 143
Enterovirus encephalitis virus, 3
Enteroviruses, 2, 8, 9, 139–150, 340
clinical diseases caused by, 142–144
nonpolio, 143, 150
Ependymal cells, 82, 185, 190, 191, 195
Ependymitis, 191
Epstein-Barr virus (EBV), 2, 41, 42, 45, 282
neuropathies, 47, 157
Experimental allergic encephalomyelitis (EAE), 6, 87, 192, 282, 283–285, 288, 291
Experimental allergic neuritis (EAN), 283, 284

Extracellular edema (spongiosis), 95

Facial palsy, 45
Far Eastern virus, 261, 262
Feline immunodeficiency lentivirus (FIV),
    94, 102, 103, 230
Feline leukemia virus, 291
Feline T-lymphotropic virus (FTLV), 102; see
    also Feline immunodeficiency
    lentivirus
Fibroblastoid cells, 65
Fibroblasts, 97
    viral protein expression, 71, 73
Fibronectin, 31
Flaviviruses, 255, 259–266
Foamy viruses, 101, 102
Forsmann antigen, 285
Fowl plague virus, 286
Friend leukemia virus, 287
Frontal cortex, 196
Frontal gyrus, 67

Galactocerebroside, 166, 284, 288, 289
Ganglia, 45
Ganglioneuronitis, 49, 50
Genital ulcerations, 43; see also Herpes
    simplex virus, type 2
Gerstman-Straussler-Scheinker syndrome,
    320
Gerstmann-Straussler syndrome, 4, 112
Giant cells, 57, 99
    multinucleated, 65, 71, 101, 103, 240
Glia cells, 193, 194
Glial cell membranes, 95
Glial cells, 158, 192, 196, 209, 210, 217, 219,
    287, 342
Glial fibrillary acidic protein (GFAP), 85,
    217
Glial nodules, 69, 265
Glioblastomas, 216
Glioma, 11, 220, 216
Gliosis, 100, 101, 103, 122, 259, 290, 304,
    308
Glycolipids, brain: see Brain, glycolipids
    viral, 286, 287–294
Granulocytes, 27
Guillain-Barre syndrome, 6, 7, 8, 49, 239,
    240, 290, 291
    animal models, 156
    viral etiology, 341

Hamster-adapted agent, 114

Hemisensory deficits, 213
Hemorrhagic fever, 263
Hepatic dysfunction, 47
Hepatitis, 260
Hepatitis B virus, 41, 87
    neuropathy, 41, 47–48, 50
Hepatitis B virus polymerase (HBVP), 282
Herpes encephalitis, 308
Herpes simplex virus (HSV), 2, 4, 41, 43–
    44, 48, 51, 156, 158, 279
    antibodies to, 166, 278, 308, 309
    association with Alzheimer's disease, 326
    latent infections, 42, 160, 168, 312
    neuropathies, 9, 42, 45, 48, 54, 156
Herpes simplex virus, type 1 (HSV-1), 2, 3,
    157, 236, 241, 282
    antibodies to, 309, 310, 311
    association with dementia, 327
    encephalitis, 42, 158, 161, 162, 166
    latent infections, 42, 155
    neuropathies, 48
    reactivation, 43
Herpes simplex virus, type 2 (HSV-2), 2, 43,
    44, 157, 236
    latent infections, 42, 44, 155
Herpes-like structures, 309
Herpesviruses, 155
    latency, 157, 326
Hippocampal cortex, 324
Hippocampic fissure, 67
Hippocampus, 186, 194, 196, 311, 324, 325,
    327
Human adenoma, 221
Human adenovirus type 12, 282
Human cytomegalovirus (HCMV), 41, 45,
    46, 51
    neuropathies, 47, 48, 54
Human immunodeficiency virus (HIV), 3,
    7, 33, 42, 93, 102, 245
    CNS involvement in disease, 48, 94, 98,
    101, 103, 104, 235, 241
    encephalitis, 99, 100
    myopathies, 57
    neurological manifestations of infection,
    229–245, 340
    neuropathies in, 8, 9, 10, 48–50, 54, 93
Human immunodeficiency virus, type 1
    (HIV-1), 7, 66, 102, 230, 318–319
    immunosuppression, 216, 229, 230, 235,
    318, 319, 321, 342
Human immunodeficiency virus, type 2
    (HIV-2), 230

Human papovavirus, 207–222, 320
Human polyomaviruses, 208, 211
Human T-lymphotropic retrovirus (HTLV), 3, 94, 96, 328
 neuropathy, 93
Human T-lymphotropic retrovirus, type 1 (HTLV-1), 3, 100, 101, 103, 104, 232, 342
 association with multiple sclerosis, 340
 myelopathy, HTLV-1-associated (HAM), 100, 101, 294, 328
 neuropathies, 10, 50, 328
Human T-lymphotropic retrovirus, type 2 (HTLV-2), 294
Hydrocephalus, 192, 194
Hypothalamic-pituitary axis, 121
Hypothalamus, 144, 191
Hypothalamus-pituitary-adrenal axis, 114

Ilheus virus, 261
Immune complexes, 48
Immune lymphocyte-mediated cytotoxicity (ILMC), 183, 188
Immune tolerance, 95
Immunoassay, 83
Immunocompromised host, 222; see also Immunosuppression
Immunocompromising disease
 in JCV infection, 215, 216
Immunoglobulin, 15, 21, 26, 27, 31, 101, 122, 232; see also Antibodies
Immunoglobulin G (IgG), 21, 26, 83, 98, 100, 122, 279, 289, 307, 308, 309, 312
Immunosuppression, 7, 85, 98, 161, 259
 HIV-1-associated, 216, 229, 230, 235, 318, 319, 321, 342
 reactivation of latent infections during, 47
 viral infection-associated, 100, 102, 157, 161, 164, 283, 340
Infectious mononucleosis, 8
Inflammatory polyradiculoneuropathy, 240
Influenza, 8, 303, 329
Influenza A, 282
 antibodies to, 278, 308
 association with schizophrenia, 307
Influenza B, 282
Influenza virus, 162, 278, 279, 286, 287, 288, 290
Inkoo virus, 266, 267
Interferon (IFN), 21, 48, 72, 122, 308
Interferon α (IFN α), 24, 25, 162, 291

Interferon β (IFN β), 24, 162, 291
Interferon γ (IFN γ), 24, 25, 36
Interleukin-1 (IL-1), 23, 24, 25, 30, 31, 32, 244
Interleukin-2 (IL-2), 23, 32, 231, 294
 receptors, 21, 25
Interleukin-3 (IL-3), 23, 24
Interleukin-4 (IL-4), 23, 24
Interleukin-5 (IL-5), 23, 24

Jamestown Canyon virus, 266, 267, 268
Japanese encephalitis, 260, 339
Japanese encephalitis virus, 3, 4, 261, 263, 265, 266
 antibodies to, 266
JC polyomavirus, 9
JC virus (JCV), 2, 207–213, 241, 321, 340, 342
 antibodies to, 215
 oncology, 216–222
 pathology, 213–216
Junin virus, 3

Kuru, 4, 95, 112, 121, 280, 319, 326
Kyasanur Forest virus, 261, 262, 263

LaCrosse encephalitis, 267
LaCrosse virus, 3, 266, 267, 268
Lactate-dehydrogenase-elevating virus, 342
Langerhans' cells, 234, 244
Lassa virus, 3
Latency-associated transcripts (LATS), 160
Lateral ventricles, 190
Lentiviruses, 63, 64, 71, 73, 74, 93, 94, 100–102, 230, 232, 319
 models for HIV infection, 232–234
Leptomeninges, 67, 82, 191, 193
Lipopolysaccharide, 117, 121
 complement activation by, 25
 B cell activation by, 26
Louping-ill virus, 3, 261, 262
LPM virus, 197
Lumbosacral neuropathy, 44
Lumbosacral polyradiculoneuritis, 43
Lumbosacral spine, 44
Lymphocytes, 67, 82, 100, 122, 193, 238, 327
 B: see B cells
 T: see T cells
Lymphocytic choriomeningitis virus (LCMV), 3
Lymphokines, 116

Lymphoproliferative disease, 10

Macroglial cells, 208, 213, 214
Macrophages, 34, 35, 86
  involvement in immune response, 15, 18,
    19, 26, 27, 28, 30, 32, 54
  involvement in viral pathogenesis, 67, 82,
    100, 101, 103, 104, 117, 118, 122, 238,
    239
  levels during arboviral infection, 259, 265
  levels during encephalitis, 36
  viral persistence in, 84, 85
  viral protein expression in, 71, 240, 241
  viral tropism for, 66, 73, 97, 162
  virus infection of, 229, 243, 318, 319
Maedi-visna virus (MVV), 63, 71, 72; see also
    Lentiviruses
  pathogenesis, 73–74
Maediviruses, 64; see also Lentiviruses
Major histocompatibility complex (MHC),
    86, 88, 100, 162, 163, 232, 291
  influence in multiple sclerosis, 278, 280,
    293, 294
  involvement in immune response, 19, 22,
    25, 28, 30, 32
Marburg virus, 3
Marek's disease, 156
Marek's-disease virus (MDV), 42, 44, 45, 47,
    157
Mason Pfizer monkey virus, 97
Measles, 8, 181, 182, 185, 187, 308, 312
  persistent infection, 193, 322
Measles inclusion body encephalitis (MIBE),
    178, 185, 186, 192
Measles virus, 3, 178, 180, 183, 184, 193,
    194, 279, 282
  antibodies to, 187, 192, 193, 194, 278,
    279, 308, 309
  association with SSPE, 321–323
  immunosuppression, 283, 340
  neuropathies, 8, 9
  recovery from cerebrospinal fluid, 181,
    182
Medulloblastomas, 216, 217
Mengo virus, 80
Meningeal irritation, 143
Meninges, 33, 71
Meningitis, 4, 67, 143, 186, 189, 191, 195,
    262
  acute, 2, 236, 341
  aseptic, 2, 6, 101, 235, 241, 263, 264
  recurrent, 9, 341

Meningitis (cont.)
  viral, 43, 44, 339
Meningoencephalitis, 157, 181, 189, 190,
    194, 236, 259
  alphavirus associated, 256
Meningoencephalomyelitis, 327
Mesencephalon, 194
Microglia, 29, 31, 82, 195, 197, 240, 249,
    291, 318
  hypertrophy, 195
Midbrain, 327
Molecular mimicry, 87, 281–283, 284, 292,
    329
Monoclonal antibody, 259, 282, 295
  immunobiological characterization using,
    182, 183, 185, 186, 189, 192, 194, 196
  neutralizing, 182
Monocyte-macrophages, 16, 231, 232, 245
  surface antigens, 21
Monocytes, 15, 21–22, 28, 31, 100, 103, 244
  role of, in the inflammatory process, 25,
    54, 70
  virus infection of, 66, 234
Mononeuritis multiplex, 47, 49, 239
Mononuclear cells, 33, 80, 124, 191, 229
Morbillivirus, 178, 179, 180, 197
Mosquitoes, as a vector in virus
    transmission, 257–262, 266–268, 286
Motor neuron disease, 54
Mouse cytomegalovirus (MCMV), 46, 47
Mouse encephalomyelitis virus, 323
Mouse teratoma cells, 47
Multiple sclerosis, 10, 11, 70, 80, 187, 193,
    277–295, 304, 306, 307, 308, 343
  epidemiology, 277–278
  pathogenesis, 277, 282
  viral etiology, 11, 277, 278–281, 288–295,
    341
Mumps, 8, 186, 187
Mumps meningitis, 186–190
Mumps virus, 3, 178, 180–182, 185, 186,
    189, 190, 191, 192, 279, 288
  antibodies to, 189, 190, 278, 295, 309,
    310
Murine leukemia virus (MuLV), 93, 100, 102
  neurotropicity, 94–97, 103, 104
Murray Valley encephalitis, 260
Murray Valley encephalitis virus, 3, 261, 262
Myalgic encephalomyelitis, 147
Myelin, 47, 84, 85, 87, 184, 195, 209, 212,
    214, 217, 238, 280, 281, 287, 289,
    292, 294, 295, 321

Myelin basic protein (MBP), 83, 87, 166, 184, 282–284, 286, 288, 292, 295, 321
Myelin breakdown, 68, 69, 70, 80, 82, 83
Myelin sheaths, 100
Myelin-associated glycoprotein (MAG), 83
Myelitis, 43, 157
Myeloneuropathies, 101
Myelopathy, 11, 241
Myocarditis, 150, 283

Natural killer (NK) cells, 27, 28, 31, 32, 34, 36, 231, 307, 328
   deficiency, 289
   levels in viral infection, 162, 163, 259
Necrotizing vasculitis, 50
Negishivirus, 261, 262
Neocortex, 324
Neoplasms, 220
Nerve conduction, 47
Neuritis, 99, 157, 282, 295
Neuroblastomas, 217
Neuroglia, 240
Neuroleukins, 101, 319
Neuromyelitis optica, 11
Neuronal cell membranes, 95
Neuronal loss, 197
Neuronophagia, 82, 259
Neurons, 46, 82, 144, 190, 194, 265, 319, 326
Neuropeptides, 101
Neurotransmitters, 325, 326
Neutralizing antibody, 71, 72, 73, 144, 164, 165, 168, 182, 195, 212, 230, 259, 265
Neutrophils, 31, 32
Newcastle disease virus (NDV), 178, 180, 196–197
Nigrostriatum, 115
Nude mice, 36, 166, 289
Null cells, 27
Numbness, 50

Occipital lobe, 67
Ockelbo virus, 259
Olfactory mucosa, 180
Olfactory nerve, 180
Oligodendrocytes, 196, 209, 212–214, 217, 279–281, 287, 291, 293, 295, 319, 321
   viral replication in, 70, 73, 82–87, 158, 193,
Oligodendroglial cells, 95

Oncoviruses, 93, 102
Onyong-nyong virus, 256
Optic chiasma, 119
Osteosarcomas, 219

Papovavirus, 9, 236; see also Human papovavirus
Parainfluenza virus, 178, 191, 279, 288
   antibodies to, 278
Paralysis, 79, 94, 95, 96, 113, 196, 309
   flaccid, 82, 263
   total, 67
Paramyxovirus-like inclusions, 279
Paramyxoviruses, 177–197, 321
   pathogenesis of infection, 180–189
   persistent infections, 182–184
Paraplegia, 67, 240
Paresis, 113, 143, 239
Parkinson's disease, 11, 329, 341
Pericytes, 71, 73
Peripheral neuropathy, 9, 42, 45, 50, 54, 236
Perithelial cells, 94
Persistent infection, 8–9, 182–184, 193, 268, 322
Picornaviruses, 80, 84, 283, 342
Pneumoviruses, 178
Polio vaccine
   inactivated (IPV), 142, 146–150
   orally administered (OPV), 142, 146–150
Polioencephalomyelitis, 195
Poliomyelitis, 4, 79, 81, 82, 141–145
   paralytic, 7, 142, 147, 148, 149
Polioviruses, 3, 139–144, 342
   antibodies to, 278
   virus recovery from cerebrospinal fluid, 140, 147
Polymerase chain reaction (PCR), 231
Polymorphonuclear cells (PMN), 28–29, 240
Polymyositis, 239, 240
Polyneuritis, 156, 157
Polyneuropathy, 49, 54, 239
Polyomaviruses, 10, 216, 320–321, 329; see also Human polyomaviruses
Postencephalitic Parkinsonism, 328–329
Posterior horn, 144
Posterior median sulcus, 67
Posterior paralysis, 192
Postherpetic neuralgia, 44, 157
Postinfectious encephalitis, 6
Postinfectious encephalomyelitis, 8, 41, 157, 341

Postinfectious measles encephalitis, 192
Postviral fatigue syndrome, 147
Postviral thyroiditis, 295
Powassan virus, 3, 261, 262
Priapism, 82
Progressive congenital encephalomyelitis, 9
Progressive inflammatory
       polyradiculoneuropathy, 51
Progressive inflammatory polyradiculopathy
       (PIP), 239
Progressive multifocal leukodystrophy, 99
Progressive multifocal leukoencephalopathy
       (PML), 9, 207–209, 213–216, 219,
       220, 222, 320–321, 329
   viral etiology in, 236, 320–321, 329, 341
Prostaglandins, 25
Protease-resistant protein (PrP), 126–128
Psychiatric disorders; see also Schizophrenia
   viral etiology of, 10
Purkinje neurons, 196
Pyramidal cell layer, 186, 196
Pyramidal tracts, 100

Rabies, 4, 7, 8, 341
Rabies virus, 3
Radiculoneuropathy, 43
Rage, 102
Ramsey-Hunt syndrome, 45
Reovirus type 1, 283
Respiratory syncytial virus, 178, 278
Reticuloendothelial system (RE system), 16,
       18, 19, 22
Retroviruses, 10, 279, 287, 328, 340, 342
   animal models of infection, 93–104
   complement activation by, 25
   nononcogenic, 63, 229
Reye's syndrome, 6, 8
Rhinovirus, 139
Rift valley fever virus, 3
Rinderpest virus (RV), 178, 180
Rocio encephalitis, 260
Rocio virus, 3, 261, 262, 265
Ross River virus, 256, 259
Rotavirus, 278
Rubella, 8, 9
Rubella virus, 3, 279, 340
   antibodies to, 278, 309, 310
Rubeola, 309, 310
Russian spring-summer encephalitis (RSSE)
       virus, 3, 262

Schizophrenia, 303–312

Schizophrenia (cont.)
   viral etiology, 11, 303–305, 341
Schwann cells, 42, 45, 46, 47, 83, 209, 240
Sciatica, 43
Scrapie, 4, 95, 111–128, 319, 326
   autoimmune response, 121
   pathogenesis, 116–121
   prion theory, 126
   transmission, 112–113
   virino theory, 126, 127, 128
Scrapie-associated fibrils (SAF), 125–127
Self antigens, viral mimicry of, 28
Semliki Forest virus (SFV), 256, 259, 288–
       290
Sendai virus, 178, 185, 186, 191, 288
Sensory ganglia, 41, 42, 161
Sensory neuropathy, 51
Shingles, 42, 44, 165
Simian immunodeficiency lentivirus (SIV),
       94, 98–104, 230, 244
Sindbis virus (SV), 29, 33, 37, 256, 286, 288
Smallpox virus, 8
Snowshoe hare virus, 266, 267
Spinal cord, 33, 67, 70, 82, 119, 144, 265,
Spinal cord lesions, 195
Spinal ganglia, 71, 186
Splanchnic nerve, 119
Spongiform encephalopathy, 9, 326
Spongiform polioencephalomyelopathy, 93,
       95, 96, 97, 100
Spongiosis, 101
Spongy degeneration, 194
Spumaviruses, 93, 101–102; see also Foamy
       viruses
St. Louis encephalitis, 260, 261
St. Louis encephalitis virus, 3, 5, 257, 261,
       263, 265
Stenosis, 190
Striatum, 186
Stromal keratitis, 166
Subacute encephalopathy, 236, 318
Subacute neurological disorders, 8–9
Subacute sclerosing panencephalitis (SSPE),
       9, 33, 178, 184, 185, 187, 188, 192,
       193, 194, 312
   viral etiology, 321–323, 329, 341
Subacute spongiform encephalopathies, 319
Subependymal inflammation, 68
Subependymal parenchyma, 191, 195
Substantia nigra, 120, 121, 328
SV 40, 208–213, 218, 282
   neurooncogenicity in hamsters, 210

SV-40-like virus, 214
Systemic lupus erythematosus, 214

T cells, 16–18, 20, 22, 34, 85, 121, 162, 163, 168, 188, 328
  deficiency, 36; *see also* Nude mice
  proliferation, 84
  receptor, 26, 27, 88, 99
  role in autoimmune disorders, 286, 289, 290, 307
  role in disease pathogenesis, 54, 117, 166, 265
  types
    cytolytic (CTL), 283, 289, 291, 294, 295
    cytotoxic (CD8+ cells), 21, 22, 26, 32, 33, 54, 72, 86, 87, 162–164, 189, 231, 232, 259, 265, 287
    helper cells (CD4+ cells), 21, 24, 25, 26, 29, 32, 34, 36, 84, 166, 244, 245
    depletion in AIDS, 229–232, 234
    levels in viral infections, 163, 259, 265
    receptors, 99
    suppressor, 34, 35, 54
T helper/T suppressor-cytotoxic cell ratios, 15
  in schizophrenics, 308
T protein, 211–213, 217–221
T suppressor cells/cytotoxic cells, 21, 24, 25, 28, 37
Tahyna virus, 3, 266, 267
Thalamus, 119, 121, 144, 191, 265, 318
Theiler's murine encephalomyelitis virus (TMEV), 79, 280, 283, 295, 342
  immunity, 83–87
  infections, 80
  neurovirulence, 81
  pathological features, 82–83
Tick-borne encephalitis, 264, 265, 287
Tick-borne encephalitis (TBE) virus, 3, 262, 289
Togaviridae, 256
Togaviruses, 2, 5
Toscana virus, 3
Transmissible mink encephalopathy, 4, 112, 319
Trigeminal complex, 158
Trigeminal ganglia, 42, 43, 46, 161, 326, 327
Trojan Horse hypothesis, 73, 117, 244, 319
Tropical ataxic neuropathy (TAN), 100
Tropical spastic paraparesis (TSP), 10, 11, 100, 294, 327–328

Tumor necrosis factor (TNF), 23, 24, 25, 163, 243, 244

Unconventional viruses, 1, 4, 320, 340
  neuropathies caused by, 8–9, 341

Vaccination, 8
Vaccinia virus, 279, 287
  antibodies to, 278
Vacuolar degeneration, 102
Vacuolar myelopathy, 101, 235, 236, 238–239
Vacuolation, 115, 119
Varicella, 8
Varicella-zoster virus (VZV), 2, 41, 44–45, 46, 155, 279
  antibodies to, 278, 310
  latent infections, 42
  neuropathy, 8, 156, 157, 236
Vasculitis, 54
Vasoactive intestinal peptide (VIP), 101, 241
Venezuelan equine encephalitis virus (VEE), 3, 256–259, 288
Ventral horns, 33
Ventricles, 69
Vertigo, 45
Vestibular nucleus, 309
Viliuisk encephalomyelitis (VE), 323–324
Viral persistence, 84, 85, 190, 193, 229, 259, 265, 222, 312, 322, 340
Visna, 111, 112
Visna virus, 33, 63, 94, 101, 102, 103, 230, 234, 319; *see also* Lentiviruses
  neurotropism, 64
  pathology, 67–71
  replication, 65–66
Von Economo's disease, 6, 8, 303, 329, 341
Von Recklinghausen's disease, 11, 342

West Nile encephalitis, 260
West Nile encephalitis virus, 3, 261
Western equine encephalitis (WEE) virus, 3, 5, 33, 256–259
  encephalitis, 4
Wiskott-Aldrich syndrome, 220

Yellow fever, 8

Zoster, 45

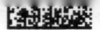